Synthetic Biology

Synthetic Biology

Edited by **Daniel McGuire**

SYRAWOOD
PUBLISHING HOUSE
New York

Published by Syrawood Publishing House,
750 Third Avenue, 9th Floor,
New York, NY 10017, USA
www.syrawoodpublishinghouse.com

Synthetic Biology
Edited by Daniel McGuire

International Standard Book Number: 978-1-68286-337-4 (Hardback)

Printed in China.

Contents

Preface

This book was inspired by the evolution of our times; to answer the curiosity of inquisitive minds. Many developments have occurred across the globe in the recent past which has transformed the progress in the field.

Synthetic biology is a rapidly evolving discipline which integrates concepts of engineering with that of biology. It is a field which has applications in a number of innovative fields, such as diagnostics, designing innovative biological systems to fabricate materials, etc. This book includes some of the vital researches being conducted across the world, on various topics related to synthetic biology, such as cell design, synthetic DNA, designed proteins, biosensors, etc. As this field is emerging at a rapid pace, the contents of the text will help the readers understand the modern concepts and applications of the subject. It is highly recommended for students and academicians pursuing synthetic biology, biotechnology and allied fields of study.

This book was developed from a mere concept to drafts to chapters and finally compiled together as a complete text to benefit the readers across all nations. To ensure the quality of the content we instilled two significant steps in our procedure. The first was to appoint an editorial team that would verify the data and statistics provided in the book and also select the most appropriate and valuable contributions from the plentiful contributions we received from authors worldwide. The next step was to appoint an expert of the topic as the Editor-in-Chief, who would head the project and finally make the necessary amendments and modifications to make the text reader-friendly. I was then commissioned to examine all the material to present the topics in the most comprehensible and productive format.

I would like to take this opportunity to thank all the contributing authors who were supportive enough to contribute their time and knowledge to this project. I also wish to convey my regards to my family who have been extremely supportive during the entire project.

Editor

1

Teaching the principle of biological optimization

Arthur T Johnson

Abstract

Among the important principles in biology that should be taught in biological engineering educational programs is the principle of optimization, what it means, why it is important, and how it comes about. This material can be presented at numerous levels throughout the curriculum. Understanding of this principle can lead biological engineers to expect it in many, if not all, biological system applications. Understanding optimization in biological systems can help understand the predictive power of evolutionary principles and what to expect from living things incorporated in designs.

Keywords: Instruction, Courses, Evolution, Biological systems

Introduction

The paradox of biology is that living things are so energy expensive, even lavish, yet display a certain energy parsimony in almost everything they do. Living things are highly organized internally and externally; they are far from the random assortment of compounds that they would become, and do become once they die, if this organization was not maintained. It takes a great amount of energy to sustain biological organization, whether we consider a single cell, organization of cells within a tissue, tissues within organs, organs comprising an organism, or different individuals interacting in complex ways in an ecosystem.

The only means to overcome the tendency for disorder and increased entropy is a constant supply of energy in form of heat, radiation (sunlight) or chemicals (food or environmental energy-rich compounds). This energy is the means that biological beings and systems survive.

But survival is only the first of two biological imperatives. The second is reproduction amid a (usually) competitive environment. There is a tendency toward domination in the world of biology to make the maximum use of finite resources. Saving operational energy where it is possible allows the organism or system to expend more energy to reproduce. This is, however, usually a zero-sum game. The biological unit that achieves better efficiency than its neighbors has a reproductive advantage and can exploit a larger share of whatever energy resources are available.

Correspondence: artjohns@umd.edu
Fischell Department of Bioengineering, University of Maryland, College Park, MD 20742, USA

Although energy is often the most limiting of necessary resources (because of the huge need for energy to sustain biological order), the same can be said for other limited resources, such as food (a source not only of energy but also of essential nutrients), minerals, space, etc.

This tendency towards energy efficiency becomes extreme when the competition extends to the living products of reproduction, the offspring themselves. Hence, the offspring incline to even greater efficiency so that they can compete among themselves.

This applies to biological units (BU) of sundry hierarchical levels and organizations, from individual genes, to cells within a tissue, to organisms, colonies, and ecosytems. Think of the case where there are two BU: one is able to perform the same functions as the second, but it can do so with one half the energy requirements. If each BU is a prey animal, which is most likely to be caught and eaten by a predator? If each BU is a predator, which is most likely to chase and catch an elusive prey, or which is most likely to survive long times between kills? If each BU is a plant, which is most likely to be able to outgrow competitors and grazing herbivores? If each BU is a microbe, which is most likely to inhabit a region with limited nutrient availability? If each BU is a bodily tissue, which is more likely to confer to the entire organism a reproductive advantage? If each BU is a colony of bees, which is more likely to produce more swarms? If each BU is an entire ecosystem, which is most likely to thrive and expand into new territory?

In each of these cases, the answer is clearly that the BU with the advantage in resource utilization is the winner of the competition. Only in the instance where competition is at a minimum, say, for example, for the first species in a virgin environment, will there be little primary reason to reduce energy costs. However, as soon as the second species arrives, or even as soon as the number of individuals of the first species increases to the point that they force significant intraspecific competition, there will be an advantage to those individuals that can make more efficient use of resources. Thus, there is a tendency to minimize dependence upon environmental sources of energy and nutrients.

Natural selection is a powerful force leading to evolution of living things. There is evidence of *convergent evolution* (where organs and tissues with different origins form identical final forms) and *allometric relationships* (scaling of different forms and functions among species). There is enough competitive pressure in biology that the benefit to cost ratio of almost every biological function must be optimized. A hummingbird, for example, needs enough strength in its wings and energy to be supplied to its wing muscles to hover. There is no advantage to be gained with excess wing strength, and so the benefit to cost ratio changes abruptly after sufficient strength is satisfied. On the other hand, animals that jump have a survival advantage if they can jump farther, faster, or higher. There is an overhead cost of supporting larger muscles or bones needed for better jumping, but the benefit to cost ratio changes gradually for jumping. The biologically-optimal solution for these cases would be expected to be different for each.

In the classroom

An essential part of biological engineering education should be the principles governing typical biological behavior, and, among these principles, biological optimization is one of the most important. Biological engineers should be on the lookout for optima to occur in all aspects of biological activity, and expect optima to define biological states. This is true for everything from locomotion to respiration to neural activity to genetic variability. Each of these will be further discussed later.

Because biological optima are the direct result of competition among living things in an environment with limited resources, and the evolutionary tendency to maximize survival and reproduction, biological engineers who understand how optima appear in biological systems will also better understand the workings of evolution and the ways in which environments are affected by living things and living things affect their environments [1]. Evolutionary principles have been used in powerful ways by biologists, physiologists, and behaviorists to predict biological behaviors. Biological engineers who intend to predict behaviors of living things involved in their designs should thoroughly understand both evolutionary principles, of which optimization is one, and interactions with the environment.

Engineers often translate efficiencies into mathematical optima. If the function to be optimized can be expressed in mathematical form, then simple differential calculus, by setting the first derivative of the function to zero, can be used to find an extremum of that function, as long as the extreme point lies within the boundaries of the problem. There may be local extremes or one global extreme point, and, when discussing these in class, the teacher should be careful to caution students that sometimes the discovery of an optimum depends on the local region around which the extremum is found. This is not too likely to happen with biological systems because then the biological system would exhibit some degree of indeterminacy and that is not very characteristic of living things. It does happen sometimes, but these are often pathological states.

Biological engineering students are characteristically exposed to mathematical models in biology soon after they have taken differential calculus, so some detailed explanations of finding optimum points using the methods they learned in previous calculus courses can be warranted. The teacher should also remind students that extrema may be located at the boundaries rather than in between. This might mean, for instance, that some threshold has been reached, and the instructor can find examples of this to present to the class. For more advanced students, some discussion of means to find optimum solutions for entire functions, such as using calculus of variations, can be included.

At the heart of many biological optimization problems are at least two processes: one that is more costly with increase of the independent variable, and another that is less costly as the independent variable increases. Taken together, there is usually some point where the sum of the outputs from both of these processes is a minimum. This is the optimum point. For biology, investigators have often been interested in energy as the output variable, and have taken as the input variable some entity that makes sense for that particular optimization problem. There are other possible output variables where the maximum becomes the optimum of interest. An example of this might be the efficacy of a certain drug as a function of the time of the day [1,2].

Optima can be broad or narrow (Figure 1). Narrow optima are very selective, and don't tolerate much variation before the cost of locating at a nonoptimum point becomes too high to be sustained. An example of this would be the wing strength of the hummingbird just mentioned. In this case, the energy penalty for maintaining wings of excessive strength would not be able to be justified for either survival or reproduction. Broad optima can still have the same optimum, but the costs of deviation from the exact optimum point do not rise significantly quickly.

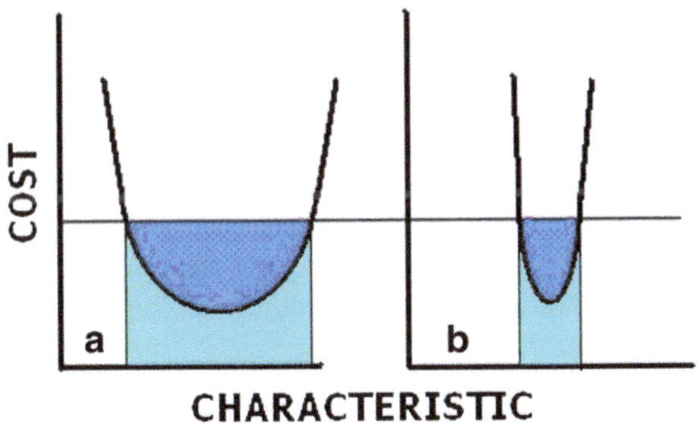

Figure 1 Illustration of broad (a) optimum and narrow (b) optimum. For the same level of cost, the broad optimum allows a greater range of the optimized characteristic. Biological systems often seem to prefer broad optima because of the flexibility that they offer.

Many biological properties and systems seem to have broad optima. In this case, the advantage is flexibility, allowing deviations that don't cost too much but that can meet new (environmental) challenges requiring responses different from those determining the original optimum.

Broad biological optima can be a consequence of several properties characteristic of living things. The first is the ability of a living system to sense its environment and to communicate in various ways. The second is the ability to respond to the immediate environment that results in a mathematically chaotic system; that is, the end result is dependent on the pathway it takes to get there. The third property is the appearance of alternate forms that differ very little in energy levels. As a result of these properties, it is easy for biological systems to have varying forms with very little additional energy costs, but which have been determined by the history of the organism.

It is sometimes easier for students to grasp these ideas if biological examples of different optima are presented to illustrate key points. A plethora of examples can also give the impression that optimization is very important in biology and is the result of generations upon generations of evolutionary tendencies. Several examples are given in a succeeding section of this paper to illustrate the range of biological activity seemingly optimized, but the instructor can often add others from personal knowledge or from the current literature.

Biological engineering students who understand the importance of optimization in biology will become more effective in the practice of their profession. They will have known that a biological optimum is located at the preferred operating point, and that deviations from this point can have serious consequences. The engineers are thus likely to make better decisions for their designs involving living things.

Examples

Optimization can be seen at all biological levels. A few of these are discussed in the following sections.

Genetic variability

There is genetic variation within a species that cannot be easily explained. The principle of survival of the fittest (natural selection) should result in the elimination of all but the most survival and reproductively successful genes. This means that genes not optimum for survival in a competitive environment should not persist over many generations, no matter how small their disadvantage. However, a few of them remain, and they give the species the possibility that, should the environment change, there would be genes already present that could be better able to allow the species to adapt correctly.

This can be considered to be an optimization problem; if so, then the unexplained genetic variation found in almost all organisms could be a consequence of the broad optima that characterize biological systems. Therefore, genetic variation could be explained by the fact that carrying nonoptimum genes does not turn out to be too expensive for the species as long as the result of those genes being present does not differ too much from the results of the fittest genes. As with many other biological optima, this means of genetic optimization turns out to be energetically less expensive (and maybe more likely for species survival) than an optimum that confers too much advantage to the best genes.

This can also be illustrated by the presence of the same set of genes in each cell of an organism. Certainly, all the genes present are not activated in every cell, especially if the organization of the organism is highly complex. There should be no need to maintain those genes that are not activated, and the cell expends some excess energy to keep them. There would also be an extra energy expense to tailor a reduced set of genes to each particular type of cell. There is apparently a benefit to cost advantage to the organism to maintain the entire genome in each cell and epigenetically regulate genetic expression.

Cellular biochemicals

The activity of many cellular compounds depends on their configurations. This is especially true of proteins and enzymes. Subtle differences in protein folding can have profound effects on effectiveness. Protein misfolding is common, but there are error-correcting schemes within the cell to recycle useless molecules, if possible. Heat shock proteins act as chaperone molecules to reform proteins damaged as a result of heat or other severe environmental stresses. All these mechanisms are designed to recover otherwise worthless molecules and turn them into functioning compounds, thus saving energy and nutrients that otherwise would have been expended by the cell to fabricate these molecules from the start.

As a particularly illustrative example of biomolecule optimization, consider actions of monoclonal and polyclonal antibodies. The former are examples of very narrow functional optima, while the latter are examples of broad functional optima. The biological engineering choice of which antibody type to attach to a biosensor, for instance, depends on how selective the antibody and biosensor is meant to be.

Central nervous system

The central nervous system is one of the most energy expensive organs of the human body, requiring about 20% of resting energy expenditure [3]. This energy cost is only indirectly related to survival and reproduction, which are determined directly by physical actions. Optimization applies to the brain, which appears to operate for maximum information transfer among neurons per unit of neural action potential energy used. The two competing functions to form the optimum solution would be: 1) energy maintenance of a number of neurons in an active state, which increases as the number of active neurons increases, and 2) the minimum number of active neurons needed to process required information, a threshold value. Thus, only one to 15 percent of the neurons in the human brain are active at any particular time. Although some would say that using only 15% of the brain's capacity is not very efficient use of resources, this saves energy for physical movement, such as hunting for food and avoiding predators, that otherwise would have been used for neural activity. Maintaining the presence of extra central nervous system neurons does not cost as much as if they were constantly active, but does result in the flexibility of possible environmental responses.

It has been hypothesized that the timing of human birth optimizes the ability of cognitive and motor neuronal development in the child by allowing the child to maximize the absorption of important cultural information (memes) in its environment [4]. This happens at the expense of the welfare of the child being completely dependent on its caregivers, a condition not shared to the same degree by many other animals. It is also possible that the timing of human birth occurs at the time when the mother's metabolism is no longer able to support both her needs and the needs of the fetus [4]. In this case, the optimum would be determined by a threshold value instead of the maximum in the case of cognitive development.

Walking and running

Walking has long been recognized as one bodily function that appears to be optimized, and appears to occur at a speed that minimizes the rate of energy expenditure of the body walking a certain distance (Figure 2). Power consumption of walking, as measured by oxygen consumption, depends on walking speed [5-9]. For any given distance walked, the speed can be higher with extra energy expenditure per unit of time over a shorter total length of time or the energy expended per unit of time can be lower for a longer length of time. The optimum speed of walking is that which minimizes the total amount of energy spent.

Walking is inherently inefficient because at least one foot is on the ground at all times. There are stages when both feet are simultaneously on the ground and pushing against each other [10]. The legs remain nearly straight, and the position of the center of mass of the body is therefore highest when the leg is vertical and the body passes over the supporting foot. Contrarily, the body is lowest when both feet are touching the ground. The body is constantly rising and falling as stepping proceeds [3]. This means that walking comprises a series of episodes of positive physical work when lifting the body mass and negative physical work when the body mass falls, each of these episodes having its own degree of inefficiency. Without optimization, walking would be more inefficient than it is, and so consume too much of the body's energy resources, making successful survival and reproduction less likely.

Running is a different mode of locomotion in which each foot is on the ground less than half the time. There

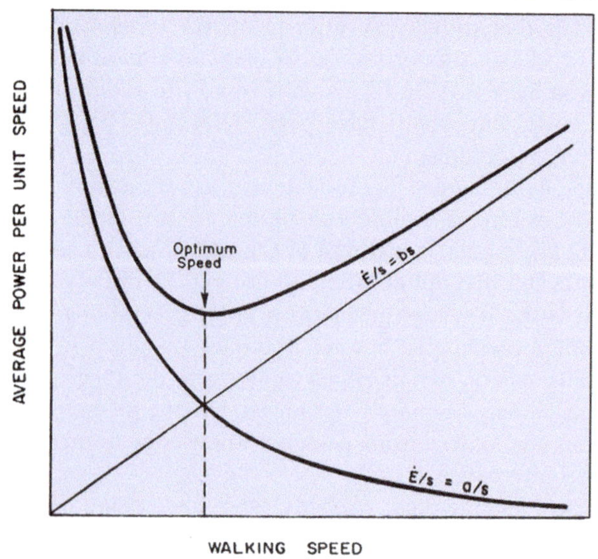

Figure 2 Walking appears to occur at a speed very close to the optimum based on average power consumption per unit walking speed (E/s). Two components of average power (E/s), one increasing with speed (bs, where b = a constant) and the other decreasing with speed (a/s, where a = a constant), make possible a minimum average power.

are stages of running during which neither foot is on the ground. The runner travels in a series of leaps, with the center of mass of the body at its highest in midleap. Its lowest point occurs when the trunk passes over the supporting foot, and the supporting leg is bent at this stage. The running body does not rise and fall as much as during walking, so running can be more efficient at higher speeds.

Caloric measurements have shown that very slow running is more energy expensive than walking at the same speeds, while walking at very high speeds is more costly than running at those same speeds. The transition between walking and running (Figure 3) occurs at a fairly predictable speed of about 2.3 m/sec (6 mi/hr) for normal-sized adults [10], and has been found to occur at the point where the energy expenditures for both walking and running are equal. Again, the energy expenditure of moving about is minimized to make survival and reproduction of the individual more likely than without the minimization.

Quadripedal animals have one mode of locomotion, besides walking and running, that humans do not: they gallop at high speed. Galloping involves bending movements of the back that briefly store leg kinetic energy fluctuations as elastic energy, contributing to overall efficiency [11]. These animals appear to have two transitional power points, one from walking to running or trotting and another from running to galloping.

Other animals use different means to reduce the energy inefficiency of locomotion. Animals such as centipedes have so many legs that their bodies do not rise and fall during locomotion. Snakes crawl without the benefit of legs, but their locomotion must overcome the friction created between the ground surface and their skin. Crocodiles use a side-to-side waddling motion to move their legs forward to propel themselves. Penguins have very short legs that do not raise and lower the body mass by very much. Birds have wings that convert forward motion into lift to improve efficiency. Fishes' bodies contain a special swim bladder that allows them to maintain a vertical position in the water without muscular effort. It has been estimated that a 1% improvement in the efficiency of a swimming fish can be expected to make 3% more energy available for growth and reproduction [12].

Breathing

Human breathing at rest consumes approximately 1-2% of total oxygen consumption of the body, whereas during exercise breathing may consume 8-10% or higher. The oxygen used to breathe is bodily maintenance overhead. Oxygen consumed by the diaphragm and other muscles involved in breathing cannot be used directly by the skeletal muscles in the legs to escape predators, so there should be an evolutionary advantage to save breathing oxygen during heavy exertion.

Breathing at normal frequencies is dominated by resistance located in the airways and lung tissue, and compliance of the lung tissues. As respiration rate increases, so does the rate of airflow and so does the pressure required to push air through the respiratory resistance. Work rate depends on both pressure and flow rate, so the rate of work required to overcome resistance increases nonlinearly with frequency. If the depth of

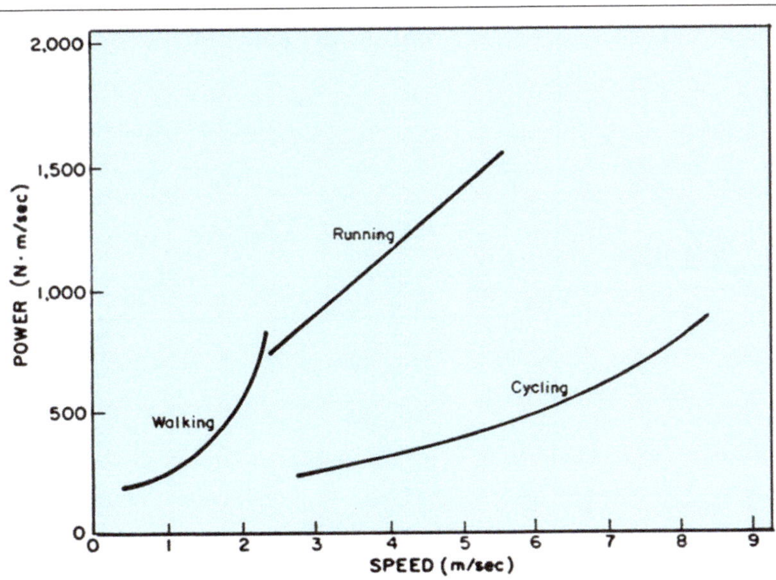

Figure 3 Power required for walking and running for an adult male human. Curves for walking and running intersect at about 2.3 m/sec and show that walking is more efficient below the intersection and running is more efficient above.

breathing were to be maintained the same at high frequencies as at lower frequencies, then the work rate due to respiratory resistance would be proportional to the square of the flow rate. However, at higher frequencies the alveolar ventilation rate increases above that which is necessary, so flow rate decreases somewhat and work rate is not quite exactly related to flow rate squared.

The pressure required to store energy in a compliance depends upon the volume stored. At higher respiration rates the lungs are not required to fill as much to deliver the same amount of oxygen to the tissues. Thus, compliance pressure, and, consequently work rate, is nearly inverse in magnitude to the frequency increase.

The result is that total respiratory work rate is composed of two components, one of which increases with frequency and the other of which decreases with frequency (Figure 4). There is an optimum breathing rate that minimizes work rate, and most published data indicates that people and some other animals breathe at a rate corresponding to the minimum [13]. Exercise breathing is faster because the minimum work rate frequency moves higher as more air is inhaled. Airway caliber, air flow waveshape, ratio of inhalation time to exhalation time, and lung midposition also appear to be adjusted to reduce energy expenditure [3].

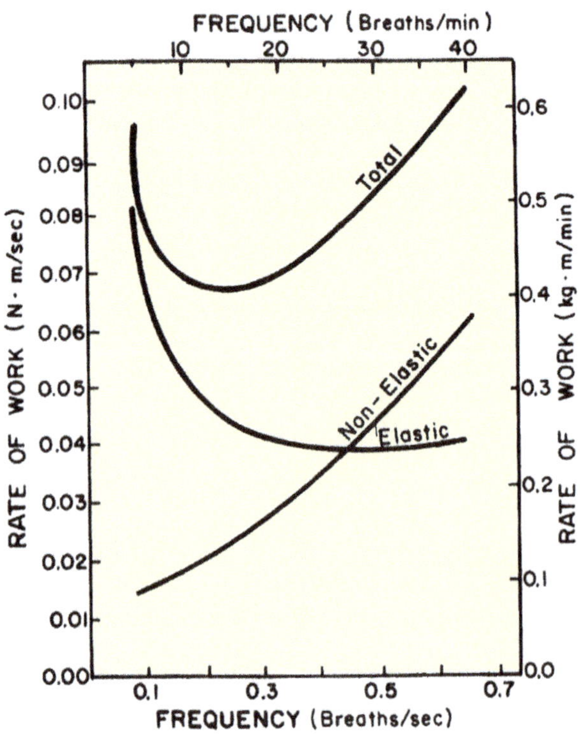

Figure 4 The work rate of breathing is the sum of resistive (nonelastic) work and elastic (compliance) work. The sum reaches a minimum at some particular frequency.

Heart rate

The work of the heart, similarly to that of the lungs, is necessary to sustain life, but does not perform the work of skeletal muscles necessary to obtain food, escape predators, and survive in other ways. Like respiration, cardiac energy expenditure, although necessary, is part of the overhead used to maintain the health of the physical body. Survival and reproduction of the animal could thus benefit from minimizing cardiovascular work rate.

Cardiovascular models have shown that it is conceivable that cardiac contraction is performed so as to minimize the rate of work necessary [14]. Variation in heart rate (Figure 5) appears to occur at the most sensitive region to regulate blood pressure [15].

Allometry

Allometry is defined as the change of proportions with increase in size of a single species or between adults of related groups [16]. Allometric relations are also known as scaling relationships, and only exist when there is similarity of structure and function between biological units of different size. If completely different mechanisms are involved (for example, locomotion of bacteria compared to horses), then no allometric relationship between them would be expected.

Scaling relationships can be used to predict the value of some biological property of one species if the value of a similar property is known for another species [1]. There appear to be universal biological principles at work in scaling relationships, although the natures of these principles have not yet been fully explored. Allometric relationships among very divergent species seem to be scaled with body mass to some simple multiple of one-quarter power ($m^{1/4}$).

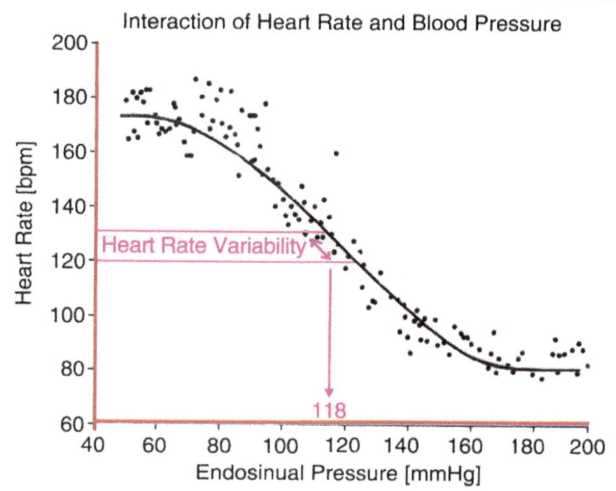

Figure 5 The heart rate is variable, but appears to be located in the most sensitive region to control blood pressure.

Natural selection seems to have led to an economy of design of structures and functions so that they just meet maximum demands. Any greater capacity would be biologically uneconomical [17]. If evolution results in allometric relationships among BU, then it is only because the benefit to cost ratio of the form or function in question has been optimized. Another way to think of allometry is to consider that, if organisms do *not* change their form as they change in size, their function *is* altered, and such functional shifts might be a source of evolutionary innovation [18].

Ecosystems

Entire ecosystems are also models of efficiency. Populations of species that cohabitate these ecosystems complement and supplement each other. Waste from some species is used as nutrition by others; some plants shade others and cool them, and keep them from drying; large trees protect less hardy species from the wind; nutrients that would ordinarily be lost to the ecosystem are stored in the bodies of organisms and recovered when they die.

Many organisms also act as ecosystems home to different varieties of cells, tissues, microbiomes, parasites, and other living things. There is an optimization in this ecosystem to conserve scarce nutrients. Iron, for instance, is recycled from worn-out red blood cells to be used by new red blood cells. The kidney reabsorbs glucose, bicarbonate, some sodium (depending on intake), potassium, and chloride in order to conserve them and reuse them. Desert animals save water by excreting a very concentrated urine.

While these are not exactly the same as optimization that lends itself to mathematical modeling, it is optimization nevertheless. If a more efficient BU is introduced, it likely has a competitive advantage over existing BU, and the less efficient BU is soon displaced. In this way, biological systems are constantly improving their utilization of scarce resources.

Summary

Understanding the importance of optimization to biological systems is an essential element of biological engineering education [19,20]. Including biological optimization in one or more undergraduate courses can result in better biological engineering graduates who are able to see broad implications for their work. Optimization is a natural result of evolutionary tendencies, so the student who understands the role of optimization in biology can see the workings of evolution as well as the complex interactions of living things with their environments.

Competing interests

The authors declare that they have no competing interests.

References

1. Johnson AT: *Biology for engineers*. Boca Raton, FL: Taylor and Francis; 2011.
2. Hoven LA, Sherman KA, Butler S, Giebultowicz JM: **Does the clock make the poison? Circadian variation in response to pesticides.** *PLoS One* 2009, **4**(7):e6469.www.plosone.org/article/info.../10.1371/journal.pone.000646.
3. Johnson AT: *Biomechanics and exercise physiology: quantitative modeling*. Boca Raton, FL: Taylor and Francis; 2007.
4. Dunsworth HM, Warrener AG, Deacon T, Ellison PT, Pontzer H: **Metabolic hypothesis for human altriciality.** *Proc Nat Acad Sci* 2012, **109**(38):15212–15216.
5. Alexander RM: **Energetics and optimization of human walking and running. The 2000 Raymond pearl memorial lecture.** *Am J of Human Biol* 2002, **14**:641–648. doi:10.1002/ajhb.10067.
6. Anderson FC, Pandy MG: **Dynamic optimization of human walking.** *J Biomech Engr* 2001, **123**(5):381–391.
7. Dean GA: **An analysis of the energy expenditure in level and grade walking.** *Ergonomics* 1965, **8**:31–47.
8. Dingwell JB, John J, Cusumano JP: **Do humans optimally exploit redundancy to control step variability in walking?** *PLoS Comput Biol* 2010, **6**(7):e1000856. http://www.ploscompbiol.org/article/info%3Adoi%2F10.1371%2Fjournal.pcbi.1000856.
9. Milsum JH: *Biological control systems analysis*. New York: McGraw-Hill; 1966.
10. Alexander RM: **Walking and running.** *Am Sci* 1984, **72**:348–354.
11. Alexander RM: **Why mammals gallop.** *Am Zoolog* 1988, **28**:237–245.
12. Alexander RM: *Principles of animal locomotion*. Princeton, NJ: Princeton University Press; 2003.
13. Christie RV: **Dyspnoea in relation to the visco-elastic properties of the lung.** *Proc Roy Soc Med* 1953, **46**:381–386.
14. Livnat A, Yamashiro SM: **Optimal control of left ventricular systolic dynamics.** *Am J Physiol* 1981, **240**:R370–R383.
15. Moser M, Frühwirth M, Kenner T: **The symphony of life.** *IEEE Eng Med Biol Soc Mag* 2008, **27**(1):29–37.
16. Li JK-J: **Scaling and invariants in cardiovascular biology.** In *Scaling in biology*. Edited by Brown JH, West GB. New York: Oxford University Press; 2000:113–128.
17. Brown JH, West GB, Enquist BJ: **Scaling in biology. Patterns and processes, causes and consequences.** In *Scaling in biology*. Edited by Brown JH, West GB. New York: Oxford University Press; 2000:1–24.
18. Koehl MAR: **Consequences of size change during ontogeny and evolution.** In *Scaling in biology*. Edited by Brown JH, West GB. New York: Oxford University Press; 2000:67–86.
19. Johnson AT: **Essential concepts for biological engineers.** *Biol Eng* 2010, **3**(1):3–15.
20. Johnson AT: **Making environmental biology central to a course in biology for engineers.** *J Ecosys Ecograph* 2011, http://dx.doi.org/10.4172/2157 – 7625.S1-002.

2

A standard vector for the chromosomal integration and characterization of BioBrick™ parts in *Escherichia coli*

Susanna Zucca[1,2], Lorenzo Pasotti[1,2], Nicolò Politi[1,2], Maria Gabriella Cusella De Angelis[2] and Paolo Magni[1,2*]

Abstract

Background: The chromosomal integration of biological parts in the host genome enables the engineering of plasmid-free stable strains with single-copy insertions of the desired gene networks. Although different integrative vectors were proposed, no standard pre-assembled genetic tool is available to carry out this task. Synthetic biology concepts can contribute to the development of standardized and user friendly solutions to easily produce engineered strains and to rapidly characterize the desired genetic parts in single-copy context.

Results: In this work we report the design of a novel integrative vector that allows the genomic integration of biological parts compatible with the RFC10, RFC23 and RFC12 BioBrick™ standards in *Escherichia coli*. It can also be specialized by using BioBrick™ parts to target the desired integration site in the host genome. The usefulness of this vector has been demonstrated by integrating a set of BioBrick™ devices in two different loci of the *E. coli* chromosome and by characterizing their activity in single-copy. Construct stability has also been evaluated and compared with plasmid-borne solutions.

Conclusions: Physical modularity of biological parts has been successfully applied to construct a ready-to-engineer BioBrick™ vector, suitable for a stable chromosomal insertion of standard parts via the desired recombination method, i.e. the bacteriophage integration mechanism or homologous recombination. In contrast with previously proposed solutions, it is a pre-assembled vector containing properly-placed restriction sites for the direct transfer of various formats of BioBrick™ parts. This vector can facilitate the characterization of parts avoiding copy number artefacts and the construction of antibiotic resistance-free engineered microbes, suitable for industrial use.

Background

Plasmids are extensively used tools to generate genetically engineered microbes for the expression of recombinant proteins or complex genetic circuits [1-4]. Even if they are very easy to manipulate and incorporate in the desired host, many disadvantages affect their use in both industrial applications and research studies. Common plasmids require the selective pressure of an antibiotic to be maintained in cells, which is costly for industrial scale recombinant protein production [5]. The spreading of antibiotics and resistance markers is also potentially unsafe for the environment [5] and some (e.g., ampicillin) should be avoided in therapeutic protein production because of the potential for human allergic reactions [6]. Selection systems without antibiotics are available, but they require mutant host strains, specific growth media or expensive reagents and in some cases they show low efficiency [5,7]. Plasmids are often replicated in multiple copy in the host cell, which enables the industrial production of a large amount of protein [6]. However, plasmid-free strains with the desired recombinant genes in single copy are required in many studies, e.g., to investigate the effect of these genes in normal physiological conditions, thus avoiding copy-number artefacts [8]. Genome integration can provide the stable insertion of the desired genes in the host chromosome without the need of any antibiotic or resistance marker. Several tools for *Escherichia coli* have been proposed which exploit homologous or site-specific recombination. Homologous recombination can be

* Correspondence: paolo.magni@unipv.it
[1]Dipartimento di Ingegneria Industriale e dell'Informazione, Università degli Studi di Pavia, via Ferrata 3, Pavia, Italy
[2]Centro di Ingegneria Tissutale, Università degli Studi di Pavia, via Ferrata 3, Pavia, Italy

used to insert the desired DNA fragment (here called *passenger*) into a specific genomic locus that must show sufficient sequence homology with a second DNA fragment (here called *guide*) used to target the locus [9]. This technology can also be used for gene knockout and mutation of native genes [10]. Integrative plasmids (that perform single- or double-crossover) [11], linear PCR fragments [12] and also single-stranded DNA [13] perform the described tasks by means of unspecific endogenous or heterologous protein machinery (e.g., the λRed system [12,13]). On the other hand, site-specific recombination uses genome insertion of bacteriophages in the host chromosome through the phage attachment site (attP) and the bacterial attachment site (attB) sequences [14]. This mechanism has been exploited to develop integrative vectors carrying the attP site (guide) and the passenger [15]. The gene expression machinery that mediates homologous or site-specific recombination can be placed on an easily curable *helper* plasmid transformed in the host strain [9,15]. Figure 1 illustrates how homologous recombination (with a single-crossover event) and site-specific recombination work.

Generally, integrant clones are selected with an antibiotic resistance marker. This marker can be removed by exploiting FRT sites: by flanking a sequence with FRT sites, it can be targeted for excision through the yeast Flp recombinase. Helper plasmids expressing the Flp recombinase have also been constructed [16].

Integrative plasmids must be easily amplified *in vivo*, and clones with a successful integration must be easily selected. To this aim, conditional-replication origins are exploited. They support plasmid replication only in specific conditions, like a specific strain or a temperature range, while the plasmid becomes non-replicative otherwise [10].

For example, the R6K replication origin can be used to propagate integrative plasmids only when the pir or pir-116 gene is present in the host strain [15].

Recent advances in the field of synthetic biology include the standardization of biological parts to facilitate the assembly of genetic circuits [17]. BioBrick™ parts in the Registry of Standard Biological Parts are a rapidly-growing collection of DNA parts that conform to a specific physical standard [18]. BioBrick™-compatible genetic circuits can be easily incorporated in a microbial host through *ad-hoc* constructed BioBrick™ plasmids [19]. Although the construction of integrative systems by using BioBrick™ parts has been reported [20], no standard and ready-to-use solution is available to produce engineered strains with BioBrick™ parts via chromosomal integration. In fact, current tools for site-directed genome integration via integrative vectors include either non-BioBrick™ plasmid vectors with specific, non-customizable DNA guide [15] or BioBrick™-compatible parts [20] that can be used to compose integrative vectors via a standard assembly procedure, but a pre-assembled ready-to-use solution is not available.

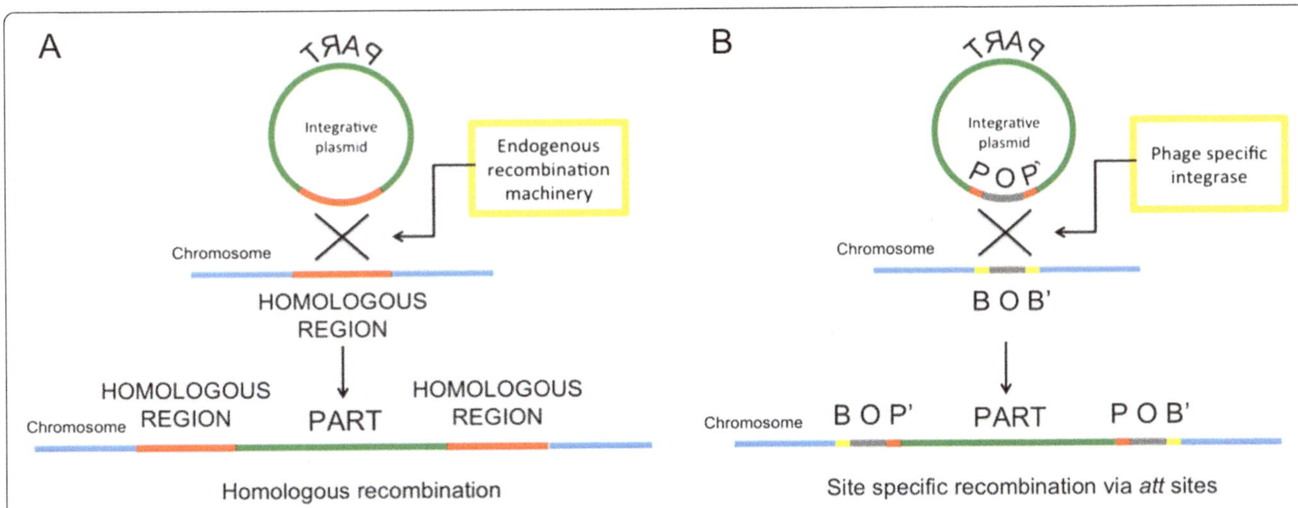

Figure 1 Main plasmid-based methodologies for chromosomal integration of parts in *E. coli*. A) Homologous recombination. The integrative plasmid (green) carries a sequence (red) that is homologous with a region (red) in the bacterial chromosome (blue). After a single-crossover event, mediated by the endogenous recombination machinery of *E. coli*, the whole plasmid sequence is integrated in the target region of the chromosome. In specific recA-knockout strains the recombination machinery can be expressed via a helper plasmid. **B**) Site-specific recombination. The integrative plasmid carries a bacteriophage attachment site (attP) that targets the whole plasmid into the specific attachment site (attB) in the host genome. This process is mediated by a specific recombinase that can be expressed via a helper plasmid. The attP and attB sites are composed by the POP′ and BOB′ sequences, respectively. They share a homologous core sequence (O) and different flanking sequences (P, P′, B and B′), so that after site-specific recombination the integrated plasmid is flanked by the BOP′ and POB′ sequences. All the bacteriophage specific *att* sequences have this common structure. In both homologous and site-specific recombinations, positive integrants are usually selected via antibiotic resistance, provided that the integrative plasmid sequence contains an antibiotic resistance marker.

In this work, we report the design of a BioBrick™-compatible integrative vector that allows the genomic integration of BioBrick™ parts and can also be specialized to target the desired integration site in the host genome by using BioBrick™ parts. We also demonstrate the usefulness of the designed tool by providing data on the modularity of promoters when characterized in a single-copy context and on plasmids.

Results and discussion

Design of the integrative base vector

Figure 2 shows the structure of the designed integrative vector, pBBintΦ. This base vector uses the default integration guide Φ80 attP, although it can be specialized according to user needs by changing guide and passenger. The cloning site is compatible with the original BioBrick standard (RFC10) and its related standards RFC23 and

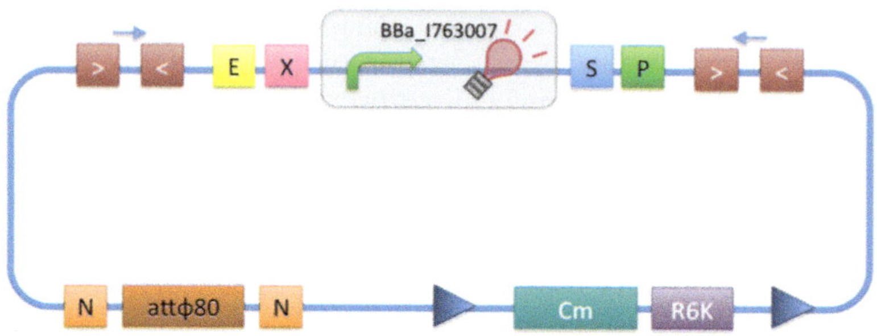

Part number	Function	Notation
BBa_G00000	BioBrick cloning site prefix	E X
BBa_G00001	BioBrick cloning site suffix	S P
BBa_I763007	PR-mRFP1 expression cassette	BBa_I763007
BBa_J72001	FRT recombination site	▶
BBa_B0053 BBa_B0054	forward transcriptional terminators	>
BBa_B0055 BBa_B0062	reverse transcriptional terminators	<
BBa_G00100	forward sequencing primer annealing site (VF2)	→
BBa_G00102	reverse sequencing primer annealing site (VR)	←
BBa_B0045	NheI restriction site	N
BBa_K300032	chloramphenicol resistance marker	Cm
BBa_J61001	R6K conditional replication origin	R6K
BBa_K300991	attϕ80 integration guide	attϕ80

Figure 2 Integrative base vector pBBintΦ structure. It is a chloramphenicol-resistant vector with a conditional replication origin (R6K) that impairs its replication in commonly used *E. coli* strains (without pir or pir-116 gene). Two FRT sites flank the resistance gene and the conditional origin, so that they can be excised via Flp-mediated recombination once integrated in the genome, thus leaving a marker-less integrant without R6K. This vector targets the Φ80 attB site in the chromosome of *E. coli* via the attP integration guide in the vector. The attP sequence is flanked by two NheI restriction sites to enable the engineering of this base vector by easily changing the guide. The cloning site is flanked by BioBrick™ Prefix and Suffix sequences, while four transcriptional terminators implement the insulation of the part when placed in the genome. Primer binding sites are also present. The default insert of the vector is an mRFP1 constitutive expression cassette, driven by the PR promoter. The glossary explains all the used symbols and lists the BioBrick™ basic parts used to compose this vector.

RFC12, as it is composed by the original BioBrick™ Prefix and Suffix [17]. The presence of illegal restriction sites (XbaI in FRT and SpeI in the Φ80 attP) prevents the usage of this backbone in the classic BioBrick™ Standard Assembly process. However, the presence of unique EcoRI and PstI sites in Prefix and Suffix fully supports the assembly of the desired BioBrick™ parts in the cloning site upon EcoRI-PstI digestion and also supports the 3A Assembly [19,21] (see Figure 3A). The two NheI restriction sites flanking the default integration guide enable the engineering of this backbone by assembling new user-defined BioBrick™ integration guides upon XbaI-SpeI digestion, if the desired guide conforms to the RFC10 or to a compatible standard (see Figure 3B). Although such assembly is non-directional, integration occurs regardless to guide orientation. Like in many other standard vector backbones (e.g., the pSB**5 vector series in the Registry of Standard Biological Parts [18]), the binding sites for standard primers VF2 and VR are present upstream and downstream of the BioBrick™ cloning site respectively. These two sequences are sufficiently distant from the cloning site to enable a good quality sequencing of the insert. The R6K conditional replication origin is necessary to avoid the extra-chromosomal maintenance of the vector during integration; specific strains with the pir or pir-116 genes are required to propagate it during the cloning steps.

The four BioBrick™ transcriptional terminators BBa_B0053, BBa_B0054, BBa_B0055 and BBa_B0062 ensure the transcriptional insulation of the integrated part from its flanking genome sequences, as it is achieved in the pSB**5 vector series. The two FRT recombination sites enable the excision of the R6K origin and the chloramphenicol resistance marker upon Flp recombinase activity. This marker excision allows users to make multiple serial integrations in the same strain in different target loci, always using the same antibiotic resistance marker. The same FRT recombination procedure can also be used, provided that essential chromosomal genes lie between the different target loci (see [12] and [15] for a detailed description of such potential problem when multiple FRT sites are introduced in the same genome).

Additional details about vector sequence, features and construction are available in the BBa_K300000 Registry page.

The engineering of the integration guide allows the integration of parts in user-defined genome positions and for this reason this vector supports the integration by exploiting bacteriophage attP-mediated integration as well as homologous recombination.

In this study, the default insert of pBBintΦ is BBa_I763007, a constitutive mRFP1 expression cassette driven by the PR promoter from lambda phage. Note that an early version of the integrative vector included a different default insert that in our experience showed problems, as discussed in Additional file 1: Additional information about integrative base vector design.

Vector performance validation in the default integration locus

The integration and marker excision capabilities of pBBintΦ were tested by integrating a number of BioBrick™ passengers in the chromosome of MG1655 and MC1061 strains. The integrative base vector targets the default integration locus Φ80 attB.

Integration: 100% of the screened colonies (N = 39) on the chloramphenicol plate lost the helper plasmid and 100% of the screened clones (N = 11) showed a correct integration position by PCR (see Additional file 1: Figure S1). 82% of the screened clones (N = 11) had at least two tandem copies of the integrated DNA, as PCR showed (see Additional file 1: Figure S2).

Marker excision: 100% of the screened colonies (N = 100) lost the helper plasmid, while 68% of them had a successful marker excision, validated via chloramphenicol sensitivity. 94% of the chloramphenicol-sensitive screened clones (N = 33) also showed a correct amplicon (see Additional file 1: Figure S3 for a representative experiment) by PCR with primers P1-P4 (see Methods section), thus validating the presence of the construct of desired size in the correct genomic position. In this case, the multiple tandem copies, previously identified by PCR, became a single copy without antibiotic resistance or R6K origin. This tandem copy loss happened because the Flp enzyme excised the entire sequence flanked by the two most distal FRT sites, thus generating a single integrant of the desired construct.

100% of the screened clones with correct P1-P4 amplicon (N = 12) showed the expected sequencing results. 83% of them also showed the expected phenotype. Additional file 1: Additional information about integrated BioBrick™ devices and phenotypes of recombinant strains reports the full list of integrated BioBrick™ devices (Additional file 1: Table S1) and additional information on integration experiments.

Characterization of a set of BioBrick™ promoters in two different genomic positions

A representative set of widely used BioBrick™ constitutive promoters with the RBS-mRFP1-terminator sequence downstream were integrated in the Φ80 or aspA genomic locus and characterized. The considered BioBrick™ promoters were BBa_J23100, BBa_J23101, BBa_J23118, BBa_I14032 and BBa_R0051, indicated in this work as J23100, J23101, J23118, PlacIQ and PR respectively, while the downstream sequence was the BioBrick™ reporter device BBa_I13507. These constructs were also characterized in the low copy vector pSB4C5, used as a term of

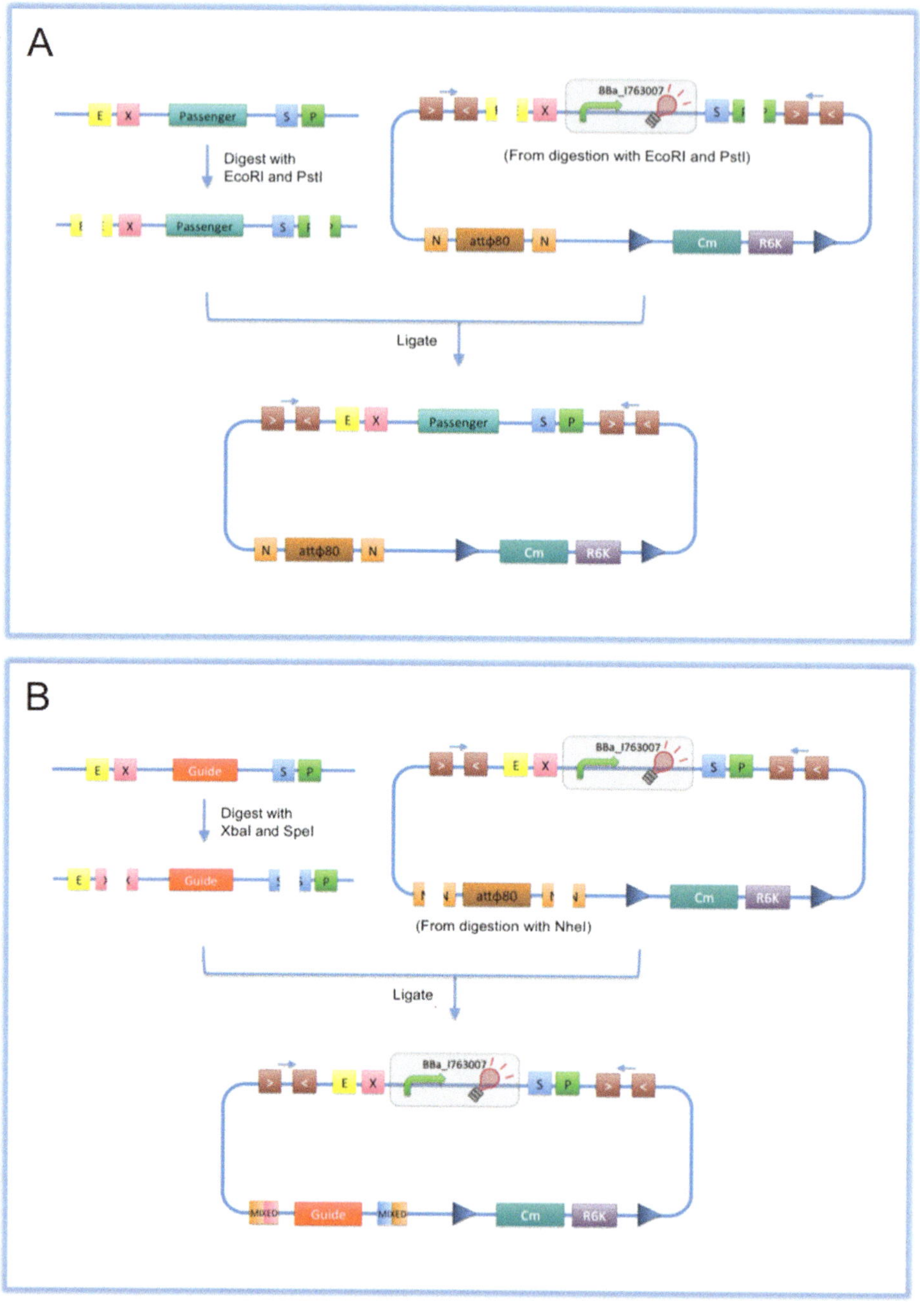

Figure 3 How to engineer the base vector. A) The desired BioBrick™ insert can be assembled as passenger in the integrative vector: the BioBrick™ passenger of interest must be digested with EcoRI-PstI, while the integrative vector is also digested with EcoRI-PstI to eliminate the default insert BBa_I763007. The two parts can be ligated and the product has to be transformed into a pir/pir-116 strain to enable the propagation of the vector. **B**) The vector can be specialized with BioBrick™ parts to target the desired locus of the *E. coli* chromosome: the BioBrick™ guide of interest must be digested with XbaI-SpeI, while the integrative vector must be digested with NheI to eliminate the default guide (Φ80 attP). The open vector must be dephosphorylated to avoid self-ligation, the two parts can be ligated and the product has to be transformed into a pir/pir-116 strain. XbaI, SpeI and NheI all have compatible protruding ends. Note that the ligation is not directional, but the guide can work in both directions. The guide can be another attP site or a part which shows a significant homology with a genomic region of the host.

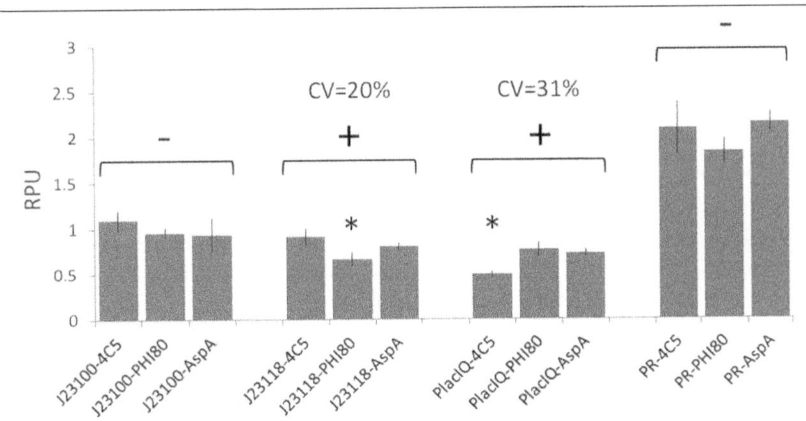

Figure 4 BioBrick™ promoters characterization, expressed in Relative Promoter Units, in single chromosomal copy (Φ80 and aspA loci) and in low-copy context (pSB4C5 vector). Promoter activity was measured via the BBa_I13507 (RBS BBa_B0034-mRFP1-double terminator) part. Grey bars represent the average values, computed on at least five clones, and the error bars represent the 95% confidence intervals. For each promoter, statistical analysis was performed via ANOVA test to compare the RPU activities measured in the three different contexts. Conditions showing a statistical difference ($P < 0.05$) in the mean activities among the three conditions are marked with a '+' sign, while promoters not showing any significant difference ($P \geq 0.05$) are marked with a '-' sign. Asterisks represent the post-hoc comparison results. They indicate the individual significantly different condition.

comparison. This set of promoters has been previously characterized in different strains and conditions [22].

Figure 4 shows the RPU values of promoters measured in the three different physical contexts. Given a promoter and a physical context, the activity of the J23101 promoter in the same context was used as a reference to compute RPUs. While the activity of both J23100 and PR is not statistically different when compared in the three contexts, J23118 and PlacIQ activities show a significant difference in one of the tested contexts with a CV of 20% and 31% respectively. The activity of PlacIQ in plasmid context is different from its activity in the chromosome. This difference could be due to the sequence upstream of the promoter that is identical for the two genomic contexts, but it is different in the pSB4C5 vector which could affect promoter activity [22,23]. The different upstream sequence is not sufficient to explain the difference observed in the J23118 activity, where the promoter shows the same activity in low copy vector and aspA locus, but a slightly lower activity in the Φ80 locus.

Figure 5 shows the absolute promoter activities (S_{cell} value, see Methods section) in the two genomic loci. The

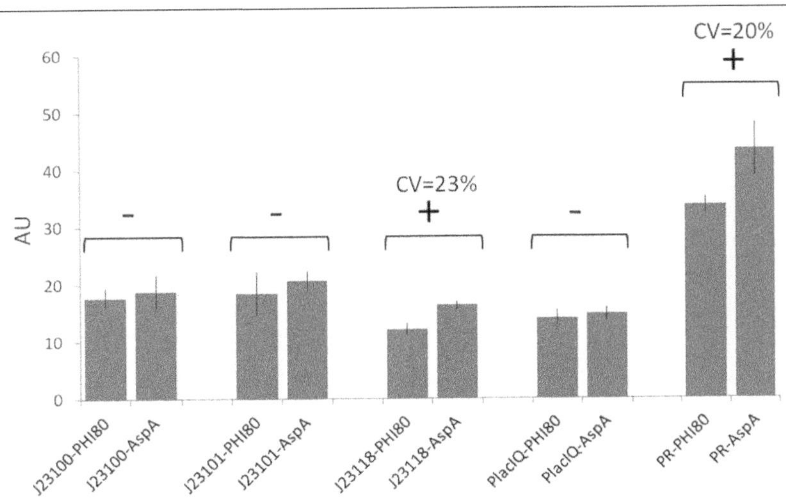

Figure 5 BioBrick™ promoters characterization, expressed in absolute units (S_{cell}), in the two single-copy contexts (Φ80 and aspA loci). Promoter activity was measured via the BBa_I13507 (RBS BBa_B0034-mRFP1-double terminator) part. Grey bars represent the average values, computed on at least five clones, and the error bars represent the 95% confidence intervals. For each promoter, statistical analysis was performed via t-test to compare the activities measured in the two different contexts. Comparisons showing a statistical difference ($P < 0.05$) in the mean activities between the two conditions are marked with a '+' sign, while promoters not showing any significant difference ($P \geq 0.05$) are marked with a '-' sign.

five promoters in the aspA locus have a systematically higher activity than in the Φ80 locus, although statistical analysis showed only two promoters with a significant activity difference. These results could be due to several reasons, such as local sequence effects and distance from the replication origin [24]. However, given the complexity of the genome context, the investigation of these effects is beyond the scope of this work.

Finally, the average estimated copy number of the low copy vector is about 7 (measured by dividing the absolute promoter activity in low copy by the one in single copy number), which is consistent with the reported copy number of the pSC101 origin [18,19] and also with previous works [25,26].

Evolutionary stability of the integrated BioBrick™ devices

All the recombinant strains tested in the previous section were also studied during continual bacterial growth for 150 generations. Figure 6 shows the average percent activity, relative to generation 0, of promoters over the generations for each investigated condition: devices integrated in the Φ80 or aspA locus or contained in a low copy vector propagated with or without antibiotic. Additional file 1: Figure S4 shows the results for each individual promoter in the above mentioned conditions. All the recombinant strains are reasonably stable after 150 generations, with only one of the J23100 samples in the aspA locus showing a significant activity loss after >80 generations (see Additional file 1: Figure S4A). Surprisingly, promoter activities are reasonably stable after 150 generations even in the low copy condition without antibiotic. These results show that the majority of the population of the integrated strains likely has not lost the desired function and their stability is comparable with the low copy plasmid condition.

In order to extend the characterization of the propagated cultures, a second experiment was performed on integrated strains and on plasmid-bearing strains grown in presence of antibiotic, at generation 0 and 150. The fluorescence of single clones isolated from glycerol stocks of 150 generation-old cultures was measured to evaluate fluorescence distribution in the population. Single clones from the original glycerol stocks (generation 0) were also assayed for each recombinant strain as a term of comparison. Results are shown in Additional file 1: Figure S5. According to these data and their interquartile ranges (see Additional file 1: Table S2), the variability of integrated strains at generation 150 is not significantly different from the variability of the same cultures at generation 0 ($p > 0.05$, Wilcoxon signed rank test). The variability of fluorescence in plasmid-bearing strains is lower than in integrated strains for both generation 0 and 150 ($p < 0.05$, Wilcoxon rank sum test). In accordance with population-based experimental results, a large portion of the clones bearing the J23100 construct (culture 2) in the aspA locus lost their activity. Finally, the averaged fluorescence of the single clones from evolved cultures is highly consistent with the fluorescence of the whole evolved populations, obtained by assaying a culture propagated from glycerol stock (see Additional file 1: Figure S6).

Conclusions

Several methods have been proposed to target the desired genetic construct into a specific chromosomal locus [27]

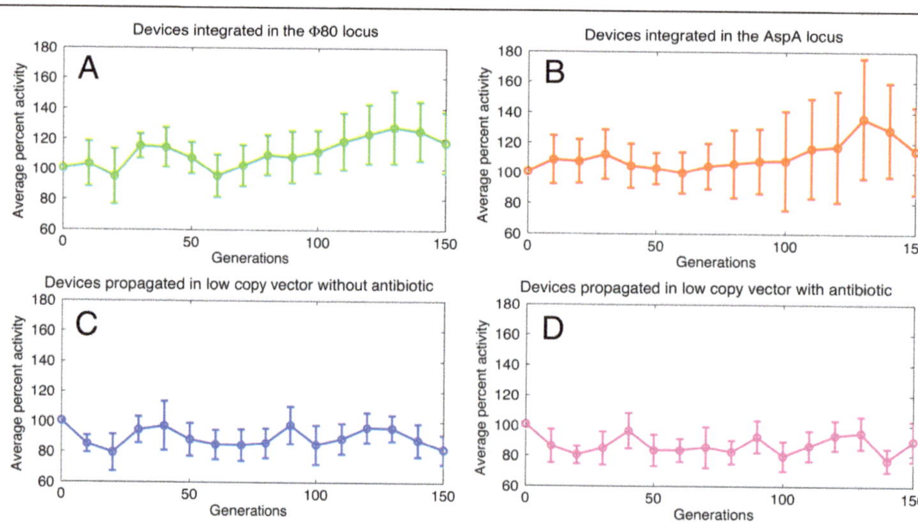

Figure 6 Evolutionary stability of the studied BioBrick™ promoters expressing RFP. The devices integrated in the Φ80 (**A**) and aspA (**B**) loci were propagated without antibiotic, while the devices carried on the low-copy vector pSB4C5 were propagated without (**C**) or with (**D**) antibiotic. The four panels show the average percent activity of all the five studied promoters over time in each condition, where 100% represents the average activity at generation 0. For each time point, error bars represent the standard deviations of the measured values.

or even randomly in the genome [28]. Many studies also focused on single-copy integrants generation [15], on the elimination of antibiotic resistance [12,28] or on the copy number amplification of the integrated device [29,30].

Although several genetic tools have been proposed in literature to perform the tasks described above, no plasmid-based standard tools are available to rapidly construct a ready-to-integrate genetic system.

This work describes a BioBrick™ integrative base vector for *E. coli*. Its standard physical interface allows the assembly of the desired parts as passenger or integration guide, both in BioBrick™ format. This vector is desirable to rapidly disclose, via quantitative characterization, if the parts under investigation are suitable in the genomic context and, if required, in single-copy.

The design of this vector was inspired by previous works. In particular, Shetty et al. [19] conceived a BioBrick™ base vector that could be specialized to construct BioBrick™-compatible vector backbones by using BioBrick™ parts as well. For example, vectors with different replication origins or antibiotic resistance markers could be easily assembled from a standard ancestor. Before the work of Shetty et al., the construction of vector backbones required specific genetic manipulations whose knowledge is normally owned by experts. On the other hand, Anderson et al. [20] used BioBrick™ parts to construct different integrative vectors to produce methylase-expressing stable strains without antibiotic resistance marker.

Our work merges such concepts by providing a ready-to-engineer base vector containing all the required features for integration (i.e., a conditional replication origin and an antibiotic resistance marker), marker excision (i.e., FRT sites flanking antibiotic marker and conditional origin), assembly of BioBrick™ parts as passenger (i.e., BioBrick™ Prefix and Suffix) or guide (i.e., NheI sites flanking the default guide). The structure of the vector enables the integration of parts properly isolated from the genomic context. In this way, all the integrated sequences have the same upstream and downstream sequences.

Whatever the integration procedure is, the compatibility of the vector with the target chassis always has to be verified: the presence of the integration locus and, when using site-specific recombination, the absence of prophages in the attB locus must be evaluated according to the declared genotype, genomic databases or by PCR.

While our aim was to provide a physical support to standardize the integrative vector construction that can be used with the desired integration procedure (site specific or homologous recombination by single crossover), this work is not intended to contribute novel integration protocols or methodologies that improve its efficiency. In fact, the efficiency depends on the specific guide, passenger and protocol used. For example, in this work the Φ80 site-specific recombination had a very high success rate while the aspA homologous recombination showed a lower efficiency (see Additional file 1: Additional information about integrative vector efficiency).

A set of BioBrick™ devices, including widely used constitutive promoters assembled with an RFP measurement system, were used as passengers in most of the integration experiments. The behaviour of these devices was quantitatively characterized in single copy in two different loci and in a low copy plasmid. The measured promoter activities were consistent with previous works [22,25]. Given a promoter, its relative activity was comparable among the tested contexts, with a maximum CV of 31% (PlacIQ promoter). Although the different sequence between the upstream regions in the single- and low-copy contexts could be responsible of this activity variation in the case of PlacIQ, the sequence difference could not explain all the observed variability among the contexts. The detected variability entity was consistent with our previous studies on context-dependent variability of these promoters when tested via different measurement systems or when assembled in different genetic circuits [22].

The absolute activity of each integrated promoter was also compared in the two chromosomal positions. Results highlighted a promoter activity difference between the aspA and the Φ80 loci, even if only two of the five promoters showed a statistically significant difference. Although a detailed investigation of such effects is beyond the scope of this work, these results demonstrated the importance of quantitative characterization of parts in different genomic contexts for the predictable design of genetic functions.

Finally, the stability of these integrated BioBrick™ devices has been evaluated via population-based assays and via variability analysis of individual clones. These experiments demonstrated that many integrated strains had comparable stability with plasmid-based ones, even though a small subset of evolved integrated strains showed activity loss. Analysis of the activity of individual clones at generation 0 and 150 showed that variability between the two generations was similar and that integrated strains had higher variability when compared to plasmid-bearing strains. As the used parts were not toxic or hard-to-express parts for the chassis, these stability experiments only represented a proof-of-concept, while the stability of parts has to be tested for each desired construct in the future and it could be dependent on the integration site, DNA sequence, host organism, metabolic burden and/or specific function implemented by the integrated device.

Methods

Strains and plasmids

All the *E. coli* strains used in this study are listed in Table 1. MG1655 [31] and MC1061 [32] were used as

Table 1 Strains used in this study

Strain	Relevant genotype	Source
MG1655	wild-type K-12, no Φ80 prophage	CGSC (#7740)
MC1061	no Φ80 prophage	CGSC (#6649)
BW23474	recA1 endA9(del-ins)::FRT uidA4(del)::pir-116	CGSC (#7838)
TOP10	recA1 endA1	Invitrogen

integration hosts. BW23474 [15] were used to propagate at high copy number the plasmids with the R6K conditional replication origin. TOP10 were heat-shock transformed according to manufacturer's protocol, while the other strains were made chemically competent as described in [33] and were heat-shock transformed at 42°C for 1 min.

Table 2 lists the plasmids used for the construction of the integrative vectors and the used helper plasmids. The genetic devices used for single copy and low copy characterization of BioBrick™ parts have been reported previously [22,25]. Briefly, all of them are composed by a promoter (BBa_J23100, BBa_J23101, BBa_J23118, BBa_I14032 or BBa_R0051) assembled to the RBS-mRFP1-Terminator sequence (BBa_I13507 reporter device) via BioBrick™ Standard Assembly [17]. For this reason, DNA junctions between the two sequences are identical (i.e. TACTAGAG) for all the constructs. Low copy characterization was performed on the genetic devices in the pSB4C5 BioBrick™ vector.

Cloning methods

Plasmids were propagated in recombinant strains grown in selective Luria-broth (LB) [33] at 37°C, except plasmids containing a temperature-sensitive replication origin that were grown at 30°C. Antibiotics were routinely used for plasmid maintenance during DNA propagation: ampicillin (100 mg/l for high copy plasmids, 50 mg/l for low copy plasmids), chloramphenicol (12.5 mg/l) and kanamycin (50 mg/l). Liquid cultures (5 ml) were incubated with shaking at 220 rpm. DNA was extracted with the NucleoSpin Plasmid kit (Macherey-Nagel) from recombinant cultures grown to saturation. Purified DNA was digested as required and gel-extracted with the NucleoSpin Extract II kit (Macherey-Nagel). DNA parts were ligated to construct all the plasmids described in this work. 1 μl of ligation mix was transformed into a proper *E. coli* strain, DNA was purified and screened via restriction digest and electrophoresis. Sequencing was performed to validate all the plasmids with primers VF2 (TGCCACCTGACGTCTAAGAA) and VR (ATTACCGCCTTTGAGTGAGC). Long-term bacterial stocks were prepared by mixing 750 μl of a saturated culture with 250 μl of 80% glycerol and stored at −80°C.

Table 2 Plasmids used for integrative vector construction

Name	BioBrick™ code	Description
pHC-attP-CS	pMK-RQ(BBa_K300983)[a]	DNA fragment including FRT-NheI-attP-NheI-CS(B0033)-FRT, flanked by AvrII sites, in high copy vector
pHC-CmR	pSB1A1(BBa_P1004)[b]	Chloramphenicol resistance gene with promoter in high copy vector
pHC-ter	pSB1AK3(BBa_B0015)[b]	Double terminator in high copy vector
pHC-R6K	pSB1A2(BBa_J61001)[b]	R6K conditional replication origin in high copy vector
pHC-CmRter	pSB1AK3(BBa_K300032)[c]	Chloramphenicol resistance cassette in high copy vector
pHC-PR-RFP	pSB1A2(BBa_I763007)[b]	RFP constitutive expression cassette, driven by the PR promoter in high copy vector
pHC-CmRter-R6K	pSB1A2(BBa_K300008)[c]	Chloramphenicol resistance cassette with R6K origin downstream in high copy vector
pBBintΦ-B0033	BBa_K300982[c]	Integrative base vector, targeting the Φ80 locus, with RBS BBa_B0033 as insert
pBBintΦ-RFP	BBa_K300000(BBa_I763007)[c]	Integrative base vector, targeting the Φ80 locus, with RFP constitutive expression cassette as insert
pHC-aspA	pSB2K3(BBa_C0083)[b]	aspA coding sequence in high copy vector
pBBintAsp-RFP	BBa_J107058(BBa_I763007)[c]	Integrative vector, specialized to target the aspA locus, with RFP constitutive expression cassette as insert
pInt80-649	BBa_J72008[d]	Ampicillin-resistant low copy helper plasmid, with temperature-sensitive pSC101 origin, containing a constitutively expressed pir-116 gene and the site-specific recombination machinery of the Φ80 phage under the control of a heat-sensitive promoter.
pCP20	not in the Registry[e]	Ampicillin- and chloramphenicol-resistant low copy helper plasmid, with temperature-sensitive pSC101 origin, containing the yeast Flp recombinase under the control of a heat-sensitive promoter.

The BioBrick™ code of the plasmids include the vector name out of brackets and insert name in brackets.
[a]: obtained from Mr. Gene DNA synthesis service. pMK-RQ is a kanamycin-resistant high copy number delivery vector from Mr. Gene.
[b]: obtained from DNA Distributions of the MIT Registry of Standard Biological Parts.
[c]: constructed in this study.
[d]: gift from the Dr. J.C. Anderson lab (UC Berkeley).
[e]: obtained from CGSC (plasmid purified from recombinant strain CGSC#7629).

NheI restriction enzyme is from Fermentas. Antarctic Phosphatase is from New England Biolabs. Platinum Taq DNA polymerase (Invitrogen) or Phusion Hot Start Flex DNA Polymerase (New England Biolabs) were used to perform PCR. All the other DNA-modifying enzymes are from Roche Diagnostics. Enzymes have been used according to manufacturer's protocols. Oligonucleotides were purchased from Primm (San Raffaele Biomedical Science Park, Milan, Italy) or from Sigma Aldrich. Sequencing was performed by BMR Genomics (Padua, Italy) DNA analysis service.

Integrative base vector construction procedure

Additional file 1: Figure S7 shows the assembly scheme followed to construct the pBBintΦ vector backbone. Briefly, pHC-CmRter was constructed via BioBrick™ Standard Assembly by assembling the insert of pHC-CmR, digested with EcoRI-SpeI, to pHC-ter digested with EcoRI-XbaI. Analogously, pHC-CmRter-R6K was obtained by assembling the insert of pHC-CmRter, digested with EcoRI-SpeI, to pHC-R6K digested with EcoRI-XbaI. The insert of pHC-attP-CS is composed by, in this order, FRT-NheI-attP-NheI-CS(BBa_B0033)-FRT, where CS is a cloning site composed by EcoRI and PstI with four transcriptional terminators flanking them and two standard primer binding sites for VF2 and VR (see Figure 2). In this plasmid, the CS contains the small BioBrick™ RBS BBa_B0033 between EcoRI and PstI. The insert of pHC-attP-CS was excised through digestion with AvrII and it was assembled with the insert of pHC-CmRter-R6K, which had been excised from its vector through digestion with XbaI-SpeI and dephosphorylated, to obtain pBBintΦ-B0033.

Finally, pBBintΦ-RFP was obtained by assembling the insert of pHC-PR-RFP, digested with EcoRI-PstI, to the vector backbone of pBBintΦ-B0033, digested with EcoRI-PstI in order to cut out and eliminate the BBa_B0033 RBS.

Assembly of the desired passenger and guide

The source or assembly procedure of all the BioBrick™ constructs used as passengers has been reported previously [22,25]. These constructs have been excised from their original vector backbone upon EcoRI-PstI double digest and they have been cloned into the pBBintΦ or pBBintAsp vector as described in Figure 3A.

The aspA coding sequence, used as DNA guide for homologous recombination, was excised from its original vector backbone upon XbaI-SpeI double digest and it has been cloned into the pBBintΦ-RFP plasmid, digested with NheI, as described in Figure 3B, and dephosphorylated.

Chromosomal integration

The protocol described in [20] was used for site-specific recombination. Briefly, competent cells were transformed with the pInt80-649 helper plasmid [20]. Transformants were grown at 30°C and were made competent again. They were transformed with the pBBintΦ integrative vector containing the part of interest as passenger and were grown on chloramphenicol plates at 30°C until colonies appeared. A single colony was propagated in LB with chloramphenicol at 37°C, 220 rpm overnight. The culture was streaked on a chloramphenicol plate and incubated overnight at 43°C. A single colony was grown in LB + chloramphenicol to yield the integrant strain. This colony was also streaked on an ampicillin plate to validate the loss of pInt80-649. Colony PCR was occasionally performed with primers P1 (CTGCTTGT GGTGGTGAAT) [15] and P2 (CTCTTACGTGCCCGA TCA), which anneal upstream of the Φ80 chromosomal attB site and in the R6K origin respectively. The P1-P2 amplicon (452 bp) indicates the correct integration position, as described in [15], while negative clones show no amplicon. It is also possible to identify clones with multiple tandem copies of the integrated part via colony PCR with primers P2 (described above) and P3 (AGACGTCAGGTGGCAAAC), which anneal in opposite directions in the R6K origin and in the upstream region of the cloning site. PCR yields a 572-bp amplicon when at least two tandem copies are present in the genome, while no amplicon is produced otherwise.

If not differently stated, the chloramphenicol concentration used in all the steps of the site-specific recombination was 12.5 mg/l.

An analogous protocol was used for homologous recombination, with the exception that the passenger was present in the pBBintAsp vector and the chloramphenicol concentration used was 8 mg/l. In this case, the pInt80-649 plasmid was used to replicate the integrative plasmid via the pir-116 gene, while the Φ80 site-specific recombination machinery was not exploited, although it was expressed during the protocol.

Other integration protocols were tested (see Additional file 1: Additional integration protocols and their relative results).

Excision of antibiotic resistance and R6K origin from an integrant strain

The protocol described in [20] was used. Briefly, the integrant strain was grown in LB and made competent. This strain was transformed with the pCP20 helper plasmid [16], plated on ampicillin plates and incubated at 30°C until colonies appeared. A single colony was propagated in non-selective LB at 37°C, 220 rpm overnight. The culture was streaked on a non-selective plate and incubated at 43°C. A single colony was propagated in non-selective LB to obtain the final single-copy integrant strain without the chloramphenicol resistance marker and the R6K origin. The loss of the antibiotic marker and the pCP20 helper were validated by streaking the obtained strain on chloramphenicol and ampicillin plates respectively. The obtained integrant strain was also assayed through colony PCR with primer pair P1 (reported above) and P4 (CTCTTACGTGCCCGATCA) for Φ80 attB targeting and with primer pair AspAFw (TGCGAGGATCGTGATGTATTTCGG) and AspARv (ATGATCTCGGGTATTCGGTCGATG) for aspA targeting. PCR products were also purified with the NucleoSpin Extract II kit and sequenced with the previously reported primer pairs or with VF2 and VR.

Characterization of promoters

MG1655 was used as chassis for the quantitative experiments. 1 ml of M9 supplemented medium (11.28 g/L M9 salts, 1 mM thiamine hydrochloride, 2 mM MgSO4, 0.1 mM CaCl2, 0.2% casamino acids and 0.4% glycerol as carbon source) [33] was inoculated with a single colony of the desired recombinant strain, grown on a streaked LB agar plate. The culture, in a 15-ml tube, was incubated at 37°C, 220 rpm for about 20 hours and then it was 500-fold diluted in 1 ml and incubated in the same conditions as before for 6 hours. 200 μl were transferred into a 96-well microplate (Greiner), the optical density at 600 nm (OD600) was measured with an Infinite F200 microplate reader (Tecan) and the background absorbance of the medium was subtracted. The culture was diluted to an OD600 of 0.05 (pathlength of the culture volume in the microplate) in 0.5 ml and it was incubated again for 45 min. 200 μl were transferred into a 96-well microplate, incubated at 37°C for 10–15 hours in the Infinite F200 and assayed via the following kinetic cycle, programmed with the i-control™ software (Tecan): linear shaking (3 mm amplitude) for 15 s, wait for 10 s, OD600 measurement, fluorescence measurement (excitation: 535 nm; emission: 625 nm; gain: 80), sampling time: 5 min. Sterile medium and a non-fluorescent culture of MG1655 were always present in order to measure the background absorbance and fluorescence. This procedure was repeated for at least five different colonies from the freshly streaked plate for each recombinant strain.

Integrant strains were tested without antibiotic, while the plasmid-bearing strains were tested in presence of 12.5 mg/l of chloramphenicol. When required, analogous procedures were performed to test recombinant MC1061 strains.

Data analysis for promoter activity measurement

Data were analyzed with Microsoft Excel and the MATLAB 2007b suite (MathWorks, Natick, MA). Raw absorbance and fluorescence time series were processed as previously reported [22,25,34] to obtain a value proportional to the average RFP synthesis rate per cell in exponential phase (called S_{cell}).

Relative Promoter Units (RPUs) were obtained as described previously [22,25,34], by dividing the S_{cell} of the promoter of interest, in a given copy number condition, by the S_{cell} of the J23101 standard reference promoter in the same copy number condition. By definition, the RPU of the J23101 promoter is 1 in all the copy number conditions. Given a physical context, (i.e. the Φ80 locus, the aspA locus or the pSB4C5 vector), the genetic device should be present at the same copy number for all the tested promoters, as the integrated strains after marker/R6K excision have a single genomic copy of the BioBrick™ passenger in the target locus and the strains with low copy plasmid should maintain the device at the same copy number because all the plasmids are equal except the promoter to be measured. All the devices should produce the same mRNA sequence, as the predicted transcription start site is very similar for all the promoters ([34,35]) and the sequence downstream of the promoters is exactly the same (BBa_I13507). Finally, given a physical context, the growth rate of strains bearing a promoter of interest was similar to the growth rate of the strain bearing the BBa_J23101 reference promoter (data not shown). For these reasons, the RPU approach can be applied to characterize the promoters of interest of this work, as the validity hypotheses of the RPU approach are assumed to be respected.

All the coefficients of variation (CVs) were corrected for small samples by multiplying them by the $(1 + 1/(4\ N))$ factor, where N is the number of samples. Hypothesis tests were performed via MATLAB.

When assessing the statistical difference among the mean values of a group composed by more than two samples, ANOVA was performed. If the test showed a significant difference, post-hoc comparisons with the Bonferroni correction were performed via individual t-tests to compare the mean values.

In order to compute the CV within a group, the non-significantly different values were averaged and the final CV was computed on the mean values of the statistically different sub-groups.

Evolutionary stability experiments

Two 2-ml tubes with 1 ml of M9 supplemented medium were inoculated with two single colonies from a freshly streaked LB agar plate and they were incubated as described above for 24 hours. Every 24 hours, the cultures grown to saturation were 1000-fold diluted in 1 ml of sterile medium and incubated under the same conditions as before, in order to achieve ~10 generations per day [36-38]. Every day, 1 μl of each saturated culture was also added to 200 μl of sterile medium, incubated in the Infinite F200 reader and assayed as described above. Data were analyzed as before to obtain S_{cell}. The activity of a promoter over time was expressed as percentage of the activity at generation 0 (i.e. the first point of the time series).

Integrant strains were propagated and tested without antibiotic. Plasmid-bearing strains were propagated and tested in presence of 12.5 mg/l of chloramphenicol at generation 0, while they were tested with and without antibiotic for the following generations.

A freshly-inoculated non-fluorescent culture of MG1655 was always included in order to perform the fluorescence background subtraction.

After 150 generations, a glycerol stock was prepared for each evolved culture by mixing 750 µl of culture with 250 µl of 80% glycerol and stocks were long-term stored at −80°C. Evolved cultures of integrant strains and plasmid-bearing strains propagated with antibiotic were streaked on LB or LB + chloramphenicol agar plates, respectively. Plates were incubated at 37°C for about 20 hours. For each culture, 28 single colonies were picked, 1 ml of M9 supplemented medium (without antibiotic for integrant strains, with chloramphenicol for plasmid-bearing strains) was inoculated in 2-ml tubes and incubated at 37°C, 220 rpm for about 20 hours. Grown cultures were then assayed and data analyzed as described above. For each culture, 1 ml of M9 supplemented medium was also inoculated with 1 µl of glycerol stock of generation 150 and then the culture was incubated, assayed and analyzed as described above, to perform the fluorescence measurement of the whole evolved population (measured in duplicate).

Individual colonies isolated from non-evolved cultures were also assayed as a term of comparison for each recombinant strain. In this case 28 colonies were isolated by streaking the original glycerol stocks of these strains on LB agar plate.

For each culture, S_{cell} data were normalized by the median of the same recombinant strain at generation 0. Interquartile range (IQR) was used to express the variability of clones in each culture. Wilcoxon signed rank test was used to compare variability between generation 0 and 150 for all the cultures (paired samples). Wilcoxon rank sum test was used to compare variability between integrated strains and plasmid-bearing strains (unpaired samples).

Endnotes

[1] BioBrick™ is a trademark of The BioBricks Foundation. (http://www.biobricks.org).

Additional file

Additional file 1: Supplementary information.

Competing interests

The authors declare that they have no competing interest.

Authors' contributions

SZ, LP and PM conceived the study, designed the experiments and analyzed the data. LP designed and constructed the integrative base vector. SZ generated the integrant strains. SZ performed all the promoter characterization experiments. LP and NP contributed to the evolutionary stability experiments. LP, MGCDA and PM wrote the manuscript. All authors read and approved the final manuscript.

Acknowledgements

The authors thank the anonymous reviewers and the editor for their useful suggestions. We thank Dr. J.C. Anderson (UC Berkeley) for the pInt80-649 helper plasmid. We thank Michela Casanova for help during quantitative experiments. We also thank Matteo Meroso, Giacomo Zambianchi and the UNIPV-Pavia iGEM 2010 Team for preliminary experiments and helpful discussions. This project was partially funded by the Italian Ministero dell'Universita' e della Ricerca through the FIRB ITALBIONET project.

References

1. Lipps G: *Plasmids: current research and future trends.* Norfolk, U.K: Caister Academic Press; 2008.
2. Cohen SN, Chang AC, Boyer HW, Helling RB: **Construction of biologically functional bacterial plasmids *in vitro*.** *PNAS* 1973, **70:**3240–3244.
3. Mandecki W, Hayden MA, Shallcross MA, Stotland E: **A totally synthetic plasmid for general cloning, gene expression and mutagenesis in *Escherichia coli*.** *Gene* 1990, **94:**103–107.
4. Wang T, Ma X, Zhu H, Li A, Du G, Chen J: **Available methods for assembling expression cassettes for synthetic biology.** *Appl Microbiol Biotechnol* 2012, **93:**1853–1863.
5. Goh S, Good L: **Plasmid selection in Escherichia coli using an endogenous essential gene marker.** *BMC Biotechnol* 2008, **8:**61.
6. Jana S, Deb JK: **Strategies for efficient production of heterologous proteins in Escherichia coli.** *Appl Microbiol Biotechnol* 2005, **67:**289–298.
7. Kroll J, Klinter S, Schneider C, Voss I, Steinbuchel A: **Plasmid addiction systems: perspectives and applications in biotechnology.** *Microb Biotechnol* 2010, **3:**624–730.
8. Boyd D, Weiss DS, Chen JC, Beckwith J: **Towards Single-Copy Gene Expression Systems Making Gene Cloning Physiologically Relevant: Lambda InCh, a Simple Escherichia coli Plasmid-Chromosome Shuttle System.** *J Bacteriol* 2000, **182:**842–847.
9. Martinez-Morales F, Borges AC, Martinez A, Shanmaugam KT, Ingram LO: **Chromosomal Integration of Heterologous DNA in Escherichia coli with Precise Removal of Markers and Replicons Used during Construction.** *J Bacteriol* 1999, **181:**7143–7148.
10. Posfai G, Koob MD, Kirkpatrick HA, Blattner FR: **Versatile Insertion Plasmids for Targeted Genome Manipulations in Bacteria: Isolation, Deletion, and Rescue of the Pathogenicity Island LEE of the Escherichia coli O157:H7 Genome.** *J Bacteriol* 1997, **179:**4426–4428.
11. Datta S, Costantino N, Court DL: **A set of recombineering plasmids for gram-negative bacteria.** *Gene* 2006, **379:**109–15.
12. Datsenko KA, Wanner BL: **One-step inactivation of chromosomal genes in Escherichia coli K-12 using PCR products.** *PNAS* 2000, **97:**6640–5.
13. Ellis HM, Yu D, DiTizio T, Court DL: **High efficiency mutagenesis, repair, and engineering of chromosomal DNA using single-stranded oligonucleotides.** *PNAS* 2001, **98:**6742–6746.
14. Campbell AM: **Chromosomal insertion sites for phages and plasmids.** *J Bacteriol* 1992, **174:**7495–7499.
15. Haldimann A, Wanner BL: **Conditional-replication, integration, excision, and retrieval plasmid-host systems for gene structure-function studies of bacteria.** *J Bacteriol* 2001, **183:**6386–6393.
16. Cherepanov PP, Wackernagel W: **Gene disruption in *Escherichia coli* : TcR and KmR cassettes with the option of Flp-catalyzed excision of the antibiotic-resistance determinant.** *Gene* 1995, **158:**9–14.
17. Knight TF: **Idempotent Vector Design for Standard Assembly of Biobricks.** *MIT DSpace* 2003. doi:1721.1/21168.
18. MIT: *Registry of Standard Biological Parts.* http://partsregistry.org.
19. Shetty RP, Endy D, Knight TF: **Engineering BioBrick vectors from BioBrick parts.** *J Biol Eng* 2008, **2:**5.

20. Anderson JC, Dueber JE, Leguia M, Wu GC, Goler JA, Arkin AP, Keasling JD: **BglBricks: A flexible standard for biological part assembly.** *J Biol Eng* 2010, **4**:1.
21. Speer MA, Richard TL: **Amplified insert assembly: an optimized approach to standard assembly of BioBrick™ genetic circuits.** *J Biol Eng* 2011, **5**:17.
22. Pasotti L, Politi N, Zucca S, Cusella De Angelis MG, Magni P: **Bottom-up engineering of biological systems through standard bricks: A modularity study on basic parts and devices.** *PLoS One* 2012, **7**:e39407.
23. Davis JH, Rubin AJ, Sauer RT: **Design, construction and characterization of a set of insulated bacterial promoters.** *Nucleic Acids Res* 2011, **39**:1131–1141.
24. Block DH, Hussein R, Liang LW, Lim HN: **Regulatory consequences of gene translocation in bacteria.** *Nucleic Acids Res* 2012, **40**:8979–92.
25. Zucca S, Pasotti L, Mazzini G, De Angelis MG, Magni P: **Characterization of an inducible promoter in different DNA copy number conditions.** *BMC Bioinforma* 2012, **13**:S11.
26. Lutz R, Bujard H: **Independent and tight regulation of transcriptional units in Escherichia coli via the LacR/O, the TetR/O and AraC/I1-I2 regulatory elements.** *Nucleic Acids Res* 1997, **25**:1203–1210.
27. Lee DJ, Bingle LEH, Heurlier K, Pallen MJ, Penn CW, Busby SJW, Hobman JL: **Gene doctoring: a method for recombineering in laboratory and pathogenic *Escherichia coli* strains.** *BMC Microbiol* 2009, **9**:252.
28. Yomano LP, York SW, Zhou S, Shanmugam KT, Ingram LO: **Re-engineering *Escherichia coli* for ethanol production.** *Biotechnol Lett* 2008, **30**:2097–2103.
29. Turner PC, Yomano LP, Jarboe LR, York SW, Baggett CL, Moritz BE, Zentz EB, Shanmugam KT, Ingram LO: **Optical mapping and sequencing of the *Escherichia coli* KO11 genome reveal extensive chromosomal rearrangements, and multiple tandem copies of the *Zymomonas mobilis pdc* and *adhB* genes.** *J Ind Microbiol Biotechnol* 2012, **39**:629–639.
30. Tyo KEJ, Ajikumar PJ, Stephanopoulos G: **Stabilized gene duplication enables long-term selection-free heterologous pathway expression.** *Nat Biotechnol* 2008, **27**:760–767.
31. Blattner FR, Plunkett G III, Bloch CA, Perna NT, Burland V, Riley M, Collado-Vides J, Glasner JD, Rode CK, Mayhew GF, Gregor J, Davis NW, Kirkpatrick HA, Goeden MA, Rose DJ, Mau B, Shao Y: **The complete genome sequence of *Escherichia coli* K-12.** *Science* 1997, **277**:1453–1462.
32. Durfee T, Nelson R, Baldwin S, Plunkett G, Burland V, Mau B, Petrosino JF, Qin X, Muzny DM, Ayele M, Gibbs RA, Csorgo B, Posfai G, Weinstock GM, Blattner FR: **The complete genome sequence of *Escherichia coli* DH10B: insights into the biology of a laboratory workhorse.** *J Bacteriol* 2008, **190**:2597–2606.
33. Sambrook J, Fritsch EF, Maniatis T: *Molecular cloning: a laboratory manual.* Cold Spring Harbor, N.Y.: Cold Spring Harbor Laboratory Press; 1989.
34. Kelly JR, Rubin AJ, Davis JH, Ajo-Franklin CM, Cumbers J, Czar MJ, de Mora K, Glieberman AL, Monie DD, Endy D: **Measuring the activity of BioBrick promoters using an *in vivo* reference standard.** *J Biol Eng* 2009, **3**:4.
35. Harley CB, Reynolds RP: **Analysis of *E. coli* promoter sequences.** *Nucleic Acids Res* 1987, **15**:2343–2361.
36. Sleight SC, Bartley BA, Lieviant JA, Sauro HM: **Designing and engineering evolutionary robust genetic circuits.** *J Biol Eng* 2010, **4**:12.
37. Yang S, Sleight SC, Sauro HM: **Rationally designed bidirectional promoter improves the evolutionary stability of synthetic genetic circuits.** *Nucleic Acids Res* 2012, **41**:e33. doi:10.1093/nar/gks972.
38. Pasotti L, Zucca S, Lupotto M, Cusella De Angelis MG, Magni P: **Characterization of a synthetic bacterial self-destruction device for programmed cell death and for recombinant proteins release.** *J Biol Eng* 2011, **5**:8.

Secretion of polyhydroxybutyrate in *Escherichia coli* using a synthetic biological engineering approach

Asif Rahman, Elisabeth Linton, Alex D Hatch, Ronald C Sims and Charles D Miller*

Abstract

Background: Polyhydroxyalkanoates (PHAs) are a group of biodegradable plastics that are produced by a wide variety of microorganisms, mainly as a storage intermediate for energy and carbon. Polyhydroxybutyrate (PHB) is a short-chain-length PHA with interesting chemical and physical properties. Large scale production of PHB is not wide-spread mainly due to the downstream processing of bacterial cultures to extract the PHB. Secretion of PHB from *Escherichia coli* could reduce downstream processing costs. PHB are non-proteinaceous polymers, hence cannot be directly targeted for secretion. Phasin, PhaP1, is a low molecular weight protein that binds to PHB, reducing PHB granule size. In this study PHB is indirectly secreted with PhaP1 from *E. coli* via type I secretion using HlyA signal peptides.

Results: This study demonstrated the successful secretion of phasin and phasin bound PHB outside of the cell and into the culture medium. The secretion of PHB was initiated between 24 and 48 h after induction. After 48 h of culturing, 36% of the total PHB produced in the secreting strain was collected in the secreted fraction and 64% remained in the internal fraction. To further support the findings of this study, the PHB secretion phenomenon was observed using SEM.

Conclusions: From this study, the ability to use type I secretion to: 1) secrete phasin and 2) successfully secrete PHB has been shown.

Keywords: Polyhydroxyalkanoates, PHA, PHB, Secretion, Phasin, Translocation, HlyA, Synthetic biology

Background

Fossil derived plastics are non-biodegradable and toxic to the environment. Based on an United States Environmental Protection Agency study in 2011, there was an increase in non-biodegradable plastic accumulation in municipal solid waste systems from 0.5% to 12.4% during 1960 to 2010 [1]. Alternative means of producing plastics in large quantities that are both economically and environmentally friendly have recently gained considerable attention [2].

Replacing traditional plastic with biodegradable plastic such as polyhydroxyalkanoates (PHAs) can potentially reduce total waste by up to 20% [3]. PHAs are produced by a variety of microorganisms as an intercellular storage medium for energy and carbon and can accumulate up to 90% of the cell dry weight [4]. PHAs are biodegradable polymers. The biodegradability can range from days [5] to months [6] with degradation either taking place extracellularly or intracellularly. Extensive review on degradation of PHAs can be found in [7,8]. There are 155 different confirmed types of PHA monomer subunits, each with varying monomer repeat number and side groups [9]. Additionally, PHAs have melting temperatures between 50-180°C and crystallinity of 30–70% [10]. Thus, PHAs have a variety of possible applications, that could replace traditional plastics derived from petroleum [11]. Some of the possible applications are highlighted in previous studies [12,13] and include: packaging, medical uses [14], agricultural uses, and in carbon nanotubes [15].

Polyhydroxybutyrates (PHB) are a short-chain-length (scl) PHA polyester with between 3–5 carbon monomers [9]. The production of PHB in recombinant systems such

* Correspondence: charles.miller@usu.edu
Department of Biological Engineering, Utah State University, 4105 Old Main Hill, Logan 84322-4105, UT, USA

as *Escherichia coli* has been made possible by the isolation of the *phaCAB* operon from *Ralstonia eutropha* (*Cupriavidus necator*) and cloning into pBluescriptSK- to generate plasmid pBHR68. pBHR68 has been widely used for recombinant production of PHB in *E. coli.* The *phaCAB* operon is a three-step enzymatic process by which acetyl-CoA is converted to PHB: *phaC* (PHA synthase), *phaA* (β-ketothiolase), and *phaB* (acetoacetyl-CoA reductase) [4,16]. After production of PHB, the PHB polymer forms spherical granules with a hydrophobic core and attached proteins at the surface, including PHA synthase and phasin, PhaP1 [17-19].

The cost of producing PHAs is approximately US$ 4-6/kg and one of the major bottlenecks in scaling up recombinant PHA production systems is the isolation of the PHAs [12,20]. Current techniques that are used to isolate PHB from *E. coli* are invasive, including mechanical, chemical, and biological treatments. These techniques involve lysing the cellular membrane prior to isolation of the PHBs. There are a variety of methods that have been suggested for recovery of PHBs from microorganisms and a detailed review of the different PHB isolation methods is outlined in [20].

The development of a secretion mechanism eliminates the need for cell disruption by mechanical or chemical means, and may lead to continuous (or semi-continuous) PHB production systems. In gram-negative bacteria like *E. coli*, compounds are exported via six major secretory pathways [21-24]. Recombinant proteins can be targeted to type I and II secretory pathways through genetic fusion with signal peptide targeting sequences.

PHBs are non-proteinaceous polyesters and therefore cannot be directly targeted for translocation by signal peptide fusion. Phasins are low-molecular weight proteins that play a role in PHB granule formation by binding to the PHB granule surface [19,25]. Phasins are structural proteins found in organisms that naturally produce PHAs and are similar in function to oleosins which are found in plants [19,26]. Currently, oleosins are used to purify various compounds such as pharmaceuticals from plants [27].

Translocation of PHBs is possible through optimization of granule size, which reportedly varies from 50 to 1,000 kDa based on growth parameters and host strain [4]. One of the functions of phasin is to increase the surface-to-volume ratio of granules so that higher accumulation levels can be achieved [25,28,29]. Therefore, the size of the PHB granule can be decreased significantly through phasin overexpression. In one such study, overexpression of phasin resulted in a decrease of PHB granules from 70–310 nm diameter to 20–60 nm [28].

Type I secretion is a simple one step secretion system that can translocate proteins from the cytoplasm to the extracellular medium without protein interaction with the periplasm [30]. Proteins of nearly 900 kDa (large adhesion protein, *LapA*) have reportedly been secreted to the extracellular milieu by the type I secretory mechanism of gram-negative bacteria [31,32]. Specifically using the Hemolysin (HlyA) secretion mechanism proteins such as β-galactosidase (117 kDa) [33], β-gal-OmpF (56 kDa) [34], and green fluorescent protein (GFP) (27 kDa) [35] have been secreted by *E. coli.* The physical characteristics of the secretion channel are approximately 3.5 nm in diameter with a length of 14 nm as reported by Fernández et al., which makes the secretion phenomena of large proteins very interesting [36].

Our group has previously used a synthetic biological engineering approach to demonstrate the feasibility of HlyA, GeneIII, PelB, and TorA secretion systems in *E. coli* with the use of GFP [35]. From the aforementioned study, the type I secretion system using the HlyA signal peptide yielded the best results for secretion of GFP outside of the cell and into the medium [35]. The objective of this study was to demonstrate that phasin can be used to secrete PHB from *E. coli* using type I secretion machinery.

Results and discussion

Initial studies were carried out to demonstrate expression and then successful translocation of phasin, PhaP1, into the extracellular medium. Once this was demonstrated, PHB secretion experiments were conducted that included: growth studies, PHB production in secreted and non-secreted fractions, and visualization with scanning electron microscopy (SEM).

Analysis of phasin translocation

For *E. coli* cells expressing pCMEL1 and pLG575, a phasin band is observed at 22–26 kDa in the cytoplasmic fraction, the periplasmic fraction, the membrane fraction, and the concentrated extracellular media (Figure 1). This polyacrylamide gel and corresponding immunoblot demonstrated: 1) the ability for *E. coli* to produce non-codon optimized Phasin (from *R. eutropha),* a protein not naturally expressed in *E. coli* and 2) translocation of phasin into different fractions of the cell. Compared to other studies that focused non-translocated phasin, the phasin band sizes in all of the different fractions are similar [19,29]. Since the translocation of phasin was successful, secretion of PHB using phasin was then attempted.

Growth studies

The PHB secreting strain consisted of pCMEL3 + pLG575, whereas the non-secreting strain consisted of pBHR68 + pLG575. From the CFU/mL vs. time graph (Figure 2) stationary phase was reached at approximately 8–12 h for both non-secreting and secreting strains. The non-secreting and secreting strains had the highest overall

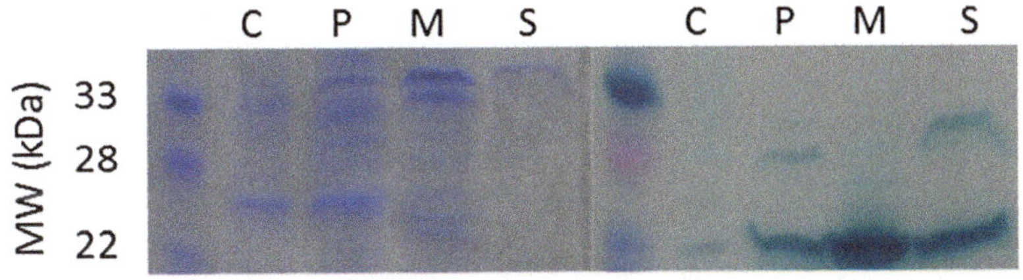

Figure 1 SDS polyacrylamide gel and corresponding immunoblot of subcellular fractions for PhaP1:HlyA. C – cytoplasmic fraction, P – periplasmic fraction, M – membrane fraction, S – concentrated supernatant (media) fraction. The position of phasin bands varies from roughly 22–26 kDa.

CFU/mL at approximately 9×10^{12} and 5×10^{12} CFU/mL, respectively, after 12 h. There was no significant difference between CFU/mL for the non-secreting and secreting strains after 24 and 48 h ($p > 0.05$). This statistical analysis on the CFU/mL demonstrated that there were no significant differences in *E. coli* growth between the two samples at times when PHB analysis was conducted (24 and 48 h). Furthermore, this suggests that the secretion of PHB does not affect cell viability.

PHB production analysis

PHB production analysis was carried out 24 and 48 because after 24 h the *E. coli* harboring the plasmid systems were in stationary phase. Previous studies demonstrated that PHB did not accumulate to significant levels during the exponential growth phase. Acetyl-CoA is required for cell synthesis during the exponential phase but is diverted to produce PHB in the stationary phase, thus, there is a delay between carbon source utilization and PHB production [37].

PHB measured inside the cell is defined as the internal fraction and PHB collected by the $CaCl_2$ precipitation method is the secreted fraction. 24 h after induction, the internal fraction of the cells demonstrated PHB production for both the secreting and non-secreting strains. While the non-secreting strain accumulated approximately 41.93 ± 13.5% of PHB in the dry cell weight at 24 h, the secreting strain accumulated approximately 28.85 ± 0.41% (Table 1). 48 h after induction the non-secreting and secreting strains had accumulated 47.24 ± 6.0% and 38.80 ± 15.5% PHB, respectively. There was no significant statistical difference seen in internally accumulated PHB in either the non-secreting or secreting strains at 24 and 48 h ($p > 0.05$) after induction. The internal levels of PHB accumulation in the non-secreting and secreting strains are comparable to those seen in other studies such as [38], where the authors showed accumulation of 42% PHB in *E. coli* DH5α harboring pBHR68 after 24–48 h [38].

PHB secreted fractions were analyzed at 24 and 48 h after induction for the secreting and non-secreting strains. PHB harvested in the secreted fraction of the non-secreting strains was 0.31 ± 0.35% and 0.72 ± 0.89%, respectively. The PHB present in the secreted phase demonstrated that some lysed cells containing PHB were found in this fraction after $CaCl_2$ precipitation. This is to be expected from a differential centrifugation technique, such as that used in this study. 24 and 48 h after induction, the

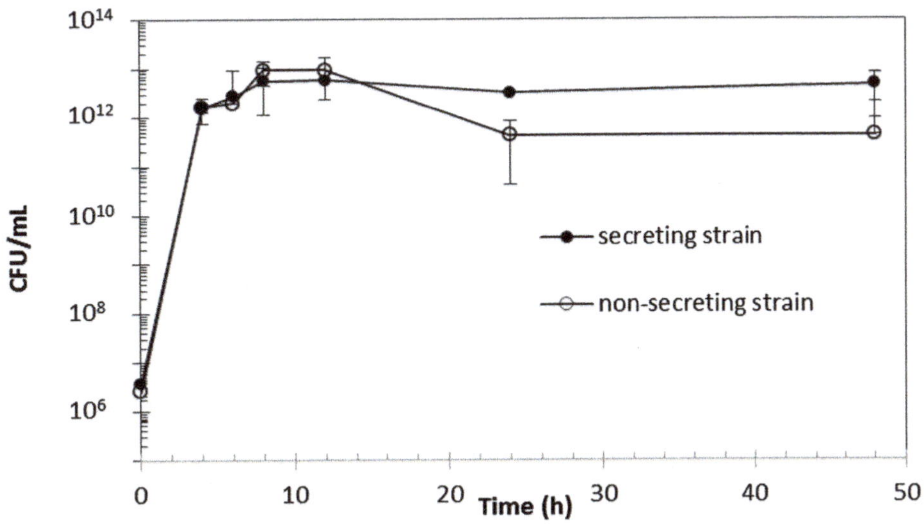

Figure 2 CFU/mL vs. Time (hours) for secreting and non-secreting strains of PHB producing *E. coli*, averaged from triplicate experiments (one standard deviation shown).

Table 1 Production of PHB in secreting (pCMEL3 + pLG575) and non-secreting (pBHR68 + pLG575) strains of *E. coli at* 24 and 48 h

Strain	Time (h)	% mass of PHB in dry mass		Production g/L PHB			
		Secreted fraction	Non- secreted fraction	Secreted fraction	Non-secreted fraction	Total production	g PHB/g Glucose
Non-secreting	24	0.31 ± 0.35	41.93 ± 13.5	0.40 ± 0.27	5.65 ± 1.1	6.05 ± 1.1	0.40 ± 0.07
	48	0.72 ± 0.89	47.24 ± 6.0	0.50 ± 0.41	5.43 ± 1.7	5.93 ± 1.8	0.40 ± 0.12
Secreting	24	0.69 ± 0.18	28.85 ± 0.41	0.40 ± 0.06	3.42 ± 0.33	3.82 ± 0.3	0.25 ± 0.02
	48	28.29 ± 7.2*	38.80 ± 15.5	2.57 ± 0.75*	4.58 ± 2.47	7.15 ± 2.6	0.48 ± 0.17

*Indicates statistical significance within column ($p < 0.05$).

secreting strain produced 0.69 ± 0.18% and 28.29 ± 7.2% respectively in the secreted fraction. The level of PHB seen in the secreted fraction of the secreting strain was statistically significant after 48 h post induction, compared to the non-secreting strain ($p < 0.05$). This increase in PHB present in the secreted phase after 48 h indicates that the secretion system is functioning and producing higher amounts of PHB, with a small amount of non-secreted PHB ending up in the secreted fraction. These results demonstrate that PHB secretion is initiated 24 h after induction.

Secretion of PHB can help in downstream processes by aiding in PHB separation from biomass. Of the total PHB produced by the secreting strain after 48 h, 36% was collected in the secreted fraction and the remaining 64% was in the internal fraction. The secreting strain had a total PHB production of 7.15 g/L compared to 5.93 g/L for the non-secreting strain after 48 h post induction.

It has been demonstrated from previous studies that PHB can accumulate in larger quantities in *E. coli* when using a bioreactor compared to a shaker flask. A fed-batch bioreactor study by Choi et al. 1999, reported accumulation of up to 77% PHB of dry cell weight [39] and another study reported accumulation of up to 80% [37]. Future studies will be performed to determine how well the secreting strain performs under similar bioreactor growth conditions.

SEM analysis

SEM is not widely used for PHB analysis since a surface topographical analysis is typically of little use. SEM has however been used to visualize PHA granules produced from recombinant *E. coli* [40] and PHA degradation from a variety of organisms [41]. In the case of secretion, SEM can provide images of what is occurring at the surface of *E. coli* during the secretion process. Figure 3A shows a PHB non-secreting *E. coli* strain harboring the pBHR68 plasmid. Figure 3B shows *E. coli* that is accumulating PHB and overexpressing phasin. Figure 3C shows the full secretion system in *E. coli* (pCMEL3 + pLG757). Figure 3C suggests that PHB is being secreted outside of the bacteria and into the medium. These observations further demonstrate the functioning of the PHB secretion system.

The SEM photo in Figure 3C suggests that secretion of PHB occurs at the polar regions of the cell. Interestingly it has been found in previous studies [42,43] that PHB granule formation occurs at the cell poles in *E. coli* when both phasin and PHB are being produced. When PHB is being produced the interpretation is that PHA synthase is active, and it has previously been suggested that PHA synthase has polar targeting information [43].

It is interesting that *E. coli* is able to secrete PHB through pore sizes of 3.5 nm [36] in diameter. From the results of this this study not all the PHB produced is

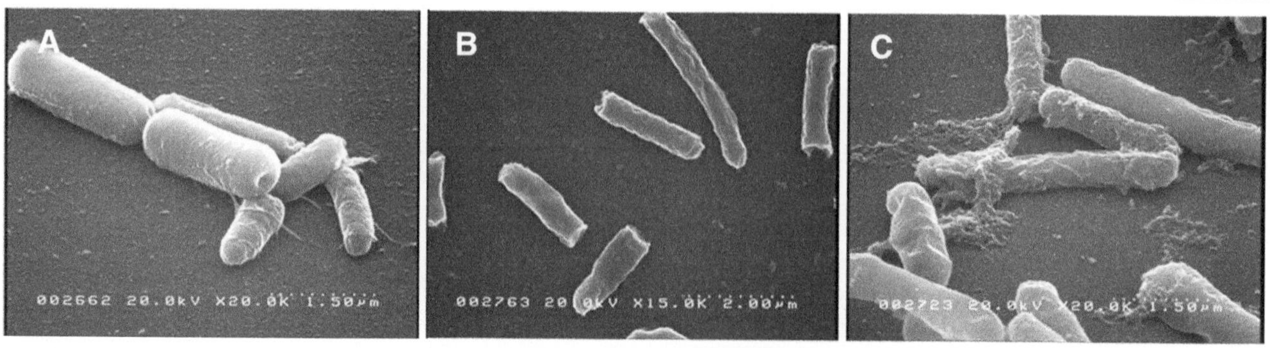

Figure 3 SEM images taken from overnight cultures of *E. coli* XL1Blue harboring different plasmid systems: A) pBHR68 (non-secreting), B) pCMEL3 (non-secreting with phasin overexpression), and C) pCMEL3 + pLG575 (complete PHB production and secretion system).

being secreted, suggesting there could be differences between the secreted and non-secreted PHB.

Conclusions

This study demonstrated the successful expression of phasin, PhaP1, and its translocation using type I secretion in *E. coli*. Once translocation of PhaP1 was successful, a recombinant synthetic biological system to secrete PHBs was designed and tested. Initiation of PHB secretion occurs between 24 and 48 h after induction of PHB synthesis. Of the total PHB produced by the secreting strain, 36% was collected in the secreted fraction and 64% remained in the internal fraction after 48 h.

Secretion of PHB should help in downstream processing whereby PHB is separated from the cell mass, which can aid in PHB recovery and purification. Future studies will include a detailed comparison of secreted versus non-secreted PHB characteristics.

Methods

Strains and plasmids

Genetic parts were constructed in accordance with the BioBrick and BioFusion technical standards [44,45]. The BioFusion standard was specifically used to create phasin and signal peptide parts because this genetic fusion system is compatible with the original BioBrick standard [45,46].

Descriptions of strains and plasmids used to study PHB and phasin production and translocation are provided as Table 2. Completed BioBrick parts were transformed in BL21-Gold (DE3) competent *E. coli* (Agilent Technologies, Santa Clara, CA) for protein expression studies. PHB production and secretion studies were carried out in *E. coli* XL1 Blue (Agilent Technologies, Santa Clara, CA). Plasmid pLG575 includes the coding regions for proteins HlyB and HlyD [47,48]. Plasmids pSB1AK3, pSB1A3, and pSB3K3 are BioBrick standard vectors for assembly and expression of BioBrick genetic devices [49].

BioBrick and plasmid construction

All restriction enzymes and related reagents were purchased from Thermo Fisher Scientific Inc. (Glen Burnie, MD). The signal peptide HlyA was made through synthetic design and construction as mention in Linton et al. 2012 (DNA 2.0, Menlo Park, CA) [35]. A BioFusion-compatible phasin BioBrick was constructed by isolating phaP1 from the genomic DNA of *R. eutropha* using PCR with primers PhaP1FOR and PhaP1REV that included the BioFusion prefix and suffix as overhanging ends (Table 2). The 620 bp PCR product was isolated by gel electrophoresis, digested with EcoRI and SpeI, and ligated into pSB3K3. A PstI site was removed from phaP1 (while conserving amino acid sequence) using a QuikChange II Site-Directed Mutagenesis Kit and the QuikChange® Primer Design Program (Agilent Technologies, Santa Clara, CA). The designed primers (g114t and g114t_antisense) for site-directed mutagenesis are shown in Table 2. The

Table 2 Strains, plasmids, and oligonucleotides used in this study

Strains	Relevant characteristics	Reference
Strains		
BL21-Gold (DE3)	E. coli B F^- ompT hsdS(rB^- mB^-) dcm^+ Tet^R gal λ(DE3) endA Hte	Agilent technologies
XL1 Blue	recA1 endA1 gyrA96 thi-1 hsdR17 supE44 relA1 lac [F′ proAB lacIqZΔM15 Tn10 (Tetr)]	Agilent technologies
Cupriavidus necator H16	Wild type, PHA producing	ATCC 17699
Plasmids		
pLG575	pACYC184 derivative, HlyBD, p15A origin, Cm^R	[47]
pBHR68	pBluescript SK–, phbCAB genes from R. eutropha	[16]
pSB1AK3	High copy BioBrick vector, pMB1 origin, Amp^R and Kan^R	[49]
pSB1A3	High copy BioBrick vector, pMB1 origin, Amp^R	[49]
pSB3K3	Medium copy BioBrick standard vector, p15A origin, Kan^R	[49]
pCMEL1	phaP1, C-terminal BioFusion with HlyA signal peptide, Lac promoter (BBa_R0010), RBS(BBa_B0034), in pSB1A3	This study
pCMEL2	phaP1, C-terminal BioFusion with HlyA signal peptide, Lac promoter (BBa_R0010), RBS(BBa_B0034), in pSB3K3	This study
pCMEL3	phaP1, C-terminal BioFusion with HlyA signal peptide, Lac promoter (BBa_R0010), RBS(BBa_B0034), in pBHR68	This study
Oligonucleotides		
PhaP1FOR	**5′**-gaattcgcggccgcttctagaatgatcctcaccccggaaca-**3′**	This study
PhaP1REV	**5′**- ctgcagcggccgctactagttcaggcagccgtcgtcttct-**3′**	This study
g114t	**5′**-cgtcgagctgaaccttcaggtcgtcaagact-**3′**	This study
g114t_antisense	**5′**-agtcttgacgacctgaaggttcagctcgacg-**3′**	This study

PstI mutation in phaP1 was successfully carried out and confirmed by sequence analysis.

Step-wise assembly of composite BioBrick devices was primarily carried out in pSB1AK3. Completed devices were subsequently ligated into pSB1A3 and pSB3K3. The lac promoter (BBa_R0010) and ribosome binding site (BBa_B0034) were used as described in Linton et al. 2012 [35]. The pCMEL1 plasmid was used for studies on phasin translocation because its origin of replication (pMB1) was compatible with the origin of replication of pLG575 (p15A). BL21-Gold (DE3) was co-transformed with pCMEL1 and pLG575. The pCMEL3 plasmid was used for studies on type I secretion of PHA. The composite part containing the promoter, RBS, coding region, and terminator were cloned into pBHR68 from pCMEL2 by digestion with EcoRI and XhoI. XL1-Blue was co-transformed with pCMEL3 and pLG575 for PHA secretion studies. A schematic of the secretion system is presented in Figure 4.

Cellular fractionation and western blotting

Overnight cultures of *E. coli* BL21-Gold (DE3) harboring pCMEL1 and pLG575 were used to inoculate (1:100 v/v dilution) 100 ml of LB media containing 25 μg/ml chloramphenicol (Acros Organics, Fair Lawn, NJ) and 50 μg/ml ampicillin (IBI Scientific, Peosta, IA) [50]. The lysozyme/EDTA/osmotic shock and chloroform-based cellular fractionation procedures and methods for analyzing supernatant fractions are previously discussed in Linton et al. 2012 [35]. Briefly, 50 mL of cell culture were centrifuged and the pellet was resuspended in a buffer and subjected to osmotic shock. Centrifugation was used to

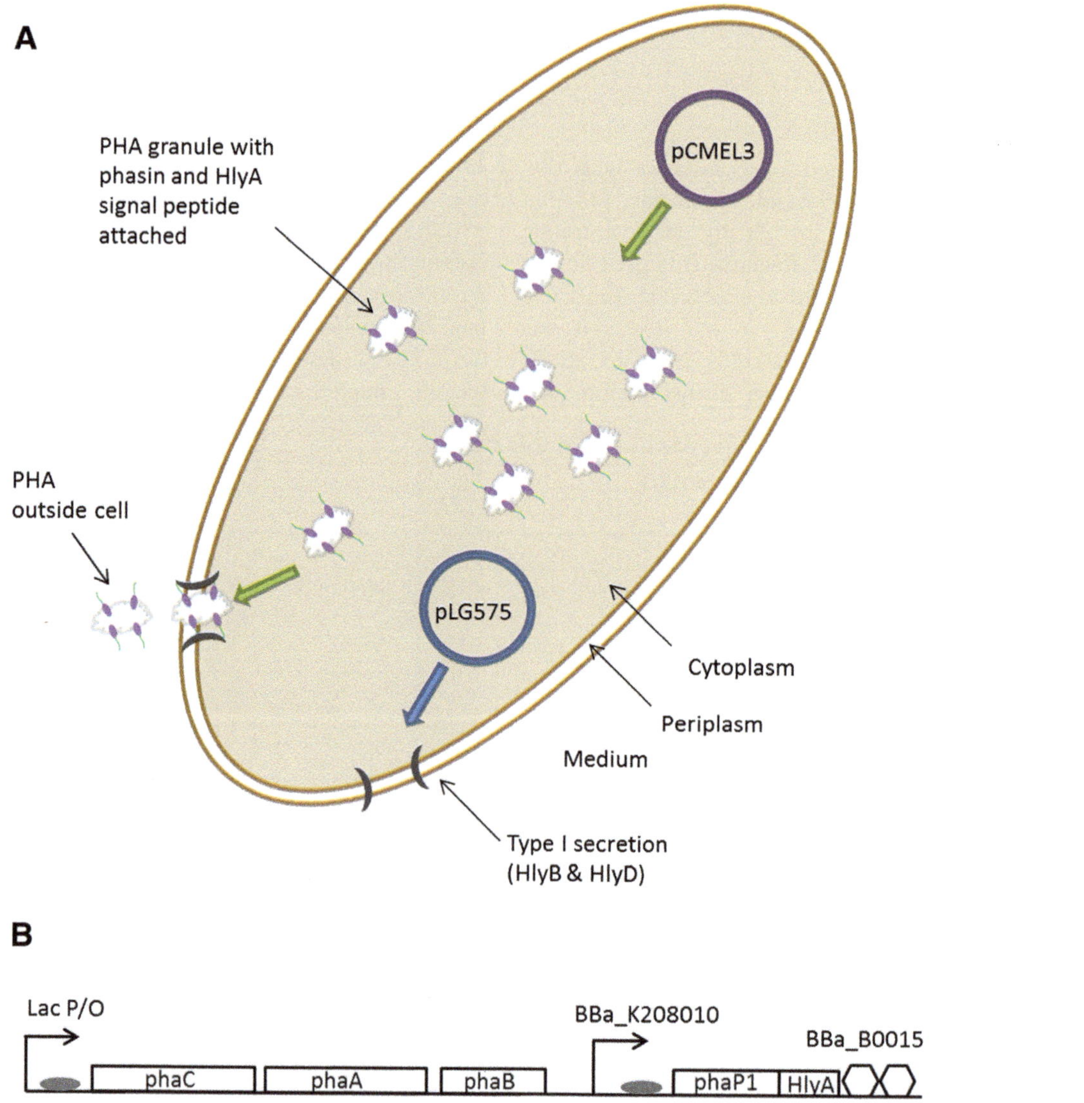

Figure 4 PHB secretion. A. Schematic for PHB secretion containing dual plasmid system pCMEL3 and pLG575. Phasin with attached signal peptide binds to PHB granule surface and the PHB-phasin-signal peptide complex is targeted for type I secretion. **B**. pCMEL3 plasmid consisting of phaC, phaC, and phaB genes from pBHR68. pCMEL3 also contains the genes needed for phasin-HylA production.

separate the periplasm from spheroplast. Cytoplasm and membranes were fractionated via ultracentrifugation.

Approximately 40 μg of protein from each of the respective fractioned samples were separated using precast 12% SDS-polyacrylamide tris-glycine gels (Jule Inc., Milford, CT). Electrophoresis operating conditions were used as specified by the manufacturer. Phasin immunoblotting was carried out using a PhaP1 specific antibody [51,52]. An Anti-Rabbit IgG HRP conjugated secondary antibody was purchased from Promega (Madison, WI). Respective primary and secondary antibody concentrations of 1:50,000 v/v and 1:2,500 v/v were used.

PHB secretion studies

All secretion studies were carried out in triplicate. Secretion studies were conducted with the secreting system (pCMEL3 + pLG575 in XL1 Blue) and non-secreting system (pBHR68 + pLG575 in XL1 Blue). The non-secreting system does not contain phasin-HlyA.

Media formulation and growth conditions

R. eutropha was cultured following the methods outlined in Linton et al. 2012 [53]. M9 salts (Becton, Dickinson and Co, Sparks, MD) supplemented with 1.5% (w/v) glucose (ACS grade, Acros Organics, Fair Lawn, NJ) and 0.2% (w/v) yeast extract (Becton, Dickinson and Co, Sparks, MD) was used for PHB secretion studies in *E. coli* [38]. Overnight *E. coli* cultures harboring specific plasmids were inoculated from freezer stocks into 5 ml of M9 media with chloramphenicol (34 μg/ml), ampicillin (50 μg/ml), and grown in an orbital shaker table at 220 rpm at 37°C. Overnight cultures were then used to seed larger 250 ml flasks (50 ml media volume) at an initial optical density (OD_{600}) of 0.05 at time 0 h. 0.1 mM Isopropyl β-D-1-thiogalactopyranoside (IPTG) (Gold Biotechnology, Inc. St. Louis, MO) was added to each flask at time 0 h. Flasks were removed at 24, and 48 h and analyzed for PHB. CFU/mL was measured at time points 0, 4, 6, 8, 12, and 24 h.

Recovery of secreted PHB

PHB granules can agglomerate together naturally and previous studies have demonstrated the use of Calcium chloride ($CaCl_2$) to enhance this process. A study by Fidler et al. used $CaCl_2$ to purify PHB from lysed cells by selectively aggregating PHB granules. It was observed that PHB granules fell to the bottom of the test tube after addition of $CaCl_2$ [54]. Another $CaCl_2$ method for PHB recovery by Resch et al. used a low speed centrifugation step to further enhance PHB recovery from cell debris [55].

In this study, techniques for secreted PHB recovery were adapted from the methods outlined in [55]. At 24, and 48 h 0.01 M $CaCl_2$ (final concentration, Avantor Performance Materials, Inc. Center Valley, PA) was added to the bacterial culture and mixed by inverting the tube several times. The tubes were then allowed to sit for 10 mins at room temperature and then centrifuged at 54 × g for 5 min. The supernatant was removed and transferred to a fresh tube and the pellet was freeze dried. The supernatant was centrifuged at 3452 × g for 10 mins and the pellet was freeze dried. The pellet from the first centrifugation contained secreted PHB with $CaCl_2$ and the pellet from the second centrifugation contained bacterial mass and non-secreted PHB. Secretion studies and PHB analysis were conducted in triplicate.

PHB concentration determination

PHB concentrations were determined based on a NMR-GC method described in Linton et al. 2012 [53]. Briefly, samples were lyophilized after which approximately 15 mg of sample was mixed with equal volumes of sodium hypochlorite and deuterated chloroform. Samples were centrifuged and 1H NMR was carried out on the PHB fraction. PHB concentration was determined from a NMR-GC standard.

Scanning electron microscopy (SEM)

To visually show PHB secretion from *E. coli*, SEM was performed. SEM protocols were used as mentioned in [56]. Briefly, secreting and non-secreting strains were grown up overnight and fixed onto glass cover slips. Samples were mounted on aluminum stubs and sputter coated with 10 nm gold. SEM was carried out using a Hitachi S4000 SEM.

Statistical analysis

All growth (CFU/mL) and PHB yield studies were carried out in triplicate to show consistency of data. Statistical analysis was conducted with Statistical Analysis Software (SAS 9.3, SAS Institute Inc., Cary, NC). A two-way analysis of variance (ANOVA) with tukey post hoc comparison performed on significant results (confidence level 95%).

Competing interests

The authors declare that they have no competing interests.

Authors' contributions

AR and EL conducted experiments and wrote the manuscript draft. CDM conceived and designed the experiments and wrote the manuscript. AH assisted with experiments and data analysis. RCS assisted with experimental design and manuscript revision. All authors read and approved the final manuscript.

Acknowledgements

The authors would also like to acknowledge FenAnn Shen and T.C. Shen in the Physics Department at Utah State University for assisting with the SEM. Marie K. Walsh in Nutrition, Dietetics and Food Sciences at Utah State University helped with phasin translocation studies. The authors would like to thank I. Barry Holland and Thorsten Jumpertz for the pLG575 plasmid. The authors would also like to acknowledge the gift of the primary phasin antibody from Anthony J. Sinskey and JoAnne Stubbe. This research was supported by the Utah State University Synthetic Bio-Manufacturing Center (SBC) and Utah Science Technology and Research initiative (USTAR).

References

1. Philp JC, Ritchie RJ, Guy K: **Biobased plastics in a bioeconomy.** *Trends Biotechnol* 2013, **31**:65–67.
2. Keshavarz T, Roy I: **Polyhydroxyalkanoates: bioplastics with a green agenda.** *Curr Opin Microbiol* 2010, **13**:321–326.
3. Dias JML, Lemos PC, Serafim LS, Oliveira C, Eiroa M, Albuquerque MGE, Ramos AM, Oliveira R, Reis MAM: **Recent advances in polyhydroxyalkanoate production by mixed aerobic cultures: from the substrate to the final product.** *Macromol Biosci* 2006, **6**:885–906.
4. Madison LL, Huisman GW: **Metabolic engineering of poly(3-hydroxyalkanoates): from DNA to plastic.** *Microbiol Mol Biol Rev* 1999, **63**:21–53.
5. Shang L, Fei Q, Zhang YH, Wang XZ, Fan DD, Chang HN: **Thermal properties and biodegradability studies of poly(3-hydroxybutyrate-co-3-hydroxyvalerate).** *Journal of Polymers and the Environment* 2012, **20**:23–28.
6. Mergaert J, Anderson C, Wouters A, Swings J: **Microbial degradation of poly(3-hydroxybutyrate) and poly(3-hydroxybutyrate-co-3-hydroxyvalerate) in compost.** *Journal of Environmental Polymer Degradation* 1994, **2**:177–183.
7. Jendrossek D, Handrick R: **Microbial degradation of polyhydroxyalkanoates.** *Annu Rev Microbiol* 2002, **56**:403–432.
8. Jendrossek D, Schirmer A, Schlegel HG: **Biodegradation of polyhydroxyalkanoic acids.** *Appl Microbiol Biotechnol* 1996, **46**:451–463.
9. Agnew DE, Pfleger BF: **Synthetic biology strategies for synthesizing polyhydroxyalkanoates from unrelated carbon sources.** *Chemical Engineering Science* 2012. In press. Available online, ISSN 0009-2509, http://dx.doi.org/10.1016/j.ces.2012.12.023.
10. Rehm BHA: **Bacterial polymers: biosynthesis, modifications and applications.** *Nat Rev Microbiol* 2010, **8**:578–592.
11. Chanprateep S: **Current trends in biodegradable polyhydroxyalkanoates.** *J Biosci Bioeng* 2010, **110**:621–632.
12. Akaraonye E, Keshavarz T, Roy I: **Production of polyhydroxyalkanoates: the future green materials of choice.** *J Chem Technol Biotechnol* 2010, **85**:732–743.
13. Gumel AM, Annuar MSM, Chisti Y: **Recent advances in the production, recovery and applications of polyhydroxyalkanoates.** *J Polym and the Environ* 2013, **21**:580–605.
14. Numata K, Doi Y: **Dual biosyntheses of poly[(R)-3-hydroxybutyric acid] and silk protein for the fabrication of biofunctional bioplastic.** *Polymer Journal* 2011, **43**:642–647.
15. Xu C, Qiu Z: **Crystallization behavior and thermal property of biodegradable poly(3-hydroxybutyrate)/multi-walled carbon nanotubes nanocomposite.** *Polymers for Advanced Technologies* 2011, **22**:538–544.
16. Spiekermann P, Rehm BHA, Kalscheuer R, Baumeister D, Steinbüchel A: **A sensitive, viable-colony staining method using Nile red for direct screening of bacteria that accumulate polyhydroxyalkanoic acids and other lipid storage compounds.** *Arch Microbiol* 1999, **171**:73–80.
17. Grage K, Jahns AC, Parlane N, Palanisamy R, Rasiah IA, Atwood JA, Rehm BHA: **Bacterial polyhydroxyalkanoate granules: biogenesis, structure, and potential use as nano-/micro-beads in biotechnological and biomedical applications.** *Biomacromolecules* 2009, **10**:660–669.
18. Rehm BHA: **Polyester synthases: natural catalysts for plastics.** *Biochem J* 2003, **376**:15–33.
19. Pötter M, Müller H, Reinecke F, Wieczorek R, Fricke F, Bowien B, Friedrich B, Steinbüchel A: **The complex structure of polyhydroxybutyrate (PHB) granules: four orthologous and paralogous phasins occur in ralstonia eutropha.** *Microbiology* 2004, **150**:2301–2311.
20. Jacquel N, Lo C-W, Wei Y-H, Wu H-S, Wang SS: **Isolation and purification of bacterial poly(3-hydroxyalkanoates).** *Biochem Eng J* 2008, **39**:15–27.
21. Pukatzki S, Ma AT, Revel AT, Sturtevant D, Mekalanos JJ: **Type VI secretion system translocates a phage tail spike-like protein into target cells where it cross-links actin.** *Proc Natl Acad Sci USA* 2007, **104**:15508–15513.
22. Henderson IR, Navarro-Garcia F, Desvaux M, Fernandez RC, Ala'Aldeen D: **Type V protein secretion pathway: the autotransporter story.** *Microbiol Mol Biol Rev* 2004, **68**:692–744.
23. Economou A, Christie PJ, Fernandez RC, Palmer T, Plano GV, Pugsley AP: **Secretion by numbers: protein traffic in prokaryotes.** *Mol Microbiol* 2006, **62**:308–319.
24. Desvaux M, Hébraud M, Talon R, Henderson IR: **Secretion and subcellular localizations of bacterial proteins: a semantic awareness issue.** *Trends Microbiol* 2009, **17**:139–145.
25. York GM, Stubbe J, Sinskey AJ: **New insight into the role of the PhaP phasin of Ralstonia eutropha in promoting synthesis of polyhydroxybutyrate.** *J Bacteriol* 2001, **183**:2394–2397.
26. Steinbuchel A, Aerts K, Babel W, Follner C, Kiebergesell M, Madkour MH, Mayer F, Pieper-Furst U, Pries A, Valentin HE, Wieczorek R: **Considerations on the structure and biochemistry of bacterial polyhydroxyalkanoic acid inclusions.** *Can J Microbiol* 1995, **41**:94–105.
27. Giddings G, Allison G, Brooks D, Carter A: **Transgenic plants as factories for biopharmaceuticals.** *Nat Biotechnol* 2000, **18**:1151–1155.
28. Maehara A, Ueda S, Nakano H, Yamane T: **Analyses of a polyhydroxyalkanoic acid granule-associated 16-kilodalton protein and its putative regulator in the pha locus of Paracoccus denitrificans.** *J Bacteriol* 1999, **181**:2914–2921.
29. Wieczorek R, Pries A, Steinbuchel A, Mayer F: **Analysis of a 24-kilodalton protein associated with the polyhydroxyalkanoic acid granules in Alcaligenes eutrophus.** *J Bacteriol* 1995, **177**:2425–2435.
30. Mergulhão FJM, Summers DK, Monteiro GA: **Recombinant protein secretion in escherichia coli.** *Biotechnol Adv* 2005, **23**:177–202.
31. Holland IB, Schmitt L, Young J: **Type 1 protein secretion in bacteria, the ABC-transporter dependent pathway.** *Mol Membr Biol* 2005, **22**:29–39.
32. Hinsa SM, Espinosa-Urgel M, Ramos JL, O'Toole GA: **Transition from reversible to irreversible attachment during biofilm formation by Pseudomonas fluorescens WCS365 requires an ABC transporter and a large secreted protein.** *Mol Microbiol* 2003, **49**:905–918.
33. Kenny B, Haigh R, Holland IB: **Analysis of the haemolysin transport process through the secretion from Escherichia coli of PCM, CAT or β-galactosidase fused to the Hly C-terminal signal domain.** *Mol Microbiol* 1991, **5**:2557–2568.
34. Mackman N, Baker K, Gray L, Haigh R, Nicaud JM, Holland IB: **Release of a chimeric protein into the medium from Escherichia coli using the C-terminal secretion signal of haemolysin.** *EMBO J* 1987, **6**:2835–2841.
35. Linton E, Walsh MK, Sims RC, Miller CD: **Translocation of green fluorescent protein by comparative analysis with multiple signal peptides.** *Biotechnol J* 2012, **7**:667–676.
36. Fernández LA, De Lorenzo V: **Formation of disulphide bonds during secretion of proteins through the periplasmic-independent type I pathway.** *Mol Microbiol* 2001, **40**:332–346.
37. Lee SY, Yim KS, Chang HN, Chang YK: **Construction of plasmids, estimation of plasmid stability, and use of stable plasmids for the production of poly(3-hydroxybutyric acid) by recombinant Escherichia coli.** *J Biotechnol* 1994, **32**:203–211.
38. Kang Z, Wang Q, Zhang HJ, Qi QS: **Construction of a stress-induced system in Escherichia coli for efficient polyhydroxyalkanoates production.** *Appl Microbiol Biotechnol* 2008, **79**:203–208.
39. Choi JI, Lee SY: **Efficient and economical recovery of poly(3-hydroxybutyrate) from recombinant Escherichia coli by simple digestion with chemicals.** *Biotechnol Bioeng* 1999, **62**:546–553.
40. Ramachander TVN, Rohini D, Belhekar A, Rawal SK: **Synthesis of PHB by recombinant E. coli harboring an approximately 5 kb genomic DNA fragment from Streptomyces aureofaciens NRRL 2209.** *Int J Biol Macromol* 2002, **31**:63–69.
41. Molitoris HP, Moss ST, De Koning GJM, Jendrossek D: **Scanning electron microscopy of polyhydroxyalkanoate degradation by bacteria.** *Appl Microbiol Biotechnol* 1996, **46**:570–579.
42. Peters V, Rehm BHA: **In vivo monitoring of PHA granule formation using GFP-labeled PHA synthases.** *FEMS Microbiol Lett* 2005, **248**:93–100.
43. Peters V, Becher D, Rehm BHA: **The inherent property of polyhydroxyalkanoate synthase to form spherical PHA granules at the cell poles: The core region is required for polar localization.** *J Biotechnol* 2007, **132**:238–245.
44. *Idempotent vector design for standard assembly of biobricks.* http://hdl.handle.net/1721.1/21168.
45. *A New biobrick assembly strategy designed for facile protein engineering.* http://hdl.handle.net/1721.1/32535.
46. Anderson JC, Dueber JE, Leguia M, Wu GC, Arkin AP, Keasling JD: **BglBricks: a flexible standard for biological part assembly.** *J Biol Eng* 2010, **4**:1.

47. Mackman N, Nicaud JM, Gray L, Holland IB: **Genetical and functional organisation of the Escherichia coli haemolysin determinant 2001.** *Molecular and General Genetics* 1985, **201**:282–288.
48. Jobling MG, Holmes RK: **Construction of vectors with the p15a replicon, kanamycin resistance, inducible lacZα and pUC18 or pUC19 multiple cloning sites.** *Nucleic Acids Res* 1990, **18**:5315–5316.
49. *Registry of standard biological parts.* http://partsregistry.org/.
50. Sambrook J, Russel DW: *Molecular cloning: a laboratory manual.* Cold Spring Harbor, NY: Cold Spring Harbor Laboratory Press; 2001.
51. York GM, Junker BH, Stubbe J, Sinskey AJ: **Accumulation of the PhaP phasin of Ralstonia eutropha is dependent on production of polyhydroxybutyrate in cells.** *J Bacteriol* 2001, **183**:4217–4226.
52. Tian J, He A, Lawrence AG, Liu P, Watson N, Sinskey AJ, Stubbe J: **Analysis of transient polyhydroxybutyrate production in Wautersia eutropha H16 by quantitative Western analysis and transmission electron microscopy.** *J Bacteriol* 2005, **187**:3825–3832.
53. Linton E, Rahman A, Viamajala S, Sims RC, Miller CD: **Polyhydroxyalkanoate quantification in organic wastes and pure cultures using a single-step extraction and 1H NMR analysis.** *Water Sci Technol* 2012, **66**:1000–1006.
54. Fidler S, Dennis D: **Polyhydroxyalkanoate production in recombinant Escherichia coli.** *FEMS Microbiol Rev* 1992, **103**:231–235.
55. Resch S, Gruber K, Wanner G, Slater S, Dennis D, Lubitz W: **Aqueous release and purification of poly([beta]-hydroxybutyrate) from Escherichia coli.** *J Biotechnol* 1998, **65**:173–182.
56. Mortensen T, Shen S, Shen F, Walsh MK, Sims RC, Miller CD: **Investigating the effectiveness of St John's wort herb as an antimicrobial agent against mycobacteria.** *Phytother Res* 2012, **26**:1327–1333.

A demonstration of the antimicrobial effectiveness of various copper surfaces

Victor K Champagne[1] and Dennis J Helfritch[2*]

Abstract

Background: Bacterial contamination on touch surfaces results in increased risk of infection. In the last few decades, work has been done on the antimicrobial properties of copper and its alloys against a range of micro-organisms threatening public health in food processing, healthcare and air conditioning applications; however, an optimum copper method of surface deposition and mass structure has not been identified.

Results: A proof-of-concept study of the disinfection effectiveness of three copper surfaces was performed. The surfaces were produced by the deposition of copper using three methods of thermal spray, namely, plasma spray, wire arc spray and cold spray The surfaces were then inoculated with meticillin-resistant *Staphylococcus aureus* (MRSA). After a two hour exposure to the surfaces, the surviving MRSA were assayed and the results compared.
The differences in the copper depositions produced by the three thermal spray methods were examined in order to explain the mechanism that causes the observed differences in MRSA killing efficiencies. The cold spray deposition method was significantly more effective than the other methods. It was determined that work hardening caused by the high velocity particle impacts created by the cold spray technique results in a copper microstructure that enhances ionic diffusion, and copper ions are principally responsible for antimicrobial activity.

Conclusions: This test showed significant microbiologic differences between coatings produced by different spray techniques and demonstrates the importance of the copper application technique. The cold spray technique shows superior anti-microbial effectiveness caused by the high impact velocity imparted to the sprayed particles which results in high dislocation density and high ionic diffusivity.

Background

Bacterial growth on surfaces is a cause of concern in many hospitals and food processing industries due to the possibility of increased risk of bacterial infection [1]. The bacterial contamination of hospital surfaces, including patient rooms, nurse stations and kitchens has been extensively documented [2-5]. Contamination of meat and vegetable preparation surfaces, including refrigerators, and conveyors have also been the subject of investigation [6-10]. In addition to topically applied disinfectants, the use of surfaces that can self-disinfect would enhance overall infection prevention.

In the last few decades, work has been done on the antimicrobial properties of copper and its alloys against a range of micro-organisms threatening public health in food processing and healthcare applications [11]. The use of copper and copper alloys for frequently touched surfaces such as door and furniture hardware, bed rails, light switches and food preparation surfaces can help limit microbial infections in hospitals and food dispensing organizations. Michels, et al. [12] show that increasing the copper content of alloys increases antimicrobial effectiveness. The contact killing is so rapid that the production of protective biofilms is not possible [13].

The specific mechanism by which copper affects cellular structures is not yet proven, but the active agent of cell destruction is generally considered to be the copper ion [11,14,15].

Recent studies showed that large amounts of copper ions were taken up by E. coli over 90 min, when cells were applied to copper coupons via an aqueous suspension (a standing drop). When cells were plated on copper using minimum liquid and a drying time of 5 seconds, the accumulation of copper ions by cells was even more dramatic, reaching a high concentration in a fraction of the time.

* Correspondence: dennis.j.helfritch.ctr@mail.mil
[2]Dynamic Science, Aberdeen, MD, USA
Full list of author information is available at the end of the article

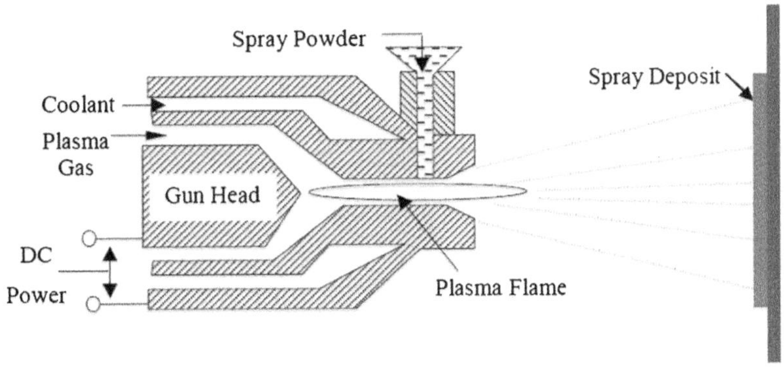

Figure 1 The plasma spray deposition system.

The copper ion level of cells remained high throughout the killing phase, suggesting that cells become overwhelmed by their intracellular copper [15]. The grain structure of the copper material affects ion diffusion and hence affects bacterial destruction by copper ions.

The US Environmental Protection Agency (EPA) registers five copper alloys with public health claims [16]. All of the alloys have minimum nominal copper concentrations of 60%. Registration of copper and certain copper alloys such as brass and bronze means that the EPA recognizes these solid materials' antimicrobial properties. Products made from any of the registered alloys are legally permitted to make public health claims relating to the control of organisms that pose a threat to human health. Laboratory studies conducted under EPA-approved protocols have proven copper's ability to kill, within 2 hours of contact time, more than 99.9% of the following disease-causing bacteria: *Staphylococcus aureus, Enterobacter aerogenes, Escherichia coli* O157:H7, *Pseudomonas aeruginosa,* Vancomycin-resistant *Enterococcus faecalis* (VRE) and MRSA.

Copper surface generation

In order to make use of the antimicrobial ability of copper, surfaces that contact skin and foods should be composed of copper or copper alloy. This can be accomplished with solid copper equipment or by means of copper surface coating. In general, cost considerations favour copper coatings over solid structural copper. Various metal spray techniques are available for the purpose of depositing a copper surface onto implements that can transmit microorganisms, and it is desired to identify an optimal deposition method. Accordingly, three metal spray techniques are evaluated with respect to the anti-microbial activity of the copper surfaces produced by each.

Plasma spray

The plasma spray process shown in Figure 1 uses a DC electric arc to generate a stream of high temperature ionized plasma gas, which acts as the spraying heat source. The coating material, in powder form and carried by an inert gas, is injected into the plasma jet where it is melted and propelled towards the substrate.

The plasma spray gun includes a copper anode and tungsten cathode, which are both water cooled. Plasma gas (argon, nitrogen, hydrogen, helium) flows around the cathode and through the anode, which is shaped as a constricting nozzle. The plasma, containing suspended

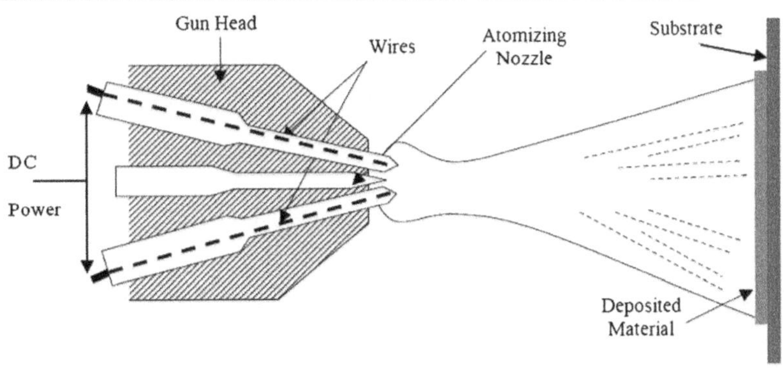

Figure 2 The wire Arc spray deposition system.

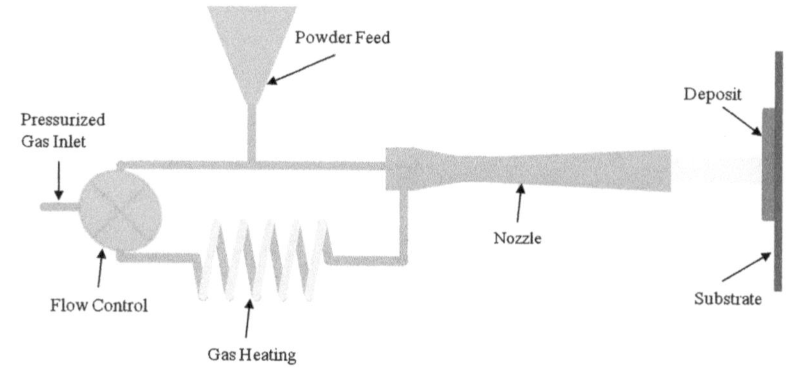

Figure 3 The cold spray deposition system.

metal droplets, exits in the anode nozzle and is directed toward a surface, where the particles deposit.

Arc spray

The arc spray process shown in Figure 2 creates an arc between two metallic wires acting as consumable electrodes. A DC voltage is applied between the wires, and an arc discharge is created at the contact of the wires. The wire electrodes are melted by the electric arc and a compressed air jet disperses the molten droplets and propels them onto a surface.

Cold spray

The cold spray process shown in Figure 3 imparts supersonic velocities to metal particles by placing them in a heated nitrogen or helium gas stream that is expanded through a converging–diverging nozzle. The powder feed is inserted at high pressure at the nozzle entrance. The particles, entrained within the gas, are directed towards a surface, where they are embedded on impact, forming a strong bond with the surface. The term "cold spray" has been used to describe this process due to the relatively low temperatures (100-500°C) of the expanded gas stream that exits the nozzle. Subsequent spray passes increase the structure thickness. The adhesion of the metal powder to the substrate, as well as the cohesion of the deposited material, is accomplished in the solid state.

The relatively low porosity of the cold spray coating results from particle packing caused by high velocity impact. Another characteristic of high velocity impacts is the creation of grain dislocations and work hardening. The low oxide content of cold sprayed deposits occurs because the particle temperature remains low and thus inhibits oxidation.

The spray techniques described each produce impacting metal particles in distinct temperature and velocity ranges. These temperatures and velocities create metal coatings with different characteristics with respect to the presence of oxides, porosity, grain dislocations, and hardness. Because of these metallurgic differences, it is reasonable to assume that the coatings will exhibit differences in antimicrobial efficiency. Table 1 gives the particle temperatures and impact velocities, as well as the porosity and oxide ranges of the resulting deposits.

Test procedure

These three surface coating techniques were used to produce copper-coated metal coupons. Approximately 1 mm thick coatings were applied to aluminum substrates. The coatings completely covered the metal substrates with an impervious seal. The copper powder feedstock used for the plasma and cold sprays is shown in Figure 4. Cross sections of the coupons produced by the three spray techniques are shown in Figures 5, 6 and 7. Differences in microstructure are clearly evident, suggesting that differences in biological activity may also occur. Evidence of particle melting is clear for the high-temperature plasma and wire arc processes. The incidence of large voids is seen in the wire arc process cross section.

The coated coupons were inoculated with MRSA. The plated samples were then held at room temperature for two hours, after which survivors were resuspended and cultured. This procedure followed the EPA Protocol [17] "Test Method for Efficacy of Copper Alloy Surfaces as a Sanitizer". The details of this procedure are given below [17-20].

Carrier surfaces and preparation

The copper coupon surfaces were utilized as the test carriers and stainless steel squares were used as control

Table 1 Typical spray gun operating parameters

Spray/ Property	Temperature °C	Velocity, m/s	Porosity, %	Oxides, %
Plasma	2500 - 3500	100 - 300	1 - 10	1 - 3
Wire Arc	2500 - 3500	50 - 100	5 - 20	10 - 20
Cold Spray	100 - 500	600 - 1000	<1	<1

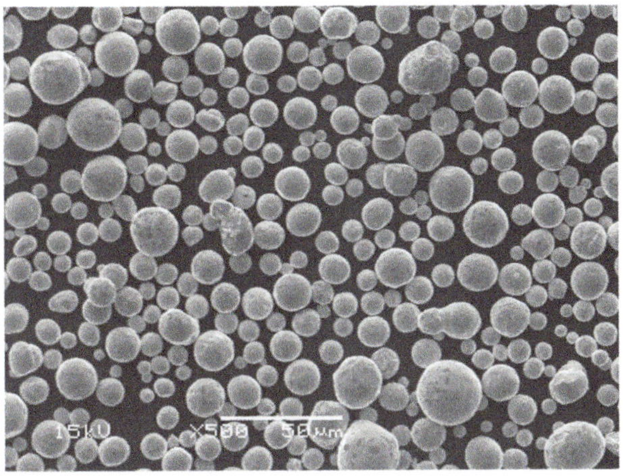

Figure 4 Feedstock copper powder for plasma and cold spray.

carriers. The carriers were dipped into ethanol, rinsed with deionized water, and allowed to air dry. The carriers were autoclaved prior to use in test. After sterilization, each carrier was placed into a Petri dish matted with two pieces of filter paper.

Preparation of test organisms

Ten (10) mL tubes of Synthetic Broth were inoculated from stock cultures and incubated for 24 hours at 36°C. Using a 4-mm inside diameter disposable sterile plastic transfer loop, at least three consecutive daily transfers of cultures were made in Synthetic Broth prior to use as test inoculum. Two loopfuls of culture were transferred to 10 ml broth medium and incubated for 48 hours.

The culture was thoroughly mixed on a "vortex" mixer and allowed to settle. The upper two thirds of this suspension was used as the inoculum for testing.

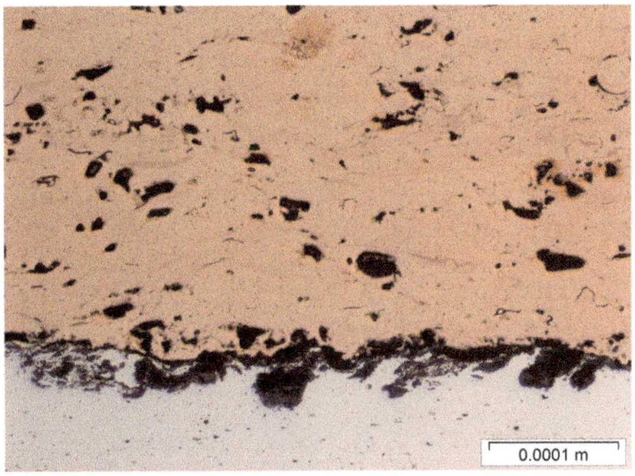

Figure 6 Cross sectional view of the wire arc sprayed copper deposit.

Addition of organic soil load

An organic soil load containing Triton X-100 (to aid in spreading of the inoculums) was added to the test culture. A 0.25 ml aliquot of fetal bovine serum and 0.05 mL aliquot of 1% Triton X-100 was added to 4.70 ml of culture to yield a 5% fetal bovine serum and 0.01% Triton X-100 soil load.

Inoculation of carriers

Each test and control carrier was inoculated at staggered intervals with 0.02 ml of 48 hour culture using a calibrated pipettor. The inoculum was spread to within 3 mm of the edges of the carrier. The lids of the Petri dishes were replaced and the carriers were held at room temperature (20°C) for 2 hours. The exposure period began immediately after inoculation.

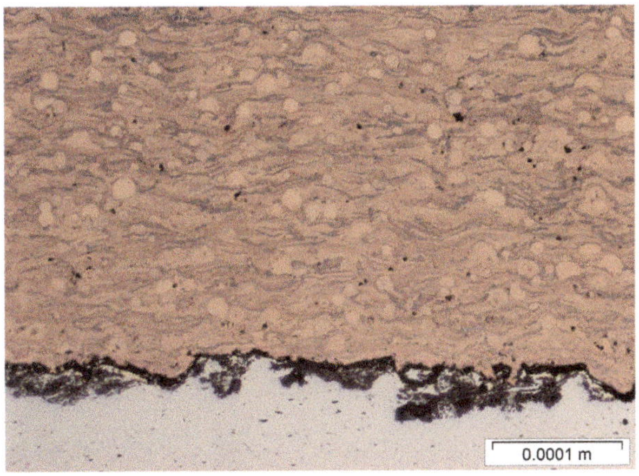

Figure 5 Cross sectional view of the plasma sprayed copper deposit.

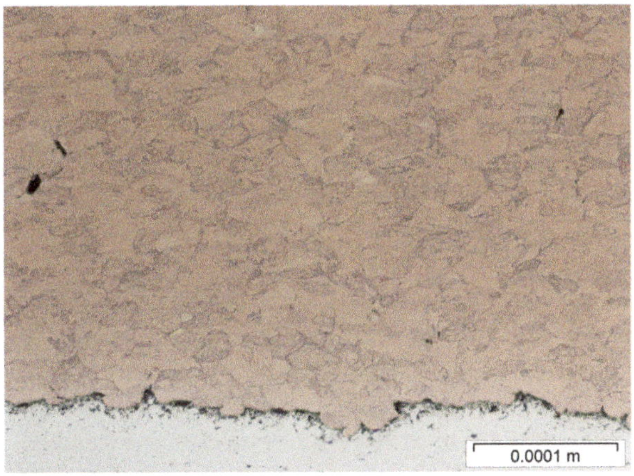

Figure 7 Cross sectional view of the cold sprayed copper deposit.

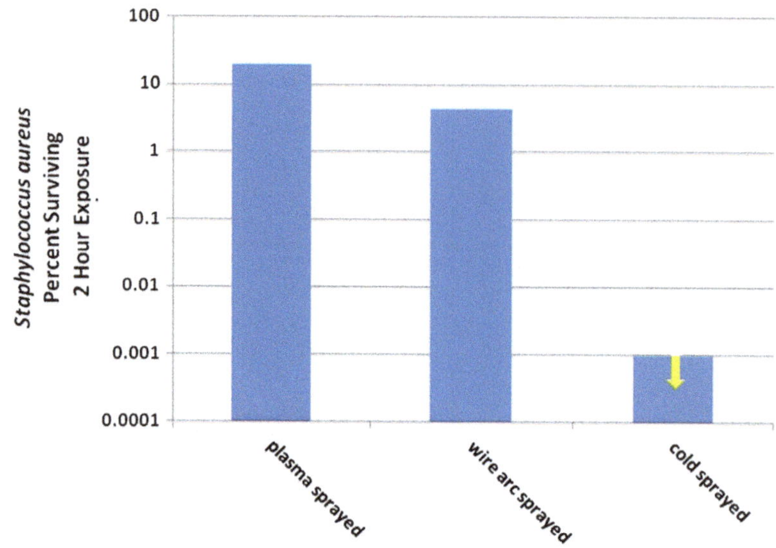

Figure 8 Percent MRSA surviving after exposure to various copper deposits.

Neutralization and subculture

Following the 2 hour exposure, the carriers were transferred to jars containing 20 mL of Letheen Broth + 0.07% Lecithin + 0.5% Tween 80 at staggered intervals. Each neutralizer jar was sonicated for five minutes to suspend any survivors and rotated to mix. Serial dilutions (10^0 – 10^{-4}) of the neutralized solution from each of the jars were prepared. One (1.0) mL aliquots of those dilutions were plated in duplicate using standard spread plate technique onto sheep blood agar plates (BAP).

Incubation and observation

The plates were incubated at 36°C for 44 hours prior to observation and enumeration. Following incubation, the plates were visually enumerated. Subcultures containing 30–300 colonies were used for calculations.

Results and discussion

The reduction of inoculated *S. aureus* was normalized by the results of the control exposure to a stainless steel surface. The results of these tests in terms of the percent of surviving *S. aureus* after two hours are shown in Figure 8. The result for cold spray was below minimal measurement thresholds and is thus reported as "less than".

The results show a greater than three order of magnitude difference in kill efficiency between the plasma and wire arc methods and the cold spray method of copper deposition. This large difference in anti-microbial effectiveness between copper spray deposition methods requires an examination of how the deposition mechanism affects the nature of the copper. The plasma and wire arc methods deposit molten particles at relatively low velocity (<200 m/s). The cold spray method deposits solid particles at high velocity (>600 m/s). Champagne, et al. [21] have shown that the high velocity impacts of cold sprayed particles lead to extreme work hardening and correspondingly high dislocation density within the deposit. For example, the Vickers Hardness values for plasma, wire arc, and cold spray deposited copper were 94, 105, and 141, respectively. Ion diffusion in metals is augmented by the presence of grain dislocations, known as "pipe diffusion", and ionic diffusion occurs principally through these dislocations. The relationship between

Figure 9 Hospital table coated with cold-sprayed copper.

dislocation density and Vickers Hardness is [22] $\rho \propto H^2$, and the relationship between diffusivity and displacement density is given by [23] $D_p \propto \rho$. Diffusivity in metals thus varies as the square of hardness and is therefore very sensitive to impact hardening by cold spray deposition. The diffusion of copper ions can therefore be significantly increased through the hardness increase produced by the cold spray process, which serves to enhance the flow of Cu^{2+} ions needed for microbial destruction.

Conclusions

The effectiveness of copper and copper alloys as antimicrobial coatings on touch-surfaces has been well documented by many researchers [5-7,10,12-15]. Except for the copper content of alloys, the effects of the metallurgical properties of the copper coatings have not been investigated by these efforts. The significant anti-microbiologic differences between coatings produced by different spray techniques, as shown here, demonstrate the importance of the copper application technique and of the resulting deposition structure. The cold spray technique showed superior anti-microbial effectiveness caused by the high impact velocity imparted to the sprayed particles which results in high dislocation density and high ionic copper diffusivity.

The cold spray process is a mature technology which is currently in use for a variety of applications requiring various metal coatings. The cold spray process can readily apply copper coatings onto touch surfaces. Figure 9 is an example of copper coating by cold spray. The hospital tray and the entire metal support structure of the hospital table have been coated with cold-sprayed pure copper. In addition to providing highly efficient antimicrobial surfaces, the cold spray technique is less likely to damage heat sensitive substrates than are high temperature thermal sprays. This work is a proof-of-concept effort, and additional, more statistically significant, work should be performed in order to justify commercialization.

Competing interests

The authors declare that they have no competing interests.

Authors' contributions

VKC conceived and directed the test program. DJH interpreted the test results and justified the mechanism. Both authors read and approved the final manuscript.

Acknowledgements

The sanitizer efficacy testing conducted against *Staphylococcus aureus* and reported in this paper was carried out under contract by ATS Labs at Eagan MN. The experiments were performed by Amy Jeske and Becky Lien.

Author details

[1]U.S. Army Research Laboratory, Aberdeen Proving Ground, MD, USA.
[2]Dynamic Science, Aberdeen, MD, USA.

References

1. Page K, Wilson M, Parkin I: **Antimicrobial surfaces and their potential in reducing the role of the inanimate environment in the incidence of hospital-acquired infections.** *J Mater Chem* 2009, **19**:3819–3831.
2. Aycicek H, Oguz U, Karci K: **Comparison of results of ATP bioluminescence and traditional hygiene swabbing methods for the determination of surface cleanliness at a hospital kitchen.** *Int J Hyg Environ Health* 2006, **209**:203–206.
3. Bernard L, Kereveur A, Durand D, Gonot J, Goldstein F, Mainardi J, Acar J, Carlet J: **Bacterial contamination of hospital physicians' stethoscopes.** *Infect Control Hosp Epidemiol* 1999, **20**:626–628.
4. Rutala WA, Katz EB, Sherertz RJ, Sarubbi FA: **Environmental-study of a methicillin-resistant staphylococcus -aureus epidemic in a burn unit.** *J Clin Microbiol* 1983, **18**:683–688.
5. White LF, Dancer SJ, Robertson C: **A microbiological evaluation of hospital cleaning methods.** *Int J Environ Health Res* 2007, **17**:285–295.
6. Faúndez G, Troncoso M, Navarrete P, Figueroa G: **Antimicrobial activity of copper surfaces against suspensions of *Salmonella enterica* and *Campylobacter jejuni*.** *BMC Microbiol* 2004, **4**:19.
7. Gounadaki S, Skandamis PN, Drosinos EH, Nychas GJ: **Microbial ecology of food contact surfaces and products of small-scale facilities producing traditional sausages.** *Food Microbiol* 2008, **25**:313–323.
8. Jackson V, Blair IS, McDowell DA, Kennedy J, Bolton DJ: **The incidence of significant foodborne pathogens in domestic refrigerators.** *Food Control* 2007, **18**:346–351.
9. Kaneko K, Hayashidani H, Takahashi K, Shiraki Y, Limawongpranee S, Ogawa M: **Bacterial contamination in the environment of food factories processing ready-to-eat fresh vegetables.** *J Food Prot* 1999, **62**:800–804.
10. Noyce J, Michels H, Keevil C: **Use of copper cast alloys to control *Escherichia coli* O157 cross-contamination during food processing.** *Appl Environ Microbiol* 2006, **72**:4239–4244.
11. Grass G, Rensing C, Solioz M: **Metallic copper as an antimicrobial surface.** *Appl Environ Microbiol* 2011, **76**:1541–1547.
12. Michels H, Wilks S, Noyce J, Keevil C: *Copper Alloys for Human Infectious Disease Control.* Pittsburgh, PA: Proceedings of the Materials Science and Technology Conference; 2005.
13. Nie Y, Kalapos C, Nie X, Murphy M, Hussein R, Zhang J: **Superhydrophilicity and antibacterial property of a Cu-dotted oxide coating surface.** *Ann Clin Microbiol Antimicrob* 2010, **9**:25.
14. Raffi M, Mehrwan S, Bhatti T, Akhter J, Hameed A, Yawar W, Masood Ul Hasan M: **Investigations into the antibacterial behavior of copper nanoparticles against Escherichia coli.** *Ann Microbiol* 2010, **60**:75–80.
15. Santo CE, Lam EW, Elowsky CG, Quaranta D, Domaille DW, Chang CJ, Grass G: **Bacterial killing by dry metallic copper surfaces.** *Appl Environ Microbiol* 2011, **77**:794–802.
16. EPA registers copper-containing alloy products. www.epa.gov/pesticides/factsheets/copper-alloy-products.htm.
17. Test method for efficacy of copper alloy surfaces as a sanitizer. www.epa.gov/oppad001/pdf_files/test_method_copper_alloy_surfaces.pdf.
18. Lien B: *Test Method for Determining the Efficacy of Antimicrobial Surfaces as Sanitizers.* Eagan: ATS Labs Report A09966; 2010.
19. Jeske A: *Custom Microbiology – Evaluation of Coated Copper Surfaces.* Eagan: ATS Labs; 2009 Report.
20. Jeske A: *Custom Microbiology – Evaluation of Cold Spray Copper Surfaces*, ATS Labs Report. Eagan: ATS Labs; 2008. Report A06255.
21. Champagne V, Helfritch D, Trexler M: **Some Material Characteristics of Cold-Sprayed Structures.** *Res Lett Mater Sci* 2007, **2007**:ID 27347.
22. Al-Rub R, Faruk NM: **Dislocation-based model for predicting size-scale effects on the micro and nano indentation hardness of metallic materials.** *Int J Mater Struct Integrity* 2010, **4**:251–277.
23. Hart EW: **On the role of dislocations in bulk diffusion.** *Acta Metall* 1957, **5**:597.

Optimized conditions for successful transfection of human endothelial cells with in vitro synthesized and modified mRNA for induction of protein expression

Meltem Avci-Adali, Andreas Behring, Timea Keller, Stefanie Krajewski, Christian Schlensak and Hans Peter Wendel*

Abstract

Background: The induction of protein synthesis by exogenous delivery of coding synthetic mRNA in desired cells is an auspicious strategy in the fields of basic cell biology, regenerative medicine, treatment of diseases, and reprogramming of cells. Here, we produced modified messenger RNA (mRNA) with reduced immune activation potential and increased stability and performed transfection experiments with different cells, HEK293 cells, BJ fibroblasts, and endothelial cells (ECs).

Results: The mRNA induced protein expression in cells was analyzed after transfection with different mRNA amounts and transfection reagents using flow cytometry. Different cell types showed different degrees of eGFP expression. HEK293 cells exhibited the highest eGFP expression compared to the BJ fibroblasts and ECs. However, the mRNA induced eGFP expression was detected in all cell types until 3 days after transfection. Already, the use of 0.5 μg of the synthesized mRNA led to the significant expression of eGFP in ECs. From all analyzed ECs approximately 87% were eGFP positive, which showed a high transfection efficiency.

Conclusions: The synthesis of stabilized mRNA and the high transfection efficiency will enable the mRNA engineering of ECs as well as other somatic cells. The delivery of synthetic exogenous mRNA into cells allows the transient expression of desired proteins, which would be normally not expressed by the cells.

Keywords: mRNA synthesis, Transfection, Protein synthesis

Background

Proper functioning of cells, differentiation and production of signaling molecules are controlled via translation of messenger RNA (mRNA) into proteins by ribosomes (Figure 1). Thus, hereditary or acquired genetic disorders followed by defects of protein expression can lead to serious diseases. In the last years, the administration of mRNA as a therapeutic agent has gained an enormous potential in the fields of disease treatment [1], regenerative medicine [2-4], and vaccination [5]. Hitherto, clinical applications of conventional mRNAs were limited due to the low stability and the strong immunogenicity. However, pioneering work from Kariko and colleagues demonstrated that the replacement of cytidine and uridine by 5-methylcytidine and pseudouridine dramatically abrogate immune activation [6,7]. They also found that pseudouridine modified mRNAs have a higher translational capacity than unmodified mRNAs and increased biological stability [8,9].

The ability to produce modified mRNAs allows the protection of naturally labile mRNA in biological fluids and greatly minimizes innate immune responses. Kormann et al. demonstrated that a combination of nucleotide modifications prevents the interaction of synthesized mRNA with TLR3, TLR7, TLR8, and RIG-1 [1]. Thus, the use of modified mRNA provides the possibility for controlled and temporary delivery of desired transcripts in vivo or in vitro to induce protein expression. Furthermore, in comparison to the viral gene therapy vectors [10], the risk of oncogenesis [11] is prevented due to lack of integration into the

* Correspondence: hans-peter.wendel@med.uni-tuebingen.de
Department of Thoracic, Cardiac, and Vascular Surgery, University Hospital Tuebingen, Calwerstr. 7/1, 72076 Tuebingen, Germany

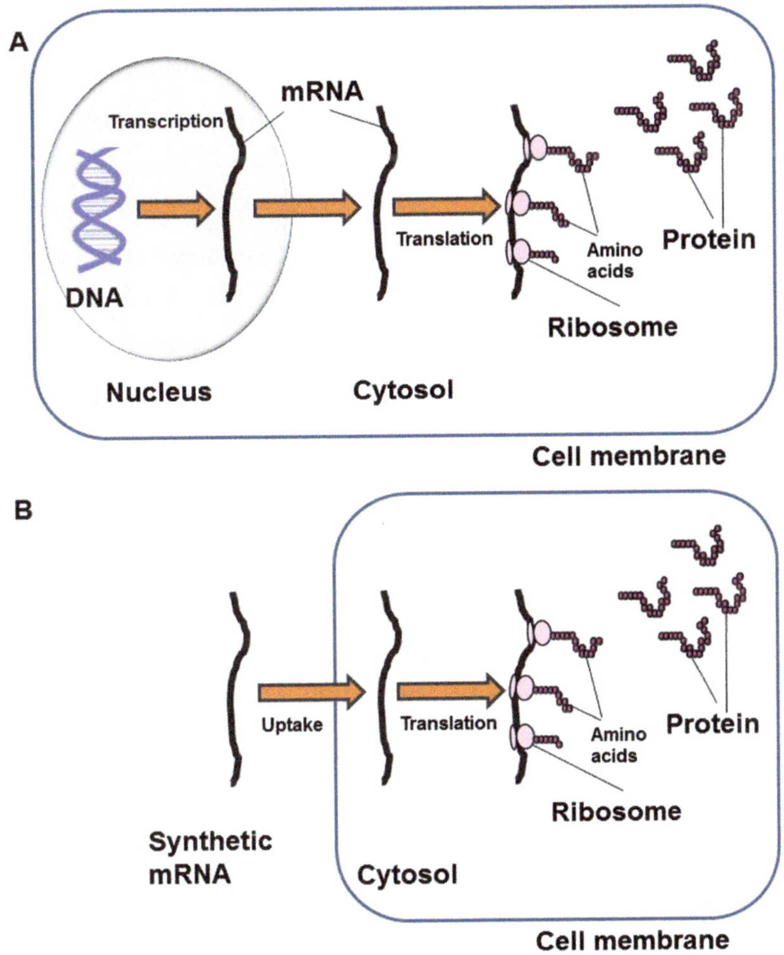

Figure 1 Synthesis of proteins in the cells. A) Natural pathway. Genetic code of the protein is transcribed into mRNA in the nucleus and transported to the cytosol. The mRNA is then translated by ribosomes into proteins. **B)** Synthetic pathway. Synthetically produced mRNA is transfected into the cells. The mRNA is then translated by ribosomes into proteins.

host cell genome. Thus, the therapy with modified synthetic mRNA will get better clinical acceptance in the future.

Here, we synthesized modified eGFP mRNA and transfected different cell types, endothelial cells (ECs), BJ fibroblasts as well as HEK293 cells, with this mRNA. Transfection efficiency and mRNA induced eGFP expression in cells was compared after the transfection with different amounts of mRNA and different transfection reagents. Furthermore, the protein expression in transfected cells was measured up to 3 days after transfection.

Materials and methods

Production of modified mRNA

Coding DNA sequence (CDS) with known flanking sequences was amplified by PCR using specific primers. PCR product was then purified and the quality of the generated DNA was determined. Using the in vitro transcription (IVT) process, mRNA was generated from the DNA product. Subsequently, the product was purified and treated with phosphatase to remove 5′-triphosphates. After the additional purification and quality control of generated mRNA, transfection experiments were performed.

Consent

ECs were isolated from residual saphenous vein biopsies of patients who had undergone coronary artery bypass graft surgery. All patients gave informed consent to participate and for the publication of this report. The study was approved by the Ethical Committee of the Faculty of Medicine of the University of Tuebingen.

Amplification of plasmid inserts and adding of polyT-tail by polymerase chain reaction (PCR)

To obtain, the DNA template for the IVT, pcDNA 3.3_eGFP Plasmid (Addgene, cat. no. 26822) containing CDS of interest, here eGFP, was amplified using PCR.

The forward primer 5′-TTGGACCCTCGTACAGAAGC TAATACG-3′ and reverse primer 5′- T_{120}-CTTCCTA CTCAGGCTTTATTCAAAGACCA-3′ for the insert amplification were purchased from ELLA Biotech (Martinsried, Germany) in HPLC grade. Thereby, a poly T-tail of 120 thymidines (T) was added to the insert by using a reverse primer with a T_{120} extension. Thus, after IVT, the generated mRNAs obtained a poly A-tail with a defined length. PCR reactions of 100 µl were performed using HotStar HiFidelity Polymerase Kit (Qiagen, Hilden, Germany) and contained 0.7 µM of each forward and reverse primer, 1× Q-solution, 1× HotStar HiFidelity PCR buffer, 50 ng plasmid DNA, 2.5 U HotStar HiFidelity DNA polymerase. PCR was run using the following cycling protocol: initial activation step at 95°C for 5 min, followed by 25 cycles of denaturation at 95°C for 45 s, annealing at 55°C for 1 min, extension at 72°C for 1 min and final extension at 72°C for 10 min. PCR reaction was purified using QIAquick PCR Purification Kit (Qiagen, Hilden, Germany) according manufacturer's instructions and the DNA was eluted using 20 µl nuclease-free water. The quality and purity of the DNA was assessed by 1% agarose gel electrophoresis.

In vitro transcription (IVT)

After the PCR, the genetic information is in vitro transcribed from DNA to mRNA using MEGAscript® T7 Kit (Life Technologies, Darmstadt, Germany). The generated mRNA transcript is then used to induce protein expression in cells. At first, 23 µl NTP/cap analog mixture containing 7.5 mM ATP, 1.875 mM GTP (both from MEGAscript® T7 Kit), 7.5 mM Me-CTP, 7.5 mM Pseudo-UTP (both from TriLink BioTechnologies, San Diego, USA), and 2.5 mM 3′-O-Me-m^7G(5′)ppp(5′)G RNA cap structure analog (New England Biolabs, Frankfurt am Main, Germany) was prepared and mixed thoroughly. The IVT reaction mixture of 40 µl was then assembled by adding 40 U RiboLock RNase inhibitor (Thermo Scientific, Waltham, USA), 1 µg PCR product, 1× reaction buffer and 1× T7 RNA polymerase enzyme mix. The IVT reaction mixture was incubated at 37°C for 3 h in a thermomixer. To remove the template DNA, 1 µl TURBO DNase (from MEGAscript® T7 Kit) was added to the IVT reaction mixture and incubated for 15 min at 37°C. Then, the reaction mixture was purified using RNeasy Mini Kit (Qiagen, Hilden, Germany) according to manufacturer's instructions. The modified mRNA was eluted from the spin column membrane twice with 40 µl nuclease-free water.

Treatment of purified mRNA with Antarctic phosphatase

The generated mRNA is treated with Antarctic phosphatase (New England Biolabs, Frankfurt am Main, Germany) to remove 5′ triphosphates, which can be recognized by RIG-1, and lead to the immune activation. Furthermore, the phosphatase treatment prevents the recircularization in a self-ligation reaction. For this purpose, 9 µl of 10× Antarctic phosphatase reaction buffer was added to 79 µl of purified mRNA solution. Subsequently, 2 µl of Antarctic phosphatase (5 U/µl) was added to the reaction mixture and incubated at 37°C for 30 min. The treated mRNA was purified using RNeasy Mini Kit (Qiagen, Hilden, Germany) according to manufacturer′s instructions. The modified mRNA was eluted from the spin column membrane twice with 50 µl nuclease-free water. The concentration was measured using ScanDrop spectrophotometer (Analytic Jena, Jena, Germany). The concentration of mRNA was adjusted to 100 ng/µl by adding nuclease-free water. The quality and purity of synthesized modified mRNA was determined by 1% agarose gel electrophoresis. The modified mRNA was aliquoted and stored at -80°C and used for transfections.

Cultivation of cells

HEK293 cells were cultivated in DMEM with high glucose (PAA Laboratories, Cölbe, Germany) supplemented with 10% FBS (Life Technologies, Darmstadt, Germany), 2 mM L-glutamine (PAA Laboratories), and 1% penicillin/streptomycin (PAA Laboratories). BJ human foreskin fibroblasts (Stemgent, Cambridge, USA) were cultivated in DMEM with high glucose containing 10% FBS, 2 mM L-glutamine, 1% penicillin/streptomycin, and 30 mM HEPES (Life Technologies, Darmstadt, Germany). Human endothelial cells (ECs) were isolated as described previously [12] and cultivated in flasks precoated with 0.1% gelatin in Vasculife® EnGS EC culture medium (CellSystems, Troisdorf, Germany) containing VascuLife EnGS LifeFactors Kit, 50 mg/ml gentamicin, and 0.05 mg/ml amphotericin B (PAA Laboratories). Cells were kept at 37°C with 5% CO_2 and media was changed every 3 days. Cells were passaged using trypsin/EDTA (0.04%/0.03%, PromoCell, Heidelberg, Germany). BJ fibroblasts at passage 7-9, endothelial cells at passage 3-5 were used for all experiments.

Preparing of cells for mRNA transfection

For performing of transfection experiments, 1.5×10^5 cells were plated per well of 12-well plate. The cells were incubated overnight at 37°C in a cell incubator. Next day, transfection experiments were performed.

Detachment of cells and performing of flow cytometry

Using trypsin/EDTA (0.04%/0.03%), cells were detached, washed, and fixed in 1× CellFix solution (BD Biosciences, Heidelberg, Germany). A total of 10,000 cells was analyzed by FACScan flow cytometer (Becton Dickinson, Heidelberg, Germany). Cells expressing eGFP were detected using the FL1 (green) channel and measuring the fluorescence at 509 nm. FACS data were displayed in histograms. For the evaluation, a marker M1 was set, which contained maximum of approximately 1.5% of control cells (cells treated with transfection medium without mRNA). Using histogram statistics of the CellQuest Pro Software (Becton Dickinson), the percentage of eGFP positive cells was determined to assess the transfection efficiency. Geo Mean (geometric mean) values show the strength of the fluorescence intensity. Higher Geo Mean values reflect stronger fluorescent intensity, indicating a higher production of eGFP.

Performing of transfections with different transfection reagents

Expression of eGFP was verified after performing of transfection with 1.5 µg mRNA, using TransIT mRNA transfection reagent (Mirus Bio LLC, Madison, USA), Lipofectamine® 2000 (Life Technologies, Darmstadt, Germany), or Stemfect RNA transfection reagent (Stemgent, Cambridge, USA). From each transfection reagent 2 µl was used for transfections. The cells were incubated for 4 h with transfection complexes, subsequently the transfection mixture was replaced by cell culture medium. The expression of eGFP was measured 24 h after transfection.

To perform the transfection with Trans-IT-mRNA transfection reagent, 1.5 µg mRNA, 2 µl TransIT-mRNA transfection reagent and 2 µl boost reagent were added to 100 µl Opti-MEM I reduced serum media (Life Technologies, Darmstadt, Germany) and incubated for 5 min at room temperature. Afterwards, transfection complexes were added for 4 h to 1 ml cell medium in 12 well plates.

Two tubes containing 25 µl Stemfect transfection buffer were prepared for assembling of the Stemfect RNA transfection mixture. In the first tube, 2 µl of Stemfect RNA transfection reagent was added. In the second tube, 1.5 µg of mRNA was added. Then the diluted transfection reagent solution was pipetted to the diluted mRNA solution, mixed, and incubated for 15 minutes at room temperature. The mRNA transfection complex was added for 4 h to 1 ml cell medium in 12 well plates.

Using the transfection reagent Lipofectamine 2000, two different transfections were performed. Lipofectamine 2000/mRNA complexes were added to the cell culture media or given only in Opti-MEM I reduced serum media to the cells. For transfection of cells only with Opti-MEM I reduced serum media, the Lipofectamine 2000 transfection mixtures were prepared by addition of 2 µl Lipofectamine 2000 and 1.5 µg mRNA to 500 µl Opti-MEM I reduced serum media and incubation for 20 min at room temperature. Cell culture medium was aspirated, cell layer was washed with 1 ml DPBS and transfection mixture was given on adherent growing cells for 4 h at 37°C. This transfection method is referred to as Lipofectamine 2000_optimized. For transfection of cells in cell culture medium, 2 µl Lipofectamine 2000 and 1.5 µg mRNA were added to 100 µl Opti-MEM I reduced serum media and incubated for 5 min at room temperature. Then, the mixture was added to 1 ml cell culture medium per well of a 12-well plate for 4 h at 37°C.

Transfection of cells with different Lipofectamine 2000 amounts

To determine the required amount of Lipofectamine 2000 for forming of lipoplexes, different amounts of Lipofectamine 2000, 1, 2, 4, 6 µl, were used to transfect the cells. For transfection of one well of 12-well plate, 500 µl Opti-MEM I reduced serum media was prepared containing 2.5 µg of eGFP mRNA and respective amount of Lipofectamine 2000. The components were gently mixed by pipetting. The transfection mixture was then incubated at room temperature for 20 min to generate lipoplexes for transfection. Cells were washed with 500 µl DPBS/well, the transfection mixture was pipetted into the well. After 4 h incubation at 37°C and 5% CO_2, the transfection mixture was replaced by 1 ml complete cell culture medium. Cells were cultivated for 24 h in the cell incubator and analyzed using flow cytometry.

Transfection of cells with different mRNA amounts

To determine the required amount of mRNA for induction of protein expression, different amounts of eGFP mRNA, 0, 0.5, 1, 1.5, 2, 2.5 µg, were used to perform the transfection of cells. For transfection of one well of 12-well plate, 500 µl Opti-MEM I reduced serum media with respective amount of mRNA and 2 µl of Lipofectamine 2000 was prepared. The components were gently mixed and incubated 20 min at room temperature. Cells were washed with 500 µl DPBS/well and the transfection mixture was added. Cells were incubated for 4 h at 37°C and 5% CO_2. Afterwards, the transfection mixture was aspirated and 1 ml complete cell culture medium was added to the cells. Cells were incubated for 24 h in the cell incubator. Using flow cytometry, the eGFP expression in the cells was verified.

Determination of eGFP expression in mRNA transfected cells over time

To evaluate the duration of protein expression in cells, eGFP expression was determined 1, 2, and 3 days after mRNA transfection. Transfection complexes were generated in 500 µl Opti-MEM I reduced serum media by adding of 2.5 µg eGFP mRNA and 2 µl Lipofectamine 2000. The components were incubated at room temperature for 20 min to form lipoplexes. Cells were washed with 500 µl DPBS/well. Subsequently, cells were incubated for 4 h at 37°C and 5%

CO_2 with the transfection mixture. Then, the mixture was aspirated and 1 ml complete cell culture medium was added to the cells. Cells were cultivated for 1, 2, and 3 days in the cell incubator and analyzed using flow cytometry.

Fluorescence microscopic analysis of eGFP transfected cells

Using fluorescence microscopy, the expression of eGFP in HEK293 cells, BJ fibroblasts, and ECs was examined. The cells were transfected with 1.5 μg eGFP mRNA

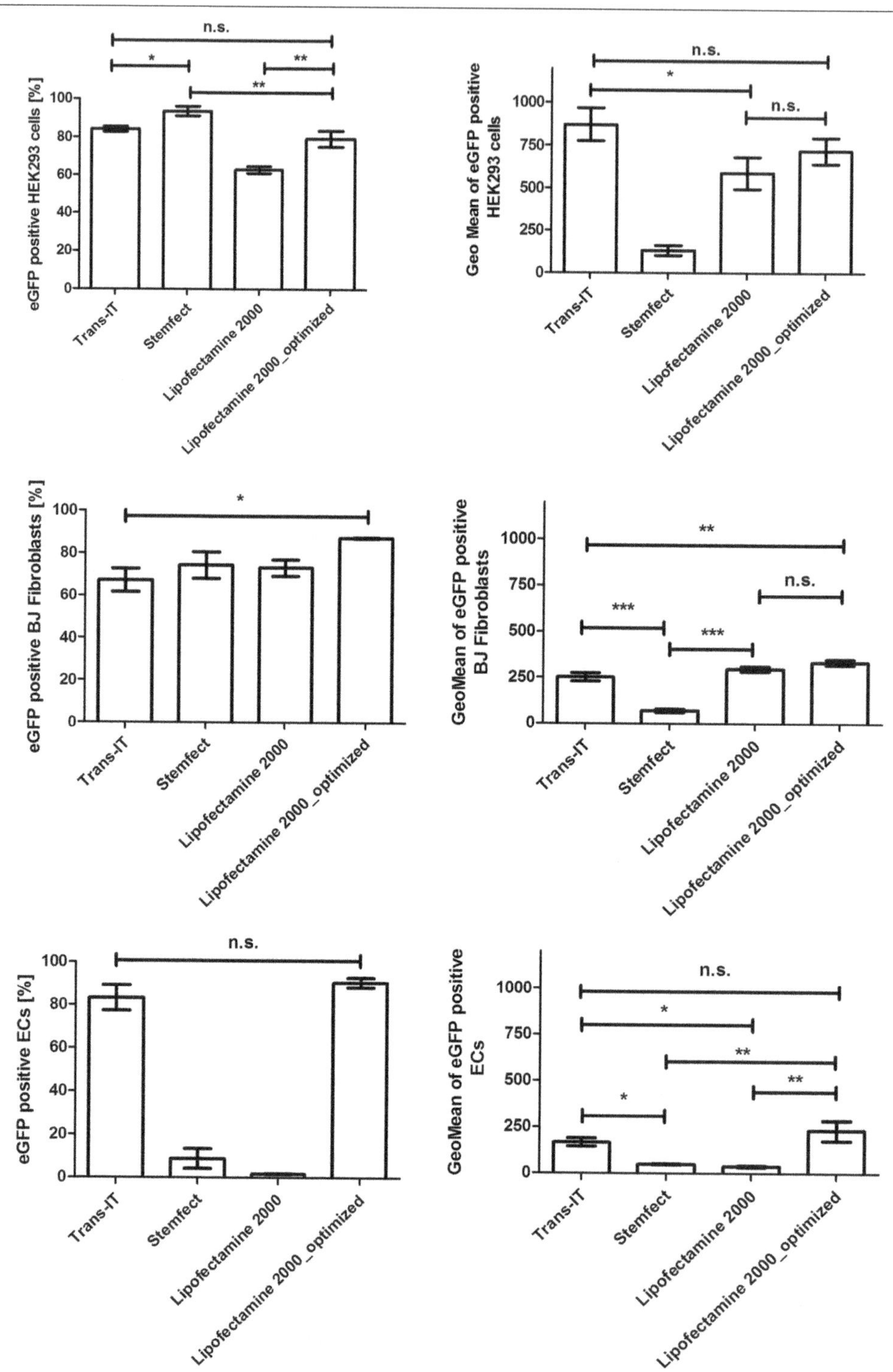

Figure 2 Transfection of HEK293 cells, BJ fibroblasts, and ECs with 1.5 μg eGFP mRNA using different transfection reagents and modifications.

using 2 µl Lipofectamine 2000. Fluorescence micrographs were taken 24 h post-transfection.

Statistical analysis

Data are shown as means ± standard deviation (SD). One-way analysis of variance (ANOVA) for repeated measures followed by the Bonferroni's multiple comparison test was performed to compare the means. All statistical analyses were performed double-tailed using GraphPad Prism version 5.01. Differences of $p < 0.05$ were considered significant.

Results

Performing of mRNA transfection with different transfection reagents

HEK293 cells, BJ fibroblasts, and ECs were transfected with 1.5 µg mRNA using TransIT mRNA transfection reagent, Lipofectamine 2000, or Stemfect RNA transfection reagent (Figure 2).

Transfection of HEK293 cells using Stemfect RNA transfection reagent led to the highest transfection efficiency, approximately 94% of cells expressed eGFP. However, the eGFP expression in cells measured by fluorescence intensity was lowest compared to the other transfection reagents. The transfection efficiency using TransIT mRNA transfection reagent and Lipofectamine 2000_optimized was similar with approximately 80% eGFP expressing cells. The fluorescence intensity in these cells was also not significantly different.

Using Stemfect RNA transfection reagent, about 74% of the analyzed BJ fibroblasts were transfected with mRNA. However, like in HEK293 cells, the protein expression in the BJ fibroblasts was lowest after the transfection with Stemfect RNA transfection reagent compared to other transfection reagents. Lipofectamine 2000_optimized resulted in best transfection efficiency with approximately 87% eGFP positive cells.

Very low amounts of ECs could be transfected using Stemfect RNA transfection reagent and Lipofectamine

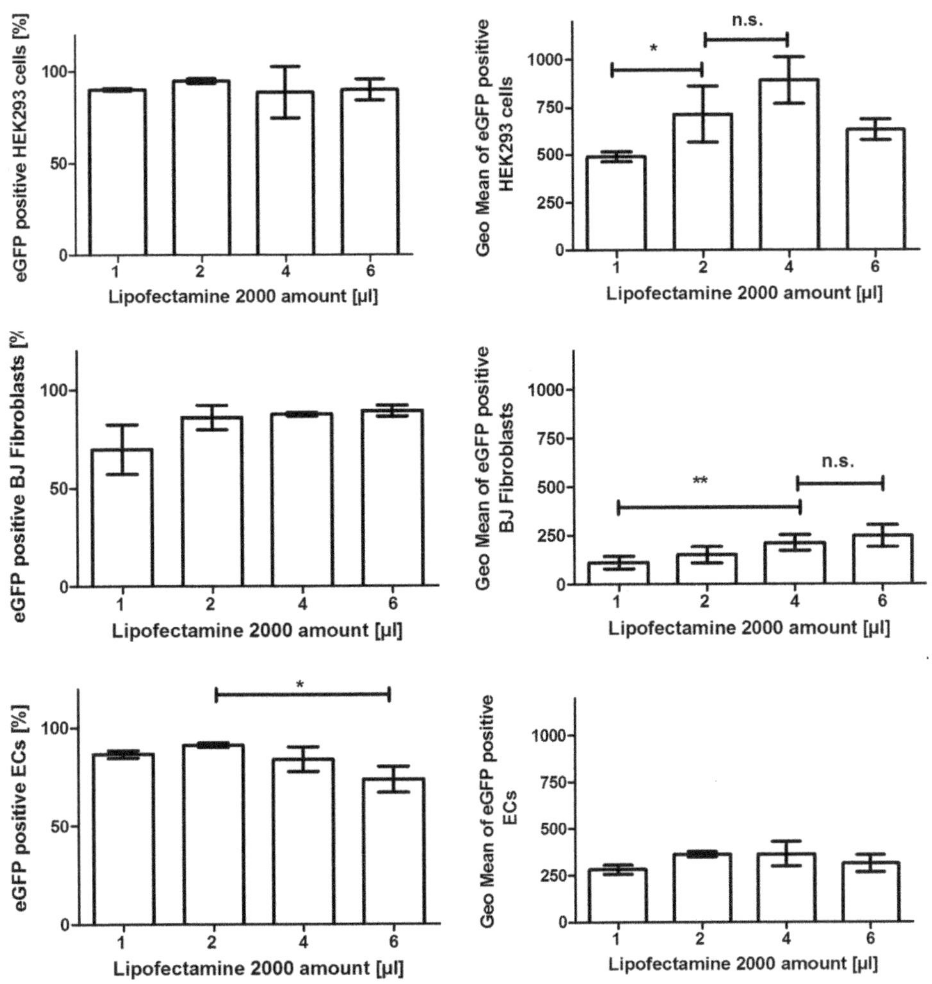

Figure 3 Transfection of HEK293 cells, BJ fibroblasts, and ECs with 2.5 µg eGFP mRNA using different amounts of Lipofectamine 2000.

2000/mRNA complexes in cell culture medium. The transfection using Trans-IT mRNA transfection reagent or Lipofectamine 2000_optimized resulted in high transfection efficiency. The eGFP expression in cells was not significantly different after transfection with both transfection reagents.

Transfection of cells with different Lipofectamine 2000 amounts

HEK293 cells, BJ fibroblasts, and ECs were transfected with 2.5 µg eGFP mRNA using 1, 2, 4, and 6 µl of Lipofectamine 2000 for generation of transfection complexes (Figure 3).

The percentage of eGFP expressing HEK293 cells was not significantly different between the transfections with different Lipofectamine 2000 amounts. Approximately 91% of the cells were eGFP positive. However, the use of 2 µl Lipofectamine 2000 resulted in significantly higher fluorescence intensity (Geo Mean) in HEK293 cells than after the transfection with 1 µl Lipofectamine 2000. The use of 4 µl Lipofectamine 2000 did not lead to significantly higher fluorescence intensity than the transfection with 2 µl. After the transfection with 2 µl Lipofectamine 2000, the Geo Mean was approximately 710.

The transfection of BJ fibroblasts with different Lipofectamine 2000 amounts did not result in significantly different percentage of eGFP positive cells. Approximately 83% of the cells were positive for eGFP expression. The use of 4 µl Lipofectamine 2000 showed significantly higher Geo Mean compared to the use of 1 µl Lipofectamine. Using 4 µl Lipofectamine 2000, a Geo Mean of 210 was detected.

The performance of EC transfection with 6 µl Lipofectamine 2000 led to decreased percentage of eGFP positive cells compared to the transfection with 2 µl Lipofectamine 2000. The transfection with 1, 2, and 4 µl Lipofectamine 2000 resulted in approximately 87% eGFP positive cells. The use of these different Lipofectamine 2000 amounts for transfection showed no significant differences. Using 2 µl Lipofectamine 2000, a Geo Mean of 362 was detected.

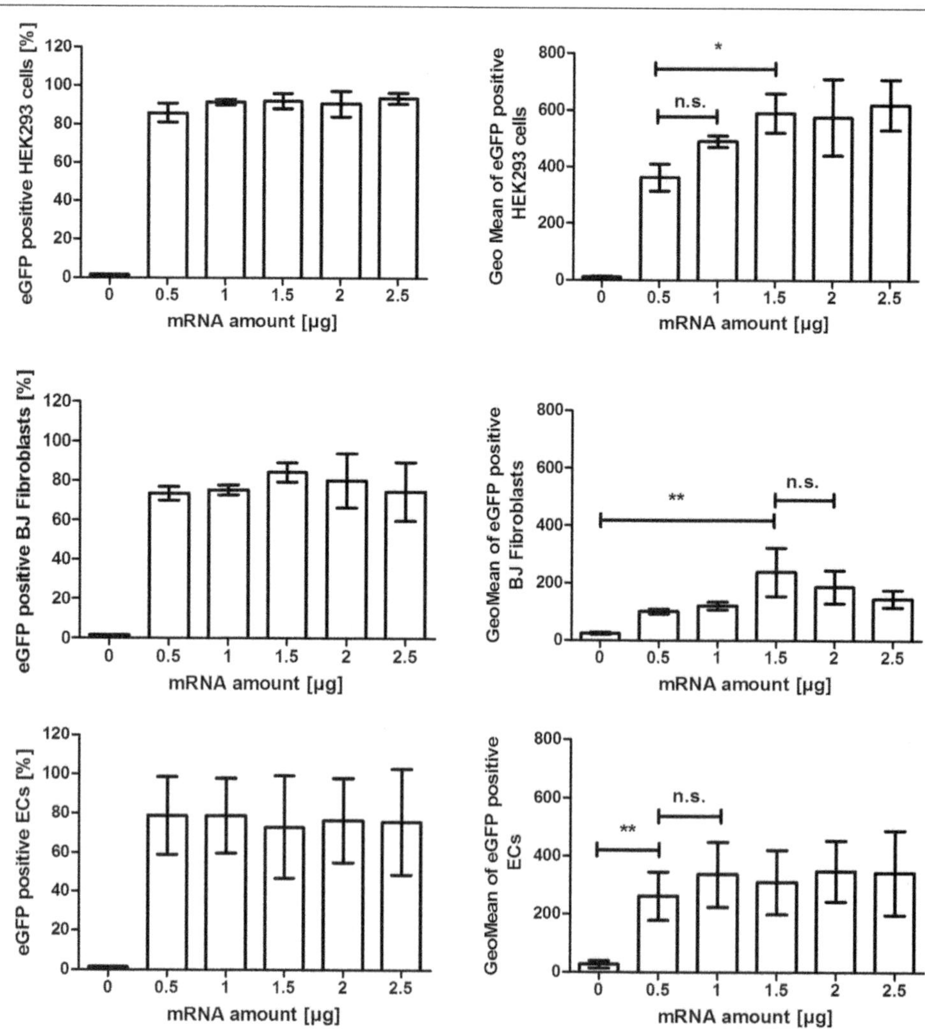

Figure 4 Transfection of HEK293 cells, BJ fibroblasts, and ECs with 2 µl Lipofectamine 2000 and different amounts of eGFP mRNA.

Transfection of cells with different mRNA amounts

The mRNA transfection complexes were generated by using 2 µl Lipofectamine 2000 and different amounts of eGFP mRNA, 0, 0.5, 1, 1.5, 2, and 2.5 µg. After the transfection of HEK293 cells, BJ fibroblasts, and ECs, the percentage of eGFP positive cells and the Geo Mean was determined using flow cytometry (Figure 4).

Already the transfection of HEK293 cells with 0.5 µg mRNA resulted in detection of 86% eGFP positive cells. The increasing of the mRNA amount did not significantly increase the percentage of transfected cells. But the use of 1.5 µg mRNA led to higher eGFP expression in cells than the use of 0.5 µg mRNA (Geo Mean of 590 versus 363). However, the transfection of cells with more than 1.5 µg mRNA did not increase the eGFP expression in the cells.

The transfection of BJ fibroblasts with 0.5 µg mRNA led to the transfection of 75% of the cells. The increasing of the mRNA amount did not result in significantly higher transfection efficiency. The transfection of 1.5 µg mRNA led to significantly higher fluorescence intensity than the negative cells. The Geo Mean was 239.

The use of only 0.5 µg mRNA led to the transfection of 79% of all ECs. The transfection efficiency was not enhanced by increasing the mRNA amount. Already 0.5 µg mRNA led to significantly higher expression of eGFP in ECs than in the negative cells. The detected Geo Mean was 262. The use of higher amounts of mRNA did not lead to increased eGFP synthesis.

Determination of eGFP expression in mRNA transfected cells over time

The duration of synthetic mRNA induced protein expression in HEK293 cells, BJ fibroblasts, and ECs was determined after the transfection of cells with 2.5 µg mRNA using 2 µl Lipofectamine 2000. The eGFP expression in cells was determined 1, 2, and 3 days after transfection using flow cytometry (Figure 5).

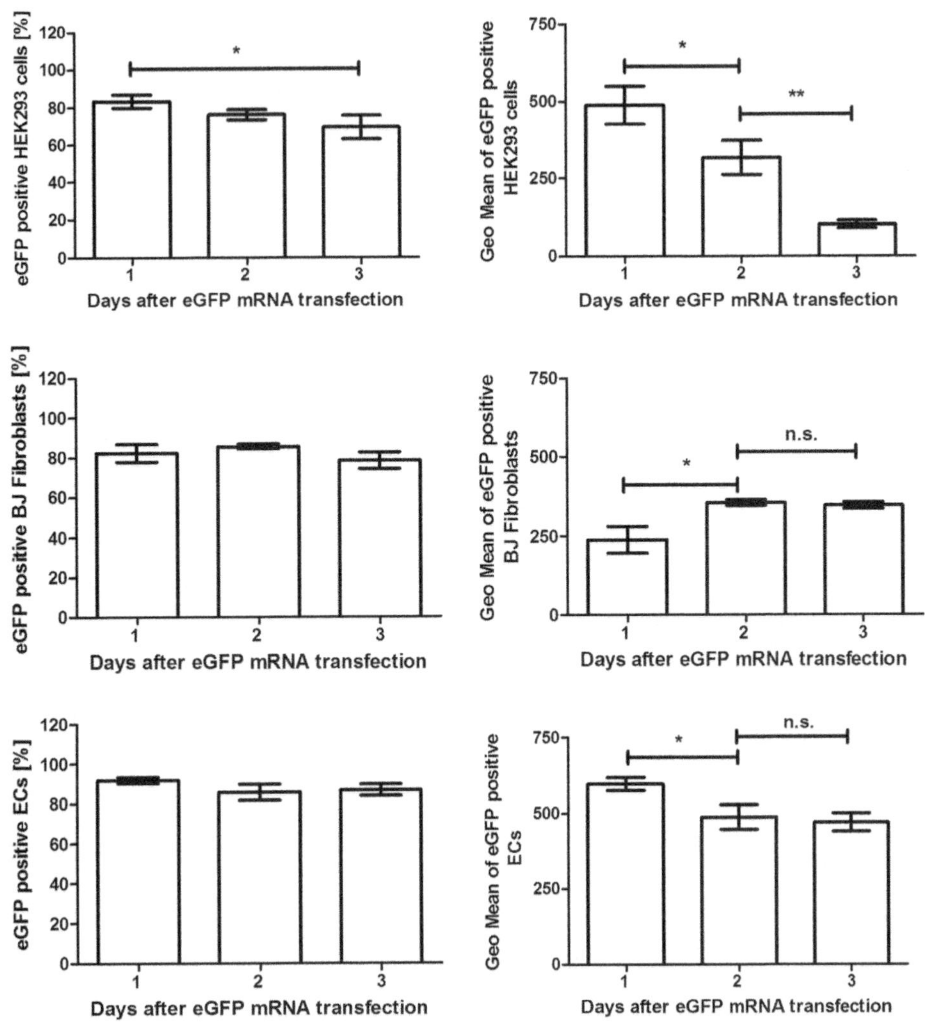

Figure 5 Transfection of HEK293 cells, BJ fibroblasts, and ECs with eGFP mRNA and determination of eGFP expression for 3 days.

Flow cytometry analysis of eGFP mRNA transfected HEK293 cells showed a significant decrease of the portion of eGFP producing cells after 3 days. However, after 3 days, 69% of analyzed cells were still positive for eGFP expression. Over time, a continuous decrease of eGFP expression in the cells was detected. After 3 days, the detected Geo Mean was approximately 4 times less than 1 day after transfection.

The percentage of eGFP positive BJ fibroblasts did not significantly change over time. But the eGFP expression in the cells was 2 days post-transfection higher than after 1 day. The fluorescence intensity of the cells did not change after 3 days.

Approximately 88% of ECs were positive for eGFP expression and the amount of eGFP positive cells did not change during 3 days. The eGFP expression peak was measured 1 day post-transfection. A significant decrease of Geo Mean was detected 2 days after transfection. However, the detected fluorescence intensity did not change in the following day.

Fluorescence microscopic analysis of eGFP transfected cells

HEK293 cells, BJ fibroblasts, and ECs were transfected with 1.5 µg eGFP mRNA using 2 µl Lipofectamine 2000. After 24 h, the eGFP expression in cells was investigated by fluorescence microscopy (Figure 6). The analyses demonstrated strongly fluorescent HEK293 cells. The fluorescence intensity of BJ fibroblasts was weaker than of the ECs.

Discussion

In recent years, the induction of protein expression in cells by delivery of synthetic modified mRNA gained in importance. Thereby, cells can be triggered to synthesize proteins for treatment of diseases caused by inability of cells to produce a key protein. For example genetic disorders can be corrected by introduction of mRNA that codes for the missing or defective protein [13]. Furthermore, mRNA therapeutics can also be used as vaccines to synthetize protein antigens for activation of the host immune system to effectively eliminate tumor cells or prevent infections [5,14]. Moreover, mRNA was also successfully used to reprogram finally differentiated somatic cells into pluripotent stem cells by the expression of reprogramming factors [3]. In a recent study, MSCs were mRNA engineered to synthetize increased amounts of proteins for enhanced regenerative potential [15].

Here, modified eGFP mRNA was synthetized and the mRNA transfection behavior of different cell types, primary cells such as ECs, BJ fibroblasts, and the HEK293 cell line, was analyzed. The determined optimal conditions can be used to achieve maximal transfection efficiency and expression of desired proteins especially in ECs. But above all, we demonstrated the successful transfection of ECs with synthetic mRNA, which resulted in production of functional protein.

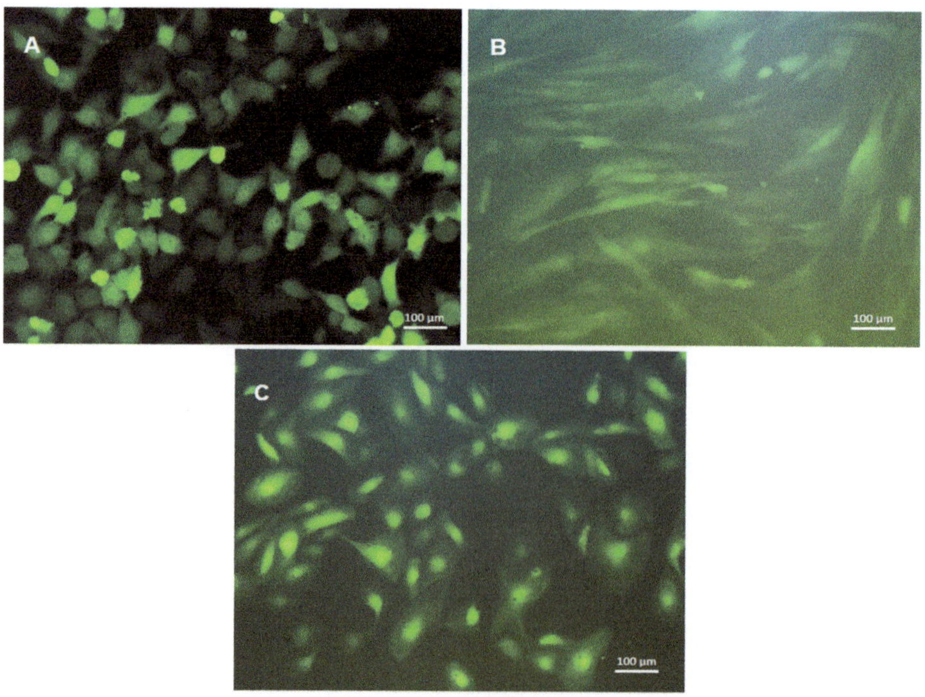

Figure 6 Fluorescence microscopic analyses of A) HEK293 cells, B) BJ fibroblasts, and C) ECs 24 h after eGFP mRNA transfection.

Firstly, the appropriate transfection reagent and the transfection conditions were determined. The transfection of ECs with Stemfect RNA transfection reagent and Lipofectamine 2000 in cell culture medium resulted in very low levels of transfection. In contrast, transfection of cells with Lipofectamine 2000/mRNA complexes in Opti-MEM I reduced serum media or the use of TransIT mRNA transfection reagent/mRNA complexes in cell culture medium led to a high transfection efficiency between 80 and 90%. The measurements also demonstrated that HEK293 cells can be successfully transfected with high efficiency using TransIT mRNA transfection reagent or Lipofectamine 2000_optimized method. BJ fibroblasts showed highest transfection efficiency using Lipofectamine 2000_optimized method.

The use of transfection reagents in low amounts can lead to reduced transfection efficiency and on the other hand when they are used in high amounts, they can induce cell toxicity. Thus, the use of optimal mRNA to lipofectamine ratio is important to maintain cell viability and high transfection efficiency. The transfection of HEK293 cells, BJ fibroblasts, and ECs was performed with different amounts of Lipofectamine 2000. The measurements showed that the transfection of BJ fibroblasts can be performed using 4 μl Lipofectamine 2000, for HEK293 cells 2 μl was sufficient and the transfection of ECs can be performed with 1 or 2 μl Lipofectamine 2000. The use of 6 μl Lipofectamine 2000 for transfection of ECs resulted in decreased transfection efficiency, which indicates a toxic effect of Lipofectamine 2000 in higher concentrations. This shows that the Lipofectamine 2000 concentration should be optimized for each cell type.

After the mRNA transfection, HEK293 cells showed higher fluorescence intensities than ECs and BJ fibroblasts, which means that they produce higher amounts of eGFP. In comparison to ECs with a Geo Mean of 362 and BJ fibroblasts with a Geo Mean of 210, the Geo Mean of HEK293 cells was 710. This was also confirmed by fluorescence microscopic analyses. However, despite different fluorescence intensities, all cell types demonstrated a high transfectability.

The transfection of cells already with 0.5 μg mRNA and 2 μl Lipofectamine 2000 resulted in a high transfection efficiency, 86% of HEK293 cells, 75% of BJ fibroblasts, and 79% of ECs were eGFP positive. However, the increasing of mRNA amount from 0.5 μg to 1.5 μg led to significantly increased eGFP synthesis in HEK293 cells. The transfection of BJ fibroblasts with 1.5 μg mRNA was needed to detect significantly higher expression of eGFP in the cells compared to the control cells. In contrast, the transfection of ECs only with 0.5 μg eGFP mRNA was sufficient to induce significantly increased protein expression compared to the control cells. The expression did not change by increasing the mRNA amount to 2.5 μg.

Only after a single transfection with eGFP mRNA, all cell types expressed eGFP for 3 days. Thus, these results showed that by using synthetic modified mRNA the protein expression can be maintained over a few days in transfected cells. The eGFP expression reached the maximum on day 2 in BJ fibroblasts and showed no change on day 3. In ECs, the eGFP expression was highest 1 day post-transfection and decreased after 2 days. However, the detected fluorescence intensity after 3 days was still high with a Geo Mean of approximately 470. Each day, the eGFP expression continuously decreased in HEK293 cells. These cells are rapidly dividing cells with a doubling time of approximately 24 h. Thus, the rapid decrease of the fluorescence intensity compared to BJ fibroblasts and ECs could be caused by halving of the eGFP amount after each division.

Conclusion

Here, we described the production of modified mRNA for exogenous delivery into cells to induce protein expression. The delivery of the generated mRNA led to the synthesis of eGFP in ECs, which was present during 3 days in the cells. However, the results also demonstrated that the transfection method, reagent, and the amount of mRNA and transfection reagent influence the successful mRNA delivery and thus the expression of the protein. Using the exogenous mRNA delivery, each cell type could be induced to temporarily produce the desired proteins. This method has great potential for therapeutic applications, but each application has to be first optimized individually.

Competing interests

The authors declare that they have no competing interests.

Authors' contributions

Conceived and designed the experiments: MAA, HPW, CS. Performed the experiments: AB, TK. Contributed reagents/materials/analysis tools: HPW, CS. Wrote the paper: MAA, HPW. Analyzed and interpreted the data: MAA, AB, SK. All authors read and approved the final manuscript.

Acknowledgements

This project was funded by the European Social Funds in Baden-Wuerttemberg, Germany, and the MWK-BW, Germany and the German Federal Ministry of Education and Research (BMBF).

References

1. Kormann MS, Hasenpusch G, Aneja MK, Nica G, Flemmer AW, Herber-Jonat S, Huppmann M, Mays LE, Illenyi M, Schams A, Griese M, Bittmann I, Handgretinger R, Hartl D, Rosenecker J, Rudolph C: **Expression of therapeutic proteins after delivery of chemically modified mRNA in mice.** *Nat Biotechnol* 2011, **29**:154–157.
2. Mandal PK, Rossi DJ: **Reprogramming human fibroblasts to pluripotency using modified mRNA.** *Nat Protoc* 2013, **8**:568–582.
3. Warren L, Manos PD, Ahfeldt T, Loh YH, Li H, Lau F, Ebina W, Mandal PK, Smith ZD, Meissner A, Daley GQ, Brack AS, Collins JJ, Cowan C, Schlaeger TM, Rossi DJ: **Highly efficient reprogramming to pluripotency and directed differentiation of human cells with synthetic modified mRNA.** *Cell stem cell* 2010, **7**:618–630.

4. Yakubov E, Rechavi G, Rozenblatt S, Givol D: **Reprogramming of human fibroblasts to pluripotent stem cells using mRNA of four transcription factors.** *Biochem Biophys Res Commun* 2010, **394:**189–193.
5. Petsch B, Schnee M, Vogel AB, Lange E, Hoffmann B, Voss D, Schlake T, Thess A, Kallen KJ, Stitz L, Kramps T: **Protective efficacy of in vitro synthesized, specific mRNA vaccines against influenza A virus infection.** *Nat Biotechnol* 2012, **30:**1210–1216.
6. Kariko K, Buckstein M, Ni H, Weissman D: **Suppression of RNA recognition by Toll-like receptors: the impact of nucleoside modification and the evolutionary origin of RNA.** *Immunity* 2005, **23:**165–175.
7. Kariko K, Weissman D: **Naturally occurring nucleoside modifications suppress the immunostimulatory activity of RNA: implication for therapeutic RNA development.** *Curr Opin Drug Discov Devel* 2007, **10:**523–532.
8. Anderson BR, Muramatsu H, Nallagatla SR, Bevilacqua PC, Sansing LH, Weissman D, Kariko K: **Incorporation of pseudouridine into mRNA enhances translation by diminishing PKR activation.** *Nucleic Acids Res* 2010, **38:**5884–5892.
9. Kariko K, Muramatsu H, Welsh FA, Ludwig J, Kato H, Akira S, Weissman D: **Incorporation of pseudouridine into mRNA yields superior nonimmunogenic vector with increased translational capacity and biological stability.** *Mol Ther* 2008, **16:**1833–1840.
10. Hacein-Bey-Abina S, Hauer J, Lim A, Picard C, Wang GP, Berry CC, Martinache C, Rieux-Laucat F, Latour S, Belohradsky BH, Leiva L, Sorensen R, Debré M, Casanova JL, Blanche S, Durandy A, Bushman FD, Fischer A, Cavazzana-Calvo M: **Efficacy of gene therapy for X-linked severe combined immunodeficiency.** *N Engl J Med* 2010, **363:**355–364.
11. Hacein-Bey-Abina S, Garrigue A, Wang GP, Soulier J, Lim A, Morillon E, Clappier E, Caccavelli L, Delabesse E, Beldjord K, Asnafi V, MacIntyre E, Dal Cortivo L, Radford I, Brousse N, Sigaux F, Moshous D, Hauer J, Borkhardt A, Belohradsky BH, Wintergerst U, Velez MC, Leiva L, Sorensen R, Wulffraat N, Blanche S, Bushman FD, Fischer A, Cavazzana-Calvo M: **Insertional oncogenesis in 4 patients after retrovirus-mediated gene therapy of SCID-X1.** *J Clin Invest* 2008, **118:**3132–3142.
12. Avci-Adali M, Kobba J, Neumann B, Lescan M, Perle N, Wilhelm N, Wiedmaier H, Schlensak C, Wendel HP: **Application of a rotating bioreactor consisting of low-cost and ready-to-use medical disposables for in vitro evaluation of the endothelialization efficiency of small-caliber vascular prostheses.** *J Biomed Mater Res B Appl Biomater* 2013, **101:**1061–1068.
13. Bangel-Ruland N, Tomczak K, Fernandez EF, Leier G, Leciejewski B, Rudolph C, Rosenecker J, Weber WM: **CFTR-mRNA delivery: a novel alternative for cystic fibrosis 'gene therapy'.** *J Gene Med* 2013, **15:**414–426.
14. Benteyn D, Anguille S, Van Lint S, Heirman C, Van Nuffel AM, Corthals J, Ochsenreither S, Waelput W, Van Beneden K, Breckpot K, Van Tendeloo V, Thielemans K, Bonehill A: **Design of an optimized Wilms' Tumor 1 (WT1) mRNA construct for enhanced WT1 expression and improved immunogenicity In Vitro and In Vivo.** *Mol Ther Nucleic Acids* 2013, **2:**e134.
15. Levy O, Zhao W, Mortensen LJ, Leblanc S, Tsang K, Fu M, Phillips JA, Sagar V, Anandakumaran P, Ngai J, Cui CH, Eimon P, Angel M, Lin CP, Yanik MF, Karp JM: **mRNA-engineered mesenchymal stem cells for targeted delivery of interleukin-10 to sites of inflammation.** *Blood* 2013, **122:**e23–e32.

Towards an integrated systems-based modelling framework for drug transport and its effect on tumour cells

Cong Liu[1*], Krishnan[1,2] and Xiao Yun Xu[1]

Abstract

Background: A systematic understanding of chemotherapeutic influence on solid tumours is highly challenging and complex as it encompasses the interplay of phenomena occurring at multiple scales. It is desirable to have a multiscale systems framework capable of disentangling the individual roles of multiple contributing factors, such as transport and extracellular factors, and purely intracellular factors, as well as the interactions among these factors. Based on a recently developed systems-based modelling framework, we have developed a coupled system in order to further elucidate the role of drug transport, and its interplay with cellular signalling by incorporating intra- and extra-vascular drug transport in tumour, dynamic descriptions of intracellular signalling and tumour cell density dynamics.

Results: Different aspects of the interaction between transport and cell signalling and the effects of transport parameters have been investigated *in silico*. Limited drug penetration is found to be a major constraint in inducing drug effect; many aspects of the interaction of transport with cell signalling are independent of the details of cell signalling. A sensitivity analysis indicates that the effect of drug diffusivity depends on the balance between interstitial drug transport and the specific requirement for triggering apoptosis (governed by highly nonlinear signalling networks), suggesting that the effect of drug diffusivity in such cases must be considered in conjunction with descriptions of cellular dynamics.

Conclusions: The modelling framework developed in this study provides qualitative and mechanistic insights into the effect of drug on tumour cells. It provides an *in silico* experimental platform to investigate the interplay between extracellular factors (e.g. transport) and intracellular factors. Such a platform is essential to understanding the individual and combined effects of transport and cellular factors in solid tumour.

Keywords: Drug transport, Drug effect, Intracellular signalling, *in silico* experimental platform, Systems-based modelling framework

Introduction

The efficacy of chemotherapy is strongly dependent on the transport of anticancer drugs to tumour cells and their responses to the administrated drug [1]; both can be compromised significantly by complexities associated with cancer, which is a disease of collective dysregulations across multiple scales.

To exert therapeutic effects, anticancer drugs must reach tumour cells with a sufficiently high concentration [2]. Limited penetration is one of the major causes of failure of chemotherapy treatment of solid tumours. The first obstacle to most blood-borne chemotherapeutic agents is posed by the abnormal and chaotic tumour vasculature, which limits tumour blood flow and consequently the supply of drugs and nutrients [3]. After crossing the capillary wall, anticancer drugs must penetrate through the tumour interstitium, where drug distribution is determined by the effectiveness of drug transport by diffusion and convection as well as drug consumption [2]. Elevated interstitial fluid pressure in solid tumours hinders convective transport, rendering diffusion the dominant mechanism for interstitial drug transport [4]. Drug diffusivity depends strongly on the

* Correspondence: cong.liu06@imperial.ac.uk
[1]Department of Chemical Engineering, Imperial College London, South Kensington Campus, London SW7 2AZ, UK
Full list of author information is available at the end of the article

physicochemical properties of the specific drug, such as molecular weight, shape, charge and solubility. Drug consumption involves drug binding, sequestration and metabolism, which can be altered by microenvironmental conditions, such as extracellular matrix composition and structure, cell packing density and the presence of tumour acidity [3,5,6].

For most anticancer drugs, it is necessary for drug molecules to transport across cell membranes to reach the target molecules and interact with them, as a consequence, triggering cellular signal transduction. Cellular signalling is one of the important characteristics of every living cell in that it governs the basic cellular activities by perceiving and correctly responding to external/internal stimuli. Anticancer drugs as stress stimuli can regulate/trigger cell signalling to kill cells or affect their cellular responses (cell apoptosis, cell proliferation, differentiation and migration), which might be directly associated with fatal consequences of cancer [7-9]. The way by which cellular signalling functions is extremely complex as cellular pathways are not isolated from each other but are interconnected through a complex network [10]. In particular, highly non-linear input-output relationships are usually displayed in cell signalling networks, with a number of emergent properties, such as adaptive responses and robust switching by positive feedback [11,12]. It is worth emphasizing that the dynamic interactions and signal transmission in these chemical networks control cellular decision making and cellular responses such as movement, apoptosis etc.

Considerable effort has been devoted to developing mathematical models to predict drug concentration and drug effect on solid tumours (for a review, see [13]). Compartmental models have been widely adopted for prediction of temporal profiles of drug concentration in designed compartments, particularly drug concentration in blood in pharmacokinetic studies [14-16]. To obtain spatio-temporal drug distributions in a given tumour geometry, it is necessary to explicitly account for drug transport (diffusion or/and convection) [17-19]. After drug concentrations are obtained, suitable pharmacodynamic models can be used to predict the effect of drug as a function of drug concentration and/or as a function of time following drug administration in a phenomenological and empirical manner, with elaboration of observed data thus neglecting detailed underlying mechanisms [20-22]. On the other hand, deterministic models can be used to describe the tumour response by assuming a drug concentration-dependent tumour growth characteristic or tumour death kinetics [23-25]. Unfortunately, mathematical models addressing the above mentioned areas have been developed separately; furthermore, they often bypass a key component that is the dynamic process of cellular signal transduction. While all these models are capable of providing certain levels of insights, none of them offers a transparent and integrated description of drug transport and drug effect accounting for the associated cellular signalling.

In this study, an integrated systems-based mathematical modelling framework is employed and extended, which captures the information flow from drug delivery to the outcome, thus including biological transport processes of drugs and cellular response and accounting for dynamics of the relevant signal transduction. This allows us to begin to probe and elucidate different aspects of the roles and the interaction of transport and intracellular signalling dynamics. In a spatially distributed system, intracellular signalling is triggered in response to heterogeneous drug stimuli delivered through transport pathways. It must be emphasized here that drug stimuli are dynamic (being delivered through complex vascular networks), while drug transport can be affected by many tissue level features and intracellular dynamics is highly nonlinear. Finally the dynamic coupling of these factors is not necessarily unidirectional. For instance the tissue scale properties and features could be a potential factor in affecting drug transport; as an example, it is found that apoptosis-inducing pretreatment enhances drug delivery [26,27]. Furthermore, with considerable progress in unravelling the intracellular and intercellular signal transduction in systems biology, mathematical modelling approaches can begin to go beyond empirical and transport centric models to integrate dynamic descriptions of transport and cellular signalling for a systematic examination of their interactions.

This paper examines some basic aspects of the interaction of transport and cellular dynamics. As a first step, the model is formulated on a simplified geometry of tumour vasculature, in which explicit coupling of blood flow between vascular and interstitial space is incorporated, along with drug transport. The effects of anticancer drugs are addressed by integrating the above with dynamics of intracellular apoptosis signalling. The integrated model is used to evaluate treatment strategies and to analyse other factors that may influence the response of tumour cells, in order to provide insights into the complex interplay between the different processes involved.

Methods

For mathematical modelling of drug transport, a commonly adopted approach is to avoid an explicit representation of the tumour vasculature which, instead, is treated as a distributed source term in the governing equations [19,28-31]. In doing so, descriptions of transport processes are incomplete without accounting for vascular transport and the spatial relationship between blood vessels and tumour interstitium. However, incorporating realistic tumour vasculature geometry is highly challenging, given the fact that the tumour vasculature is abnormal, irregular and heterogeneous. Further complexities in evaluating drug effects are added when dynamic intracellular signalling processes are incorporated, which are triggered in response to spatio-temporal drug stimuli and exhibit highly non-linear dynamics. To obtain clear-cut and transparent insights into transport mechanisms, cellular signalling and their interaction, we employ the modelling framework as an *in silico* experimental platform which describes a well-defined tumour-drug system with minimal essential elements, definite information flow and a controlled source of variability and heterogeneity. The *in silico* experimental platform depicts an idealized tumour with no heterogeneity, in a simplified geometry. This setup is designed as an initial effort to contain the minimal components necessary for understanding the effects of drugs on tumours and elucidating the effects of transport and cellular factors in a transparent manner without consideration of other factors.

Computational geometry

The model consists of a single blood vessel surrounded by the tumour interstitium, which is a simplified representation also employed in previous studies [32,33]. Although the geometrical configuration is similar to that of a tumour cord model [17,18,34], they differ in size in that the present model mimics the entire transport domain in the tumour tissue. While the equivalent vessel geometry and the corresponding parameters adopted in the model may be an over-simplification, it is practical and tractable to start with this simple geometry for the purpose of understanding the essential interplay between drug transport and drug effect. Figure 1 depicts the computational geometry and the flow of information in the model.

Mathematical models

The current modelling framework includes basic descriptions of blood flow, drug transport, intracellular apoptosis signalling and tumour cell density dynamics. The main assumptions are as follows: (1) blood is an incompressible, Newtonian fluid and the blood vessel is straight and rigid; (2) the tumour interstitium is homogeneous, with a uniform distribution of nutrients and pH; (3) tumour cells are stationary, leading to the assumption that the tumour interstitium has a fixed outer boundary; (4) all tumour cells are distributed uniformly, identical (i.e. ignoring cellular variability, stochasticity, and the effects of cell-to-cell interactions and cell-cycle) and alive initially. These assumptions are made with the understanding that the key ingredients are reasonably represented in the initial model and that individual assumptions may be relaxed in subsequent studies. Detailed descriptions of each of the elements are given in the following sections. A brief overview of the mathematical equations for different regions in the computational domain is presented in Figure 2, with symbols and values of parameters defined in Tables 1 and 2.

Tumour blood flow

The model accounts for the coupling between vascular, transmural and interstitial fluid flow since tumour blood vessels are highly permeable. Blood flow is assumed to be steady, which is acceptable here as it is capable of serving as a fundamental platform to investigate the dynamic behaviour of drug transport and intracellular events without imposing further complexities related to pulsatile blood flow.

Flow within the blood vessel is governed by the Navier-Stokes equations:

$$\rho \mathbf{u}_v \cdot \nabla \mathbf{u}_v = -\nabla \boldsymbol{P}_v + \nabla \cdot \left[\boldsymbol{\mu}\left(\nabla \mathbf{u}_v + (\nabla \mathbf{u}_v)^T\right)\right] \quad (1a)$$

$$\nabla \mathrm{u}_v = 0 \quad (1b)$$

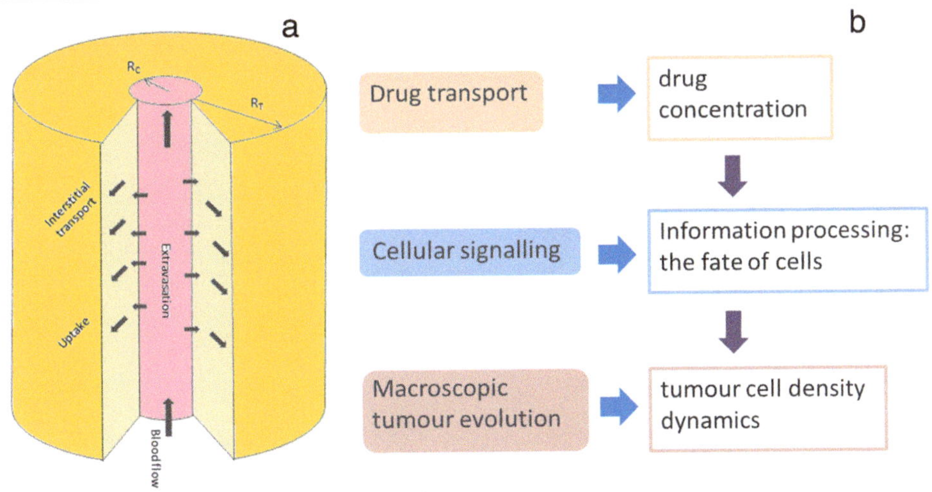

Figure 1 Schematic overview of information flow involved in the modelling framework. (a) Schematic illustration of the simplest tumour vascular geometry and **(b)** diagrammatical representation of the levels of description incorporated in the model.

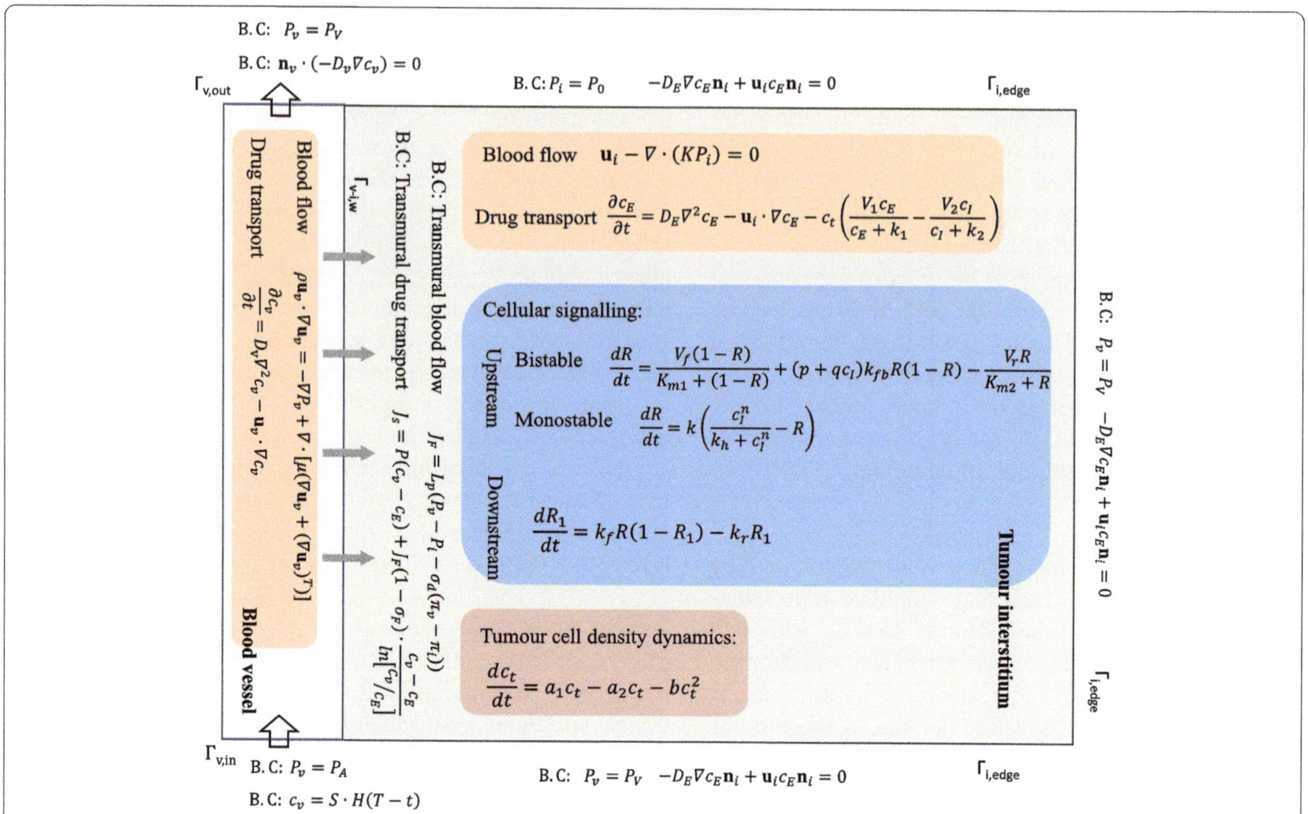

Figure 2 Overview of mathematical descriptions involved in each domain with corresponding boundary conditions. Details of biological transport (blood flow as well as drug transport), coarse-grained cellular apoptosis signal transduction and tumour cell density dynamics refer to following subsections.

Where ρ is blood density, μ is blood viscosity, $\mathbf{u}$ is blood velocity vector with subscript v denoting vascular space, and P_v is vascular blood pressure.

Eqn. 1a and Eqn. 1b are solved subject to the following boundary conditions:

$$P_v = P_A \quad on\ \Gamma_{v,in} \tag{2a}$$

$$P_v = P_V \quad on\ \Gamma_{v,out} \tag{2b}$$

$$\mathbf{n}_v \cdot \mathbf{u}_v = J_F \quad on\ \Gamma_{v-i,w} \tag{2c}$$

Fluid motion in the interstitial space is described by Darcy's law:

$$\mathbf{u}_i - \nabla \cdot (KP_i) = 0 \tag{3a}$$

$$\nabla \mathbf{u}_i = 0 \tag{3b}$$

where K represents interstitial hydraulic conductivity, and $\mathbf{u}_i$ and P_i are blood velocity vector and fluid pressure in the interstitium.

The boundary conditions for Eqn. 3a and Eqn. 3b are:

$$\text{-}\mathbf{n}_i \cdot \boldsymbol{u}_i = J_F \quad on\ \Gamma_{v-i,w} \tag{4a}$$

$$P_i = P_0 on\ \Gamma_{i,edge} \tag{4b}$$

The boundary conditions (BC) described by Eqn. 2a and Eqn. 2b specify a constant arterial and venous pressure at the inlet and outlet of the vessel, respectively; BC (2c) assumes a transmural velocity in the normal direction (outward) of the blood vessel. BC (4b) assigns the ambient pressure at all boundaries of the interstitium (except the vessel wall); BC (4a) prescribes a transmural velocity normal (inward) to the vessel wall. The transmural velocity, J_F can be calculated using Starling's law:

$$J_F = L_p(P_v - P_i - \sigma_d(\pi_v - \pi_i)) \quad \text{on } \Gamma_{\text{v-i,w}} \tag{5}$$

Where L_p is vascular hydraulic conductivity, σ_d is osmotic reflection coefficient, and π_v and π_i are osmotic pressure in the vascular and insterstitial space, respectively. As the contribution of osmotic pressure difference is negligible in solid tumours, Starling's law is reduced to

$$J_F = L_p(P_v - P_i) \quad \text{on } \Gamma_{\text{v-i,w}} \tag{6}$$

Drug transport

Three variables are considered in the drug transport processes: the intravascular drug concentration (c_v), the interstitial extracellular free drug concentration (c_E), and intracellular drug concentration (c_I). Within the blood

Table 1 Parameters and values used in the mathematical models

Parameter	Symbol	Value	Reference
Tumour blood flow			
Tumour capillary radius	R_C	100 (μm)	[33]
Tumour interstitium radius	R_T	1200 (μm)	Estimated
Tumour length	L	1 (cm)	[33]
Blood density	ρ	1000 (kg/m^3)	[32]
Blood viscosity	μ	0.004 (Pa·s)	[35]
Vascular hydraulic conductivity	L_p	0.36-2.8 × 10^{-6} (cm/mmHg·s)	[36]
Tissue hydraulic conductivity	K	0.4-2.0 × 10^{-7} (cm^2/mmHg·s)	[36]
Vascular inlet pressure	P_A	15 (mmHg)	[37]
Vascular outlet pressure	P_V	0	[36]
Ambient pressure	P_O	0	[36]
Drug transport (Doxorubicin)	D_E	1.5778 × 10^{-6} (cm^2/s)	[17]
DOX diffusion coefficient			
Rate of transmembrane transport	V_1, V_2 ($V_1 = V_2$)	0.28 (ng/(10^{5}cells)/min)	[14]
Diffusive permeability for DOX	P_E	2.778 × 10^{-4} (cm/s)	[17]
Michaelis constant for transmembrane transport	k_1	0.219 (μg/ml)	[14]
Michaelis constant for transmembrane transport	k_2	1.37 (ng/(10^{5}cells))	[14]
Tumour cell density			
Initial tumour cell density	$c_{t,0}$	10^6 (cells/mm^3)	[17]
Tumour cell growth rate	a_1	0.5 (day^{-1})	Estimated
Saturation constant in logistic equation	b	0.02592 (mm^3/(10^{5}cells)/day)	Estimated
Tumour cell natural decay rate	a_2	0.24 (day^{-1})	Estimated
Intracellular signalling:			
Bistable switch			
Michaelis Menten constants	V_f	27 (hr^{-1})	[38]
Michaelis Menten constants	V_r	0.459 (hr^{-1})	[38]
Michaelis Menten constants	K_{m1}	100	[38]
Michaelis Menten constants	K_{m2}	0.01	[38]
Kinetic parameter mediating feedback strength	k_{fb}	2.927 (hr^{-1})	[38]
Basal level in the bistable switch	p	0.7	Estimated
Parameter mediating input regulation in the bistable switch	q	0.3(ng/(10^{5}cells))$^{-1}$	Estimated
Monostable switch			
Kinetic parameter reflecting the time scale of the response	k	0.432 (hr^{-1})	Estimated
Associated constant	k_h	1 (ng/(10^{5}cells))	Estimated
Hill coefficient	n	10	Estimated
R_1 protein activation rate	k_f	3.6 (hr^{-1})	Estimated
R_1 protein degradation rate	k_r	0.144 (hr^{-1})	Estimated
R_1 Threshold for apoptosis switch	$R_{1,th}$	0.9	Estimated

vessel, the drug is transported mainly by convection with negligible reactions (e.g. metabolism, binding/unbinding, uptake). Once extravasated into the tumour interstitium, drug particles penetrate through the intersititum via diffusion and convection and at the same time they may be taken up by tumour cells. The drug is treated as a blood-borne solute and its transport is governed by a diffusion-convection-reaction equation.

Table 2 Variables and their definitions in the mathematical models

Symbol	Definition	Symbol	Definition
$\mathbf{u}_v$	Blood velocity vector in vascular domain	c_v	DOX concentration in vascular domain
$\mathbf{u}_i$	Blood velocity vector in interstitial domain	c_E	Extracellular DOX concentration
P_v	Vascular blood pressure	c_I	Intracellular DOX concentration
P_i	Interstitial fluid pressure	R	Hypothetical downstream intermediate protein in apoptosis signalling cascade
J_F	Transmural fluid velocity, determined by Starling's law	R_1	Hypothetical protein responsible for triggering apoptosis
J_s	Transmural drug flux, determined by Kedem-Katchalsky equation	c_t	Tumour cell density

Solute dynamics in the blood vessel (Ω_v) is governed by:

$$\frac{\partial c_v}{\partial t} = D_v \nabla^2 c_v - \mathbf{u}_v \cdot \nabla c_v \tag{7}$$

Where c_v refers to drug concentration in the vascular space, and D_v is drug diffusivity in the vascular space. The boundary conditions are:

$$c_v = S \cdot H(T-t), \ t>0 \quad \text{on } \Gamma_{\text{v,in}} \tag{8a}$$

$$\mathbf{n}_v \cdot (-D_v \nabla c_v) = 0 \quad \text{on } \Gamma_{\text{v,out}} \tag{8b}$$

$$-D_v \nabla c_v \mathbf{n}_v + \mathbf{u}_v c_v \mathbf{n}_v = -J_s \quad \text{on } \mathit{\Gamma}_{\text{v-i,w}} \tag{8c}$$

BC (8a) prescribes a pulse injection at the vessel inlet with a constant intensity S and infusion time T, in which Heaviside term $H()$ indicates infusion occurs during the period of t = 0 to t = T. BC (8b) defines a convective flux at the outlet; BC (8c) sets an outward solute flux across the leaky wall boundary.

Solute dynamics in the interstitium (Ω_i) is governed by the extracellular and intracellular drug transport.

Extracellular drug concentration:

$$\frac{\partial c_E}{\partial t} = D_E \nabla^2 c_E - \mathbf{u}_i . \nabla c_E - c_t \left(\frac{V_1 c_E}{c_E + k_1} - \frac{V_2 c_I}{c_I + k_2} \right) \tag{9a}$$

Where c_E and c_I refer to the extracellular and intracellular drug concentration, respectively, D_E is diffusion coefficient of drug in the interstitium, while c_t is tumour cell density. V_1, V_2, k_E and k_I are constants that describe transport across the cell membrane, in which V_1 and V_2 are the maximum rates of transmembrane transport, while k_1 and k_2 are the Michaelis-Menten constants for transmembrane transport. Eqn. 9a describes diffusion, convection of extracellular drugs and their uptake/pumping out by tumour cells with the last terms expressed by Michaelis-Menten kinetics. The boundary conditions are:

$$-D_E \nabla c_E \mathbf{n}_i + \mathbf{u}_i c_E \mathbf{n}_i = 0 \quad \text{on } \Gamma_{\text{i,edge}} \tag{10a}$$

$$-D \nabla c_E \mathbf{n}_i + \mathbf{u}_i c_E \mathbf{n}_i = J_S \quad \text{on } \mathit{\Gamma}_{\text{v-i,w}} \tag{10b}$$

BC (10a) describes a no flux condition at the other boundaries of the interstitium. In the current study, neither the surrounding host tissue nor functional lymphatics in the host tissue are explicitly described. However, it is possible to incorporate the effect of lymphatic drainage in the current set up by imposing a mixed boundary condition at the outer surface. Simulations with both BCs gave essentially identical results for the stimuli used here (results not shown). The effect of the host tissue will be explicitly examined in a future study. BC (10b) assigns an inward solute flux into the interstitium, which can be determined by the Kedem-Katchalsky equation:

$$J_s = P(c_v - c_E) + J_F(1 - \sigma_F) \triangle c_{lm} \cdots \triangle c_{lm} = \frac{c_v - c_E}{ln \left[{}^{c_v}/_{c_E} \right]} \quad \text{on } \mathit{\Gamma}_{\text{v-i,w}} \tag{11}$$

Where P is drug diffusive permeability across the vessel wall, σ_f is osmotic reflection coefficient, Δc_{lm} is the log-mean concentration across the vessel wall, and J_F is the fluid flux across the vessel wall, which is determined by Starling's law.

Intracellular drug concentration

The intracellular drug concentration depends on the corresponding extracellular drug concentration according to transmembrane transport. This is given by

$$\frac{\partial c_I}{\partial t} = \frac{V_1 c_E}{c_E + k_1} - \frac{V_2 c_I}{c_I + k_2} \tag{12}$$

It is noted that drug binding to plasma proteins (mainly albumin) is neglected in the current study. In many cases, the binding/unbinding process is described simply by first-order kinetics; therefore its effect is relatively simple in that it reduces the amount of free extracellular drug available to tumour cells. Incorporation of drug binding would not alter the system output qualitatively, but would dramatically increase computational burden. However, it is essential to take this into account when quantitative modelling is required or sophisticated binding mechanisms are involved.

Intracellular apoptosis signalling

The modelling of intracellular signalling processes, a substantial core of systems biology, plays a very important role in the entire modelling framework as decisions on cell fate are processed and determined by molecular signalling networks. It is thus very important to have a dynamic representation of this in the modelling framework. Given the complexity of signalling which contains many missing biological details and unknown parameters, it is essential to choose an appropriate level of description in the model to start with, so that the most important known aspects of the signalling and cellular decision making are included. Thus, the strategy adopted here is to start with coarse grained descriptions of the cellular signalling dynamics, which are capable of representing correctly the nature of the information flow, and ensuring that the qualitatively important features of detailed models are accounted for. We believe that this is more appropriate than starting with detailed models containing many unknown factors and other details whose correctness may be difficult to establish. At the same time, this provides a platform that allows more detailed mechanistic models to be incorporated in the future.

When modelling the intracellular processes in response to chemotherapy, the main process of interest is apoptosis (programmed cell death) following the administration of anticancer agents. Based on systems biology investigations and existing models [39-45], we recognize that two key features must be reflected by any model regardless of its complexity. Firstly, some threshold effect must be present; secondly, the "switch" to apoptosis must be realized in an irreversible way.

The apoptosis models adopted here are based on the two types of switches commonly observed in cellular signalling: bistable and monostable apoptosis switches. As bistable switches can exhibit irreversibility intrinsically, they have been used in modelling irreversible cell fate decision-making in apoptosis [40,43,45]. Generally, positive feedback (biologically existing in the intracellular caspase network) and cooperativity (in apoptosome formation) are regarded as sources of bistability. However, as far as apoptosis is concerned, it is not obvious whether representing the irreversible fate (cell death) as a steady state is necessary or even reasonable (at the cell fate decision level). It is possible that an irreversible decision is made when critical cellular events are triggered, from which there is no turning back. Thus the irreversibility could result from a simple irreversible reaction, which is kept under tight control and triggered only under very special circumstances [39,41,42]. Noting this, and the fact that both mechanisms have been discussed in the context of apoptosis, the two models employed in this study are: (1) a bistable switch with self-contained threshold behaviour and irreversibility; (2) a sequential interconnection of a monostable switch and a downstream irreversible reaction effect [34].

Bistable switch

$$\frac{dR}{dt}=\frac{V_f(1-R)}{K_{m1}+(1-R)}+(p+qc_I)k_{fb}R(1-R)-\frac{V_rR}{K_{m2}+R} \tag{13a}$$

Where K_{m1} and K_{m2} are the Michaelis-Menten parameters. k_{fb} is a kinetic parameter which parametrizes the feedback strength. The constants p and q serve to set the basal level and dynamic range of the module.

Irreversible monostable switch

$$\frac{dR}{dt}=k\left(\frac{c_I^n}{k_h+c_I^n}-R\right) \tag{13b}$$

Where n denotes the Hill coefficient, k_h an associated constant in the Hill term, and k is a parameter representing the time scale of the response.

In both models, R acts as a typical downstream intermediate element along the signal transduction while R_1 represents the output responsible for directly triggering apoptosis.

$$\frac{dR_1}{dt}=k_fR(1-R_1)-k_rR_1 \tag{14}$$

Once a threshold of R_1 is crossed, cell death is triggered. Normalised concentrations of molecules R and R_1 are adopted. Two additional points are worth emphasizing here. Firstly, the actual choice of monostable/bistable model has a very minor effect, as analysis with other model variants has led to very similar results. Secondly, detailed characterization and comparison of monostable and bistable models demonstrate their similarities and differences, which provide a basis for their use in contexts such as this [46].

Tumour cell density dynamics

The equation governing the tumour cell density is described by a continuous logistic equation [47]:

$$\frac{dc_t}{dt}=a_1c_t-a_2c_t-bc_t^2 \tag{15}$$

Where a_1 is tumour growth rate, a_2 is tumour natural decay rate and b is a saturation constant in the logistic tumour growth equation.

Eqn. 15 naturally describes the growth and death of cells with saturating growth effect leading to a finite steady state (models with some variations have very similar net results: see [46]). These equations provide an explicit representation of the key features of interest. Although the population balance formalism provides a more comprehensive description of birth and death of cells, it is computationally highly demanding and can be difficult to handle if additional cellular complexities are included.

When tumour cells are perturbed by anticancer drugs, the intracellular apoptosis signalling is initiated, resulting in cell death at the population level. Since cell density is described in a continuous, rather than discrete form, triggering of the intracellular threshold is represented by a sharp fall in growth rate at the population level. Clearly, the dynamics of the logistic model imply that if the growth rate becomes sufficiently low, the zero steady state ($c_t = 0$) would be the only biologically relevant state, indicating that all cells will eventually die. It should be noted that computational implementation of this threshold effect is achieved in a reversible way in the bistable model, but in a unidirectional (irreversible) way in the monostable model.

Initial conditions

Except for tumour cell density, all other variables are set to be zero initially. A uniform tumour cell density is assumed prior to drug injection.

Numerical methods

Computational procedures

All numerical simulations presented in this paper are implemented in the finite element based software Comsol Multiphysics. The simulation of blood flow is decoupled from that of drug transport and tumour cell density by assuming the velocity field is independent of the drug concentration field and tumour cell density distribution. Steady-state simulation of blood flow is carried out first. Upon obtaining the pressure and velocity fields, drug transport is resolved by solving the diffusion-convection-reaction equation. The boundary conditions at the vessel wall are implemented in accordance with the physical settings in Comsol Multiphysics. With regard to fluid flow, transmural velocity (J_F) is positive in both the vessel and interstitial domain as it points away from the surface in the vessel domain, and at the same time, it represents the inflow to the interstitial domain. J_F is set as a variable in accordance with Starling's law, which enables the coupling of the vascular fluid pressure (P_v) to the interstitial fluid pressure (P_i). For drug transport, an inflow flux is set by default, which means a negative transmural flux (J_S) in the vessel domain, and a positive flux in the interstitial domain. Intracellular signal transduction is triggered by the local intracellular drug concentration, the response of which is manifested through tumour cell density owing to decreased tumour growth rate or increased tumour death rate. The tumour cell density, in turn, affects drug transport.

The equations are discretised and solved on a pre-generated computational mesh. Mesh sensitivity study is carried out first to provide mesh independent solutions. The final mesh consists of 75,000 and 325,000 mapped meshes for drug transport and tumour cell density, respectively.

Model parameters

Values of all parameters as well as variables used in the integrated model are defined in Tables 1 and 2, respectively. These are extracted from a variety of sources as they span multiple scales of description and are not available in a single tumour-drug system. Values for blood flow related parameters are mainly extracted from similar mathematical models found in the literature. Doxorubicin (DOX) is one of the anticancer drugs commonly used in clinics and a large amount of experimental data is available on its physical and pharmacokinetics properties; therefore it is chosen as a representative anticancer drug for parameterization purpose. Values for tumour growth parameters are chosen to reflect both the range of steady state, as well as the appropriate time scale. In the intracellular dynamics, parameters are not generally available and their values are determined based on appropriate reflection of the time scales involved in apoptosis signalling. Further, the downstream threshold ($R_{1,th}$) is specially chosen to ensure complete R_1 activation while maintaining the upstream signal for a sufficient time period. This in turn directly reflects the cellular behaviour to exposure of drugs for a sufficient time. Thus our intracellular descriptions are parameterized so that they are capable of triggering apoptosis decisions for comparable signals as are seen experimentally. It should be emphasized that most of the essential conclusions drawn from this study are not strongly dependent on the particular numerical choice of parameters.

Results

In this section, numerical results of blood flow are presented first, followed by results of drug transport and distribution. The effect of drug examined in terms of tumour cell density distribution by considering both bistable and irreversible monostable intracellular apoptosis models under various pulse drug injections. In addition, a sensitivity analysis is performed on parameters involved in drug transport. Results for drug concentration and tumour cell density are presented in the dimensionless form, which are normalised by their corresponding reference values: 0.001 $\mu g/mm^3$ for vascular and extracellular drug concentrations, 1 $ng/10^5$cells for intracellular drug concentration and 10^6 $cells/mm^3$ for tumour cell density.

Blood flow

Axial pressure profiles on the lumen and tissue side of the blood vessel wall together with the transmural velocity are presented in Figure 3. It is seen that vascular pressure falls linearly along the flow direction while interstitial fluid pressure experiences a sharp rise near the vessel inlet,

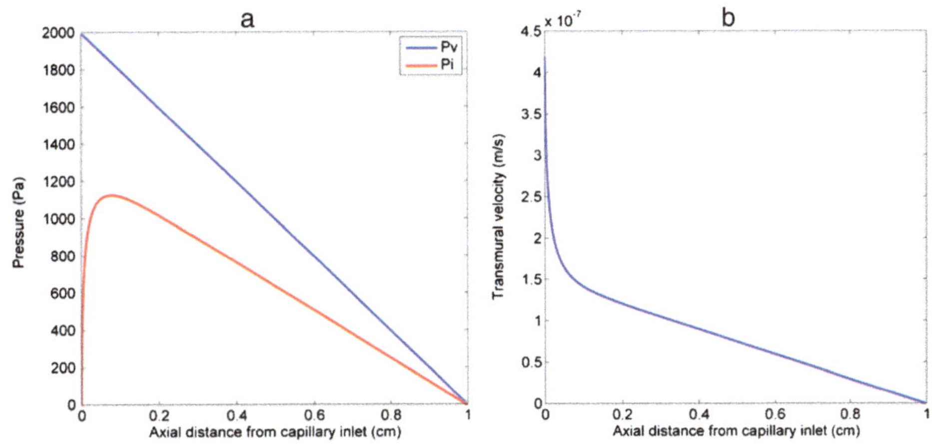

Figure 3 Axial distribution of (a) vascular (P_v) and interstitial fluid pressure (P_i) on the capillary wall, (b) transmural velocity.

followed by a gradual reduction. This is due to the entrance effect caused by the discontinuity of pressure conditions imposed at the vessel inlet and the adjacent lower boundary of the interstitium. The simulation results demonstrate that interstitial fluid pressure is strongly coupled to vascular fluid pressure due to elevated hydraulic conductivity normally found in tumour tissues. Since the pressure difference across the capillary wall determines the filtration velocity of blood, the latter (as shown in Figure 3(b)) experiences a sharp fall near the vessel inlet before declining gradually along the flow direction as the transmural pressure difference (*Pv-Pi*) diminishes.

Drug transport and distribution

Drug transport

Drugs are transported in tumour tissues by diffusion and convection. The relative importance of diffusion and convection is measured by Peclet number (*Pe*) which is defined as:

$$Pe = \frac{Lu}{D}$$

where L is the characteristic length, u is a representative velocity, and D is the diffusion coefficient for a given anticancer drug. Since D is 1.578×10^{-10} m^2/s for doxorubicin (Table 1), the corresponding *Pe* in the blood vessel is on the order of 10^6, suggesting that intravascular drug transport is dominated by convection.

Drugs extravasate into the tumour interstitium by diffusion and convection, determined by diffusive permeability P and blood filtration velocity J_F. The filtration velocity shown in Figure 3(b) is of the order of 10^{-7} m/s, which is an order of magnitude lower than diffusive permeability (2.78×10^{-6} m/s) across the vessel wall for doxorubicin, suggesting that the total transmural flux is dominated by the diffusive flux in this case.

With regard to interstitial drug transport, variations of *Pe* (based on local radial velocity) along the radial direction at three different axial locations are displayed in Figure 4(a). It shows that *Pe* is generally of the order of 10^{-1} and decreases along both the radial and axial directions, which suggests that diffusion is becoming more dominant than convection and the role of convection may be limited and confined to a region close to the wall. Furthermore, with the use of a simple tumour vascular geometry, the effect of convection can be quantified by comparing cross-sectional profiles of intracellular drug concentration. As shown in Figure 4(b), the difference is almost negligible, which confirms that diffusion plays a dominant role in interstitial drug transport.

Drug distribution

The anticancer drug is assumed to be directly injected into the blood stream at the inlet of the blood vessel in the form of a pulse, which is an appropriate type of signal as it represents the time-dependent nature without introducing further complexity to the analysis of dynamic interactions between tumours and drugs. For a systemic administration, a more realistic drug input expressed as an exponentially decaying function of time [48,49] or based on injection details could be readily incorporated in the future. Snapshots of spatial profiles of drug concentration are displayed in Figure 5, where t = 1.5 h corresponds to the end of pulse injection and t = 2 h is 0.5 h after drug injection. In Figure 5(a) drug concentration is uniform in the core region, while a concentration boundary layer is seen near the wall with drug concentration on the inner vessel wall decreasing along the direction of blood flow (z direction). Figure 5(c) shows that a steep extracellular drug concentration gradient is established close to the vessel wall while little drug reaches beyond $5R_C$ (R_C is the equivalent vessel radius). As intracellular drug concentration is dependent on the local extracellular drug concentration, it follows

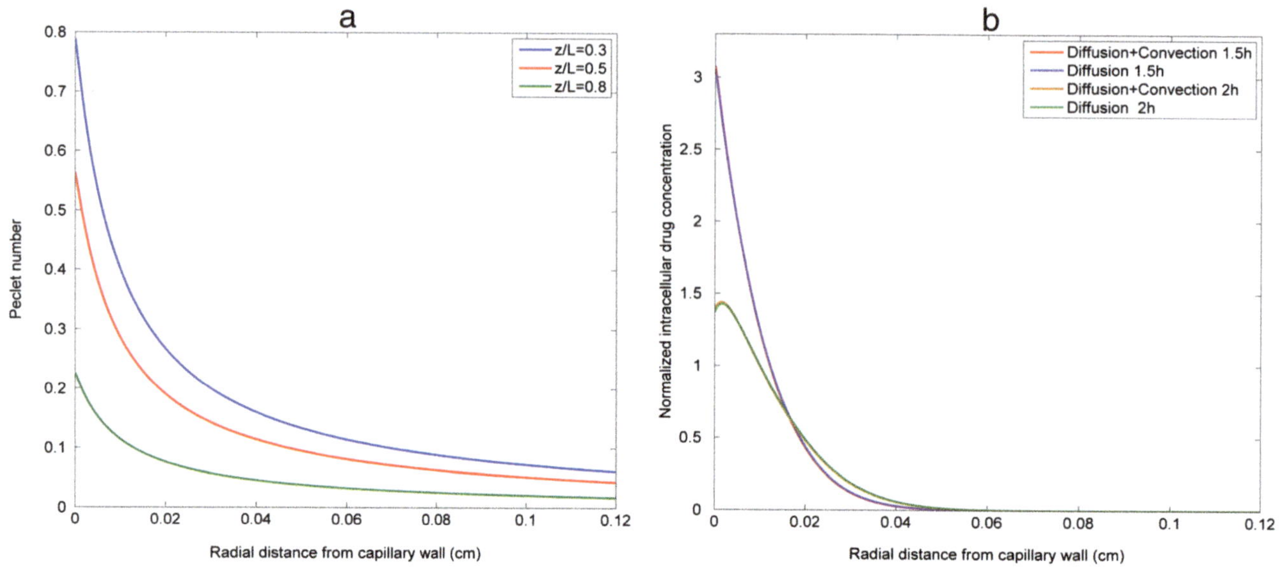

Figure 4 Role of diffusion and convection in drug transport. (a) cross-sectional profiles of *Pe* number at various axial positions in tumour interstitium, **(b)** quantitative comparison of normalized intracellular drug concentration at axial location z/L = 0.5 with and without convection under baseline pulse injection at t = 1.5 h (end of injection) and at t = 2 h (0.5 h after injection). Pulse characteristics of baseline pulse: pulse intensity $S = 1$, infusion time $T = 1.5$ h.

the same trend as shown in Figure 5(e) but with a larger value due to the kinetics of transmembrane transport.

Displayed in Figure 5(b,d,f) are snapshots of vascular, extracellular and intracellular drug concentrations at t = 2 h, half an hour after drug injection. In response to the sudden termination of drug input, reversal of concentration gradient is observed in the near wall region inside the blood vessel and in the interstitium. In this context, the interstitium acts as a reservoir, from which drugs are transported back to the blood vessel and eventually leave the blood vessel by convection. The reverse transport of drugs is confined to a thin layer close to the vessel wall, while drugs outside this layer are transported outward in the radial direction by diffusion and convection. Therefore, the extracellular drug concentration profile experiences a rise and reaches a peak before falling off; the same is observed for the intracellular drug concentration profile.

Drug effect–Tumour cell density distribution

Baseline case

As cell apoptosis is triggered by effective intracellular drug concentration (above the threshold) which shows little variation in the axial direction, tumour cell density at any cross section away from the inlet and outlet effects (e.g. z/L = 0.5) would be sufficiently representative and can be used to compare results obtained with different simulation parameters. Displayed in Figure 6(a) and (b) is the distribution of tumour cell density under the same pulse injection for bistable and irreversible monostable switches respectively. As effective drug penetration is confined to the close vicinity of the vessel wall, cell death only occurs in a small region, leaving most of the tumour interstitial space unaffected. Similar results are predicted by both bistable and monostable apoptosis switches. Although intracellular drug concentration is above its threshold transiently for a pulse injection, irreversibility of the intracellular apoptosis module ensures that tumour cell density continues to decrease in the cell killing region, implying that a single pulse can only kill cells in a confined region and the width of tumour cell death region at a specific time point (e.g. the instant when triggering of apoptosis is completed) is capable of reflecting the drug effect. Therefore, in the following results, relative tumour cell death regions are presented for bistable and irreversible monostable apoptosis switches to illustrate the effects of pulse properties, different pulse fractionations and other influencing parameters.

Effects of pulse infusion time and intensity

Figure 7(a) shows the effect of infusion time on the width of tumour cell death region for bistable and irreversible monostable switches. For both cases, an increase in infusion time causes a moderate extension of the cell killing region, as a result of improved extracellular drug transport. A similar effect is found with increased pulse intensity, as shown in Figure 7(b).

The predicted tumour cell density shown here has a weaker non-linear response to increases in infusion time or pulse intensity compared to our previous study [34]. Although the *in silico* experiments in both studies use the same basic modelling elements, they differ in the

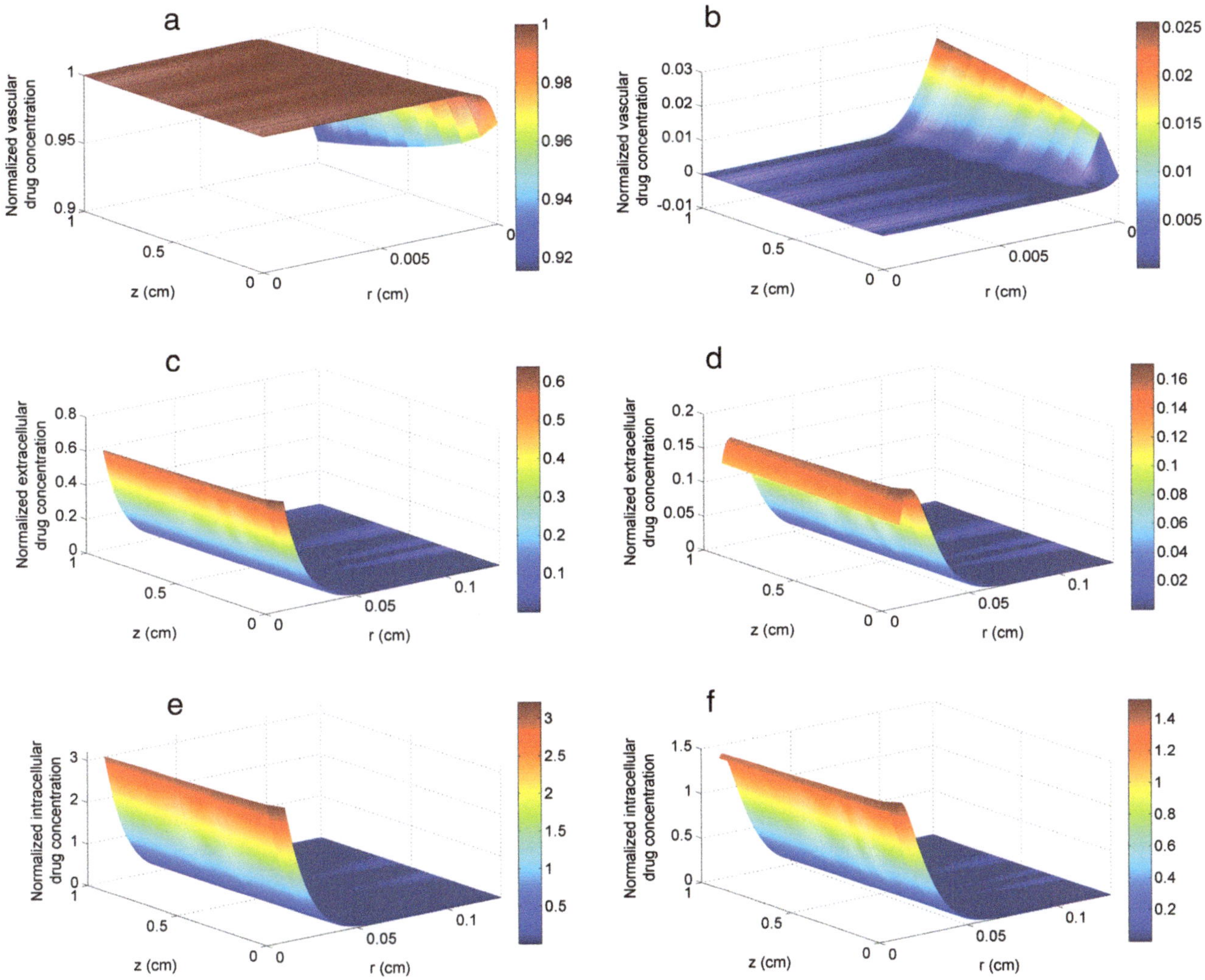

Figure 5 Snapshots of spatial profiles of drug concentration under baseline pulse injection for bistable switch. Upper panel: vascular drug concentration at **(a)** t = 1.5 h, **(b)** t = 2 h. Middle panel: extracellular drug concentration at **(c)** t = 1.5 h, **(d)** t = 2 h. Bottom panel: intracellular drug concentration at **(e)** t = 1.5 h, **(f)** t = 2 h.

geometrical scale of interest and the level of integration between different elements; the former may significantly affect the role of drug transport and intracellular signalling in cell killing. In our previous study [34], the time scale for transport (diffusion) in a microscopic tumour cord geometry is of the order of seconds, and the results suggest that transport over such a small time scale plays a less important role, which results in a more obvious nonlinear response. However, in the present *in silico* simulations, a relatively large tumour interstitium is adopted to mimic drug penetration through the entire tumour tissue. Within this context, the time scales for drug transport (dominated by diffusion) and intracellular signalling are comparable; therefore, the transport limitation poses a major constraint in inducing the effect of anticancer drugs since reactions involved in cellular signalling are triggered by intracellular drug concentration. It is concluded that limitations in transport can act as a buffer to reduce the sensitivity of cell-killing region to changes in the characteristics of stimuli. This is further demonstrated through the sensitivity analysis on the size of tumour domain (presented in sensitivity analysis section).

Effects of pulse fractionations

Also examined is the width of tumour cell death region in response to different pulse fractionations for a fixed product of pulse strength and pulse duration. Figure 8 shows similar results for different pulse fractionations except for the case with $S = 0.5$, $T = 4$ h, where contrasting results are observed for the bistable and monostable switch. At this pulse fractionation, the monostable switch predicts a much wider cell death region than that by the

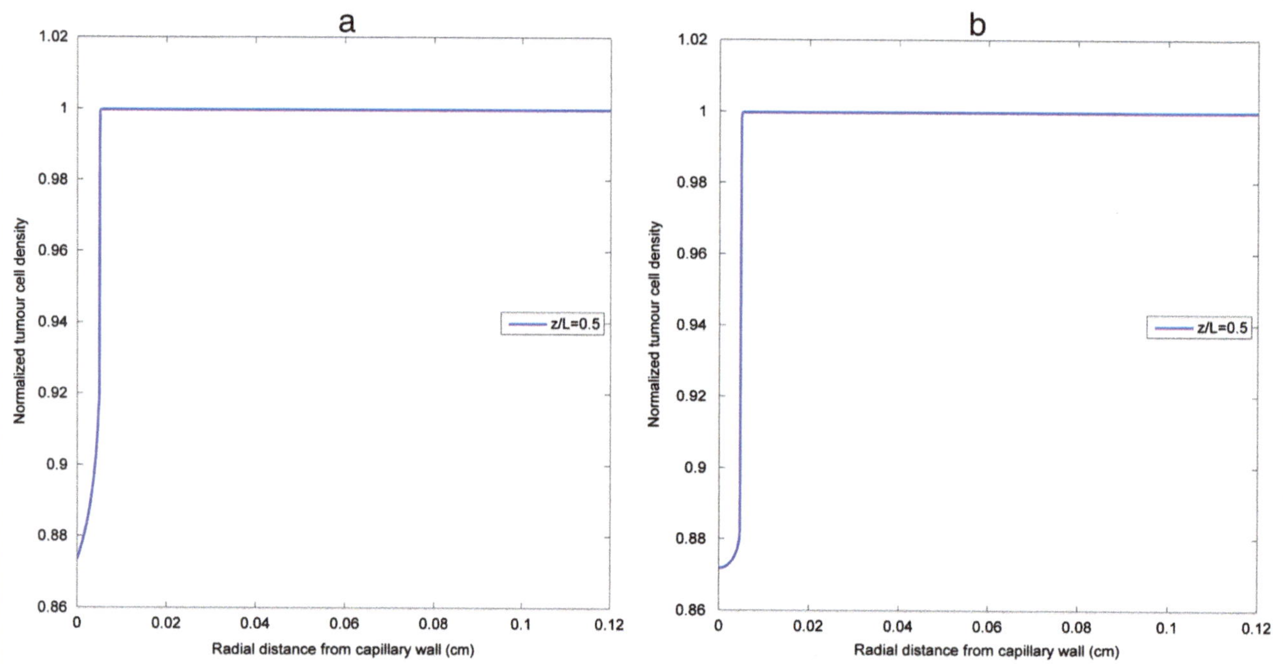

Figure 6 Tumour cell density under baseline pulse injection. Cross-sectional profiles at z/L = 0.5 of tumour cell density at 10 h display similar trends for **(a)** bistable switch, **(b)** irreversible monostable switch.

bistable switch. The difference between bistable and irreversible monostable switches may be due to the following reasons: (i) different intracellular apoptosis signalling dynamics; (ii) different intracellular drug concentrations determined by interstitial drug transport and reactions or; (iii) a combination of both. These are examined further as detailed below.

Examined first are the intracellular drug concentrations which act as upstream stimuli to trigger cell apoptosis. Intracellular drug concentrations at two time points during the injection are shown in Figure 9. At an earlier time (t = 2.5 h in Figure 9(a)), intracellular drug concentration profiles are identical for both cases, but at the end of injection (t = 4 h in Figure 9(b)), intracellular drug

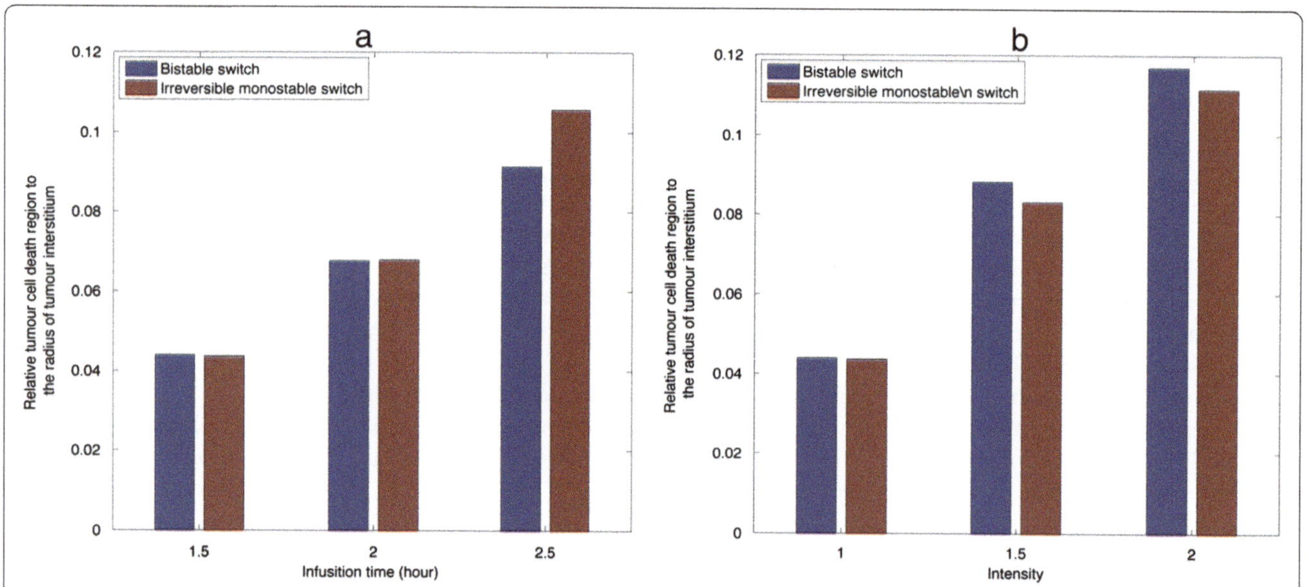

Figure 7 Effect of pulse characteristics on the relative tumour cell death region at 10 h. (a) Effect of infusion time under baseline pulse intensity. **(b)** effect of pulse intensity under baseline infusion time.

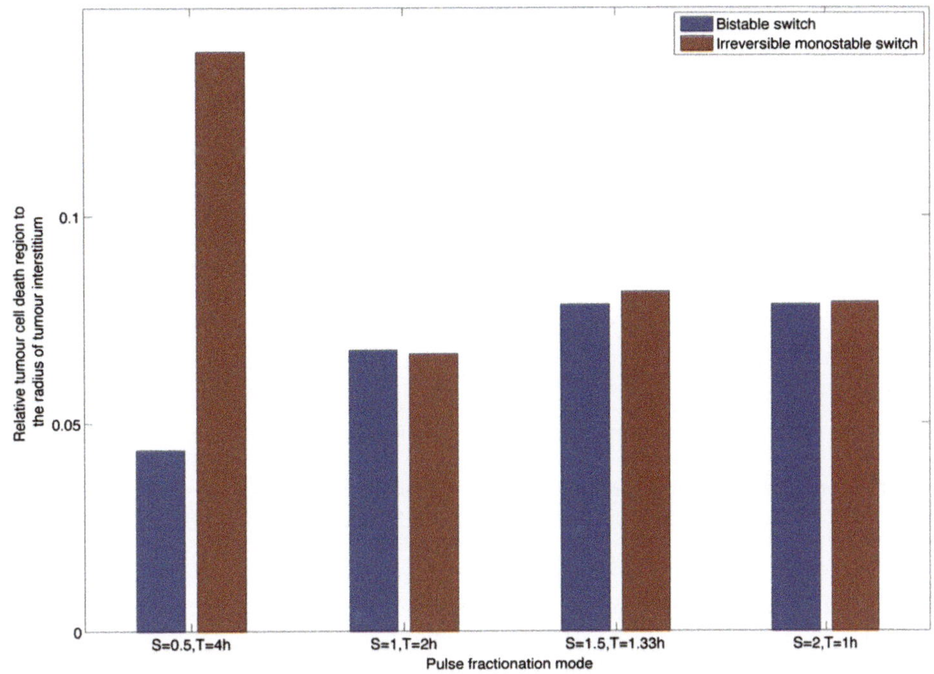

Figure 8 Effect of pulse fractionations on the relative tumour cell death region at 10 h.

concentrations with the monostable switch are apparently higher than those with the bistable switch, and the region where intracellular drug concentrations are above the threshold is wider. Temporal snapshots of tumour cell density are displayed for both cases in Figure 10. Obviously, there is no sign of cell death during the injection period for the bistable switch, but for the irreversible monostable switch cell density starts to decrease at 3 h. The falling tumour cell density triggers the feedback loop: decreased cell density leads to reduced drug consumption, which allows further drug transport and accumulation inside the cells, thus leading to cell death in the further

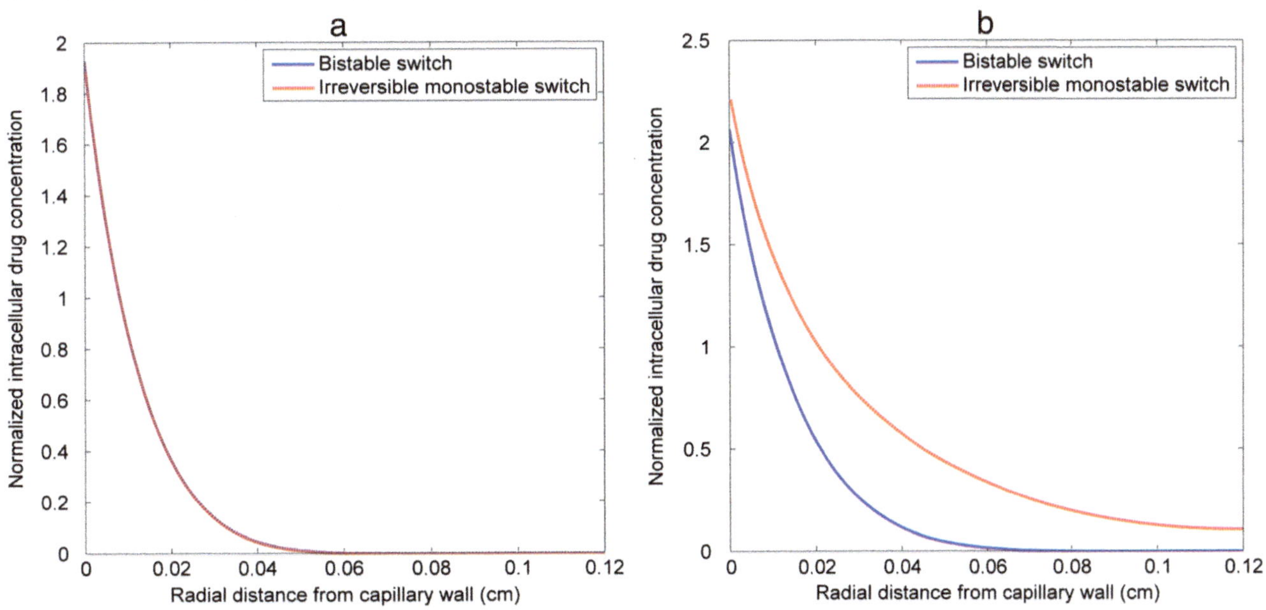

Figure 9 Time evolution of intracellular drug concentration under one specific pulse fractionation. Snapshots of cross-sectional profiles at z/L = 0.5 of intracellular drug concentration for bistable and irreversible monostable switches under pulse injection $S = 0.5$, $T = 4$ h at **(a)** t = 2.5 h, **(b)** t = 4 h.

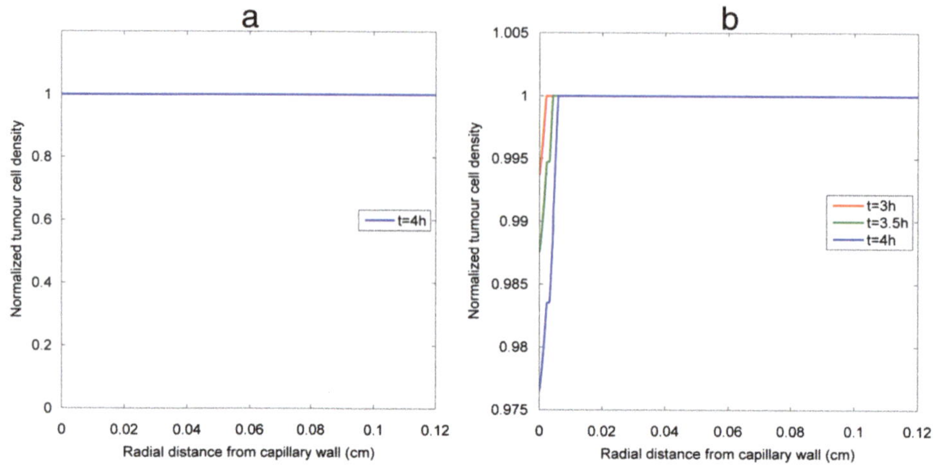

Figure 10 Time evolution of tumour cell density under the same pulse fractionation. Snapshots of cross-sectional profiles at z/L = 0.5 of tumour cell density under the same pulse fractionation as in Figure 9 for **(a)** bistable switch, **(b)** irreversible monostable switch.

region. The strong coupling between drug transport and tumour cell density is not only demonstrated by the numerical results shown here, but can also be shown by performing a non-dimensional analysis of the extracellular drug transport equation. In addition, Zheng *et al.* [27] have provided experimental evidence that drug-induced apoptosis facilitates drug penetration in solid tumours. Overall, we can conclude that there may exist small differences in response due to differences in intracellular dynamics (this was also encountered in [46]), and that the simulations reveal how transport and cellular effect may be coupled bidirectionally.

Sensitivity analysis

Parameters used in the mathematical models are either related to drug transport or involved in drug effect. With regard to drug transport, parameters can be further divided into three groups: (i) diffusion related parameters, namely drug diffusivity (D), diffusive permeability (P), (ii) convection related parameters, such as hydraulic permeability (L_p), hydraulic conductivity in the interstitium (K), (iii) tumour geometric parameters, for example, the radius of blood vessel (R_C) and tumour interstitium (R_T).

Based on the analysis presented above (in drug transport section), it is deduced that convection plays a minor role in the transmural and interstitial drug transport. Since convection is dependent on hydraulic conductivity L_p and tissue hydraulic conductivity K, their effects on blood flow are examined, and both are found to have a marginal effect on enhancing the transmural velocity (J_F) (results not shown here).

Parameters involved in the intracellular signalling models are estimated to reflect a time scale and threshold value in a reasonable range, which agree qualitatively with those obtained from a cascade of signal transduction. Overall, transparent effects of these parameters are observed as expected from the coarse-grained intracellular signalling models. A slower kinetic rate or an elevated threshold would make it more difficult to trigger apoptosis while in the opposite scenario, relieving the constraints of apoptosis may exert a further effect on improving interstitial drug transport. As the study is oriented towards an integrative understanding of drug effect with account for mechanistic drug action, the intracellular signalling models together with the estimated parameter values adopted in the present study are sufficient to serve the purpose.

Therefore, the sensitivity study presented here is focused on how interstitial drug transport may be perturbed by altering diffusion related parameters, namely drug diffusivity and diffusive permeability, and geometric parameter (the size of tumour interstitium).

Effect of drug diffusivity

The effect of increasing drug diffusivity (1D, 2D and 10D, where D is the value of diffusivity in the baseline case) is studied for two different pulse intensities. It is worth noting that the influence of drug diffusivity on the tumour cell density also depends on the intensity of pulse injection. As shown in Figure 11, increasing drug diffusivity can lead to reduced cell killing for pulse injection, with failure to trigger apoptosis at 10 D at normal intensity. Higher drug diffusivity allows the drug to transport further beyond the immediate vicinity of the vessel wall, and may help to establish a homogeneous concentration profile. However, it is not as simple when examined together with the specific requirement for apoptosis that the intracellular drug concentration needs to be sustained above its threshold for a sufficient length of time. For a smaller amount of drug, increased drug diffusivity somewhat dilutes the drug concentration, making it more difficult to satisfy

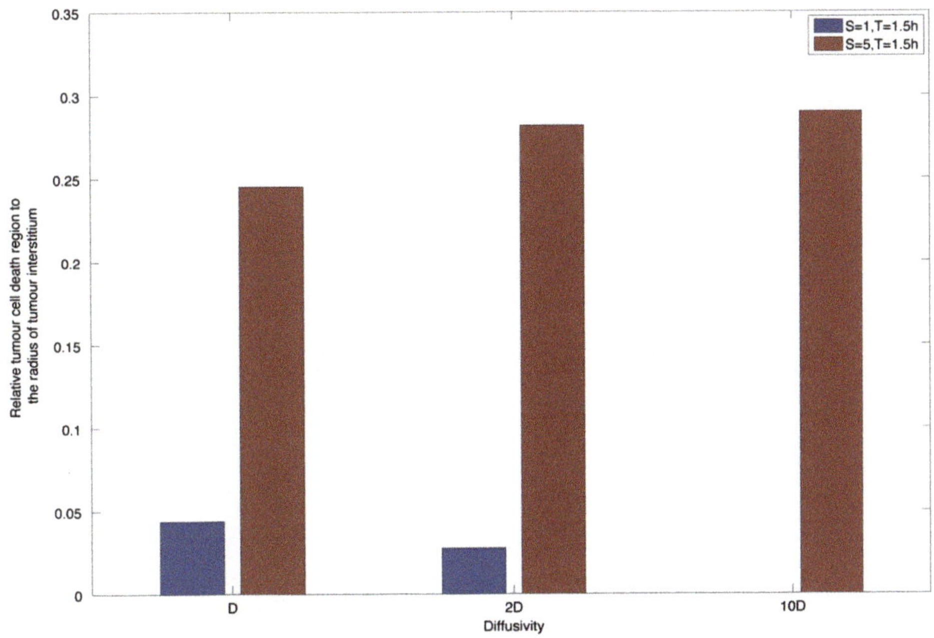

Figure 11 Effect of drug diffusivity on the relative tumour cell death region at 10 h for bistable switch.

the condition for apoptosis; as a consequence, the cell killing region is reduced. In this context, it would be favourable if the drug is concentrated in a limited region to exercise its effect locally. For a higher dosage, the opposite scenario occurs where the cell death domain is extended due to improved drug distribution as illustrated in Figure 11. An obvious increase in the width of cell death region is found when drug diffusivity is doubled, but further increase in diffusivity produces little change. The simulation results indicate that the effect of drug diffusivity needs to be examined by considering the balance between interstitial drug transport and the specific requirement of intracellular apoptosis dynamics, and it is dependent on the dosage applied. Similar trends are observed when the monostable apoptosis switch is employed. Overall, we see from the integrated study that the effect of drug diffusivity on the outcome is not necessarily as simple as may be expected by analysing diffusion in isolation.

Effect of diffusive permeability

Drugs extravasate across the blood vessel wall by diffusion and convection with the former being the dominant mode. The diffusive transmural flux is determined by the diffusive permeability of the drug and the concentration gradient across the vessel wall. Examined here is the distribution of the tumour cell density for the baseline diffusive permeability (P) and a much higher permeability ($10P$) under a given pulse. As shown in Figure 12, an increase in diffusive permeability results in an extension of the cell death region, but it is still limited to a narrow region close to the vessel wall even when the diffusive permeability is increased by ten-fold. This may be attributed to the following: (1) interstitial drug transport represents an obstacle in transporting excessive drugs away from the vessel wall; (2) more drugs are transported back to blood vessel due to the reversal concentration gradient caused by the termination of pulse injection.

Effect of the size of tumour interstitium

When the size of the tumour interstitium is reduced by half, the region of tumour cell death is still confined to the proximal region to the vessel wall with a marginal increase as shown in Figure 13. This is explained as follows. Overall, the reduced size (over a broad spectrum of tumour size) has negligible effect on interstitial drug transport during the injection phase. However, during the post-injection phase, the effect of a reduced tumour size can be seen in terms of the enhanced convective transmural flux (as a result of increased transmural velocity), which partially compensates for the negative diffusive flux back to the blood vessel, and allows more drugs to be retained in the vicinity, thus leading to enhanced penetration in the interstitium. With even smaller tumour sizes, the effect of transport limitation is gradually removed. Analysis of varying tumour sizes demonstrates that the effect of drugs is determined by the interaction between multiple drug transport processes and intracellular signalling dynamics.

Discussion

In this study, the first steps have been taken towards developing an *in silico* experimental platform with integration of

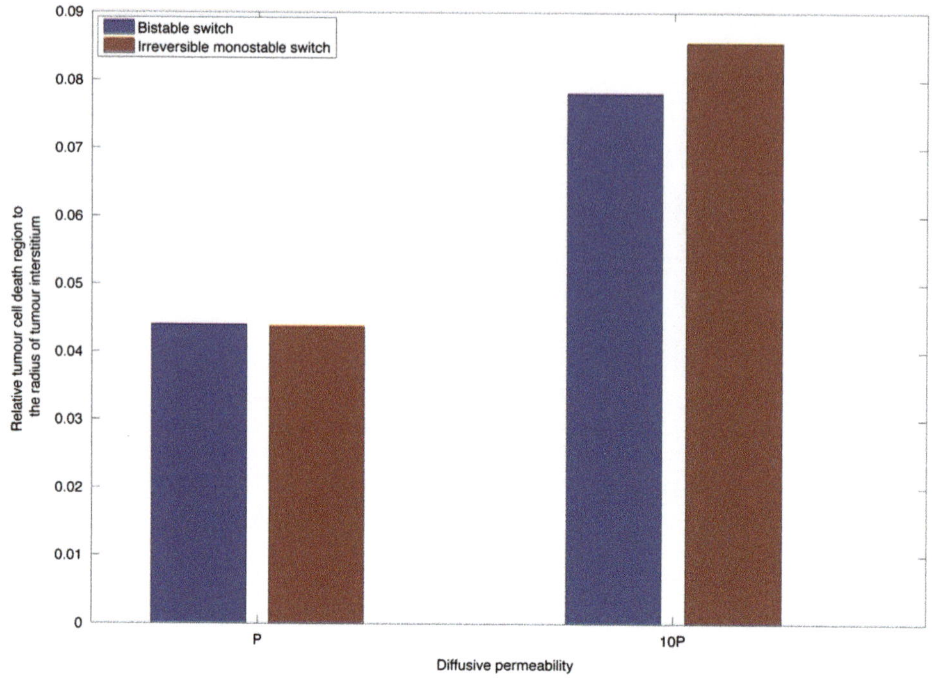

Figure 12 Effect of diffusive permeability on the relative tumour cell death region at 10 h.

blood flow, drug transport (both vascular and interstitial) and cellular signalling dynamics to provide an overall framework to systematically evaluate the effect of anti-cancer drugs on tumour cells. This platform, described by our model, contains the minimal essential elements to understand the basic aspects of the interplay between drug transport and cellular effects. The model set-up represents an idealized tumour with a well controlled environment, which is essential for understanding the interactions of many complicating factors. The tumour cell density (the output) is determined by the combined action of drug input (determined by tumour blood flow and drug transport) and intracellular signalling (involving non-linear dynamics), when subject to a particular form of drug stimulus (the input), which is itself determined by the mode of delivery (not modelled in detail). Basic

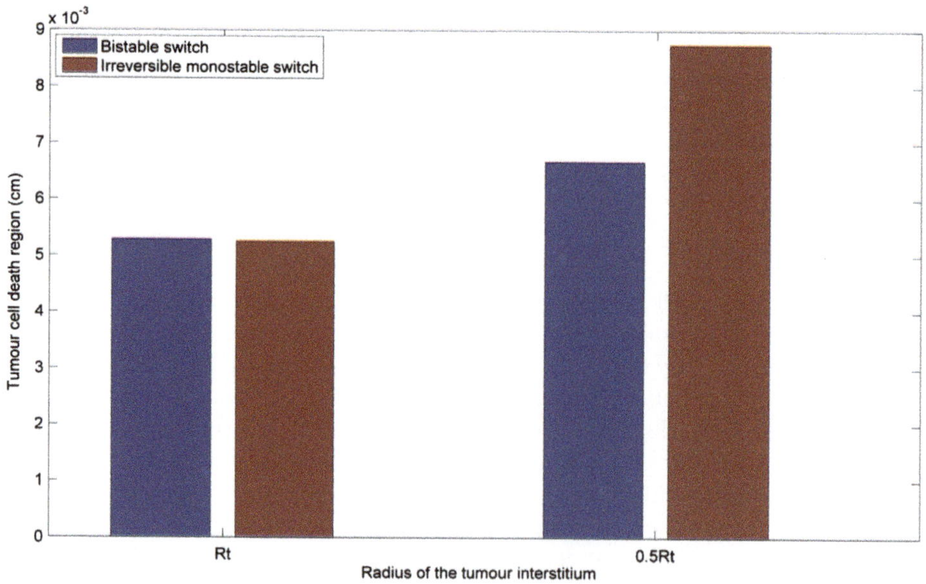

Figure 13 Effect of the size of tumour interstitium on the tumour cell death region at 10 h.

descriptions of the processes involved are incorporated with an aim to create a model system which captures the information flow from drug delivery to causal effect, and is capable of serving as a platform to understand the interplay between transport and extracellular factors on one hand and cellular features on the other hand.

Drug transport

Rather than treating the tumour blood as a homogeneous source term in the transport equation, the models describe tumour blood flow explicitly by coupling vascular blood flow with the interstitial fluid flow through an elevated tumour vascular hydraulic conductivity. Drug delivery is subsequently examined following the same transport routes. From the simulation results, few differences can be observed between the drug concentration profiles at different axial locations, indicating that intravascular transport (convection dominant) is not a rate limiting step in the drug transport processes. Of course it must be borne in mind that this finding is based on the highly simplified geometry adopted in this study – a single vessel similar to the Krogh model. It is acknowledged that a single vessel model is not always the best model; it fails to capture the heterogeneous distribution of functional blood vessels in tumours that result in large avascular regions. However, the present study focuses on providing essential and clear-cut insights into drug distribution and drug effect by conceptualizing the information flow in an integrated manner. Potentially important factors, such as the complex tumour vascular geometry (or even non-Newtonian blood rheological properties) may significantly increase resistance to tumour blood flow, thus leading to insufficient drug delivery through the tumour vasculature and even more heterogeneous drug distribution in the tumour interstitium. A description of more realistic tumour vascular network, based on advanced imaging techniques, could potentially be incorporated into the modelling framework for more realistic prediction of drug concentration distribution in specific tumours. We also note that to bring new insights into drug delivery through a complex vascular network in the tumour, it is necessary to first understand what the effect is in a simpler vasculature.

The higher vascular permeability normally observed in tumour tissues facilitates transmural transport of drugs into the tumour interstitium. However, it is noted that drug penetration is restricted to the region close to the blood vessel, with poor drug distribution in the interstitium. With regard to drug transport into tumour cells, in the case of the specific anticancer drug selected (doxorubicin), it is found that the drug is preferentially sequestered in tumour cells. Therefore, limited drug penetration in the interstitium is a major difficulty to overcome in order to improve drug efficacy. Compared to drug diffusion in the interstitium, drug consumption by tumour cells determined by the tumour cell density plays a dominant role in impairing interstitial penetration (discussed later). On the other hand, the heterogeneous drug distribution found in the spatially distributed system also implies that homogeneous compartment models may not be sufficient for accurate predictions of drug efficacy. It is worth pointing out that the modelling framework developed is not only applicable to chemotherapeutic agents, but (with minor modifications) also to oxygen and other endocrine signals, which share the same transport pathways. Further, interstitial drug transport may be implicitly affected by cellular responses, such as the presence of low pH due to the adaptive metabolism towards oxygen starvation, and detoxifying anticancer drugs due to acquirement of drug resistances. Understanding the interplay of such factors is beyond the scope of this study, but can be built on the existing framework presented here.

Drug effect

Drug response is evaluated broadly by empirical or mechanistic approaches. This study is motivated by the desire to provide a mechanistic understanding of drug effect by taking into account the interactions between drugs and targets, the downstream signalling reactions and ultimately cell fate decision making. In the current model, apoptosis is the primary cellular response triggered by anticancer drugs and dynamic analysis of drug effect is performed at both cellular and tissue level.

It is important to choose an appropriate level of cellular effects to be included in the model given the fact that many of the biochemical details are still unknown (or may be questionable or may vary between cell types). Coarse grained descriptions of intracellular process are adopted here, with a view to qualitatively capture the nature of signal transduction in the cell and to retain the relevant input-output signalling characteristics. Future work to include more biochemical details of the signalling network (including the detailed dynamics of the caspase network and its regulation) is needed in order to obtain a more detailed depiction of the cellular signalling and understand the roles of multiple intracellular regulatory mechanisms.

In the current work, two types of apoptosis modules with qualitatively distinct dynamics of signal transduction are included and examined separately: a bistable switch and an irreversible monostable switch. The two models are examined to address whether such different dynamic characteristics at the cellular level would result in different drug effects at the population level. It has been found that in most respects the two models predict broadly similar effects. A related point to be made is that when cellular signalling is included, one should be very cautious about claims of validating models from scanty data. The extent to which signalling models can be distinguished in such a context is discussed elsewhere [46]. It is worth emphasizing that our intracellular models are essentially minimal models but have the features which would be expected of all apoptosis models.

Drug transport vs. drug effect

Our analysis reveals that in pulse injections, the drug killing effect is typically confined to the proximal region due to limited drug penetration for both types of apoptosis models (Figures 5,6,7). Our analysis reveals in multiple ways, the need for incorporating the dynamics of intracellular signalling in mechanistic predictions of drug effect and investigating drug transport and drug effect in an integrated manner. Within the present modelling framework, it is found that interstitial drug transport and drug effect are strongly coupled: drug-induced pharmacological effect (apoptosis) can enhance drug penetration in the interstitium, which constrains the exertion of drug action. This indicates that an optimal staged treatment schedule might help to result in a rapid penetration and the subsequent cell killing in regions further away from blood vessels.

It is natural to expect that greater diffusivity enhances drug penetration in the interstitium, resulting in a more homogeneous distribution of drug concentration. However, as our analysis shows, it is not necessarily the case that greater diffusivity would improve drug effect on tumour cells for a given stimulus. This is due to the specific requirement for apoptosis being triggered and the interplay between intracellular dynamics and interstitial drug transport. Drug diffusivity in tumour tissues is not only related to the physiological properties of anticancer agents, i.e. their charge, size or shape, but also to the structures of tumour tissues [50]. Therefore, the analysis presented here may help to refine drug infusion strategies by noting the different effects of drug diffusivity in a drug- and tumour-specific context.

Conclusions

In this paper, an *in silico* experimental platform is employed, which describes the flow of information from drug delivery to drug effect combining tumour blood flow, anticancer drug transport and cellular dynamics. Within the simplified model setting, a series of investigations on different drug stimuli and parameters is presented, providing explicit insights into the effect of drug and the interplay between multiple transport processes and intracellular signalling dynamics. Although the modelling platform in the current study is coarse-grained and qualitative in nature, it is capable of accommodating other mathematical models and allowing for fine graining and augmented descriptions of individual subprocesses systematically. Quantitative understanding of the contributing factors could be achieved for specific tumour types and specific anticancer drugs, in conjunction with the relevant experimental data at multiple scales. The modelling platform can serve as a computational tool to perform a thorough sensitivity analysis for the control and optimization of chemotherapeutic processes.

Competing interests

The authors declare that they have no competing interests.

Authors' contributions

CL, JK and XYX planned the paper, CL performed the computational work, CL wrote the paper with input by XYX and JK. All authors read and approved the final manuscript.

Author details

[1]Department of Chemical Engineering, Imperial College London, South Kensington Campus, London SW7 2AZ, UK. [2]Centre for Process Systems Engineering and Institute for Systems and Synthetic Biology, Imperial College London, South Kensington, London SW7 2AZ, UK.

References

1. Venkatasubramanian R, Henson MA, Forbes NS: **Integrating cell-cycle progression, drug penetration and energy metabolism to identify improved cancer therapeutic strategies.** *J Theor Biol* 2008, **253**:98–117.
2. Minchinton AI, Tannock IF: **Drug penetration in solid tumours.** *Nat Rev Cancer* 2006, **6**:583–592.
3. Trédan O, Galmarini CM, Patel K, Tannock IF: **Drug resistance and the solid tumor microenvironment.** *J Natl Cancer Inst* 2007, **99**:1441–1454.
4. Boucher Y, Baxter LT, Jain RK: **Interstitial pressure gradients in tissue-isolated and subcutaneous tumors: implications for therapy.** *Cancer Res* 1990, **50**:4478–4484.
5. Grantab R, Sivananthan S, Tannock IF: **The penetration of anticancer drugs through tumor tissue as a function of cellular adhesion and packing density of tumor cells.** *Cancer Res* 2006, **66**:1033–1039.
6. Netti PA, Berk DA, Swartz MA, Grodzinsky AJ, Jain RK: **Role of extracellular matrix assembly in interstitial transport in solid tumors.** *Cancer Res* 2000, **60**:2497–2503.
7. Kreeger PK, Lauffenburger DA: **Cancer systems biology: a network modeling perspective.** *Carcinogenesis* 2010, **31**:2–8.
8. Kaufmann SH, Earnshaw WC: **Induction of apoptosis by cancer chemotherapy.** *Exp Cell Res* 2000, **256**:42–49.
9. Kohn KW, Jackman J, Oconnor PM: **Cell-cycle control and cancer-chemotherapy.** *J Cell Biochem* 1994, **54**:440–452.
10. Martin GS: **Cell signaling and cancer.** *Cancer Cell* 2003, **4**:167–174.
11. Cloutier M, Wang E: **Dynamic modeling and analysis of cancer cellular network motifs.** *Integr Biol* 2011, **3**:724–732.
12. Tyson JJ, Baumann WT, Chen C, Verdugo A, Tavassoly I, Wang Y, Weiner LM, Clarke R: **Dynamic modelling of oestrogen signalling and cell fate in breast cancer cells.** *Nat Rev Cancer* 2011, **11**:523–532.
13. Liu C, Krishnan J, Stebbing J, Xu XY: **Use of mathematical models to understand anticancer drug delivery and its effect on solid tumors.** *Pharmacogenomics* 2011, **12**:1337–1348.
14. El-Kareh AW, Secomb TW: **A mathematical model for comparison of bolus injection, continuous infusion, and liposomal delivery of doxorubicin to tumor cells.** *Neoplasia* 2000, **2**:325–338.
15. Kuh HJ, Jang SH, Wientjes MG, Au JL: **Computational model of intracellular pharmacokinetics of paclitaxel.** *J Pharmacol Exp Ther* 2000, **293**:761–770.
16. Chappell MJ, Evans ND, Errington RJ, Khan IA, Campbell L, Ali R, Godfrey KR, Smith PJ: **A coupled drug kinetics-cell cycle model to analyse the response of human cells to intervention by topotecan.** *Comput Methods Programs Biomed* 2008, **89**:169–178.
17. Eikenberry S: **A tumor cord model for doxorubicin delivery and dose optimization in solid tumors.** *Theor Biol Med Model* 2009, **6**:16.
18. Thurber GM, Weissleder R: **A systems approach for tumor pharmacokinetics.** *PLoS One* 2011, **6**:e24696.
19. Goh Y-MF, Kong HL, Wang C-H: **Simulation of the delivery of doxorubicin to hepatoma.** *Pharm Res* 2001, **18**:761–770.

20. El-Kareh AW, Secomb TW: **A mathematical model for cisplatin cellular pharmacodynamics.** *Neoplasia* 2003, **5**:161–169.
21. El-Kareh AW, Secomb TW: **Two-mechanism peak concentration model for cellular pharmacodynamics of Doxorubicin.** *Neoplasia* 2005, **7**:705–713.
22. El-Kareh AW, Labes RE, Secomb TW: **Cell cycle checkpoint models for cellular pharmacology of paclitaxel and platinum drugs.** *AAPS J* 2008, **10**:15–34.
23. Simeoni M, Magni P, Cammia C, De Nicolao G, Croci V, Pesenti E, Germani M, Poggesi I, Rocchetti M: **Predictive pharmacokinetic-pharmacodynamic modeling of tumor growth kinetics in xenograft models after administration of anticancer agents.** *Cancer Res* 2004, **64**:1094–1101.
24. Kozusko F, Chen P, Grant SG, Day BW, Panetta JC: **A mathematical model of in vitro cancer cell growth and treatment with the antimitotic agent curacin A.** *Math Biosci* 2001, **170**:1–16.
25. Jackson TL: **Intracellular accumulation and mechanism of action of doxorubicin in a spatio-temporal tumor model.** *J Theor Biol* 2003, **220**:201–213.
26. Jang SH, Wientjes MG, Au JL: **Enhancement of paclitaxel delivery to solid tumors by apoptosis-inducing pretreatment: effect of treatment schedule.** *J Pharmacol Exp Ther* 2001, **296**:1035–1042.
27. Zheng JH, Chen CT, Au JL, Wientjes MG: **Time- and concentration-dependent penetration of doxorubicin in prostate tumors.** *AAPS PharmSci* 2001, **3**:E15.
28. Baxter LT, Jain RK: **Transport of fluid and macromolecules in tumors. I. Role of interstitial pressure and convection.** *Microvasc Res* 1989, **37**:77–104.
29. Baxter LT, Jain RK: **Transport of fluid and macromolecules in tumors. II. Role of heterogeneous perfusion and lymphatics.** *Microvasc Res* 1990, **40**:246–263.
30. Baxter LT, Jain RK: **Transport of fluid and macromolecules in tumors. III. Role of binding and metabolism.** *Microvasc Res* 1991, **41**:5–23.
31. Teo CS, Hor Keong Tan W, Lee T, Wang C-H: **Transient interstitial fluid flow in brain tumors: effect on drug delivery.** *Chem Eng Sci* 2005, **60**:4803–4821.
32. Netti PA, Roberge S, Boucher Y, Baxter LT, Jain RK: **Effect of transvascular fluid exchange on pressure-flow relationship in tumors: a proposed mechanism for tumor blood flow heterogeneity.** *Microvasc Res* 1996, **52**:27–46.
33. Pozrikidis C, Farrow DA: **A model of fluid flow in solid tumors.** *Ann Biomed Eng* 2003, **31**:181–194.
34. Liu C, Krishnan J, Xu XY: **A systems-based mathematical modelling framework for investigating the effect of drugs on solid tumours.** *Theor Biol Med Model* 2011, **8**:45.
35. Jain RK: **Determinants of tumor blood flow: a review.** *Cancer Res* 1988, **48**:2641–2658.
36. Baish JW, Gazit Y, Berk DA, Nozue M, Baxter LT, Jain RK: **Role of tumor vascular architecture in nutrient and drug delivery: an invasion percolation-based network model.** *Microvasc Res* 1996, **51**:327–346.
37. Peters W, Teixeira M, Intaglietta M, Gross JF: **Microcirculatory studies in rat mammary carcinoma. I. Transparent chamber method, development of microvasculature, and pressures in tumor vessels.** *J Natl Cancer Inst* 1980, **65**:631–642.
38. Seaton D, Krishnan J: **Modular systems approach to understanding the interaction of adaptive and monostable and bistable threshold processes.** *IET Syst Biol* 2011, **5**:81–94.
39. Albeck JG, Burke JM, Spencer SL, Lauffenburger DA, Sorger PK: **Modeling a snap-action, variable-delay switch controlling extrinsic cell death.** *PLoS Biol* 2008, **6**:2831–2852.
40. Legewie S, Bluthgen N, Herzel H: **Mathematical modeling identifies inhibitors of apoptosis as mediators of positive feedback and bistability.** *Plos Comput Biol* 2006, **2**:1061–1073.
41. Gu C, Zhang J, Chen Y, Lei J: **A trigger model of apoptosis induced by tumor necrosis factor signaling.** *BMC Syst Biol* 2011, **5**(Suppl 1):S13.
42. Zhang T, Brazhnik P, Tyson J: **Computational analysis of dynamical responses to the intrinsic pathway of programmed cell death.** *Biophys J* 2009, **97**:415–434.
43. Eissing T, Conzelmann H, Gilles ED, Allgower F, Bullinger E, Scheurich P: **Bistability analyses of a caspase activation model for receptor-induced apoptosis.** *J Biol Chem* 2004, **279**:36892–36897.
44. Bentele M, Lavrik I, Ulrich M, Stosser S, Heermann DW, Kalthoff H, Krammer PH, Eils R: **Mathematical modeling reveals threshold mechanism in CD95-induced apoptosis.** *J Cell Biol* 2004, **166**:839–851.
45. Bagci EZ, Vodovotz Y, Billiar TR, Ermentrout GB, Bahar I: **Bistability in apoptosis: roles of bax, bcl-2, and mitochondrial permeability transition pores.** *Biophys J* 2006, **90**:1546–1559.
46. Krishnan J, Liu C: **An investigation of signal transduction and irreversible decision making through monostable and bistable switches.** In *Systems and Synthetic Biology: A Systematic Approach.* Edited by Kulkarni V, Stan G-B, Raman K. Springer Verlag; 2012. In press.
47. Edelstein-Keshet L: *Mathematical Models in Biology.* SIAM, 3600 Market Street, Floor 6, Philadelphia, PA 19104: Society for Industrial and Applied Mathematics; 1988.
48. Eksborg S, Andersson M, Domellöf L, Lönroth U: **A pharmacokinetic study of adriamycin and 4'epi-adriamycin after simultaneous intra-arterial liver administration.** *Med Oncol Tumor Pharmacother* 1986, **3**:105–110.
49. Robert J: **Pharmacokinetics of adriamycin in patients with breast cancer: correlation between pharmacokinetic parameters and clinical short-term response.** *Eur J Canc Clin Oncol* 1982, **18**:739–745.
50. Jain RK: **Transport of molecules in the tumor interstitium: a review.** *Cancer Res* 1987, **47**:3039–3051.

7

Development of growth selection systems to isolate a-type or α-type of yeast cells spontaneously emerging from *MAT*a/α diploids

Nobuo Fukuda and Shinya Honda*

Abstract

Background: Manufacture of *MAT*a and *MAT*α yeast cells is required for crossbreeding, a procedure that permits hybridization and the generation of new heterozygous strains. Crossbreeding also can be performed with a- and α-type of cells, which have the same mating abilities as *MAT*a and *MAT*α haploid cells, respectively.

Results: In this work, we describe a method to generate a- and α-type of cells via the naturally-occurring chromosomal aberration in parental *MAT*a/α diploids. We successfully designed suitable genetic circuits for expression of the *URA3* selection marker gene to permit isolation of a- and α-type of cells, respectively, on solid medium lacking uracil. Furthermore we succeeded in generation of zygotes by mating of both the manufactured a- and α-type of yeast cells.

Conclusions: This process does not require exposure to mutagens such as UV irradiation, thereby avoiding the accumulation of undesirable mutations that would detract from the valuable traits that are under study. All the genetic modifications in the current study were introduced into yeast cells using plasmids, meaning that these traits can be removed without altering the genome sequence. This approach provides a reliable and versatile tool for scientific research and industrial yeast crossbreeding.

Keywords: Yeast, Biotechnology, Gene regulation, Transcription factors, Promoters

Background

Crossbreeding is an effective approach used to improve and combine traits of yeast strains by zygosis of cells of opposite mating types (*MAT*a and *MAT*α) [1-3]. Generally, parental diploid and polyploid strains used for industrial application are never able to mate directly, and hence isolation of mating strains (typically, haploid strains) via sporulation is a prerequisite for crossbreeding. Yet classical crossbreeding can be problematic because numerous industrial yeast strains sporulate poorly or not at all [4-8].

The *MAT* genes locate on chromosome III that is the most unstable one among 16 chromosomes of *Saccharomyces cerevisiae* [9]. Naturally, a- and α-type of yeast cells emerge from *MAT*a/α diploids with quite low frequency, due to chromosomal aberration during mitotic division. These cells can be used as alternative mating strains, possessing the same mating ability as *MAT*a and *MAT*α haploids generated via sporulation, respectively.

There are several kinds of chromosomal aberration such as loss of heterozygosity (LOH) and mitotic chromosome loss. LOH is a natural event that generates homozygous loci via chromosomal rearrangement in heterozygous loci [10-13], and LOH occurring at the *MAT* locus within a *MAT*a/α cell produces either a *MAT*a/a or a *MAT*α/α cell. The spontaneous frequency of LOH is below 1×10^{-4} [14]. Mitotic chromosome loss is also a naturally-occurring event that polyploid cells lose single or multiple chromosomes [15]. The frequency of loss of chromosome III in yeast diploid cells was reported to be 5×10^{-5} [9], and yeast cells having lost one of two chromosomes III containing the *MAT*a or *MAT*α gene acquire either a- or α-type of mating ability. Unfortunately, however, it can be difficult to isolate the generated a- and α-type of cells, since the spontaneous frequencies of such events are quite low.

Ultraviolet (UV) irradiation of yeast diploid cells has been used successfully to increase the frequency of LOH

* Correspondence: s.honda@aist.go.jp
Biomedical Research Institute, National Institute of Advanced Industrial Science and Technology (AIST), Higashi, Tsukuba, Ibaraki 305-8566, Japan

to about 30% [16]. According to other report, chemicals such as benzonitrile and methyl ethyl ketone increase the frequency of chromosome loss [17]; however, these treatments have the potential to randomly induce undesirable mutations at loci other than the *MAT* locus or undesirable loss of chromosomes other than chromosome III, which might prevent the generated a- or α-type of cells from inheriting the desired properties of the parental strains when used in crossbreeding.

Here we designed genetic circuits to express the *URA3* gene as a selection marker, permitting isolation of the rare a- or α-type of cells arising via spontaneous chromosomal aberration (Figure 1). Furthermore, we introduced a previously constructed expression system to prevent autopolyploidization between a- and α-type of cells generated from the same parental strain. Specifically, we used the a1-α2 complex [18], a repressor of haploid-specific genes (*hsg*) that endow yeast cells with the ability to mate [19]. In this work, we show the feasibility of this approach and its potential to provide mating strains for yeast crossbreeding.

Results

General strategy

An outline of the experimental design for production of a- and α-type of cells is shown in Figure 1. Spontaneous chromosomal aberration during cultivation provide low numbers of both a- and α-type of cells in cultures containing large numbers of parental *MAT*a/α cells. The inclusion in our system of the a1-α2 complex (which represses expression of *hsg*) suppresses mating by the rare a- and α-type of cells, which therefore can grow independently in the cell mixture without forming non-hybrid (autopolyploid) cells [18]. Here we used *URA3* as the selection marker gene for a- (Figure 1A) or α-type of cells (Figure 1B), permitting isolation by selection on solid medium lacking uracil (SD – Ura plates).

All the genetic modifications in the current study were introduced into yeast cells using plasmids, permitting complete removal of the modification following isolation of the target cells. Final plasmid sets suitable for isolation of a- or α-type of cells were defined by sequential

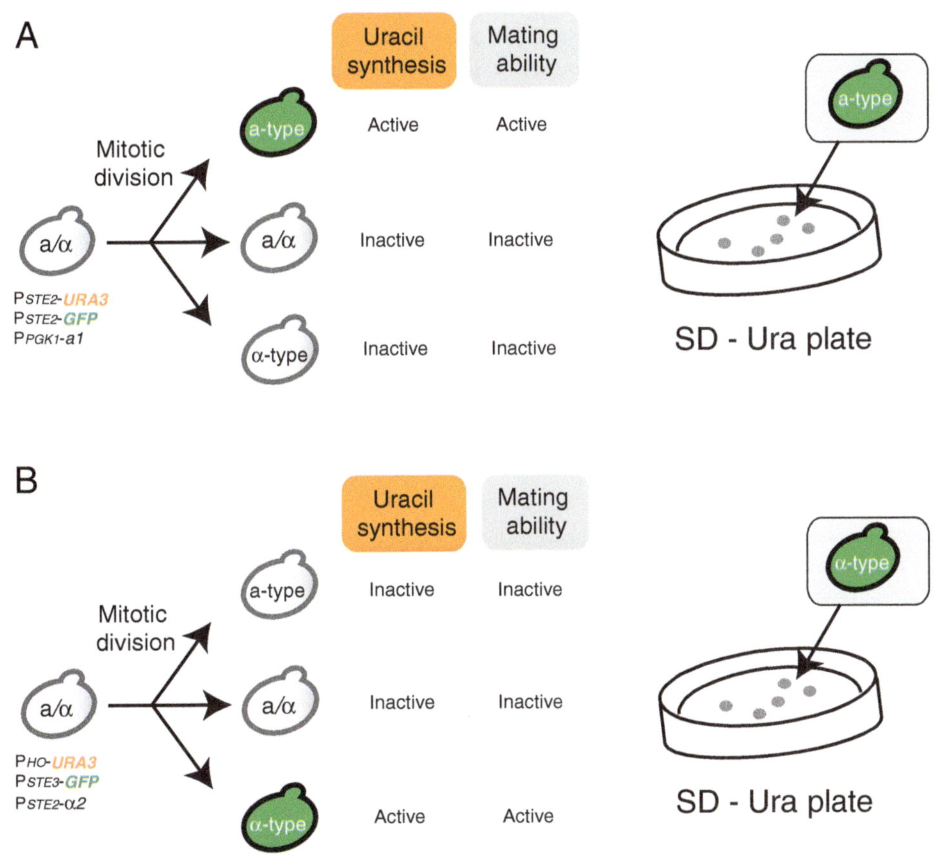

Figure 1 Schematic outline of experimental design. (A) Isolation of a-type of cells generated via spontaneous chromosomal aberration. P_{PGK1}-a1 suppresses mating ability of α-type of cells to prevent autopolyploidization (mating between a- and α-type of cells derived from the same parental strain). The *URA3* gene is expressed only in a-type cells, and a-type of cells would be selected on SD solid medium lacking uracil (SD – Ura plates). **(B)** Isolation of α-type of cells generated via spontaneous chromosomal aberration. P_{STE2}-α2 suppresses mating ability of a-type cells to prevent autopolyploidization. The *URA3* gene is expressed only in α-type cells, and α-type of cells would be selected on SD – Ura plates. The *GFP* reporter gene is used to detect yeast cells with target mating-type.

trial and error. The process is described below in a step-by-step manner.

Trial 1: expression of the *URA3* gene via YEp-type plasmids

Two kinds of YEp-type (multi-copy) plasmids were constructed to express the *URA3* gene in a- or α-type of cells (Figure 2). Plasmid pHY-2U (Figure 2A) places the selection marker under control of P_{STE2}, a promoter of a-type-specific genes (*asg*), so that only *MAT*a cells can express the *URA3* gene. On the other hand, plasmid pLY-3U (Figure 2C) places the selection marker under control of P_{STE3}, a promoter of α-type-specific genes (*αsg*), so that only *MAT*α cells can express the *URA3* gene. Growth of each transformant was evaluated by monitoring the OD_{600}.

Only *MAT*a cells possessing pHY-2U grew without uracil at the same level as with uracil (Figure 2B). Although there were obvious differences in growth ability between *MAT*a cells and others, *MAT*α and *MAT*a/α cells harboring pHY-2U exhibited some growth in the absence of uracil. As shown in Figure 2D, *MAT*a and *MAT*a/α cells harboring pLY-3U exhibited levels of growth equivalent to that of *MAT*α cells harboring pLY-3U when cultured on SD - Ura. These results suggest that leaky expression of the *URA3* gene via YEp-type of plasmids would preclude isolation of target cells on SD - Ura plates.

Trial 2: expression of the *URA3* gene via YCp-type plasmids

To reduce leaky *URA3* gene expression, DNA fragments containing P_{STE2}-*URA3* or P_{STE3}-*URA3* were transferred into YCp-type (single-copy) plasmids to yield pLS-2U (Figure 3A) or pHS-3U (Figure 3C), respectively. To make detailed examination of changes in growth ability of yeast cells, we tracked OD_{600} values over time, and then plotted growth curves for each transformant with or without uracil. Whereas pLS-2U permitted *MAT*a cells to grow without uracil, *MAT*α and *MAT*a/α cells harboring pLS-2U exhibited only minimal growth in the absence of uracil (Figure 3B). These results suggested that pLS-2U is suitable for isolation of a-type of cells.

Among cells transformed with pHS-3U, *MAT*a and *MAT*a/α grew poorly without uracil compared to *MAT*α (Figure 3D). Although the leakiness of *URA3* gene expression was significantly reduced by altering plasmid type from YEp to YCp, further optimization was needed to establish an isolation method for α-type of cells.

Trial 3: use of the a1-α2 repressor complex combined with P_{HO}, a promoter of *hsg*, for α-type-specific *URA3* gene expression

Two plasmids shown in Figure 4A were used in an improved strategy for α-type-specific *URA3* gene expression. The plasmid pL3G-2α that contains P_{STE2}-α2

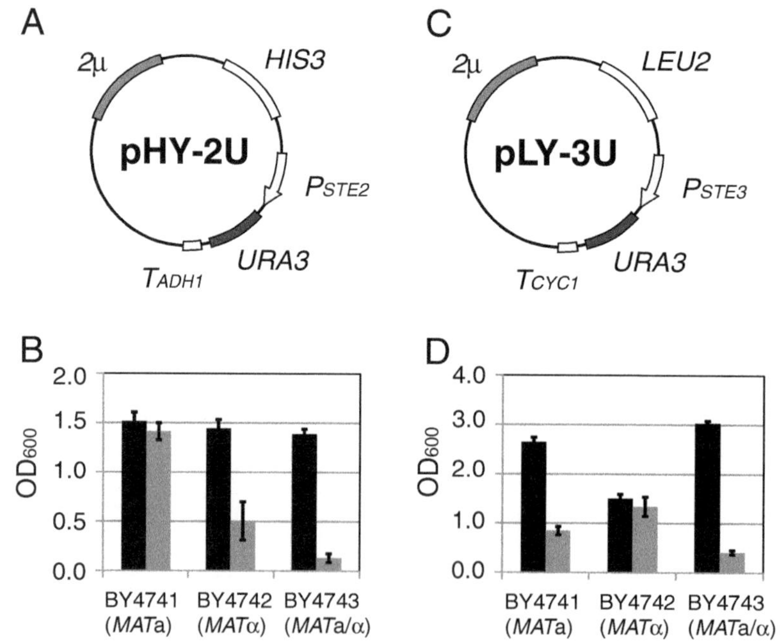

Figure 2 Growth of yeast transformants harboring YEp-type plasmids. (A) Plasmid map of pHY-2U containing *2 μ* origin of replication (providing cellular retention of multicopy plasmids) and P_{STE2}-*URA3* construct (activated in a-type yeast cells). **(B)** The OD_{600} values of cultures of pHY-2U transformants at 18 h cultivation. Black bars indicate cultivation with uracil, and gray bars indicate cultivation without uracil. Values are presented as means ± standard deviations from three independent experiments. **(C)** Plasmid map of pLY-3U containing *2 μ* origin of replication and P_{STE3}-*URA3* construct (activated in α-type yeast cells). **(D)** The OD_{600} values of cultures of pLY-3U transformants at 18 h cultivation. Symbols are as in B.

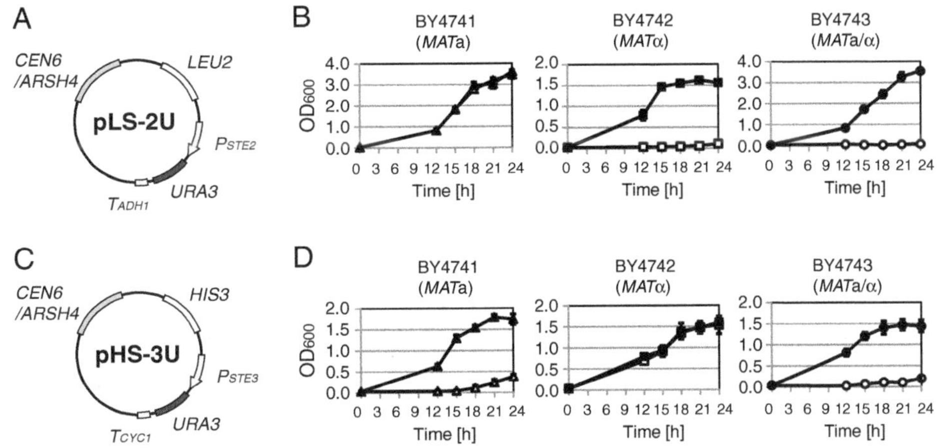

Figure 3 Growth of yeast transformants harboring YCp-type of plasmids. **(A)** Plasmid map of pLS-2U containing *CEN6/ARSH4* origin of replication (providing cellular retention of single-copy plasmids) and P_{STE2}-*URA3* construct (activated in a-type yeast cells). **(B)** The growth curves of pLS-2U transformants. Closed symbols indicate cultivation with uracil, and open symbols indicate cultivation without uracil. Values are presented as means ± standard deviations from three independent experiments. **(C)** Plasmid map of pHS-3U containing *CEN6/ARSH4* origin of replication and P_{STE3}-*URA3* contruct (activated in α-type yeast cells). **(D)** The growth curves of pHS-3U transformants. Symbols are as in B.

suppresses the mating ability of *MAT*a cells by artificially forming the a1-α2 repressor complex [18]. Because the a1-α2 complex can directly repress gene expression under the control of promoters of *hsg*, we selected P_{HO} (a *hsg* promoter) as an alternative promoter to express the *URA3* gene. Hence, as shown in Figure 1B, P_{HO}-*URA3* is expected to be activated only in *MAT*α cells, where P_{STE2}-α2 can drive expression off the *HO* promoter.

Figure 4B shows growth curves of transformants in the improved strategy. As expected, while *MAT*α transformants grew without uracil, *MAT*a and *MAT*a/α transformants exhibited minimal or no growth without uracil. These results suggest that use of pHS-HoU

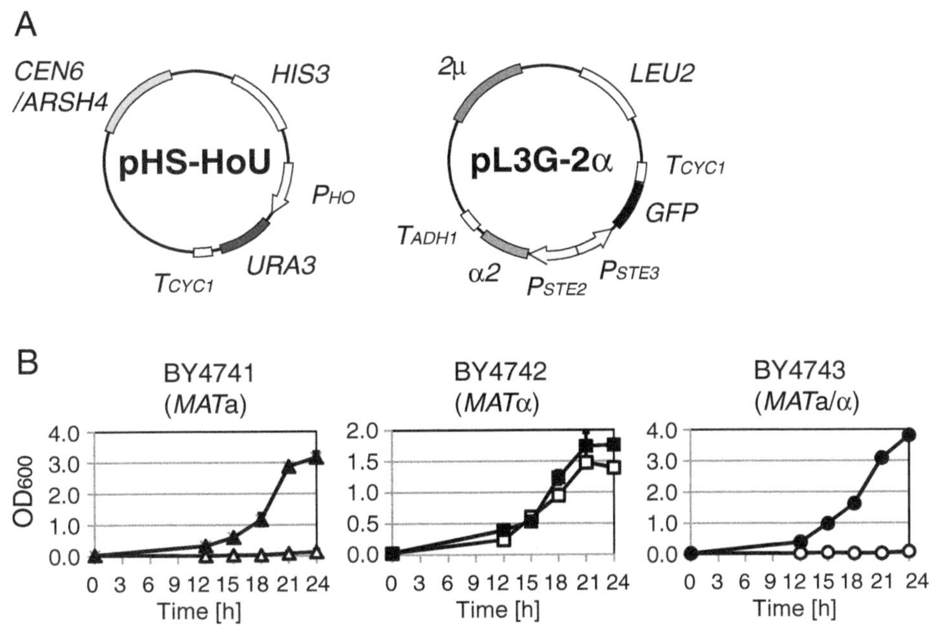

Figure 4 Improved growth selection system for α-type of cells. **(A)** Plasmids used for α-type-specific *URA3* gene expression. The plasmid pHS-HoU contains *CEN6/ARSH4* origin of replication (providing cellular retention of single-copy plasmids) and P_{HO}-*URA3* construct (activated in yeast haploid cells). The plasmid pL3G-2α contains P_{STE2}-α2, a construct that represses *URA3* gene expression in a-type yeast cells. **(B)** The growth curves of double plasmid (pHS-HoU and pL3G-2α) transformants. Closed symbols indicate cultivation with uracil, and open symbols indicate cultivation without uracil. Values are presented as means ± standard deviations from three independent experiments.

combined with pL3G-2α is suitable for isolation of α-type of cells.

Artificial regulation network designed for isolation of a- and α-type of cells

As described above, the plasmid pLS-2U (constructed in trial 2) was preferred for isolation of a-type of cells, and the combination of plasmids pHS-HoU and pL3G-2α (as described in trial 3) was preferred for isolation of α-type of cells. A model that describes how our artificial gene network can function in isolation of a- or α-type of cells is shown in Figure 5.

For isolation of a-type of cells, we introduced plasmid pLS-2U into the parental strain in combination with plasmid pH2G-Pa1 (see Additional file 1: Figure S1A and B), which prevents autopolyploidization (Table 1 and Figure 5A). The a1-α2 complex represses expression of *hsg* that are required for mating signal transduction in α-type of cells generated via chromosomal aberration, as well as in parental *MAT*a/α cells. In this system, only a-type of cells can survive on SD – Ura plates, because of a-type-specific *URA3* gene expression. To confirm the mating-type of yeast cells isolated on SD – Ura plates using this method, we used a *GFP* reporter gene under the control of the a-type-specific promoter [15]. GFP-fluorescence was observed in these yeast isolates, verifying the a-type, and confirming that this system can distinguish positives from false-positives in the scheme proposed in Figure 1A.

Next, a plasmid set composed of pHS-HoU and pL3G-2α was introduced into the parental strain for isolation of α-type of cells. pL3G-2α also provides the autopolyploidization prevention function (Table 1 and Figure 5B). The resulting a1-α2 complex represses expression of *hsg* in a-type of cells as well as in the parental *MAT*a/α cells. In this system, only α-type of cells can survive on SD – Ura plates, because of α-type-specific *URA3* gene expression. To confirm the mating-type of isolated yeast

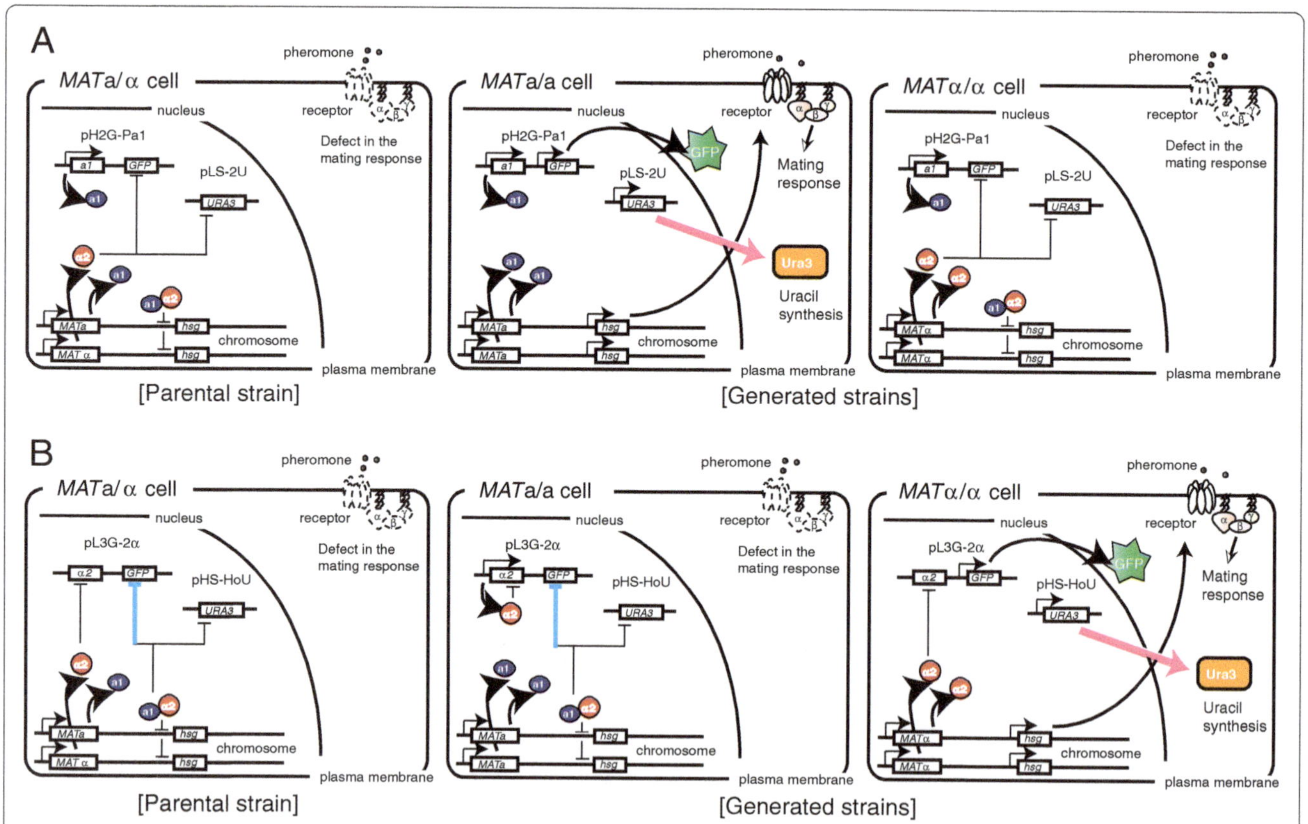

Figure 5 Schematic outline of yeast mating-type regulation and our engineered approach. (A) Engineering for isolation of a-type of cells. The *MAT*a/α parental strain indicates BY4743 harboring pH2G-Pa1 and pLS-2U (see Table 1). **(B)** Engineering for isolation of α-type of cells. The *MAT*a/α parental strain indicates BY4743 harboring pL3G-α2 and pHS-HoU (see Table 1). Chromosomal aberration regarding the *MAT* locus induces generation of two kinds of yeast cells, *i.e.*, a- and α-type of cells. Here we described yeast cells possessing two sets of *MAT* genes (generated via LOH) as an example. Then, spontaneous or engineered formation of the a1-α2 complex suppresses the mating response by repressing expression of haploid-specific genes (*hsg*), which encode components of the mating signaling pathway. Target cells expressing the *URA3* selection marker gene can be isolated on SD – Ura plates in the absence of autopolyploidization. The target cells also express the *GFP* reporter gene. Regulation colored with cyan indicates indirect gene repression; the a1-α2 complex prevents expression of transcriptional activator α1 that is required for gene expression under the control of P_{STE3}. Note that *URA3* gene is directly repressed by a transcriptional repressor in non-target cells.

Table 1 Yeast strains and plasmids used in this study

Name	Description	Reference source
Yeast strains		
BY4741	*MAT***a** *his3Δ1 ura3Δ0 leu2Δ0 met15Δ0*	Brachmann *et al.*, [24]
BY4742	*MAT***α** *his3Δ1 ura3Δ0 leu2Δ0 lys2Δ0*	Brachmann *et al.*, [24]
BY4743	*MAT***a/α** *his3Δ1/his3Δ1 leu2Δ0/leu2Δ0 LYS2/lys2Δ0 met15Δ0/MET15 ura3Δ0/ura3Δ0*	Brachmann *et al.*, [24]
BY4743A	*MAT***a** *his3Δ1 leu2Δ0 LYS2 MET15 ura3Δ0*	Present study
BY4743AL	*MAT***α/α** *his3Δ1/his3Δ1 leu2Δ0/leu2Δ0 LYS2/lys2Δ0 met15Δ0/MET15 ura3Δ0/ura3Δ0*	Present study
Plasmids		
pRS313	Yeast expression vector; *CEN6/ARSH4* ori and *HIS3* marker	NBRP*
pRS315	Yeast expression vector; *CEN6/ARSH4* ori and *LEU2* marker	NBRP
pRS316	Yeast expression vector; *CEN6/ARSH4* ori and *URA3* marker	NBRP
pHY-2GA	Yeast expression vector; *2 μ* ori, *HIS3* marker and P_{STE2}-*EGFP*	Fukuda *et al.*, [18]
pLY-3GC	Yeast expression vector; *2 μ* ori, *LEU2* marker and P_{STE3}-*EGFP*	Fukuda *et al.*, [18]
pH2G-Pa1	P_{PGK1}-*a1* in pHY-2GA	Fukuda *et al.*, [18]
pL3G-2α	P_{STE2}-*α2* in pLY-3GC	Fukuda *et al.*, [18]
pHY-2U	Yeast expression vector; *2 μ* ori, *HIS3* marker and P_{STE2}-*URA3*	Present study
pLY-3U	Yeast expression vector; *2 μ* ori, *LEU2* marker and P_{STE3}- *URA3*	Present study
pLS-2U	P_{STE2}-*URA3* in pYO315	Present study
pHS-3U	P_{STE3}-*URA3* in pYO313	Present study
pHS-HoU	P_{HO}-*URA3* in pYO313	Present study

*Resources were provided by the National Bio-Resource Project (NBRP) of the MEXT, Japan.

cells, we used a *GFP* reporter gene under the control of the α-type-specific promoter [18]. GFP-fluorescence was observed in these yeast isolates, verifying the α-type, and confirming the validity of the scheme proposed in Figure 1B.

Isolation of a- and α-type of cells generated via spontaneous chromosomal aberration

As shown in Figure 1A, *MAT*a/α cells possessing both plasmids pLS-2U and pH2G-Pa1 were washed and spread on SD – Ura plates after overnight cultivation with uracil. From approximately 1×10^7 cells spread on the SD – Ura plate, 28 colonies were formed (Figure 6A). Ten of these colonies were randomly selected, and fluorescence intensity was measured. Values were compared to that of the parental *MAT*a/α cells (harboring pH2G-Pa1), which were used as a negative control (labeled as "Ctrl" in Figure 6B). All of the isolates derived from the SD – Ura plates exhibited higher fluorescence than Ctrl, suggesting that the proposed selection system for a-type of cells rarely produced false-positives. We selected the colony (No. 3) with the highest fluorescence intensity and performed another round of colony isolation to remove the cells of all plasmids. The resulting strain was designated BY4743A (Table 1).

Next, *MAT*a/α cells possessing both plasmids pHS-HoU and pL3G-2α were washed and spread on the SD – Ura plates after overnight cultivation with uracil, as shown in Figure 1B. From approximately 1×10^6 cells spread on the SD – Ura plate, 23 colonies were formed (Figure 6C). Ten of these colonies were randomly selected, and fluorescence intensity was measured. Values were compared to that of the parental *MAT*a/α cells (harboring pL3G-2α), which were used as a negative control (labeled as "Ctrl" in Figure 6D). All of the colonies isolated on the SD – Ura plates exhibited higher fluorescence than Ctrl, suggesting that the proposed selection system for α-type of cells rarely produced false-positives. We selected the colony (No. 4) with the highest fluorescence intensity and performed another round of colony isolation to remove the cells of all plasmids. The resulting strain was designated BY4743AL (Table 1).

To verify the mating ability of BY4743A and BY4743AL, mating assays [20,21] were carried out as follows. Because BY4743A and BY4743AL have the same auxotrophy as BY4743, pH2G-Pa1 (containing *HIS3* marker) was introduced into BY4743A (a-type) and BY4743 (*MAT*a/α; parental strain), and pL3G-2α (containing *LEU2* marker) was introduced into BY4743AL (α-type) and BY4743. This allowed us to select zygotes using solid medium lacking histidine and leucine. The mating assays revealed successful mating between BY4743A and BY4743AL, but not between the two kinds of BY4743 transformants (Figure 6E). These results confirmed the validity of our proposed scheme. We believe that our approach will find application in manufacturing mating strains for yeast crossbreeding.

Discussion

The aim of this study was to establish a versatile method, using growth-based selection, for the isolation of mating strains for yeast crossbreeding (Figure 1). Although mating strains are traditionally prepared by sporulation from *MAT*a/α yeast strains used for industrial purposes, there are numerous strains that have extreme difficulty sporulating [7]. Alternative approaches using HO endonuclease have been developed in order to provide mating strains without sporulation [18,22]. Unfortunately, however, these approaches cannot be applied to yeast strains containing the "stuck" mutation (a single

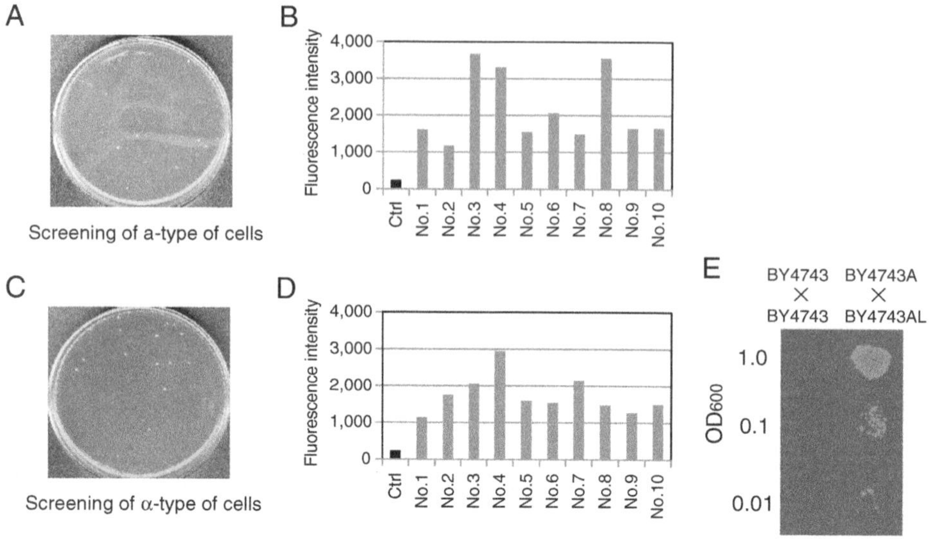

Figure 6 Isolation of a- or α-type of cells generated via spontaneous chromosomal aberration. (A) Direct images of colony formation in isolation of a-type of cells. Cell suspensions containing approximately 1×10^7 yeast cells (1 mL of suspension at OD_{600} 1.0) were spread on selective solid medium (SD – Ura plate). **(B)** Fluorescent reporter assay to quantify expression of the *GFP* reporter gene in yeast transformants possessing plasmid pH2G-Pa1 (P_{STE2}-*GFP*). Ctrl indicates the parental *MAT*a/α transformant (negative control). **(C)** Direct images of colony formation in isolation of α-type of cells. Cell suspensions containing approximately 1×10^6 yeast cells (1 mL of suspension at OD_{600} 0.1) were spread on selective solid medium. **(D)** Fluorescent reporter assay to quantify expression of the *GFP* reporter gene in yeast transformants possessing plasmid pL3G-2α (P_{STE3}-*GFP*). Ctrl indicates the parental *MAT*a/α transformant (negative control). **(E)** Mating assays to investigate the mating abilities of yeast cells. Each spot corresponds to 10 μL of suspension at the indicated OD_{600}.

base substitution) at the *MAT* locus [23]. Here we focused on chromosomal aberration such as LOH and chromosome loss, alternative processes that produce a- and α-type of cells from parental *MAT*a/α cells. Although chromosomal aberration is remarkably tolerant of slight differences in the base sequences at the *MAT* locus, spontaneous frequencies of such events are less than 1×10^{-4}.

Hashimoto *et al.* used UV irradiation to increase the frequency of LOH [16], and Whittaker *et al.* used 12 kinds of chemicals to increase the frequency of chromosome loss [17]; however, these treatments are likely to randomly induce additional mutations and loss of other chromosomes. Because the purpose of yeast crossbreeding is to combine favorable traits of parental strains, these undesirable changes would be a concern for the resulting a- or α-type of cells. To isolate a- and α-type of cells generated via spontaneous chromosomal aberration in the absence of mutagens, we describe here the development of mating-type-dependent *URA3* gene expression systems (Figure 1).

Generally, chromosomal stability becomes diminished by an increase in ploidy (*e.g.* less stable in triploid and tetraploid than in diploid cells). According to past report, the frequency of losing chromosome VII was approximately 10,000-fold higher in tetraploid than in diploid cells of *S. cerevisiase* [15]. In principle, mating-type-dependent screening cannot exclude triploid, tetraploid and other higher polyploid cells that may emerge via chromosome aberration following autopolyploidization (mating between the a- and α-type of cells generated from the same parental strain). Although these cells would presumably be available for yeast crossbreeding, chromosome stability of the hybrids should inevitably become diminished due to unneeded increase in ploidy. Hence, we utilized artificial formation of the a1-α2 complex [18] to remove the potential risk of autopolyploidization.

To develop growth selection systems for mating strains, we initially adopted P_{STE2} (for isolation of a-type of cells) and P_{STE3} (for isolation of α-type of cells as the promoters to express the *URA3* selection marker gene. By the use of YCp-type of vector, P_{STE2}-*URA3* successfully permitted only a-type of yeast cells to grow without uracil. Compared to P_{STE2}, regulation of gene expression under the control of P_{STE3} was leaky, a fact that had not been noted in a previous study using the *GFP* reporter gene [18]. Although we attempted to reduce leakiness of the *URA3* gene expression by inclusion of the artificially formed a1-α2 complex (see Additional file 1: Figure S1C and D), little change was seen in the OD_{600} value of *MAT*a and *MAT*a/α cells after 24 h cultivation (see Additional file 1: Figure S1D; compare to Figure 3D).

The difference in leakiness between P_{STE2} and P_{STE3} may reflect the distinct regulatory mechanisms controlling *asg* and *αsg* promoters. In *MAT*α and *MAT*a/α cells, expression of *asg* is suppressed by the α2 repressor encoded by the *MATα* allele [24]. The *MATα* allele also encodes the α1 activator, which facilitates the expression of *αsg*. Whereas *MAT*a cells rarely express *αsg* due to absence of the α1-coding gene within the *MAT* locus, *MAT*a/α cells are unlikely to express *αsg* because the α1-coding gene is repressed by the a1-α2 complex (a1 is encoded by the *MAT*a allele) [19]. Unlike promoters of *asg*, there is no component that can directly repress promoters of *αsg*, which might permit a low level of basal (leaky) expression. In the present work, we succeeded in tight α-type-specific *URA3* gene expression by utilizing P_{HO}, a promoter of *hsg* that is directly repressed by the a1-α2 complex (Figure 4).

While use of auxotrophic markers has an advantage in transformation efficiencies, the utility of them is limited within laboratory strains or genetically modified strains because industrial yeast strains usually do not have auxotrophic mutations such as *ura3*, *his3* and *leu2*. However, in principle, our system can adopt drug-resistance markers which are available for industrial yeast strains. Instead of the *URA3* marker, we attempted to use *kanMX4* marker (see Additional file 2) that gives resistance for geneticin (G418), and succeeded in growing only target cells in cultivation medium containing G418 (see Additional file 3: Figure S2).

To evaluate the availability of the established growth selection systems, we manufactured a- and α-type of cells as shown in Figure 1. The appearance frequencies of a- and α-type of cells were 2.8×10^{-6} and 2.3×10^{-5}, respectively. Because the frequency of chromosomal aberration generally varies by yeast strain, genetic modifications might have some effect on the frequency of chromosomal aberration in the parental BY4743 strain. Next, we verified the mating-type of colonies isolated on SD – Ura plates using *GFP* reporter gene expression. All colonies had the expected mating-type, suggesting the utility of isolating a- and α-type of cells via the constructed growth-based selection systems.

We performed another round of colony isolation to remove all plasmids from yeast cells, yielding BY4743A and BY4743AL strains, respectively. To identify the event generating BY4743A or BY4743AL strain, we carried out quantitative analysis for the number of chromosome III (containing *MAT* genes) using real-time PCR (see Additional file 4: Figure S3). BY4743A strain has only 1 set of chromosome III, suggesting that it was generated via chromosome loss. On the other hand, BY4743AL strain has 2 sets of chromosome III, suggesting that it was generated via LOH.

Furthermore, to confirm the stability of their mating ability, we carried out mating assays after serial passage in culture (see Additional file 5: Figure S4). Even after three passages, BY4743A and BY4743AL continued to exhibit equivalent levels of mating. We expect that the mating abilities of the generated a- and α-type of cells will remain extremely stable, given that additional chromosomal rearrangements are unlikely to occur at the *MAT* loci.

Conclusion

We have established a new approach to manufacture a- and α-type of cells from parental *MAT*a/α cells using growth selection systems. Use of spontaneous chromosomal aberration is quite beneficial in acquisition of mating strains inheriting desirable properties of industrially-used strains. We showed that yeast strains generated via spontaneous chromosomal aberration have stable mating ability, providing hybrid strains for use in yeast crossbreeding. Use of our method should help promote further advances of yeast-based biosynthesis approaches and in other experimental areas of research such as quantitative trait locus analysis and genome-wide association studies that permit the linking of phenotypic traits and genotypic data.

Methods

Strains and media

Detailed information about *Saccharomyces cerevisiae* strains BY4741, BY4742, and BY4743 [25], as well as other strains used in this study, is shown in Table 1. Yeast cells were grown in YPD medium (1% yeast extract, 2% peptone and 2% glucose) or in SD medium (0.67% yeast nitrogen base without amino acids (Becton Dickinson and Company, Franklin Lakes, NJ, USA) and 2% glucose). A final concentration of 2% agar was added to make solid media.

Construction of plasmids

The sequences of oligonucleotides used in this study are shown in Table 2. The plasmids shown in Table 1 were made as follows. Using pRS316 (provided by the National BioResource Project (NBRP) of the MEXT, Japan) as a template, the *URA3* gene was amplified with oligonucleotide pair o1 and o2, and inserted in place of the *EGFP* at the *Not*I-*Bam*HI sites of pHY-2GA [18], yielding a YEp-type plasmid designated pHY-2U. Similarly, using pRS316

Table 2 Sequences of oligonucleotides used to construct plasmids

Number	Sequence
1	5′-gaatcaaaaGCGGCCGCatgtcgaaagctacatat-3′
2	5′-ccccagtttgGGATCCttagttttgctggccgcat-3′
3	5′-aaaattttcGCGGCCGCatgtcgaaagctacatat-3′
4	5′-aattggagctccaCCGCGG-3′
5	5′-cgggccccccCTCGAG-3′
6	5′-aattggagctccaCCGCGGcatttttgtttcttttggac-3′
7	5′-tttcgacatGCGGCCGCtttaaagtatagatagaa-3′

as a template, the *URA3* gene was amplified with oligonucleotide pair o3 and o2, and inserted in place of the *EGFP* at the *Not*I-*Bam*HI sites of pLY-3GC [18], yielding a YEp-type plasmid designated pLY-3U.

The YCp-type plasmids used to express the *URA3* gene were constructed as follows. DNA fragments containing P_{STE2}-*URA3* were amplified from pHY-2U using oligonucleotide pair o4 and o5, and inserted into the *Sac*II-*Xho*I sites of pRS315 (provided by NBRP), yielding the plasmid pLS-2U. DNA fragments containing P_{STE3}-*URA3* were amplified from pLY-3U using oligonucleotide pair o4 and o5, and inserted into the *Sac*II-*Xho*I sites of pRS313 (provided by NBRP), yielding the plasmid pHS-3U. The haploid-type-specific promoter, P_{HO}, was amplified using oligonucleotide pair o6 and o7 from genomic DNA derived from strain BY4741, and inserted in place of P_{STE3} at the *Sac*II-*Not*I sites of pHS-3U, yielding the plasmid pHS-HoU. Each plasmid was introduced into yeast cells using the lithium acetate method [26].

Investigation of cell growth characteristics

Each yeast transformant was grown in 500 μL of SD medium without or with 20 mg/L uracil at 30°C, setting initial optical density at 600 nm (OD_{600}) at 0.03. The OD_{600} values of cultures were monitored using a UV/visible spectrophotometer (Ultrospec 3100 pro; GE Healthcare Japan Corporation, Tokyo, Japan).

Fluorescent reporter assay

The *EGFP* gene was used as a fluorescent reporter to indicate mating type. Reporter-containing cells were incubated at 30°C for 18 h, harvested and washed with distilled water. The cells then were resuspended in 100 μL of distilled water to an OD_{600} of 5.0. GFP fluorescence intensities were measured using an Infinite 200 fluorescence microplate reader (Tecan Japan Co., Ltd., Kawasaki, Japan). For detection of the GFP signal, the excitation wavelength was set at 485 nm with a bandwidth of 20 nm, and the emission wavelength was set at 535 nm with a bandwidth of 25 nm. The gain was set at 50.

Mating assay

Evaluation of mating ability was performed as follows. Yeast diploid cells were cultivated with the mating partner in 1 ml of YPD medium at 30°C for 1.5 h, with an initial OD_{600} of 0.1. After cultivation, yeast cells were harvested, washed, and resuspended in distilled water. Starting from an initial OD_{600} of 1, 0.1, or 0.01, 10 μl of cell suspensions were spotted on SD solid medium without the appropriate amino acids for growth selection of zygotes. After incubation at 30°C for 2 days, the image data was recorded for colonies on solid medium.

Isolation of yeast cells with target mating-type

Parental *MAT*ȧ/α cells were grown in 500 μL of SD media with 20 mg/L uracil at 30°C for 18 h, and then harvested and washed with distilled water. The cells then were resuspended in distilled water. Cell suspensions were spread on SD − Ura plates.

Additional files

Additional file 1: Figure S1. Alternative growth selection systems for isolation of a-type or α-type yeast cells by formation of the a1-α2 complex. (A) Plasmids used for a-type-specific *URA3* gene expression. The plasmid pLS-2U was used in combination with pH2G-Pa1, which suppresses the mating ability of α-type cells. (B) The OD_{600} values of cultures of double transformants (harboring both plasmids pLS-2U and pH2G-Pa1) at 24 h cultivation. Black bars indicate cultivation with uracil, and gray bars indicate cultivation without uracil. (C) Plasmids used for α-type-specific *URA3* gene expression. The plasmid pHS-3U was used in combination with pL3G-2α, which is required for suppressing the mating ability of a-type cells. (D) The OD_{600} values of cultures of double transformants (harboring both plasmids pHS-3U and pL3G-2α) at 24 h cultivation. Black bars indicate cultivation with uracil, and gray bars indicate cultivation without uracil.

Additional file 2: Supporting information for Materials and Methods. Table S1. Sequences of oligonucleotides used to construct plasmids.

Additional file 3: Figure S2. Growth of yeast transformants harboring *kanMX4* selection marker gene. (A) Plasmid map of pLS-2 K containing *CEN6/ARSH4* origin of replication (providing cellular retention of single-copy plasmids) and P_{STE2}-*kanMX4* construct (activated in a-type yeast cells). (B) The growth curves of pLS-2 K transformants. Closed symbols indicate cultivation without G418, and open symbols indicate cultivation with G418. (C) Plasmid map of pHS-HoK-2α containing *CEN6/ARSH4* origin of replication and P_{HO}-*kanMX4* construct combined with P_{STE2}-α2 construct (activated in α-type yeast cells). (D) The growth curves of pHS-HoK-2α transformants. Symbols are as in B.

Additional file 4: Figure S3. Ploidy analysis using real-time PCR. The normalized copy number of the *PGK1* gene is an indicator of ploidy (A) for BY4743A and (B) for BY4743AL strains. Standard deviations of three replicates are presented.

Additional file 5: Figure S4. Investigation of stability of the mating abilities of yeast cells after serial passage of cultures. Up to three serial passages were carried out, and then the resulting BY4743, BY4743A, and BY4743AL transformants were used for mating assays.

Abbreviations

EGFP: Enhanced green fluorescent protein; LOH: Loss of heterozygosity; *MAT* locus: Mating-type locus; *hsg*: Haploid-specific genes; *asg*: a-type-specific genes; *asg*: α-type-specific genes.

Competing interests

We declare that all authors are inventors on a pending patent using aspects of this system.

Authors' contributions

NF designed the study, conducted experiments, analyzed data, and co-wrote the manuscript. SH analyzed data and co-wrote the manuscript. Both authors read and approved the final manuscript.

Acknowledgements

The plasmids pRS313, pRS315, and pRS316, were provided by the National BioResource Project (NBRP) of the MEXT, Japan. This work was supported by JSPS KAKENHI Grant Number 25820406.

References

1. Higgins VJ, Bell PJL, Dawes IW, Attfield PV: **Generation of a novel *Saccharomyces cerevisiae* strain that exhibits strong maltose utilization and hyperosmotic resistance using nonrecombinant techniques.** *Appl Environ Microbiol* 2001, **67**:4346–4348.
2. Kishimoto M: **Fermentation characteristics of hybrids between the cryophilic wine yeast *Saccharomyces bayanus* and the mesophilic wine yeast *Saccharomyces cerevisiae*.** *J Ferment Bioeng* 1994, **77**:432–435.
3. Shinohara T, Mamiya S, Yanagida F: **Introduction of flocculation property into wine yeasts (*Saccharomyces cerevisiae*) by hybridization.** *J Ferment Bioeng* 1997, **83**:96–101.
4. Benitez T, Gasent-Ramirez JM, Castrejon F, Codon AC: **Development of new strains for the food industry.** *Biotechnol Prog* 1996, **12**:149–163.
5. Guijo S, Mauricio JC, Salmon JM, Ortega JM: **Determination of the relative ploidy in different *Saccharomyces cerevisiae* strains used for fermentation and 'flor' film ageing of dry sherry-type wines.** *Yeast* 1997, **13**:101–117.
6. Maráz A: **From yeast genetics to biotechnology.** *Acta Microbiol Immunol Hung* 2002, **49**:483–491.
7. Gunge N: **Breeding of bakers' yeast-determination of the ploidy and an attempt to improve practical properties.** *Japan J Genet* 1966, **41**:203–214.
8. Tsuboi M, Takahashi T: **Genetic analysis of the nonsporulating phenotype of brewer's yeasts.** *J Ferment Technol* 1988, **66**:605–613.
9. Kumaran R, Yang SY, Leu JY: **Characterization of chromosome stability in diploid, polyploid and hybrid yeast cells.** *PLoS One* 2013, **8**:e68094.
10. Andersen MP, Nelson ZW, Hetrick ED, Gottschling DE: **A genetic screen for increased loss of heterozygosity.** *Saccharomyces cerevisiae* 2008, **179**:1179–1195.
11. Takagi Y, Akada R, Kumagai H, Yamamoto K, Tamaki H: **Loss of heterozygosity is induced in *Candida albicans* by ultraviolet irradiation.** *Appl Microbiol Biotechnol* 2008, **77**:1073–1082.
12. Alvaro D, Sunjevaric I, Reid RJ, Lisby M, Stillman DJ, Rothstein R: **Systematic hybrid LOH: a new method to reduce false positives and negatives during screening of yeast gene deletion libraries.** *Yeast* 2006, **23**:1097–1106.
13. Daigaku Y, Endo K, Watanabe E, Ono T, Yamamoto K: **Loss of heterozygosity and DNA damage repair in *Saccharomyces cerevisiae*.** *Mutat Res* 2004, **556**:183–191.
14. Hiraoka M, Watanabe K, Umezu K, Maki H: **Spontaneous loss of heterozygosity in diploid *Saccharomyces cerevisiae* cells.** *Genetics* 2000, **156**:1531–1548.
15. Mayer VW, Aguilera A: **High levels of chromosome instability in polyploids of Saccharomyces cerevisiae.** *Mutat Res* 1990, **231**:177–186.
16. Hashimoto S, Aritomi K, Minohara T, Nishizawa Y, Hoshida H, Kashiwagi S, Akada R: **Direct mating between diploid sake strains of *Saccharomyces cerevisiae*.** *Appl Microbiol Biotechnol* 2006, **69**:689–696.
17. Whittaker SG, Zimmermann FK, Dicus B, Piegorsch WW, Resnick MA, Fogel S: **Detection of induced mitotic chromosome loss in Saccharomyces cerevisiae–an interlaboratory assessment of 12 chemicals.** *Mutat Res* 1990, **241**:225–242.
18. Fukuda N, Matsukura S, Honda S: **Artificial conversion of the mating-type of *Saccharomyces cerevisiae* without autopolyploidization.** *ACS Synth Biol* 2013. in press.
19. Gelfand B, Mead J, Bruning A, Apostolopoulos N, Tadigotla V, Nagaraj V, Sengupta AM, Vershon AK: **Regulated antisense transcription controls expression of cell-type-specific genes in yeast.** *Mol Cell Biol* 2011, **31**:1701–1709.
20. Fukuda N, Ishii J, Tanaka T, Kondo A: **The competitor-introduced Gγ recruitment system, a new approach for screening affinity-enhanced proteins.** *FEBS J* 2010, **277**:1704–1712.
21. Fukuda N, Ishii J, Kondo A: **Gγ recruitment system incorporating a novel signal amplification circuit to screen transient protein-protein interactions.** *FEBS J* 2011, **278**:3086–3094.
22. Kanai K, Kobayashi O: *Method for imparting mating ability to yeast.* Japanese Kokai Tokkyo: Koho patent publication number JP 2010-220481. Japanese Kokai Tokkyo Koho Application number JP 2009–068013, Date 2009.03.19.
23. Ray BL, White CI, Haber JE: **Heteroduplex formation and mismatch repair of the "stuck" mutation during matingtype switching in Saccharomyces cerevisiae.** *Mol Cell Biol* 1991, **11**:5372–5380.
24. Botstein D: **Ira Herskowitz: 1946–2003.** *Genetics* 2004, **166**:653–660.
25. Brachmann CB, Davies A, Cost GJ, Caputo E, Li J, Hieter P, Boeke JD: **Designer deletion strains derived from Saccharomyces cerevisiae S288C: a useful set of strains and plasmids for PCR-mediated gene disruption and other applications.** *Yeast* 1998, **14**:115–132.
26. Gietz D, St Jean A, Woods RA, Schiestl RH: **Improved method for high efficiency transformation of intact yeast cells.** *Nucleic Acids Res* 1992, **20**:1425.

8

Design and analysis of a tunable synchronized oscillator

Brendan M Ryback[1†], Dorett I Odoni[1†], Ruben GA van Heck[1], Youri van Nuland[1], Matthijn C Hesselman[1], Vítor AP Martins dos Santos[1*], Mark WJ van Passel[1,4] and Floor Hugenholtz[2,3]

Abstract

Background: The use of *in silico* simulations as a basis for designing artificial biological systems (and experiments to characterize them) is one of the tangible differences between Synthetic Biology and "classical" Genetic Engineering. To this end, synthetic biologists have adopted approaches originating from the traditionally non-biological fields of Nonlinear Dynamics and Systems & Control Theory. However, due to the complex molecular interactions affecting the emergent properties of biological systems, mechanistic descriptions of even the simplest genetic circuits (transcriptional feedback oscillators, bi-stable switches) produced by these methods tend to be either oversimplified, or numerically intractable. More comprehensive and realistic models can be approximated by constructing "toy" genetic circuits that provide the experimenter with some degree of control over the transcriptional dynamics, and allow for experimental set-ups that generate reliable data reflecting the intracellular biochemical state in real time. To this end, we designed two genetic circuits (basic and tunable) capable of exhibiting synchronized oscillatory green fluorescent protein (GFP) expression in small populations of *Escherichia coli* cells. The functionality of the basic circuit was verified microscopically. High-level visualizations of computational simulations were analyzed to determine whether the reliability and utility of a synchronized transcriptional oscillator could be enhanced by the introduction of chemically inducible repressors.

Results: Synchronized oscillations in GFP expression were repeatedly observed in chemically linked sub-populations of cells. Computational simulations predicted that the introduction of independently inducible repressors substantially broaden the range of conditions under which oscillations could occur, in addition to allowing the frequency of the oscillation to be tuned.

Conclusions: The genetic circuits described here may prove to be valuable research tools for the study of synchronized transcriptional feedback loops under a variety of conditions and experimental set-ups. We further demonstrate the benefit of using abstract visualizations to discover subtle non-linear trends in complex dynamic models with large parameter spaces.

Keywords: Synchronized tunable oscillator, Genetic circuit, Transcriptional feedback, Delay differential equation

Background

Synthetic genetic circuits as research tools

In order for synthetic genetic circuits to be technologically useful and modularly composable in higher order systems, their properties must be subject to formal mathematical descriptions that capture the salient features of a given circuit [1]. One of the aims of synthetic biology is to develop models that are sufficiently accurate and comprehensive to provide a basis for predicting the emergent properties of newly built genetic circuits under varying conditions [2,3]. Modeling approaches based on either *a priori* mechanistic descriptions (e.g. delay differential equations) [4], and to a lesser extent, data driven "black-box" model structure identification methods (e.g. NARMAX [5]) have become increasingly prominent. More recently, approaches attempting to consolidate models operating at varying levels of biological abstraction have also been proposed [6].

One way to improve the accuracy of models is to construct "toy" circuits which have externally controllable

* Correspondence: vitor.martinsdossantos@wur.nl
†Equal contributors
[1]Systems and Synthetic Biology, Wageningen University, Wageningen, Netherlands
Full list of author information is available at the end of the article

parameters. Such systems facilitate the rapid generation of a wide range of experimental conditions, which can be modeled in order to gain insights into potentially interesting dynamic behaviors. These systems should be complex enough to provide useful insights into the nonlinear dynamics of multi-component systems, without being so complex as to create indeterminable and/or intractable parameter spaces. Simple bi-stable memory switches have been studied extensively [7] as toy circuits for model building and parameter estimation methods, as they generally fulfill the aforementioned criteria and are amenable to study with conventional fluorescent plate-reader and FACS-based experimental set-ups. While bi-stable switches can provide insights into some nonlinear regulatory interactions, they are - by their very design specifications - stable.

In contrast, genetic oscillators exhibit unstable time-variant expression dynamics which can potentially provide insights into more complex (and subtle) emergent properties [8]. Thus, genetic oscillators are excellent objects for the study of biological nonlinear dynamics, as is evidenced by the abundance of published theoretical work [9-12]. A drawback of genetic oscillators is that their experimental implementation poses non-trivial practical difficulties. Measuring the gene expression of individual bacterial cells in real time over time-spans relevant to transcriptional oscillators is technically challenging [13]. Synchronization of populations, e.g. via quorum sensing, allows populations to be studied instead of individual cells, but imposes new constraints. The fact that systems governed by more unknown parameters than measurable (or controllable) variables are inherently underdetermined confounds the improvement of mechanistic models. Therefore, it is likely that the elucidation of the oscillatory dynamics emerging from (transcriptional) regulatory feedback loops could be facilitated by the introduction of simple control elements; assuming they remain orthogonal to the system's basic circuitry and do not increase the complexity of the nonlinear interactions (or significantly affect the host cell's metabolism). If implemented successfully, such control elements could substantially expand the range of experimental conditions used to characterize the circuit's oscillatory dynamics.

Combining synchronization and tunability

To this end, we have redesigned a genetic circuit capable of producing synchronized oscillatory GFP expression in *Escherichia coli* cells. The topology and components of the basic synchronized oscillator circuit (Figure 1) were derived from a design first published by Danino et al. [14]. The circuit consists of genes encoding products that either synthesize, degrade, or respond to the presence of the quorum sensing molecule N-(3-oxohexanoyl)-homoserine lactone (hereafter referred to as AHL) and associated regulatory elements. Oscillations emerge from the coupling of positive and negative feedback loops, of which the positive feedback is based on the AHL synthase LuxI found in *Vibrio fischeri* and the negative feedback is mediated by the AHL degrading enzyme AiiA from *Bacillus thuringiensis* [14].

Transcription of the *luxR* gene is regulated by a constitutive promoter, resulting in constant levels of the AHL-dependent transcription factor LuxR. Transcription of *luxI* and *aiiA* occurs at a basal rate when intracellular AHL concentrations are below the activation threshold of LuxR. As the cell density increases in the course of normal cell divisions in a constrained space, so too does the intracellular AHL concentration. AHL diffuses freely between the cells, which effectively synchronizes their internal states. When the AHL concentration reaches the activation threshold of LuxR, the rate of transcription of *luxI* and *aiiA* is greatly increased, initially giving rise to higher levels of the enzyme LuxI due to differences in transcription and maturation time as well as ribosomal saturation [15].

This positive feedback loop results in an exponential increase in AHL synthesis, and in turn, maximal expression of LuxI and AiiA. As catalytically active AiiA accumulates, AHL is rapidly degraded (negative feedback) to sub-LuxR activation levels, and transcription of *aiiA* and *luxI* recedes to the basal rate. Oscillations arise from the delayed interaction between these coupled positive and negative feedback loops and are contingent on the ability of the system to rapidly reset to its initial state. This property is dependent on the inclusion of LVA-degradation tags in all non-constitutively expressed proteins [16]. Changes in the intracellular AHL concentration are visualized semi-quantitatively via the expression of a fluorescent reporter gene under control of a LuxR-AHL dependent promoter. Synchronization across a population of cells results from the rapid diffusion of AHL. This onset of synchronization via quorum sensing is not gradual, but sudden and a function of varying cell densities [17].

The designs used in this study differ from previously published work [14] due to three substantial modifications: (i) elimination of redundant regulatory and coding sequences, (ii) introduction of tunable hybrid promoters and (iii) consolidation of the circuit into a single DNA sequence conforming to the BioBrick assembly standard (i.e. BioBrick device) [18].

The introduction of these tuners is intended to provide an additional set of control variables that allow the kinetics of the circuit's feedback loops to be influenced independently of one another by varying the inducer molecule concentrations within a dynamic range. This additional control may be exploited to compensate for external conditions that would prevent the basic, non-tunable circuit from producing oscillations, effectively increasing the oscillator's robustness towards variations in cell density, and

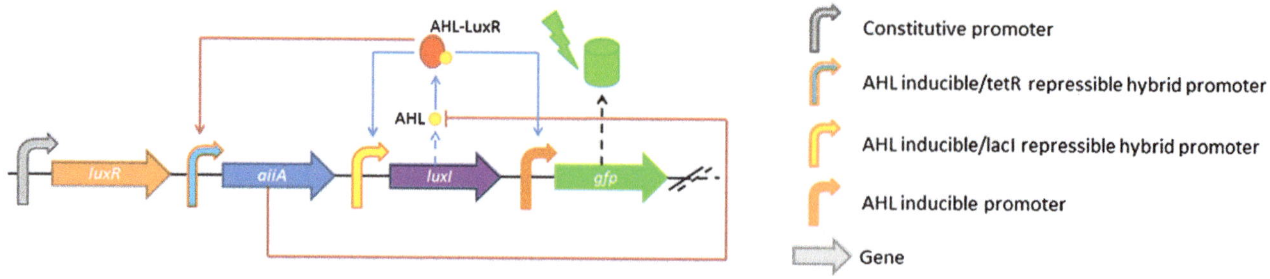

Figure 1 Scheme of the basic synchronized oscillator consisting of modified lux quorum sensing machinery, aiiA, and GFP. LuxI produces AHL, which forms a complex with the constitutively expressed LuxR. This complex induces further expression of LuxI (positive feedback loop depicted in blue) as well as AiiA, which in turn degrades AHL (negative feedback loop depicted in red). The expression of GFP is dependent on the AHL concentration present in the system and thus serves as a reporter of the oscillating AHL levels. The three AHL inducible promoters are derived from the lux promoter. Those depicted in blue and yellow are also repressible by corresponding transcription factors, but behave identically in absence of them.

by extension, expand the range of experimental set-ups under which the circuit could be employed. The functionality of the basic, non-tunable circuit in *E. coli* was verified experimentally using a custom microbial growth chamber [19] in conjunction with a fluorescence microscope.

The dynamics of both circuits are described by a set of delay differential equations which served as the basis for deterministic simulations. A broad range of input values was chosen in order to elucidate the extent to which changes in the inducer molecule concentrations influence the cell-density dependent expression dynamics.

Results and discussion

Circuit design

In the basic circuit described before, cell density is one of only two system parameters which can be manipulated, the other being the AHL-removal rate (i.e. medium replacement). All other parameters, including the maximum transcription rates of *luxI* and *aiiA*, are inaccessible barring the introduction of further regulatory elements. This constraint was circumvented by implementing "tuners", that is, constitutively expressed LacI and TetR transcription factors, which respectively repress the transcription of *luxI* and *aiiA*. The extent of the repression is dependent on the intracellular concentration of the corresponding inducer molecules isopropyl β-D-1-thiogalactopyranoside (IPTG) for LacI, and anhydrotetracycline (aTc) for TetR. The expanded circuit including the tuners is schematically represented in Figure 2.

The circuit represented in Figure 2 shows the oscillatory circuit with the aforementioned modifications: (i) elimination of two redundant copies of the sequences encoding the transcriptional regulator LuxR, (ii) replacement of natural bidirectional quorum sensing promoters with synthetic hybrid promoters containing repressor binding sites, the activity of which can be independently controlled via chemical inducers and (iii) consolidation of all the circuit's components, including the tuner module, into a single device conforming to the BioBrick assembly standard.

Basic oscillator tested in a flow device

To validate the functionality of the basic synchronized oscillator construct depicted in Figure 1, fluorescence microscopy measurements were taken of *E. coli* cells harbouring the plasmid grown in a microdish [19,20]. Time spans of 5 hours over which oscillations in GFP intensity were observed are depicted in Figure 3, showing the functionality of the oscillatory expression dynamics as compared to *E. coli* cells containing a construct with constitutively expressed GFP, as previously reported [19]. While synchronized oscillatory GFP expression was repeatedly observed, there were substantial differences in the measured frequencies and amplitudes between experiments. The measurements furthermore appear to be noisier than previously published results in which a similar genetic circuit was tested using a microfluidic chip. The specific geometry of this platform may lead to less consistent gene expression than observed in set-ups with more defined fluid control and smaller cell retention spaces [14]. Since our measurements were performed under zero-flow conditions (no media replacement), non-enzymatic AHL-removal only occurred via diffusion, resulting in a net accumulation of AHL. This eventually led to a steady state in which the enzymatic degradation was not sufficient to reset the system. It is reasonable to assume that once a critical AHL threshold is reached, further fluctuations in AHL concentration no longer significantly affect transcriptional dynamics due to saturation of LuxR. This may explain both the relative shallowness and shorter duration of the oscillations compared to a setup in which AHL was actively removed via controlled medium replacement [14]. However, another result of the accumulation of AHL was that, surprisingly, synchronization was not limited to populations within individual wells, but was observed among all of the wells within the measured area (approx. 0.63 × 0.9 mm). Since AHL is a small molecule, it is likely that it can diffuse through the porous aluminium oxide matrix [19]. This allows wells of different cell densities to be synchronized, as the AHL concentration in the medium below mimics a

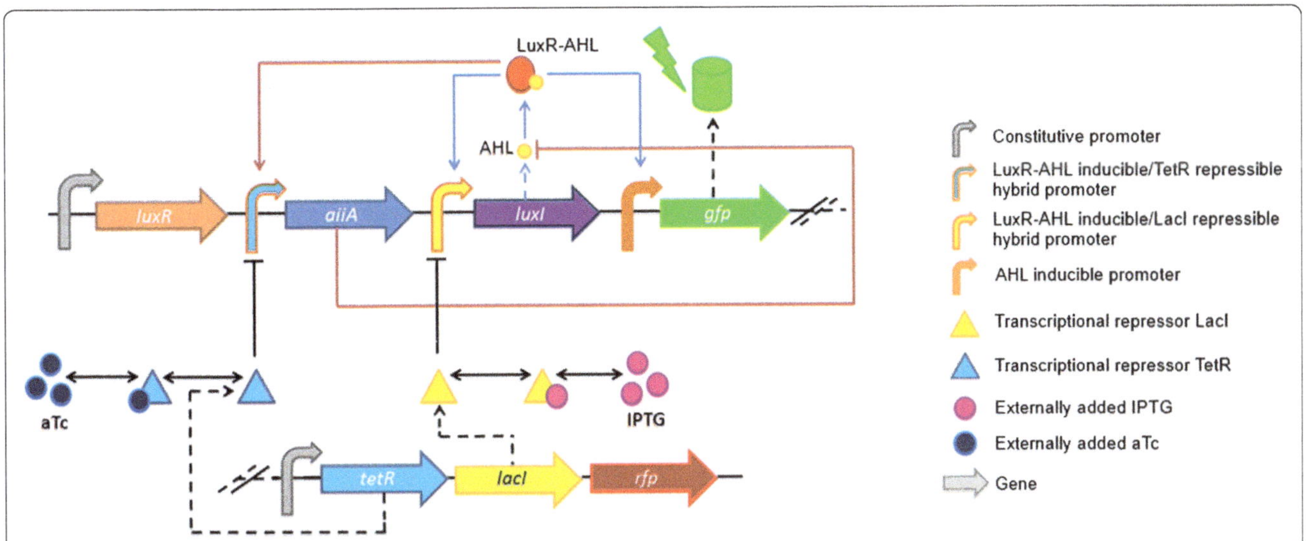

Figure 2 Circuit expanded with repressors enables tuning of the feedback kinetics. The transcriptional repressors TetR and LacI and the red fluorescent reporter mCherry are expressed as a contiguous transcript under control of the araC/pBAD promoter. TetR and LacI repress the transcription of aiiA and luxI, respectively. The presence of the inducer molecules aTc and IPTG relieve the repression as a function of their intracellular concentrations, effectively modulating the strength of the oscillating components' transcriptional feedback.

high cell density even in wells of lower cell densities. However, it is unclear whether the AHL concentration in the medium below the matrix is also homogenous.

Work published by Prindle et al. [21] demonstrated synchronization of cells trapped in microfluidic chambers across a distance of 5 mm via diffusion of H_2O_2. Our findings are consistent with this to the extent that synchronization was observed across various spatially separated sub-populations. However, an important difference is that in our system populations are solely coupled by AHL diffusion, and therefore dependent on a fluid medium to travel. The diffusion kinetics of different signaling molecules therefore need to be taken into account when designing chemically coupled regulatory systems.

Computational simulations illustrating differential regulatory dynamics

Deterministic simulations based on a dynamic model consisting of five delay differential equations were performed to illustrate the differences between the tunable and non-tunable circuits. In absence of the repressors (LacI and TetR both set to 0), the inducer molecules IPTG and aTc have no effect on gene expression. In this case the system's dynamics are governed solely by the quorum sensing machinery, which operates as a function of cell density via the inducer molecule AHL, which diffuses freely throughout the system and thus synchronizes the population. At cell densities below 0.65, where 0.88 is the theoretical maximum constraint

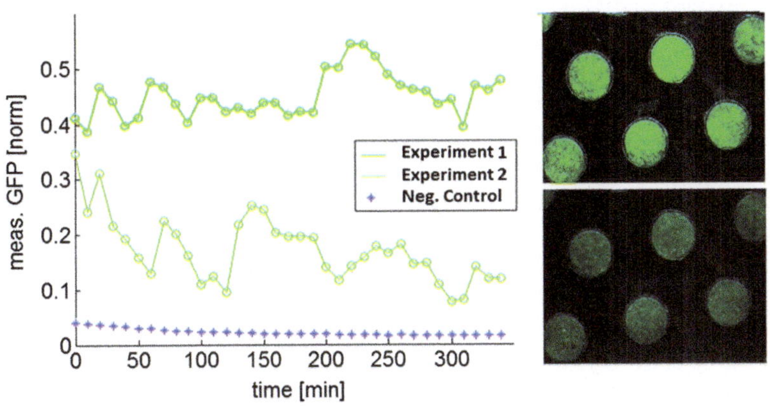

Figure 3 Oscillatory GFP expression measured in flow device. Left, measured oscillations in GFP expression of cells containing oscillator without the repressors and cells containing construct for constitutively expressed GFP as negative control. The plotted data represents the average intensity of a single well in the focus area per experiment. Right, cells grown in the wells of the microdish at high (top) and low (bottom) GFP expression levels.

given by the model, there is not enough AHL present to induce synchronization and thus no sustained oscillatory GFP expression is observed. The result is a single high peak before settling into a steady state. After passing a cell density of 0.65, the cells are able to produce enough AHL to cause damped oscillations in GFP expression. At cell densities greater than 0.75, the AHL threshold value is reached and sustained oscillations across the population of cells can emerge. The oscillations increase in frequency as they approach the maximum cell density. In all cases, the waveform is that of a relaxation oscillator, which is consistent with the previously described experimental results. The relationship between the onset of oscillations and the cell density is depicted in Figure 4.

When the repressors LacI and TetR are present, the system's expression dynamics are influenced by the concentrations of the inducer molecules IPTG and aTc. The results of 2178 simulations over 1000 time points are represented as a 5-dimensional scatterplot in Figure 5. The Cartesian x,y and z coordinates represent IPTG, aTc, and cell density, respectively. The size and color of the circular markers are each determined by scalars representing metrics that capture the essential properties of the waveform generated by the simulated GFP expression. An increase in marker size corresponds to an increase in frequency. The color change from blue to red corresponds to an increase in the normalized amplitude metric.

It is clear from this representation that the relationships between the system inputs and the resulting waveforms are non-linear, giving rise to a number of localized trends within the 5-dimensional space. The most obvious feature is the relative sparseness of the space for IPTG values greater than 0.3 and cell density values smaller than 0.6. This space is populated exclusively by damped oscillations with few peaks. It is noteworthy that for small values of IPTG the frequency steadily increases as a function of cell

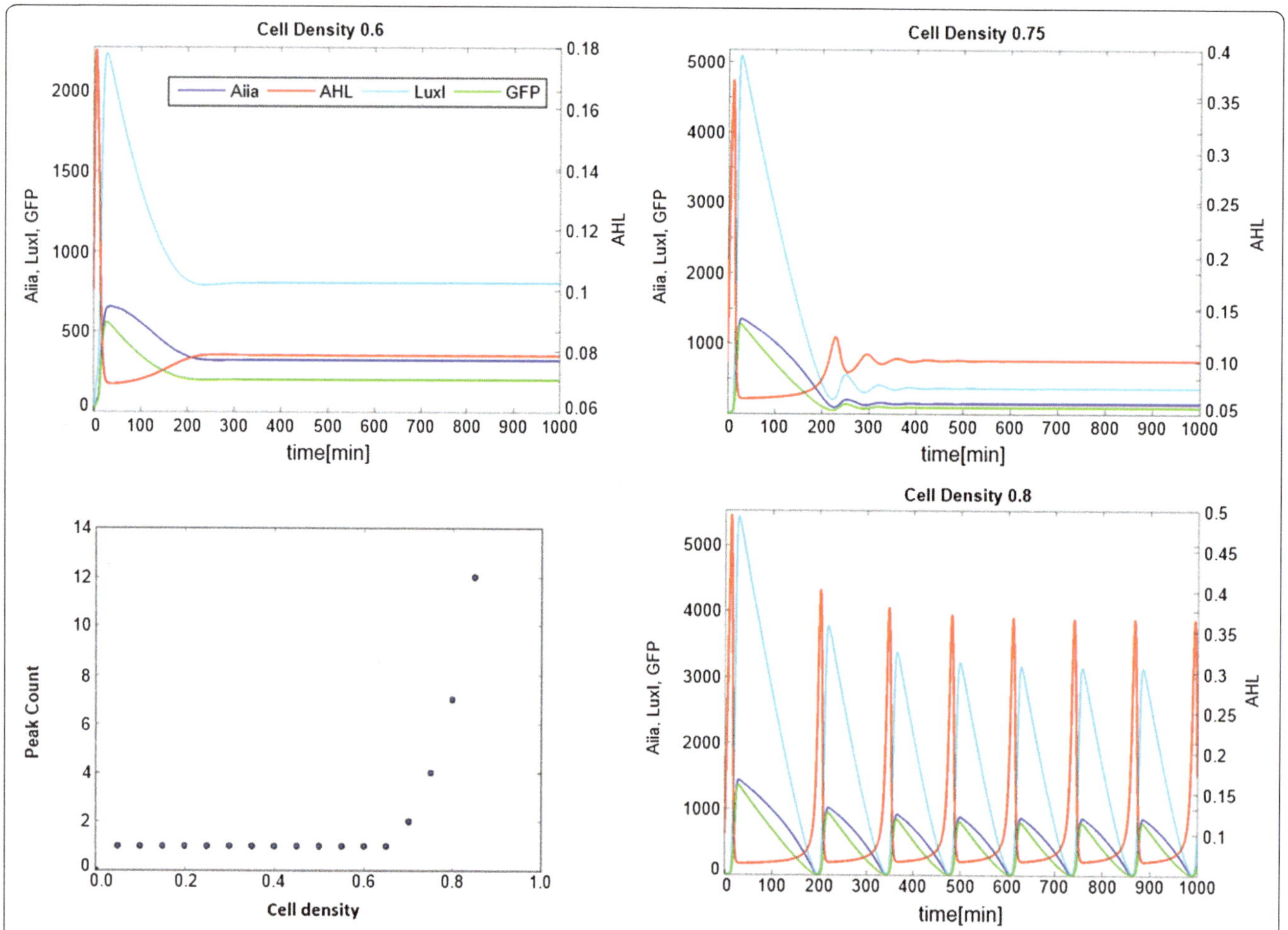

Figure 4 Correlation between the cell density and the onset of synchronized oscillatory behavior across a population of cells. As LuxI (cyan) is produced more rapidly than AiiA (dark blue), AHL (red) will initially peak to a high level before being degraded and reaching a low basal level. At low cell densities, the AHL level is not enough to induce synchronized oscillatory behavior across a population of cells. At higher cell densities however there will be enough AHL in the system to coordinate the dynamic expression of the oscillatory construct within the cells, thus sustained oscillations in GFP levels will be observed.

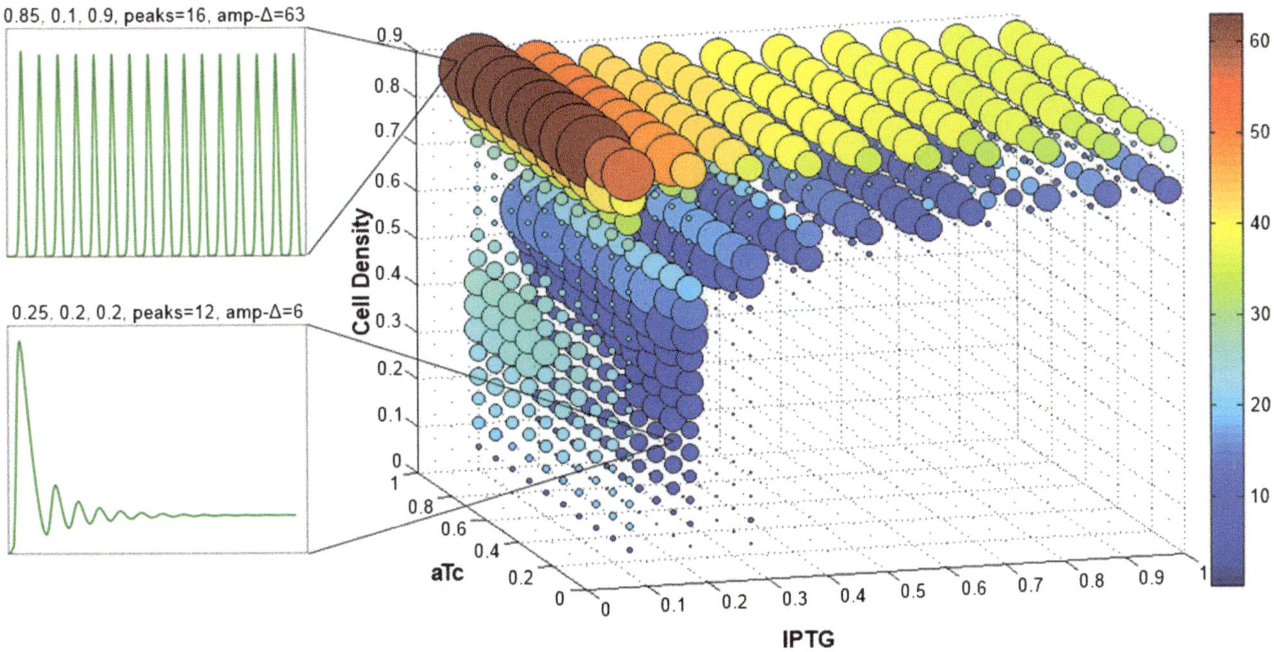

Figure 5 5-dimensional scatterplot representing the result of 2178 simulations. Inducer molecule concentrations and cell density are the input variables. The radius of each circular marker is proportional to 2^n, where n is the number of peaks in a given simulation and can be interpreted as a measure of frequency. This nonlinear scaling emphasizes the differences between high-frequency oscillations, and prevents them from being occluded by the far more numerous low-frequency oscillations. The color of each marker represents the sum of the differences between all maxima and minima normalized against the average of all values in a given simulation. Thus the color serves as a measure of amplitude.

density before decreasing (around 0.6 for IPTG = 0.2) and subsequently increasing again.

Furthermore, cell density is strongly correlated to the amplitude metric. This is related to the fact that sustained oscillations score far higher in this metric than damped oscillations, and sustained oscillations only occur outside the sparse parabolic region that spans the majority of the volume.

The relationship between the amplitude metric and the input variables is best seen as a 2-dimensional scatterplot (Figure 6, top). Values below 20 (dark blue in the 5-dimensional plot) correspond to damped oscillations in all cases. The distribution is inhomogeneous when the amplitude is plotted against the cell density. It is more homogeneous when plotted against the IPTG level, and even more so when plotted against the aTc level. The more homogeneous the plot, the less effect changes in the variable have on the amplitude. It might seem counter-intuitive that increasing levels of IPTG do not correspond to an increasing range of cell densities in which oscillations occur, as increased IPTG results in decreased repression of AHL production. We propose that the predicted dynamics are due to the difference in the time it takes for the enzymes to become active [14]. Since the maturation of the AHL-degradase takes longer than that of the AHL-synthase, AHL can accumulate and reach a saturation threshold which will effectively produce a steady state. This can be seen in the comparison between mirrored ratios of IPTG and aTc (Figure 6, bottom). When IPTG exceeds aTc, the system requires a higher cell density to oscillate than in the reverse case.

A comparison between the tunable and non-tunable system is shown in Figure 7. The 5-dimensional scatterplots show that the non-tunable system can only oscillate at cell densities above 0.6, whereas the tunable system oscillates at all simulated cell densities when IPTG is below 0.3. The maximum amplitude is around 60 for the tunable, and around 30 for the non-tunable circuit. Frequency variability is visualized by histograms, which show that the tunable system can produce a waveform with any number of peaks between 0 and 16, whereas the non-tunable system is comparatively limited. These simulations indicate that while damped oscillations (dark blue) still make up the vast majority of waveforms produced by the tunable circuit, this system is capable of a significantly more diverse behavior than the original non-tunable design. It is clear that the changes made to the original circuit have a substantial effect on the resulting protein expression dynamics. The introduction of mathematical expressions representing chemical inducer molecules and corresponding orthogonal transcription factors revealed an

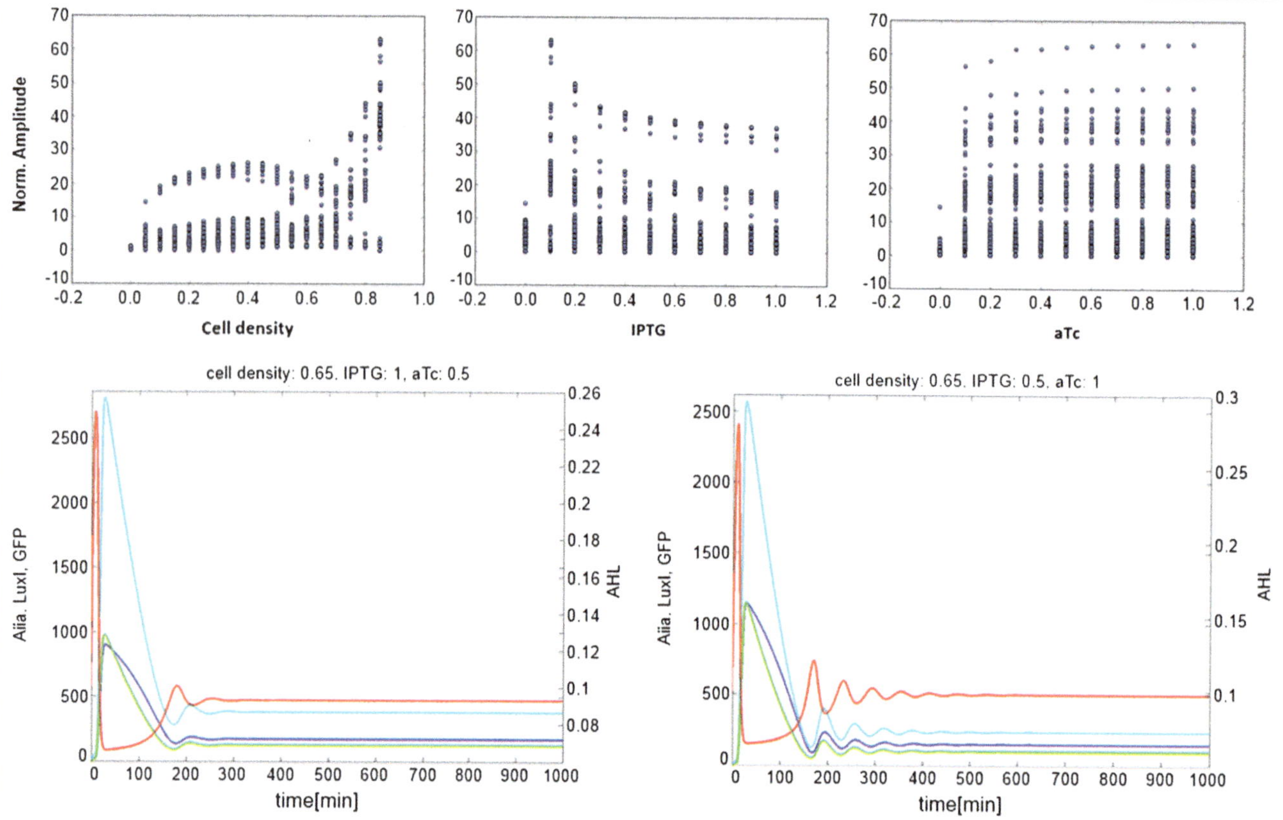

Figure 6 Top: 2-dimensional scatterplots separately comparing the normalized mean amplitudes to the 3 input variables. The amplitude metric varies the most as a function of cell density, and the least as a function of aTc. Bottom: Simulations comparing the effects of different ratios of IPTG and aTc.

unexpected range of non-obvious relationships between the system components.

Conclusion

The aim of this study was to test the functionality of a refactored synchronized transcriptional oscillator and to investigate whether its reliability and utility could be enhanced by the introduction of chemically inducible repressors. The functionality of the basic circuit, assembled from BioBrick parts, was verified experimentally using a custom experimental platform. These experiments revealed synchronization at an unexpected scale between spatially separated but chemically linked populations of bacteria. Computational simulations of the tunable circuit design revealed a rich landscape of non-linear relationships between the oscillatory behavior of the circuit and the control variables. The simulations suggested that, while cell density is the primary determinant of gene expression dynamics in this system, the ability to tune transcriptional feedback kinetics via inducer molecules substantially broadens the range of waveforms that this circuit can generate. Assuming that the model upon which the simulations were based capture the actual dynamics, the tunable oscillator design described here should be highly versatile. While fluorescent plate reader experiments aimed at characterizing this circuit's tunability repeatedly demonstrated dynamic gene expression, a lack of consistency between replicates was confounded by a low signal to noise ratio, ultimately yielding inconclusive results (data not shown).

These results offer a cursory glance at the type of methods that could be employed to study nonlinear transcriptional regulatory dynamics using this circuit. Future work on this system should aim to validate the model before exploring more rigorous analytical methods.

Due to its efficient single-plasmid design it also lends itself to the investigation of expression dynamics as a function of varying copy numbers using different plasmid backbones, or the effect of genomic integration. Such approaches could very well yield reliable, quantitative data if combined with advanced experimental platforms, such as fluorescence microscopy combined with microfluidics, fluorescence-based cell-sorting methods, or milliliter-scale continuous stirred-tank bioreactors. It is our hope that in the future, this circuit may be used by others as a tool for developing, and possibly benchmarking increasingly refined modeling approaches that shed light on the intricate and elusive properties of complex genetic circuits.

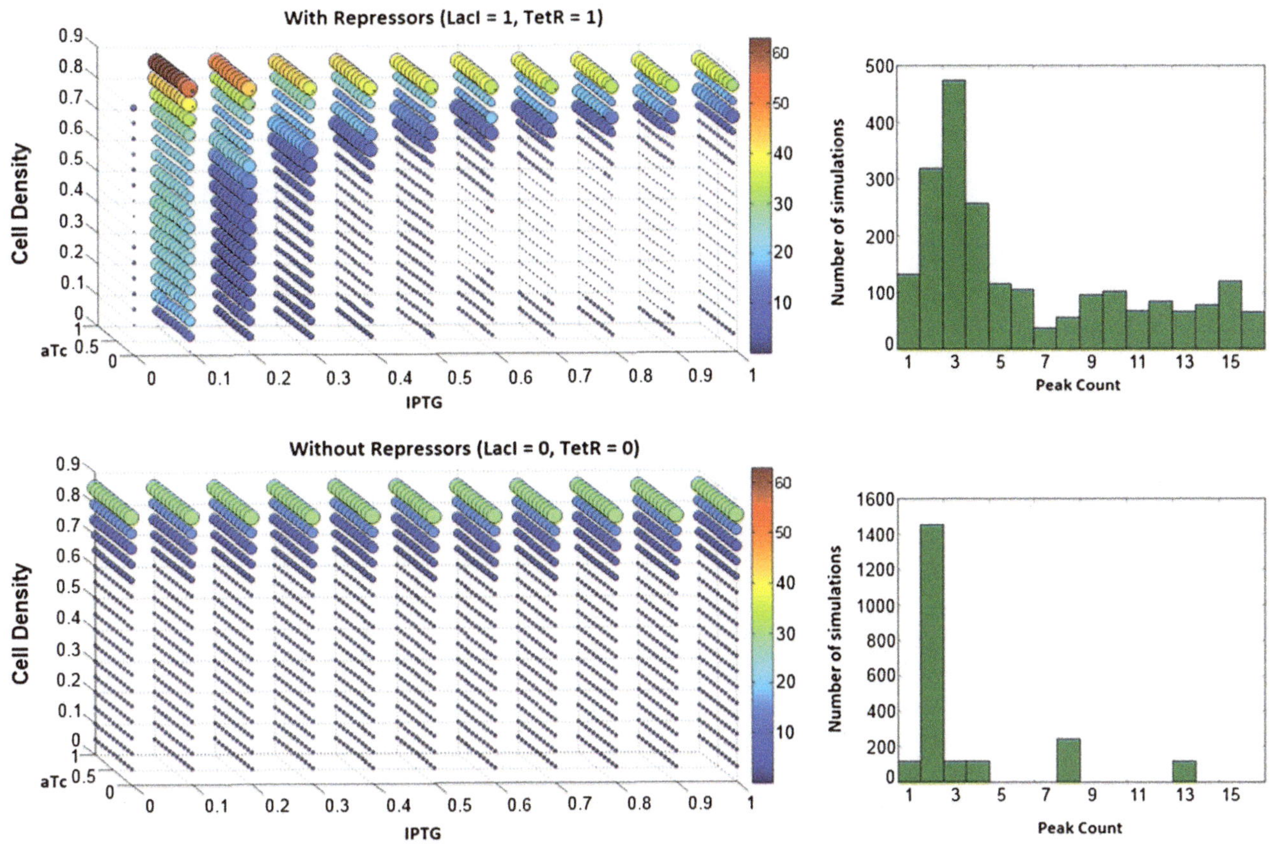

Figure 7 Comparison between the simulated circuit with and without repressors. Simulated dynamics of the tunable and non-tunable circuits demonstrates that the introduction of tuners greatly enhances the range of cell densities under which sustained oscillations can occur. Maximum achievable amplitude is also substantially higher, as is the range of frequencies that can be generated.

Methods

Circuit assembly

Both the basic and tunable oscillator circuits were assembled from BioBrick parts obtained from the Registry of Standard Biological Parts (www.partsregistry.org), listed in Table 1. Parts were assembled hierarchically, two at a time using the BioBrick standard assembly method [18,22]. The receiving plasmid with a pSB1A2 backbone was cut with either EcoRI and XbaI or with SpeI and PstI restriction enzymes. The insert was liberated from the donor plasmid by digestion with either EcoRI and SpeI or XbaI and PstI restriction enzymes. After digestion the fragments were separated via gel electrophoresis and subsequently isolated with a Qiagen Gel Extraction kit. The purified fragments were then ligated using T4 Ligase and used to transform chemically competent *E. coli* TOP10 cells using a heat shock protocol. After a recovery in SOC medium the cells were plated on LB agar containing

Table 1 BioBrick parts

Component		Function	Transcriptional control		Source
LuxR	C0062	AHL-dependent TF	Constitutive	J23101	*Vibrio fischeri*
LuxI-LVA	C0061	AHL-synthase	Induced by AHL-LuxR Repressed by LacI	I751502	*Vibrio fischeri*
AiiA-LVA	C0060	AHL-degradase	Induced by AHL-LuxR, Repressed by TetR	K176000	*Bacillus thuringiensis*
GFP-LVA	J04031	Reporter molecule	Induced by AHL-LuxR	R0062	*Aequeora victoria*
TetR	C0040	Repressor of *aiiA*	araC/pBAD	I0500	*Escherichia coli*
LacI	C0012	Repressor of *luxI*	araC/pBAD	I0500	*Escherichia coli*
mCherry	J06504	Constitutive reporter molecule	araC/pBAD	I0500	*Discosoma sp.*
Backbone	pSB1A2	Cloning & expression vector	N/A	N/A	*Escherichia coli*

All components that were used to make the genetic circuits are listed with a generic name, BioBrick ID, and function, as well as their transcriptional regulation, promoter BioBrick ID, and the biological source of the part.

either chloramphenicol or ampicillin and grown overnight at 37°C. Colony PCR was used to screen for successful transformants which were then used for the inoculation of 10 mL liquid cultures of LB and grown overnight at 37°C. The resulting composite BioBrick part was then isolated from these liquid cultures using a Qiagen miniprep kit.

Fluorescence measurements using microdish

Liquid cultures were made from single colonies which had grown on LB agar plates with ampicillin (50 μg/mL). The single colonies were grown over night at 37°C in 10-15 ml of LB ampicillin medium. The cultures were spun down and resuspended in 0.9% phosphate buffered saline (PBS), before inoculation in a custom made flow device [19] equipped with a microdish made from porous aluminium oxide containing 40 μm deep wells with a diameter of 180 μm [20]. Since LB-amp medium was supplied from below the microdish, the growth of bacteria was restricted to the wells there nutrients could be obtained via diffusion through the porous material at the base. The visual output was measured using an Olympus fluorescence microscope BX41 with an exposure time of 200 ms and 100 × magnification. Measurements were taken in a time interval of 10 minutes by a Mindstorms Lego robot (http://mindstorms.lego.com). Data analysis and processing were done with ImageJ 1.45 (http://rsbweb.nih.gov/ij/index.html) and MATLAB (http://www.mathworks.com).

DDE model

The genetic circuit described above can be represented as a system of delay differential equations, which was adapted from Danino et al. [14] and expanded with Hill functions to represent the effect of the tuner repressors and their inducer molecules on the maximum expression level of the dynamically expressed components. The final model is presented as Equations 1, 2, 3, 4, 5.

The terms proportional to C_A (1) and C_L (2) represent the dependency of AiiA and LuxI expression on the cell density d. The hybrid promoters regulating *luxI* and *aiiA* were assumed to have the same response kinetics to LuxR-AHL as the natural lux promoter in the absence of the repressor proteins. To take possible differences between the hybrid promoters into account, the leakage constants for *aiiA* and *luxI* expression were replaced by the new leakage constants δ_1 and δ_2. The terms containing these leakage constants are the history functions present in the original model which consolidate the time delay resulting from gene transcription and translation into a single parameter τ. The term after that in equations (1) and (2) is the actual tuner term and represents the influence of the repressors on the system, which in turn is dependent on the presence of either IPTG for LacI or aTc for tetR, respectively. Equations (3) and (4) contain an additional term proportional to D, which shows the diffusion of AHL throughout the cells. Finally, the terms proportional to γ show the degradation of AiiA, LuxI and GFP in equations (1), (2) and (5), respectively.

In accordance with observations of most naturally occurring regulatory elements, all promoters are assumed to be "leaky", and exhibit a basal expression level in the absence of activating TFs [13]. The model is applicable to both the basic circuit and the tunable one, as the model for the latter can be reduced to represent the former simply by setting the concentration of the repressors to 0.

Deterministic simulations were performed using the MATLAB dde23 solver in order to elucidate the relationship between inducer molecule concentrations and their effect on GFP expression relative to cell density. The input values were chosen to cover the entirety of the controllable input space, ranging from full repression (IPTG and aTc set to 0) to full induction (both IPTG and aTc set to 1) in steps of 0.1. The cell density was also iterated from 0 to 0.85 in steps of 0.05, resulting in a total of 2178 simulated conditions.

Technically the expression of TetR and LacI is not constitutive due to regulation by the AraC/pBAD promoter and its corresponding inducer molecule arabinose. However, it is treated as such for the purposes of this study.

Appendix

$$\frac{dA}{dt} = C_A \cdot \left[1 - \left(\frac{d}{d_0}\right)^4\right] \cdot \frac{\delta_1 + a \cdot Hi_{t-\tau}^2}{1 + k_1 \cdot Hi_{t-\tau}^2} \cdot \frac{1}{1 + \frac{\beta_1 \cdot R_A}{1+\beta_2 \cdot I_A}} - \gamma A \cdot \frac{A}{1 + f(A + L + G)} \tag{1}$$

$$\frac{dL}{dt} = C_L \cdot \left[1 - \left(\frac{d}{d_0}\right)^4\right] \cdot \frac{\delta_2 + a \cdot Hi_{t-\tau}^2}{1 + k_1 \cdot Hi_{t-\tau}^2} \cdot \frac{1}{1 + \frac{\beta_3 \cdot R_L}{1+\beta_4 \cdot I_L}} - \gamma L \cdot \frac{L}{1 + f(A + L + G)} \tag{2}$$

$$\frac{dHi}{dt} = \frac{b \cdot L}{1 + k \cdot L} - \gamma H \cdot \frac{A \cdot Hi}{1 + g \cdot A} + D \cdot (He - Hi) \tag{3}$$

$$\frac{dHe}{dt} = -\frac{d}{1-d} \cdot D \cdot (He - Hi) \tag{4}$$

$$\frac{dG}{dt} = C_G \cdot \left[1 - \left(\frac{d}{d_0}\right)^4\right] \cdot \frac{\delta + \alpha \cdot Hi_{t-\tau}^2}{1 + k_1 \cdot Hi_{t-\tau}^2} - \gamma_G \cdot \frac{G}{1 + f(A + L + G)} \tag{5}$$

EQUATIONS 1, 2, 3, 4, 5. Set of delay differential equations representing the interactions between the circuit's components. The terms representing time and cell-density dependent changes in AiiA (1), LuxI (2), and GFP (5) all have the same basic features. The main difference is that the maximum expression levels of AiiA and LuxI are limited by Hill functions that take the concentrations of their respective repressors and corresponding inducer molecules into account. In contrast, the expression of GFP is only dependent on the cell density and intracellular AHL concentration. Changes in intracellular AHL (3) are a function of LuxI and AiiA levels as well as a diffusion term. A = AiiA, L = LuxI, H_i = internal AHL, H_e = external AHL, C = production constant, d = cell density, δ = promoter leakage, γ = degradation constant, D = diffusion rate, τ = time delay.

Abbreviations
GFP: Green fluorescent protein; aTc: Anhydrotetracycline; IPTG: Isopropyl-β-D-thiogalactopyranoside; FACS: Fluorescence-activated cell sorting; AHL: N-(3-oxohexanoyl)-homoserine lactone.

Competing interests
The authors declare that they do not have any competing interests.

Authors' contributions
BMR and MCH designed the constructs. BMR, MCH, RvH and DIO assembled the constructs. BMR, DIO, MvP and FH designed the experiments. BMR, DIO, RvH and YvN performed the experiments and analysed data. DIO and RvH developed the model and performed the simulations. BMR and DIO analysed and visualized the simulation data. FH, MvP and VMDS supervised the project. BMR and DIO wrote the paper. All authors read and approved the final manuscript.

Acknowledgements
We gratefully acknowledge the contributions of the Wageningen University 2011 iGEM team. Furthermore, we would like to thank Maria Suarez Diez for critically reading the manuscript. MWJvP is funded by the Netherlands Organization for Scientific Research (NWO) via a VENI grant. FH is funded by the Netherlands Consortium for Systems Biology (NCSB) which is part of the Netherlands Genomics Initiative and the NWO. MicroDish BV provided the cultivation chips used in the microscopy experiments. Additional resources were provided by Wageningen University.

Author details
[1]Systems and Synthetic Biology, Wageningen University, Wageningen, Netherlands. [2]Laboratory of Microbiology, Wageningen University, Wageningen, Netherlands. [3]Netherlands Consortium for Systems Biology, University of Amsterdam, Amsterdam, Netherlands. [4]Centre for Zoonoses and Environmental Microbiology, Centre for Infectious Disease Control, National Institute for Public Health and the Environment (RIVM), Bilthoven, Netherlands.

References
1. De Jong H: **Modeling and simulation of genetic regulatory systems: a literature review.** *J Comput Biol* 2002, **9**(1):67–103.
2. McAdams HH, Arkin A: **Gene regulation: towards a circuit engineering discipline.** *Curr Biol* 2000, **10**(8):318–320.
3. Endy D: **Foundations for engineering biology.** *Nature* 2005, **438**(7067):449–453.
4. Hasty J, McMillen D, Isaacs F, Collins JJ: **Computational studies of gene regulatory networks: in numero molecular biology.** *Nat Rev Genet* 2001, **2**(4):268–279.
5. Krishnanathan K, Anderson SR, Billings SA, Kadirkamanathan V: **A data-driven framework for identifying nonlinear dynamic models of genetic parts.** *ACS Synthetic Biol* 2012, **1**(8):375–384.
6. Chandran D, Sauro HM: **Hierarchical model synthetic biol.** *ACS Synthetic Biol* 2012, **1**(8):353–364.
7. Gardner TS, Cantor CR, Collins JJ: **Construction of a genetic toggle switch inEscherichia coli.** *Nature* 2000, **403:**339–342.
8. Atkinson MR, Savageau MA, Myers JT, Ninfa AJ: **Development of genetic circuitry exhibiting toggle switch or oscillatory behavior in escherichia coli.** *Cell* 2003, **113**(5):597–607.
9. Zhou T, Zhang J, Yuan Z, Chen L: **Synchronization of genetic oscillators.** *Chaos: Interdiscipl J Nonlinear Sci* 2008, **18**(3):20.
10. Kim JR, Shin D, Jung SH, Heslop-Harrison P, Cho KH: **A design principle underlying the synchronization of oscillations in cellular systems.** *J Cell Sci* 2010, **123**(4):537–543.
11. O'Brien EL, Van Itallie E, Bennett MR: **Modeling synthetic gene oscillators.** *Math Biosci* 2012, **236**(1):1–15.
12. Purcell O, Savery NJ, Grierson CS, di Bernardo M: **A comparative analysis of synthetic genetic oscillators.** *J Roy Soc Interface* 2010, **7**(52):1503–1524.
13. Elowitz MB, Leibler S: **A synthetic oscillatory network of transcriptional regulators.** *Nature* 2000, **403**(6767):335–338.
14. Mondragón-Palomino O, Tsimring L, Hasty J: **A synchronized quorum of genetic clocks.** *Nature* 2010, **463**(7279):326–330.
15. Nandagopal N, Elowitz MB: **Synthetic biology: integrated gene circuits.** *Science* 2011, **333**(6047):1244–1248.
16. Sayut DJ, Sun L: **Slow activator degradation reduces the robustness of a coupled feedback loop oscillator.** *Mol BioSystems* 2010, **6**(8):1469–1474.
17. Garcia-Ojalvo J, Elowitz MB, Strogatz SH: **Modeling a synthetic multicellular clock: Repressilators coupled by quorum sensing.** *Proc Natl Acad Sci U S A* 2004, **101**(30):10955–10960.
18. Shetty RP, Endy D, Knight TF: **Engineering BioBrick vectors from BioBrick parts.** *J Biol Eng* 2008, **2**(1):1–12.
19. Hesselman MC, Odoni DI, Ryback BM, de Groot S, van Heck RGA, Keijsers J, Kolkman P, Nieuwenhuijse D, van Nuland YM, Sebus E, Spee R, de Vries H, Wapenaar MT, Ingham CJ, Schroën K, Santos VAP M d, Spaans SK, Hugenholtz F, van Passel MWJ: **A multi-platform flow device for microbial (co-) cultivation and microscopic analysis.** *PloS One* 2012, **7**(5):e36982.
20. Ingham CJ, Sprenkels A, Bomer J, Molenaar D, van den Berg A, van Hylckama Vlieg JE, de Vos WM: **The micro-Petri dish, a million-well growth chip for the culture and high-throughput screening of microorganisms.** *Proc Nat Acad Sci* 2007, **104**(46):18217–18222.
21. Prindle A, Samayoa P, Razinkov I, Danino T, Tsimring LS, Hasty J: **A sensing array of radically coupled genetic 'biopixels'.** *Nature* 2012, **481**(7379):39–44.
22. Knight T: **Idempotent vector design for standard assembly of biobricks.** *MIT Synthetic Biology Working Group* 2003.

A fibrin/hyaluronic acid hydrogel for the delivery of mesenchymal stem cells and potential for articular cartilage repair

Timothy N Snyder[1,2†], Krishna Madhavan[1†], Miranda Intrator[1], Ryan C Dregalla[2] and Daewon Park[1*]

Abstract

Background: Osteoarthritis (OA) is a degenerative joint disease affecting approximately 27 million Americans, and even more worldwide. OA is characterized by degeneration of subchondral bone and articular cartilage. In this study, a chondrogenic fibrin/hyaluronic acid (HA)-based hydrogel seeded with bone marrow-derived mesenchymal stem cells (BMSCs) was investigated as a method of regenerating these tissues for OA therapy. This chondrogenic hydrogel system can be delivered in a minimally invasive manner through a small gauge needle, forming a three-dimensional (3D) network structure *in situ*. However, an ongoing problem with fibrin/HA-based biomaterials is poor mechanical strength. This was addressed by modifying HA with methacrylic anhydride (MA) (HA-MA), which reinforces the fibrin gel, thereby improving mechanical properties. In this study, a range of fibrinogen (the fibrin precursor) and HA-MA concentrations were explored to determine optimal conditions for increased mechanical strength, BMSC proliferation, and chondrogenesis potential *in vitro*.

Results: Increased mechanical strength was achieved by HA-MA reinforcement within fibrin hydrogels, and was directly correlated with increasing HA-MA concentration. Live/dead staining and metabolic assays confirmed that the crosslinked fibrin/HA-MA hydrogels provided a suitable 3D environment for BMSC proliferation. Quantitative polymerase chain reaction (qPCR) of BMSCs incubated in the fibrin/HA-MA hydrogel confirmed decreased expression of collagen type 1 alpha 1 mRNA with an increase in Sox9 mRNA expression especially in the presence of a platelet lysate, suggesting early chondrogenesis.

Conclusion: Fibrin/HA-MA hydrogel may be a suitable delivery method for BMSCs, inducing BMSC differentiation into chondrocytes and potentially aiding in articular cartilage repair for OA therapy.

Keywords: Osteoarthritis, Fibrin, Hyaluronic acid, Mesenchymal stem cell, Hydrogel, Cartilage, Stem cell delivery, Regenerative medicine, Tissue engineering

Introduction

Over 30 million Americans suffer from arthritis and other rheumatic (affecting joints and connective tissue) conditions, and by 2030 nearly 25% of the American population is expected to be affected by such conditions [1]. Osteoarthritis (OA) represents the majority of these cases, and is characterized by loss of articular cartilage and subchondral bone. This condition causes severe pain, and is the leading cause of chronic disability in the U.S. [1-3]. Articular cartilage damage in particular is difficult to treat due to the lack of nerves and vasculature in the cartilage [4]. Some progress has been made in the past two decades, and regeneration of injured or degenerated cartilage has been achieved using surgical techniques such as debridement [5], microfracture [6-8], mosaicplasty [9], perichondral grafting [10,11] and periosteal grafting [12,13]. More recently, research has focused on the use of synthetic implants [14,15] for cartilage regeneration. However, about 20% of synthetic implants are known to fail approximately a decade after implantation [16]. Furthermore, this method introduces other surgery-associated risks due to the need for surgical implantation [17]. The delivery of autologous

* Correspondence: daewon.park@ucdenver.edu
†Equal contributors
[1]Bioengineering Department, University of Colorado, Anschutz Medical Campus, Mail Stop 8607, 12700 East 19th Avenue, Aurora, CO 80045, USA
Full list of author information is available at the end of the article

chondrocytes [18-21] or mesenchymal stem cells (MSCs) [22-25] for cartilage regeneration has shown some promising results. MSCs in particular have received much attention in cartilage regeneration, as they are multipotent cells capable of differentiating into cartilage, bone, muscle, fat and marrow stroma in response to appropriate signaling pathways [26]. However, because autologous chondrocytes and MSCs are highly environment-sensitive, their use has been severely hampered by the lack of suitable delivery methods [27]. For example, unfavorable conditions of the delivery matrix have led to autologous chondrocyte differentiation into fibroblast-like cells, which form fibrous tissues that can prevent healing [28].

Hyaluronic acid (HA) scaffolds have been well established as a biomaterial [29] for MSC delivery. HA hydrogels have been used to induce MSC osteogenesis [30,31], adipogenesis [32] and keratinogenesis [32] *in vitro*. Chung et al reported that delivery of MSCs in HA hydrogels promoted chondrogenic gene expression [33]. HA is known to directly interact with the fibrin precursor fibrinogen through reversible complex ionic interactions [34], yet it provides a sufficiently non-adhesive surface that allows cells to move more freely [35]. Fibrin scaffolds have also been investigated for MSC delivery both *in vitro* [23,24] and *in vivo* [24,36-38]. A recent study concluded that a system comprising bone marrow-derived MSCs (BMSCs) embedded within fibrin gels with TGFβ3 was able to stimulate appropriate differentiation and articular cartilage regeneration both *in vitro* and *in vivo* [23,39].

In general, these chondrogenic hydrogels should be biocompatible, possess sufficient mechanical strength, allow good adhesion of MSCs, and permit diffusion of nutrients and differentiation factors to induce chondrogenesis and hence cartilage repair [40]. Pure unmodified fibrin hydrogels possess excellent biocompatibility since they contain native arginine-glycine-aspartic acid (RGD) sites for MSC attachment [41]; however, without reinforcements or crosslinking, they possess low mechanical strength [42-45]. Although HA hydrogels possess better mechanical strength, they lack adhesive motifs [31] and thus deter cell proliferation at higher concentrations (>1 mg/mL) [30,46].

In this study, to exploit the advantages of both fibrin and HA, we fabricated a fibrin gel reinforced by cross-linked hyaluronic acid methacrylate (HA-MA). This fibrin/HA-MA hydrogel was examined to determine its *in vitro* mechanical and biological properties, and chondrogenesis was examined by BMSCs in the hydrogel. We hypothesized that the combination of fibrin and HA-MA would promote cell attachment and proliferation, and subsequently induce chondrogenesis, and thus cartilage regeneration and repair.

Results and discussion

BMSC phenotype verification

Phenotypical surface protein expression of BMSCs was confirmed by flow cytometry (Figure 1) after incubating BMSCs with the antibodies unlabeled with fluorescent dyes (Figure 1A, B and C), anti-CD34 antibody (Figure 1D, E and F) and fluorescent-labeled antibodies, CD105-PE, CD73-PerCP, CD-90APC, CD44-FITC & CD34-FITC (Figure 1G, H and I). The first graph of each case (Figure 1A, D and G) is the side scatter channel (SSC) vs. forward scatter channel (FSC). SSC is a measure of cell granularity, while FSC represents overall cell size. The next two graphs of each case represent the fluorescent intensity through channels FL1 (fluorescein isothiocyanate, FITC) and FL2 (phycoerythrin, PE) (Figure 1B, E and H), and FL3 (peridinin chlorophyll protein complex, PerCP) and FL4 (allophycocyanin, APC) (Figure 1C, F and I). Based on the fluorescent tag of the antibodies, positive markers display a shift in their respective fluorescent channel. In the four-color graphs (Figure 1G, H and I), there is a shift to the upper right quadrant of both fluorescence graphs, which signifies a positive result for all four markers. All markers were run individually with positive results. The unlabeled cell population was used to determine the placement of the quadrants. Thus, the BMSCs were verified to possess the typical phenotype necessary for them to undergo chondrogenesis.

HA-MA characterization

The successful conjugation of MA to HA was confirmed by proton nuclear magnetic resonance (^{1}H NMR) spectra (Figure 2). The NMR spectrum of native HA (Figure 2A) confirmed the presence of HA methyl protons at 1.9 ppm [29]. After conversion, the protons in methacrylate vinyl groups appeared at 5.6 ppm and 6.1 ppm, and the MA methyl protons were confirmed at 1.8 ppm (Figure 2B). Degree of methacrylation was determined, as described previously [47], by the ratio of the integral of the HA methyl proton peak (at ~1.9 ppm) to that of the protons of MA (at ~5.6 ppm and ~6 ppm). The degree of methacrylation was calculated as 95 ± 13%, which represents almost one methacrylate group per disaccharide unit of HA.

BMSC viability & proliferation

BMSC activity was examined using a metabolic assay after 2, 4, and 6-day culture in different fibrin/HA-MA hydrogel formulations (Figure 3). No significant statistical differences between 4 mg/mL (Figure 3A) and 6 mg/mL (Figure 3B) fibrinogen groups were observed. However, the HA-MA concentration in each group influenced cellular activity. Within the two groups, there was significant statistical difference ($p < 0.05$) as calculated by one-way ANOVA. We, further, calculated the p values between individual group-pairs by student's t-Test. With

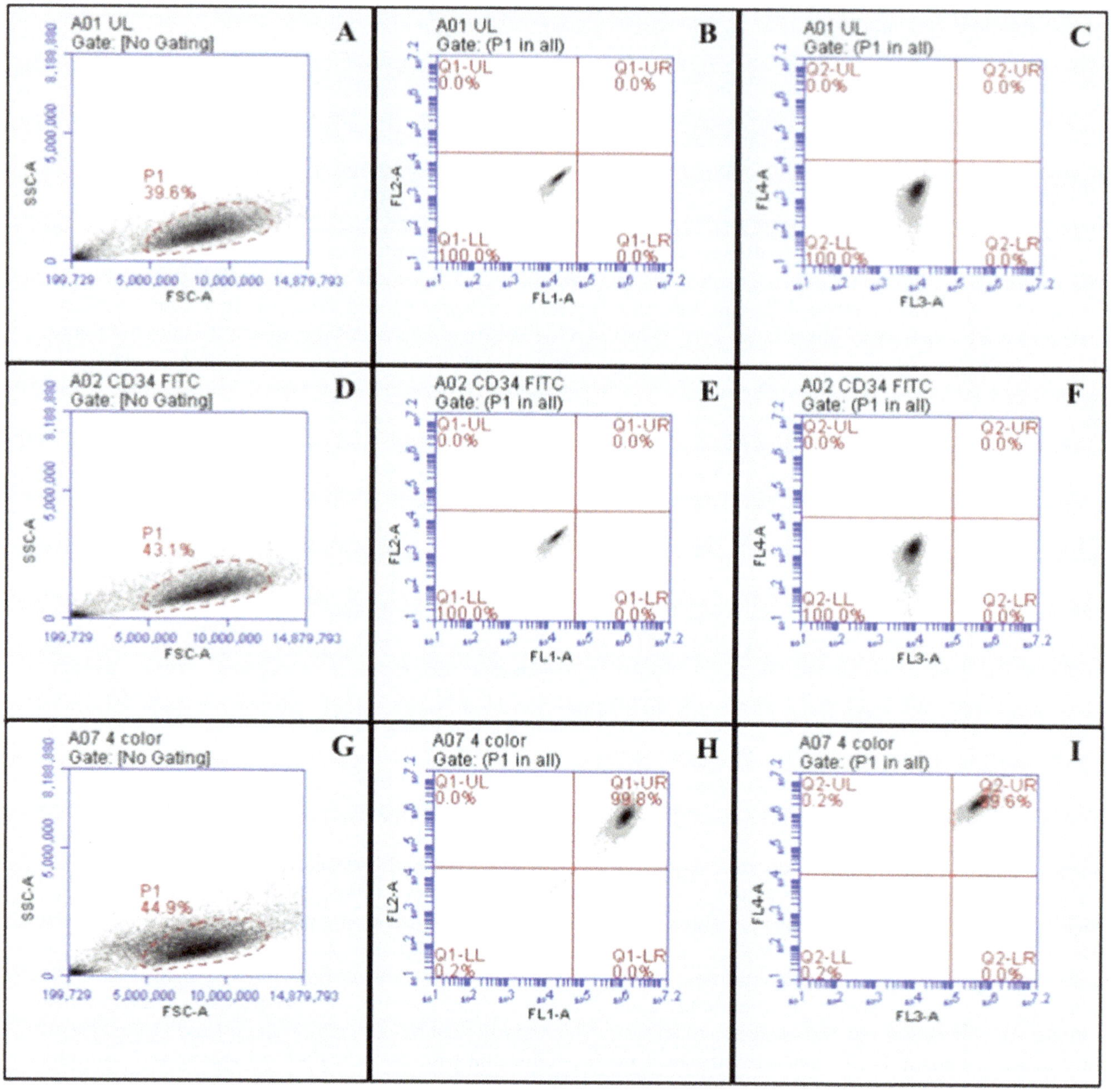

Figure 1 Flow cytometry verifies BMSC phenotype via positive and negative surface markers. Unlabeled BMSCs: **(A)** FSC vs. SSC and **(B-C)** Absence of fluorescence detected through any channels. BMSCs treated with CD34-FITC indicate no CD34 expression: **(D)** FSC vs. SSC, **(E)** all cells were FITC-negative as detected through channels FL1 and FL2 (100.0% in Q1-LL) and **(F)** through channels FL3 and FL4 (100.0% in Q2-LL). BMSCs treated with CD105-PE, CD73-PerCP, CD90-APC, CD44-FITC indicate expression of all four surface proteins: **(G)** FSC vs. SSC, **(H)** 99.8% of cells (Q1-UR) were PE, PerCP, APC and FITC-positive as detected through FL1 and FL2 and **(I)** 99.6% of cells (Q2-UR) were detected positive through FL3 and FL4.

the introduction of HA-MA into the hydrogels, both 4 mg/mL and 6 mg/mL fibrinogen concentrations supported substantial cell activity with HA-MA concentrations below 1 mg/mL. Gel compaction occurred due to the stresses imposed by the BMSC contraction in the gel matrix [48]. Gel compaction could have been increased due to the presence of TGFβ [49,50]. Based on preliminary observations (data not shown), fibrinogen concentrations less than 3 mg/mL resulted in gel compaction after only two days in culture, after which cells migrated from inside the gel to the bare plate. This suggests that the hydrogels with fibrinogen concentrations below 3 mg/mL may be mechanically unreliable. 6 mg/mL was selected as the upper limit due to the fact that higher concentrations

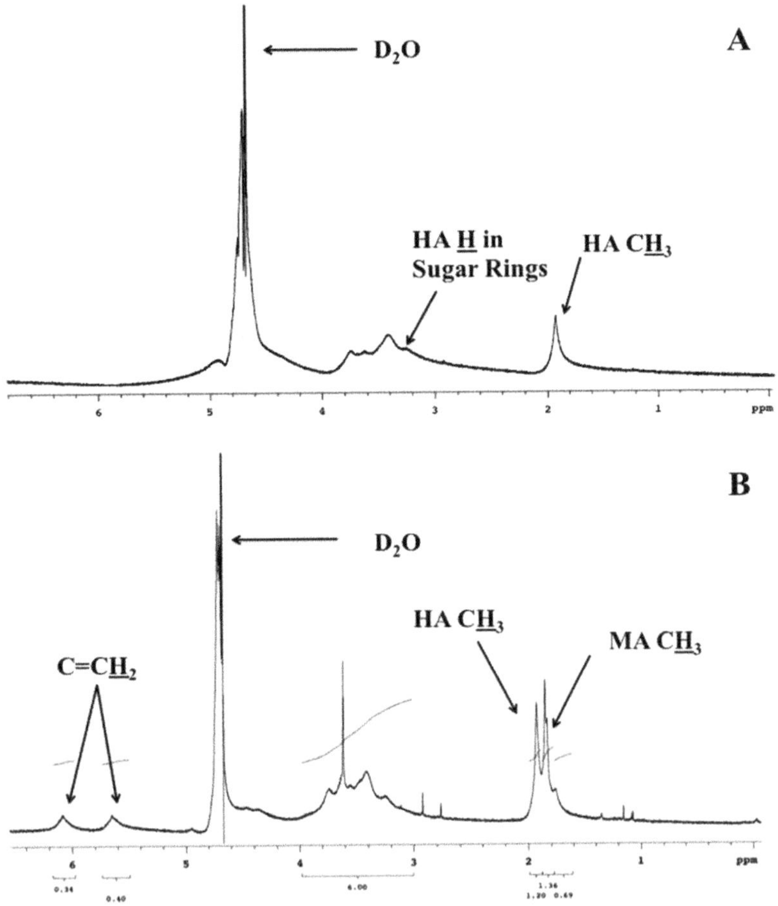

Figure 2 ^1H NMR spectra of HA (A) and HA-MA (B) with proton peaks in the methyl groups as well as the sugar rings labeled. ^1H for relevant peaks in each figure have been underlined.

became too viscous to easily mix with the HA-MA solution. The BMSC viability was also visualized by live/dead staining (Figure 4), which showed increasing numbers of viable cells and cell density at successive time points for both 4 mg/mL and 6 mg/mL fibrinogen concentrations. It was evident that there were a greater number of cells outlined in day 6 compared to day 2 in both the 4 mg/mL and 6 mg/mL conditions. The use of both the metabolic assay and live/dead staining is a strong indication of cellular proliferation. Metabolic activity alone is not reliable enough to ensure cell division; however, the quantitative measure of cellular activity with the increase in cell density as determined by live/dead staining strongly suggests cellular proliferation. A decrease in cellular proliferation as a function of mechanical strength has been recently shown in similar hydrogels. Fibrin constructs containing HA modified with tyramine capable of crosslinking in the presence of hydrogen peroxide revealed a decrease in proliferation of endothelial cells with increased mechanical strength [51].

Mechanical and structural characterization of fibrin/HA-MA hydrogels

To examine mechanical stiffness of the fibrin/HA-MA hydrogels, unconfined compression tests were performed to determine the compressive modulus at 20% strain (Figure 5), as described previously [47]. One-way ANOVA comparison of the data confirmed that the groups were statistically different ($p < 0.05$). Student's t-Test was used to find the p value between two groups to determine any statistically significant difference ($p < 0.05$). Allowing HA-MA concentrations to vary from 0 to 1.5 mg/mL, the compressive modulus of the fibrin/HA-MA hydrogel with 4 mg/mL fibrinogen varied from 1.62 ± 0.6kPa to 4.19 ± 0.28kPa, and the compressive modulus of the fibrin/HA-MA hydrogel with 6 mg/mL fibrinogen varied from 3.39 ± 0.91kPa to 6.76 ± 0.52kPa. Since there was no statistically significant difference between the two conditions 4 mg/mL of fibrinogen with 0 mg/mL of HA-MA and 4 mg/mL of fibrinogen with 0.5 mg/mL of HA-MA ($p = 0.1916$), we decided to ignore the small

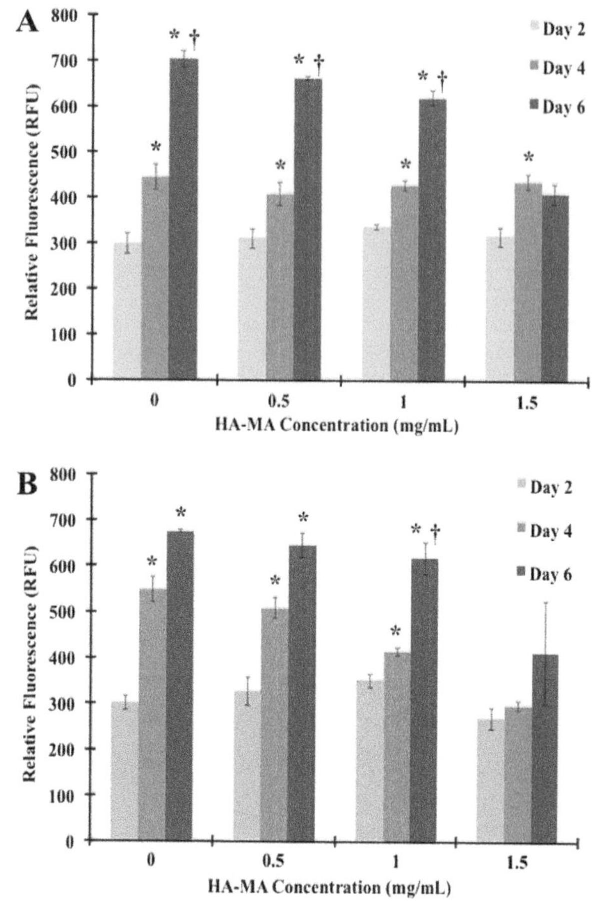

Figure 3 BMSC metabolic activity in fibrin/HA-MA hydrogels with varying HA-MA concentrations. (A) 4 mg/mL and **(B)** 6 mg/mL fibrinogen. * – Statistically significant difference between successive days for a given HA-MA concentration ($p < 0.05$). † – Statistically significant difference between successive HA-MA concentrations for a given day ($p < 0.05$).

apparent difference in the compressive moduli. The 6 mg/mL fibrinogen hydrogel possessed the higher compressive modulus at each HA-MA concentration. Increasing concentrations of both fibrinogen and HA-MA was directly correlated to an increased compressive modulus. Reinforcement with HA-MA crosslinking marginally improved mechanical strength over previously demonstrated pure fibrin hydrogels [42-45]. Since the gel was not stable below 3 mg/mL concentration, we could not measure the compressive strength at the lower concentrations.

Fibrinogen (fibrin-precursor) and HA-MA form a gel with an interpenetrating network of fibers through ionic and chemical interactions [34,51,52]. The hydrogel was then reinforced using UV photocrosslinking, which was achieved by photocrosslinking through the polymerizable vinyl group (–C = C–) present on the methacrylate side chain. The fibrin/HA-MA hydrogel even after crosslinking would still be completely biodegradable since the degradation of the hydrogel could be achieved through the hydrolysis of the ester bond present in the methacrylate side chain and the ether group present in the main hyaluronic acid backbone. To gain further insight into the structure of the hydrogels, field emission scanning electron microscopy (FESEM) was performed (Figure 6). The fibrin gels without HA-MA (Figures 6A and B) showed fibrous structure with pore sizes on the order of 1 μm. Addition of HA-MA resulted in a more sheet-like morphology with pore sizes in the range of 10-100 μm (Figure 6C). Magnifying further into the fibrin/HA-MA construct showed that fibrin fibers were still embedded within the structure (Figure 6D), indicating that the gelation occurred as intended by design: formation of fibrin gel followed by HA-MA reinforcement and crosslinking. Since 6 mg/mL fibrinogen with 1 mg/mL HA-MA led to well-controlled proliferation and an increase in mechanical strength, this fibrin/HA-MA hydrogel formulation was used for further study *in vitro* to determine mRNA expression of key chondrogenic markers.

mRNA Expression analysis

To determine chondrogenic potential, the cells were incubated within as-selected hydrogel for 12 days, and, the gene expression of collagen type 1 alpha 1 (COL1A1), collagen type 2 alpha 1 (COL2A1), SOX9 and aggrecan (ACAN) was assayed using qPCR (Figure 7). A platelet lysate (PL) was used as a supplementation to mimic real-life applications using BMSCs, since it is known to contain the growth factors such as transforming growth factor (TGFβ) that promote chondrogenesis [25]. While latent TGFβ was not quantified, active TGFβ concentration was quantified at 8.97 ng/mL using the Quantikine ELISA kit and other researchers have reported supplementation in the range of 5-20 ng/mL [53,54]. In the compared data set, one-way ANOVA confirmed that the data was statistically different ($p < 0.05$). The fold increase of mRNA transcripts was normalized to cells within fibrin/HA-MA gels supplemented with 10% fetal bovine serum (FBS). There was a downregulation of COL1A1 in all conditions with a significant downregulation in BMSCs in the fibrin/HA-MA containing PL (Figure 7A). SOX9 upregulation was observed for all conditions with a statistically significant difference in the fibrin/HA-MA with PL (Figure 7B). COL2A1 and ACAN genes were also upregulated in certain cases yet these genes were not expressed as strongly as SOX9 or COL1A1 (Refer to Additional file 1: Figure S1). A decrease in COL1A1 expression and an increase in SOX9 are good indicators of chondrogenesis [55]. Previous research has also indicated that the expressions of SOX9 and COL1A1 are very strong in comparison to the expression of COL2A1 and ACAN when BMSCs are cultured in a fibrin-based or hyaluronic acid-based scaffold [54,56-58]. Furthermore, SOX9 is by itself a favorable marker that indicates the production of COL2A1 and

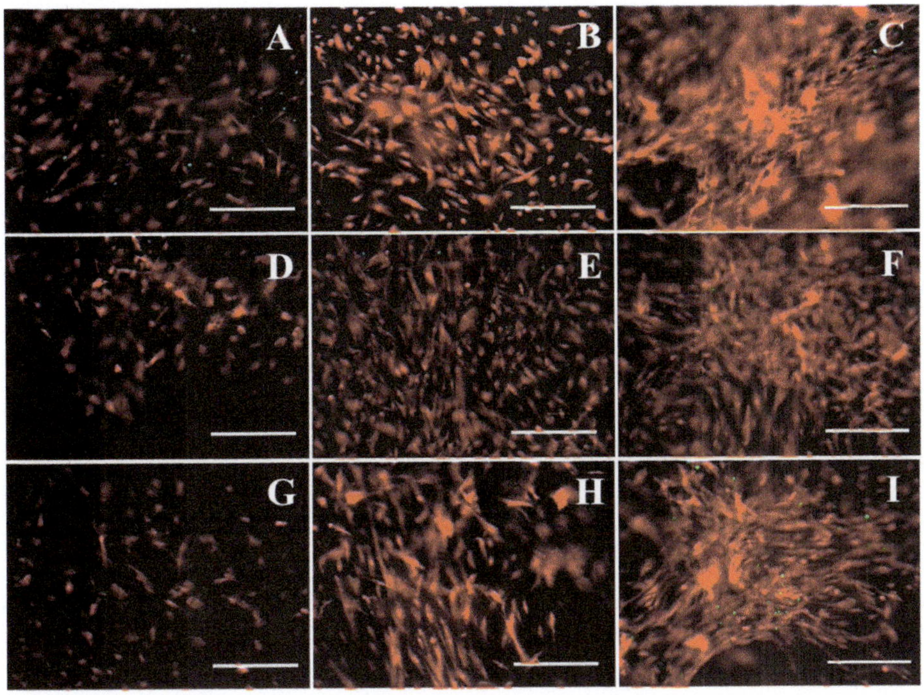

Figure 4 Representative images of live/dead staining of BMSCs in fibrin/HA-MA hydrogels. Live cells were stained with Calcein-AM (Red) and dead cells with EthD-1 (Green). Formulation with 4 mg/mL fibrinogen with 0 mg/mL HA-MA at **(A)** day 2, **(B)** day 4, and **(C)** day 6. Formulation with 4 mg/mL fibrinogen with 1 mg/mL HA-MA at **(D)** day 2, **(E)** day 4, and **(F)** day 6. Formulation with 6 mg/mL fibrinogen with 1 mg/mL HA-MA at **(G)** day 2, **(H)** day 4, and **(I)** day 6. Scale bar: 100 μm.

ACAN since it promotes the production of these extracellular components during chondrogenesis *in vivo* [59].

Conclusion

The use of a composite hydrogel formulated with 6 mg/mL fibrinogen and 1 mg/mL HA-MA promoted optimal BMSC proliferation. Crosslinking reinforcement of the fibrin scaffold with HA-MA improved the compressive modulus of fibrin/HA-MA hydrogel. Increased expression of Sox9 and decreased expression of type I collagen demonstrated the chondrogenic potential of the fibrin/HA-MA composite hydrogel, and the fibrin/HA-MA system may indeed be a promising cell delivery vehicle capable of localizing BMSCs to the site of cartilage injury and promoting chondrogenesis.

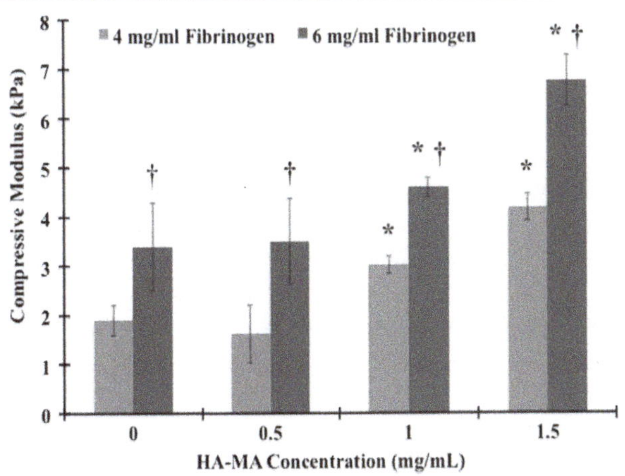

Figure 5 Compressive modulus of fibrin/HA-MA hydrogels at 20% strain. *Statistically significant difference between successive HA-MA concentrations for a given fibrinogen concentration ($p < 0.05$). † – Statistically significant difference between different fibrinogen concentrations for a given HA-MA concentration ($p < 0.05$).

Materials and methods

Materials

Dulbecco's Modified Eagle Medium (DMEM), phosphate buffered saline (PBS), trypsin and alpha minimum essential media (AMEM) were obtained from Gibco Life Technologies (Grand Island, NY). Fetal bovine serum (FBS) was purchased from Atlas Biologicals (Fort Collins, CO), and doxycycline was purchased from APP Pharmaceutical (Schaumburg, IL). The antibodies CD105-Phycoerythrin (CD105-PE), CD73-Peridinin chlorophyll protein complex (CD73-PerCP), CD90-Allophycocyanin (CD-90A PC), CD44-Fluorescein isothiocyanate (CD44-FITC) and CD34-FITC were purchased from BD Biosciences (Franklin Lakes, NJ). Bovine serum albumin (BSA), methacrylic anhydride (MA), hyaluronic acid (HA) sodium salt from Streptococcus equi, sodium hydroxide, fibrinogen from human plasma (50-70% protein, >80% clottable),

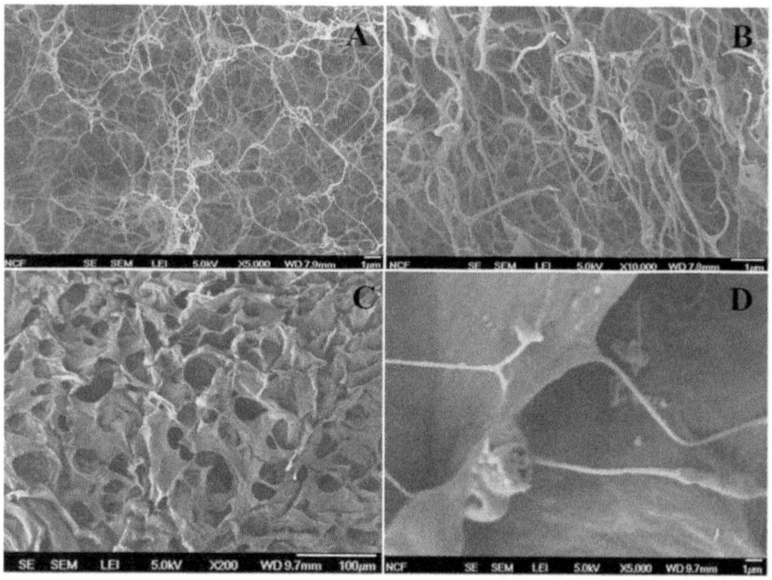

Figure 6 Scanning electron micrographs of (A) 6 mg/mL fibrinogen with no HA-MA at 5000×, (B) 6 mg/mL fibrinogen with no HA-MA at 10,000×, (C) 6 mg/mL fibrinogen with 1 mg/mL HA-MA at 200x, and (D) 6 mg/mL fibrinogen with 1 mg/mL HA-MA at 5000×. Scale bar: 1 μm.

aprotinin (3900-10400KIU/mg solid), Evithrom/Thrombin (800-1200 IU/mL) and ethanol (>99.5%) were obtained from Sigma Aldrich (St. Louis, MO). Dialysis membrane (MWCO 3500 Da) was purchased from VWR International (Denver, CO). For the cell live/dead assay, CellTrace™ Calcein Red-Orange, Sytox® Green and resazurin (PrestoBlue®) were purchased from Molecular Probes (Eugene, OR). The RealTime Ready Cell Lysis Kit, RNA Master Hydrolysis Probes kit, probes for aggrecan (ACAN), SOX9, collagen type 2 alpha 1 (COL2A1) and collagen type I alpha 1 (COL1A1) and left and right primers and the LightCycler II 480 polymerase chain reaction (PCR) platform were obtained from Integrated DNA Technologies (Coralville, IA). The Quantikine ELISA kit for quantifying TGFβ was obtained from R&D Systems (Minneapolis, MN).

Equipment

Flow cytometry was performed on an Accuri C6 Flow Cytometer (BD Biosciences). Imaging for the live/dead assay was performed on a fluorescent microscope (AMG EVOS FL, Advanced Microscopy Group). To quantify cell growth, a Fluoroskan Ascent FL 2.5 fluorescent micro-plate reader with excitation/emission filters 544/590 (Thermo Scientific) was used. Hydrogel samples were uniaxially compressed on an Insight mechanical tensile system (MTS Systems). Proton nuclear magnetic resonance (^{1}H NMR) spectroscopy was performed on a Varian INOVA 500 MHz spectrometer (International Equipment Trading Ltd.) with 5 mm triple resonance 1H/13C/15 N probe and sensitivity of 850:1. For field emission scanning electron microscopy (FESEM), the samples were sputter-coated with gold in a Cressington 108 Auto Sputter Coater (Cressington Scientific Instruments) and imaged under a JEOL JSM-7401 F field emission scanning electron microscope (JEOL USA Inc.). A LightCycler II 480 PCR platform (Integrated DNA Technologies) was used to analyze mRNA expression profiles.

BMSC isolation and culture

Age, sex and number of donors are not disclosed since it is not a non-human subject research as determined by Colorado Multiple Institutional Review Board. This information was not available to us to ensure these donors cannot be traced. However, all donors undergoing a bone marrow nucleated cell procedure consented to providing bone marrow for research purposes. Bone marrow was fractionated by centrifugation and the buffy coat layer was isolated and plated at a cell seeding density of 1×10^6 cells/cm^2 in tissue culture flasks with DMEM supplemented with 10% FBS and 5 μg/mL doxycycline. Cells were incubated under 5% CO_2 and 5% O_2 at 37°C. Media was changed every two days, washing twice with PBS before adding fresh media. Between days 6 and 10, colony formation was assessed and cells were harvested using trypsin. Cells were counted and re-plated at 6000 cells/cm^2 in AMEM with 10% FBS and 5 μg/mL doxycycline. Flasks were transferred to normoxic conditions with 21% O_2 and 5% CO_2 at 37°C. Media was changed every two days until cells were approximately 80% confluent.

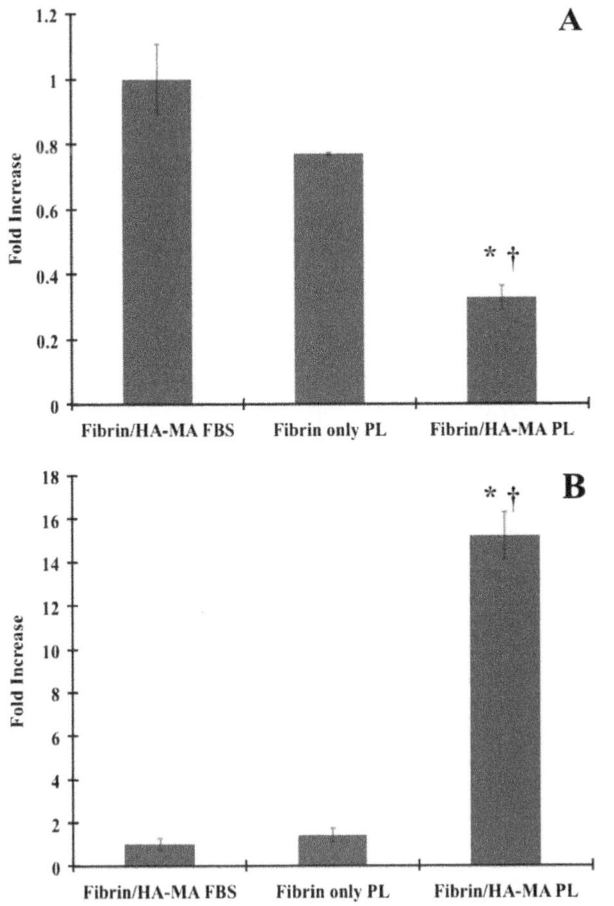

Figure 7 mRNA expression for BMSCs. (A) collagen type 1 alpha 1 gene and **(B)** SOX9 gene. All data has been normalized to fibrin/HA-MA with 10% FBS condition. Glyceraldehyde-3-Phosphate Dehydrogenase (GAPDH) is the housekeeping gene. * – Statistical significance from the fibrin/HA-MA with 10% FBS condition ($p < 0.05$) and † – Statistical significance from the Fibrin with PL condition ($p < 0.05$).

BMSC Verification

The human BMSC phenotype was verified in accordance with the Mesenchymal and Tissue Stem Cell Committee of the International Society for Cellular Therapy criteria: the surface protein expression must be positive for CD105, CD73 and CD90 and negative for at least one of the following: CD45, CD34, CD14 or CD11b [60]. CD44 has also been documented as a phenotypical MSC marker [61-63] and was also assayed for. CD105-PE, CD73-PerCP, CD90-APC, CD44-FITC and CD34-FITC surface markers were assessed using an Accuri C6 Flow Cytometer. Approximately 50,000 cells per surface protein, plus another 50,000 for four-color analysis, were centrifuged to a pellet and washed in PBS containing 2% BSA. The antibody surface markers (CD105-PE, CD73-PerCP, CD90-APC, CD44-FITC and CD34-FITC), at a concentration of 5 μg/mL, were added with cells and incubated in 2% BSA/PBS solution for 1 hour. Unbound antibody was removed; the cells were re-suspended in PBS to be run through the flow cytometer, gated at 10,000 events. An unlabeled population was run as a control to determine gating for the detection of positive and negative shifts in fluorescence; gating remained constant for all runs.

Modification of HA

The modification of HA was performed as described previously [40]. Briefly, HA sodium salt from Streptococcus equi (MW 1.5-1.8 MDa) was suspended to a final 1% (w/v) solution in diH_2O. The solution was placed in a round bottom flask and cooled to 5°C in a water bath with stirring. A 10 molar excess of MA was added to the flask and the solution was maintained at pH 10 using 5 N NaOH. After 24 hours, the reaction mixture was slowly added to excess ethanol for precipitation. The precipitate was centrifuged and the supernatant was decanted. The ethanol precipitation was repeated once more to further remove any unreacted MA. The HA-MA was transferred to a round bottom flask to vacuum dry on a rotary evaporator to remove excess ethanol. After drying, the sample was dissolved in diH_2O and dialyzed for three days against diH_2O with refreshment of water every day. The sample was then transferred to a 50 mL conical tube and lyophilized under vacuum. The MA conversion was determined using ^{1}H NMR in D_2O by the ratio of the integral values between the MA proton peaks at 5.6 ppm and 6.1 ppm to the methyl proton peak of HA at 1.9 ppm.

Fibrin/HA-MA hydrogel preparation with BMSCs

For BMSC culture, fibrinogen (final concentration: 4 mg/mL or 6 mg/mL), aprotinin (30 KIU/mL), doxycycline (5 μg/mL) and HA-MA (0.5, 1.0 or 1.5 mg/mL) were mixed thoroughly in 200 μL AMEM with 1% (w/v) photoinitiator (Irgacure 2959) and BMSCs (10,000 cells/well) in 96-well plate. To form fibrin gel, thrombin was added at 1 U/mL. Upon complete gelation, the gels were exposed to UV light (10 mW/cm^2) at, approximately, 350 nm for 5 minutes to crosslink HA-MA. The samples were limited to a 5-minute exposure to minimize the negative effects of UV on the hydrogel. BMSCs were cultured at 37°C under 5% CO_2 using AMEM with 10% FBS and 5 μg/mL doxycycline. Exposure to UV light was limited to 5 minutes since it could have a negative effect on proteins and DNA within the cells.

BMSC viability and proliferation

BMSC viability and proliferation within fibrin/HA-MA gels was investigated using live/dead staining and metabolic activity assay at 2, 4 and 7 days. CellTrace™ Calcein Red-Orange was added to a final concentration of 2 μM, with respect to the total volume of gel and media, and incubated for 15 minutes prior to observation. Similarly,

Sytox® Green was added to a final concentration of 2.5 μM and incubated for 5 minutes. After incubation, BMSC viability was observed by fluorescence microscopy. To quantify cell growth, cultures were exposed to PrestoBlue®. A 10-fold dilution of PrestoBlue® with respect to the total gel and media volume was added to the media and incubated for 2 hours on an orbital shaker. The media was removed and transferred to a 96-well plate, and fluorescence was measured by microplate reader.

Mechanical testing

Hydrogels were prepared in 24-well plate to a final volume of 1 mL as described above without the addition of BMSCs. PBS was added to allow the final gels to be fully hydrated for 24 hours. Unconfined compression testing with a 5 N load cell was performed using a 5kN MTS Insight load frame material testing system. Compression testing was performed at a 2 mm/min strain rate to 80% strain. A stress-strain curve was plotted and an exponential regression model was fitted to the data. The modulus was calculated in the initial linear region at 20% strain.

Field emission scanning electron microscopy

Hydrogels were prepared as described above to a final volume of 1 mL in 3 mL syringes with the tip removed. Fibrin-only gels with a final concentration of 6 mg/mL fibrinogen were compared to the same gels containing 1 mg/mL HA-MA. The samples were lyophilized until dry and then cross-sectioned. The samples were placed in a vacuum chamber to sputter-coat a thin layer of gold evenly across the sample surface for 40s at an operating current of 40 mA. The sample was then loaded into the field emission scanning electron microscope for morphological imaging. The sample was imaged under an operating voltage of 5 kV.

mRNA expression analysis

To determine potential chondrogenesis, BMSCs were cultured in 6 mg/mL fibrinogen gels with/without 1 mg/mL HA-MA for 12 days with the supplementation of 10% Platelet lysate (PL) obtained using osmotic stress from the platelet rich fraction of human whole blood. PL was prepared from the same blood draw and donor as to eliminate discrepancies in growth factor concentrations between conditions. Gels were harvested and washed with PBS, centrifuged to a pellet and the supernatant decanted. The probes for aggrecan (ACAN), SOX9, collagen type 2 alpha 1 (COL2A1) and collagen type 1 alpha 1 (COL1A1), and left and right primers were mixed according to manufacturer's instructions with template RNA from the samples. Solutions were added to each well of a 96-well plate and transferred to the LightCycler II 480 PCR platform. A one-step reaction protocol was programmed into the machine according to the manufacturer's instructions. RealTime ready Cell Lysis Kit contained a cell lysis solution and RNase inhibitor that was added to each gel in volumes of 39.5 μL and 0.5 μL, respectively. This was incubated on ice for 5 minutes and mixed thoroughly to extract mRNA from cells within the gel. Two master mixes were prepared before lysis, one containing primers and probes and the other containing Tth deoxyribonucleic acid (DNA) polymerase (capable of RNA transcription), deoxyribonucleotide triphosphates (dNTPs), an activator and an enhancer. The primers and probes were mixed to final concentrations of 5 μM probe and 5 μM total left and right primer. Template, PCR grade water and both master mixes were combined in ratios based on the manufacturer's instructions yielding a final primer and probe concentration of 0.325 μM. Final reaction volume was 20 μL/well run in triplicates for each probed mRNA. Glyceraldehyde-3-Phosphate Dehydrogenase (GAPDH) was used as a housekeeping gene with primers and probes provided within the qPCR kit. The relative expression of each mRNA was determined by comparing the critical threshold values of the target genes, i.e. SOX9, COL2A1, COL1A1 & ACAN, to that of the housekeeping gene, GAPDH. Table 1 indicates the sequences of primers for the various target genes. Fold expression of each condition was determined by normalizing the relative expression of the target sample with a reference sample of BMSCs cultured in fibrin/HA-MA hydrogels substituted with 10% FBS. Active TGFβ concentration was determined using the Quantikine ELISA Kit.

Table 1 Gene sequences for mRNA expression during chondrogenesis

Chondrogenic marker	Symbols	mRNA Sequence	
		Left primer	Right primer
Glyceraldehyde-3-Phosphate Dehydrogenase†	GAPDH	tagtagccgggccctacttt	tcctcctgtttcatccaagc
Aggrecan	ACAN	cctccccttcacgtgtaaaa	gctccgcttctgtagtctgc
Collagen Type 1 Alpha 1	COL1A1	gggattccctggacctaaag	ggaacacctcgctctcca
Collagen Type 2 Alpha 1	COL2A1	agggccaggatgtccatt	aggagagggcccacagag
SRY-box containing gene 9	SOX9	gtacccgcacttgcacaac	tctcgctctcgttcagaagtc

† - Glyceraldehyde-3-Phosphate Dehydrogenase (GAPDH) is the housekeeping gene.

Statistical analysis

All experiments were conducted with a sample size of five unless otherwise noted with values reported as the mean ± the standard deviation. One-way ANOVA was used to calculate p-values to determine over all statistical significance between all groups ($p < 0.05$). Unpaired Student's t-Tests were used to calculate p-values to determine statistical significance ($p < 0.05$).

Additional file

Additional file 1: Figure S1. mRNA expression for BMSCs. (A) aggrecan gene and (B) collagen type 2 alpha 1 gene. All data has been normalized to fibrin/HA-MA with 10% FBS condition. Glyceraldehyde-3-Phosphate Dehydrogenase (GAPDH) is the housekeeping gene. * – Statistical significance from the Fibrin with 10% FBS condition ($p < 0.05$).

Abbreviations

AC: Autologous chondrocytes; BMSC: Bone marrow-derived mesenchymal stem cells; HA: Hyaluronic Acid; HA-MA: Hyaluronic Acid Methacrylate; MSC: Mesenchymal stem cells; OA: Osteoarthritis; PE: Phycoerythrin; PerCP: Peridinin chlorophyll protein complex; APC: Allophycocyanin; FITC: Fluorescein isothiocyanate; BSA: Bovine serum albumin; PBS: Phosphate buffered saline; PL: Platelet lysate; SSC: Side scatter channel; FSC: Forward scatter channel; FL: Fluorescent channel; mRNA: Messenger Ribonucleic Acid; qPCR: Quantitative Polymerase Chain Reaction; TGFβ: Transforming growth factor *beta*.

Competing interests

The authors declare that they have no competing interests.

Authors' contributions

TS carried out all experiments, contributed to experimental designs and protocols and drafted the manuscript. KM helped with data analysis, writing and finalizing manuscripts. MI provided guidance on discussion and helped with finalizing the manuscripts. RD provided guidance with experimental designs and protocols and provided conceptual input and troubleshooting. DP oversaw the project providing conceptual development, advice in experimental designs and protocols, troubleshooting and finalizing the manuscript. All authors read and approved the final manuscript.

Acknowledgements

This work was partially supported by a University of Colorado start-up grant. The authors would like to thank Dr. Donghwa Yun for his help with the material synthesis.

Author details

[1]Bioengineering Department, University of Colorado, Anschutz Medical Campus, Mail Stop 8607, 12700 East 19th Avenue, Aurora, CO 80045, USA. [2]Regenerative Sciences, 403 Summit Blvd, Suite 201, Broomfield, CO 80021, USA.

References

1. Lawrence RC, Felson DT, Helmick CG, Arnold LM, Choi H, Deyo RA, Gabriel S, Hirsch R, Hochberg MC, Hunder GG, Jordan JM, Katz JN, Kremers HM, Wolfe F: **Estimates of the prevalence of arthritis and other rheumatic conditions in the United States. Part II.** *Arthritis Rheum* 2008, **58:**26–35.
2. CDC: **Prevalence of disabilities and associated health conditions among adults - United States, 1999.** *JAMA* 2001, **285:**1571–1572.
3. Spiller KL, Maher SA, Lowman AM: **Hydrogels for the repair of articular cartilage defects.** *Tissue Eng Part B Rev* 2011, **17:**281–299.
4. Stockwell RA: **Chondrocytes.** *J Clin Pathol Suppl (R Coll Pathol)* 1978, **12:**7–13.
5. Hubbard MJS: **Articular debridement versus washout for degeneration of the medial femoral condyle - A five-year study.** *J Bone Joint Surg (Br)* 1996, **78B:**217–219.
6. Steadman JR, Rodkey WG, Rodrigo JJ: **Microfracture: Surgical technique and rehabilitation to treat chondral defects.** *Clin Orthop Relat Res* 2001, (391 Suppl):S362–S369.
7. Steadman JR, Briggs KK, Rodrigo JJ, Kocher MS, Gill TJ, Rodkey WG: **Outcomes of microfracture for traumatic chondral defects of the knee: Average 11-year follow-up.** *Arthroscopy* 2003, **19:**477–484.
8. Steadman JR, Rodkey WG, Briggs KK: **Microfracture chondroplasty: indications, techniques, and outcomes.** *Sports Med Arthrosc* 2003, **11:**236–244.
9. Matsusue Y, Yamamuro T, Hama H: **Arthroscopic multiple osteochondral transplantation to the chondral defect in the knee associated with anterior cruciate ligament disruption.** *Arthroscopy* 1993, **9:**318–321.
10. Homminga GN, Bulstra SK, Bouwmeester PS, van der Linden AJ: **Perichondral grafting for cartilage lesions of the knee.** *J Bone Joint Surg (Br)* 1990, **72:**1003–1007.
11. Bouwmeester SJ, Beckers JM, Kuijer R, van der Linden AJ, Bulstra SK: **Longterm results of rib perichondrial grafts for repair of cartilage defects in the human knee.** *Int Orthop* 1997, **21:**313–317.
12. Hoikka VEJ, Jaroma HJ, Ritsila VA: **Reconstruction of the patellar articulation with periosteal grafts: 4-year follow-up of 13 cases.** *Acta Orthop Scand* 1990, **61:**36–39.
13. Ritsila VA, Santavirta S, Alhopuro S, Poussa M, Jaroma H, Rubak JM, Eskola A, Hoikka V, Snellman O, Osterman K: **Periosteal and perichondral grafting in reconstructive surgery.** *Clin Orthop Relat Res* 1994, **302:**259–265.
14. Messner K, Gillquist J: **Synthetic implants for the repair of osteochondral defects of the medial femoral condyle - a biomechanical and histological-evaluation in the rabbit knee.** *Biomaterials* 1993, **14:**513–521.
15. Hunziker EB: **Articular cartilage repair: basic science and clinical progress. A review of the current status and prospects.** *Osteoarthritis Cartilage* 2002, **10:**432–463.
16. Ahmed N, Stanford WL, Kandel RA: **Mesenchymal stem and progenitor cells for cartilage repair.** *Skeletal Radiol* 2007, **36:**909–912.
17. Kalunian K: *Patient Information: Osteoarthritis treatment (Beyond the Basics)* [http://www.uptodate.com/contents/osteoarthritis-treatment-beyond-the-basics]
18. Peterson L, Vasiliadis HS, Brittberg M, Lindahl A: **Autologous chondrocyte implantation: A long-term follow-up.** *Am J Sports Med* 2010, **38:**1117–1124.
19. Vasiliadis HS, Danielson B, Ljungberg M, McKeon B, Lindahl A, Peterson L: **Autologous chondrocyte implantation in cartilage lesions of the knee. Long-term evaluation with magnetic resonance imaging and delayed gadolinium-enhanced magnetic resonance imaging technique.** *Am J Sports Med* 2010, **38:**943–949.
20. Vasiliadis HS, Lindahl A, Georgoulis AD, Peterson L: **Malalignment and cartilage lesions in the patellofemoral joint treated with autologous chondrocyte implantation.** *Knee Surg Sports Traumatol Arthrosc* 2011, **19:**452–457.
21. Minas T, Peterson L: **Autologous chondrocyte transplantation.** *Oper Techn Sport Med* 2012, **20:**72–86.
22. Galle J, Bader A, Hepp P, Grill W, Fuchs B, Kas JA, Krinner A, Marquass B, Muller K, Schiller J, Schulz RM, von Buttlar M, von der Burg E, Zscharnack M, Loffler M: **Mesenchymal stem cells in cartilage repair: State of the art and methods to monitor cell growth, differentiation and cartilage regeneration.** *Curr Med Chem* 2010, **17:**2274–2291.
23. Park JS, Yang HN, Woo DG, Jeon SY, Park KH: **Chondrogenesis of human mesenchymal stem cells in fibrin constructs evaluated *in vitro* and in nude mouse and rabbit defects models.** *Biomaterials* 2011, **32:**1495–1507.
24. Park SH, Choi BH, Park SR, Min BH: **Chondrogenesis of rabbit mesenchymal stem cells in fibrin/hyaluronan composite scaffold *in vitro*.** *Tissue Eng Part A* 2011, **17:**1277–1286.
25. Danisovic L, Varga I, Polak S: **Growth factors and chondrogenic differentiation of mesenchymal stem cells.** *Tissue Cell* 2012, **44:**69–73.
26. Mauck RL, Yuan X, Tuan RS: **Chondrogenic differentiation and functional maturation of bovine mesenchymal stem cells in long-term agarose culture.** *Osteoarthritis Cartilage* 2006, **14:**179–189.
27. Richter W: **Mesenchymal stem cells and cartilage *in situ* regeneration.** *J Intern Med* 2009, **266:**390–405.
28. Pelttari K, Lorenz H, Boeuf S, Templin MF, Bischel O, Goetzke K, Hsu HY, Steck E, Richter W: **Secretion of matrix metalloproteinase 3 by expanded articular chondrocytes as a predictor of ectopic cartilage formation capacity *in vivo*.** *Arthritis Rheum* 2008, **58:**467–474.

29. Schante CE, Zubera G, Herlinb C, Vandammea TF: **Chemical modifications of hyaluronic acid for the synthesis of derivatives for a broad range of biomedical applications.** *Carbohydr Polym* 2011, **85:**469–489.
30. Bian L, Hou C, Tous E, Rai R, Mauck RL, Burdick JA: **The influence of hyaluronic acid hydrogel crosslinking density and macromolecular diffusivity on human MSC chondrogenesis and hypertrophy.** *Biomaterials* 2013, **34:**413–421.
31. Jha AK, Xu X, Duncan RL, Jia X: **Controlling the adhesion and differentiation of mesenchymal stem cells using hyaluronic acid-based, doubly crosslinked networks.** *Biomaterials* 2011, **32:**2466–2478.
32. Espandar L, Bunnell B, Wang GY, Gregory P, McBride C, Moshirfar M: **Adipose-derived stem cells on hyaluronic acid-derived scaffold a new horizon in bioengineered cornea.** *Arch Ophthalmol* 2012, **130:**202–208.
33. Chung C, Burdick JA: **Influence of three-dimensional hyaluronic acid microenvironments on mesenchymal stem cell chondrogenesis.** *Tissue Eng Part A* 2009, **15:**243–254.
34. LeBoeuf RD, Raja RH, Fuller GM, Weigel PH: **Human fibrinogen specifically binds hyaluronic acid.** *J Biol Chem* 1986, **261:**12586–12592.
35. Nehls V, Drenckhahn D: **A novel, microcarrier-based in vitro assay for rapid and reliable quantification of three-dimensional cell migration and angiogenesis.** *Microvasc Res* 1995, **50:**311–322.
36. Dozza B, Di Bella C, Lucarelli E, Giavaresi G, Fini M, Tazzari PL, Giannini S, Donati D: **Mesenchymal stem cells and platelet lysate in fibrin or collagen scaffold promote non-cemented hip prosthesis integration.** *J Orthop Res* 2011, **29:**961–968.
37. Anitua E, Sanchez M, Nurden AT, Nurden P, Orive G, Andia I: **New insights into and novel applications for platelet-rich fibrin therapies.** *Trends Biotechnol* 2006, **24:**227–234.
38. Hale BW, Goodrich LR, Frisbie DD, McIlwraith CW, Kisiday JD: **Effect of scaffold dilution on migration of mesenchymal stem cells from fibrin hydrogels.** *Am J Vet Res* 2012, **73:**313–318.
39. Choi JW, Choi BH, Park SH, Pai KS, Li TZ, Min BH, Park SR: **Mechanical stimulation by ultrasound enhances chondrogenic differentiation of mesenchymal stem cells in a fibrin-hyaluronic acid hydrogel.** *Artif Organs* 2013, **37:**648–655.
40. Seo S, Na K: **Mesenchymal stem cell-based tissue engineering for chondrogenesis.** *J Biomed Biotechnol* 2011, **2011:**806891.
41. Kim IL, Khetan S, Baker BM, Chen CS, Burdick JA: **Fibrous hyaluronic acid hydrogels that direct MSC chondrogenesis through mechanical and adhesive cues.** *Biomaterials* 2013, **34:**5571–5580.
42. Hunter CJ, Mouw JK, Levenston ME: **Dynamic compression of chondrocyte-seeded fibrin gels: effects on matrix accumulation and mechanical stiffness.** *Osteoarthritis Cartilage* 2004, **12:**117–130.
43. Hokugo A, Takamoto T, Tabata Y: **Preparation of hybrid scaffold from fibrin and biodegradable polymer fiber.** *Biomaterials* 2006, **27:**61–67.
44. Haugh MG, Thorpe SD, Vinardell T, Buckley CT, Kelly DJ: **The application of plastic compression to modulate fibrin hydrogel mechanical properties.** *J Mech Behav Biomed Mater* 2012, **16:**66–72.
45. Ahearne M, Kelly DJ: **A comparison of fibrin, agarose and gellan gum hydrogels as carriers of stem cells and growth factor delivery microspheres for cartilage regeneration.** *Biomed Mater* 2013, **8:**035004.
46. Camci-Unal G, Cuttica D, Annabi N, Demarchi D, Khademhosseini A: **Synthesis and characterization of hybrid hyaluronic acid-gelatin hydrogels.** *Biomacromolecules* 2013, **14:**1085–1092.
47. Seidlits SK, Khaing ZZ, Petersen RR, Nickels JD, Vanscoy JE, Shear JB, Schmidt CE: **The effects of hyaluronic acid hydrogels with tunable mechanical properties on neural progenitor cell differentiation.** *Biomaterials* 2010, **31:**3930–3940.
48. Fernandez P, Bausch AR: **The compaction of gels by cells: a case of collective mechanical activity.** *Integr Biol (Camb)* 2009, **1:**252–259.
49. Bian W, Liau B, Badie N, Bursac N: **Mesoscopic hydrogel molding to control the 3D geometry of bioartificial muscle tissues.** *Nat Protoc* 2009, **4:**1522–1534.
50. Schutte SC, Chen Z, Brockbank KG, Nerem RM: **Tissue engineering of a collagen-based vascular media Demonstration of functionality.** *Organogenesis* 2010, **6:**204–211.
51. Lee F, Kurisawa M: **Formation and stability of interpenetrating polymer network hydrogels consisting of fibrin and hyaluronic acid for tissue engineering.** *Acta Biomater* 2013, **9:**5143–5152.
52. LeBoeuf RD, Gregg RR, Weigel PH, Fuller GM: **Effects of hyaluronic acid and other glycosaminoglycans on fibrin polymer formation.** *Biochemistry* 1987, **26:**6052–6057.
53. Worster AA, Brower-Toland BD, Fortier LA, Bent SJ, Williams J, Nixon AJ: **Chondrocytic differentiation of mesenchymal stem cells sequentially exposed to transforming growth factor-beta 1 in monolayer and insulin-like growth factor-I in a three-dimensional matrix.** *J Orthop Res* 2001, **19:**738–749.
54. Lisignolia G, Cristino S, Piacentini A, Toneguzzi S, Grassi F, Cavallo C, Zini N, Solimando L, Maraldi NM, Facchini A: **Cellular and molecular events during chondrogenesis of human mesenchymal stromal cells grown in a three-dimensional hyaluronan based scaffold.** *Biomaterials* 2005, **26:**5677–5686.
55. Bosnakovski D, Mizuno M, Kim G, Takagi S, Okumur M, Fujinag T: **Gene expression profile of bovine bone marrow mesenchymal stem cell during spontaneous chondrogenic differentiation in pellet culture system.** *Jpn J Vet Res* 2006, **53:**127–139.
56. Jakobsen RB, Shahdadfar A, Reinholt FP, Brinchmann JE: **Chondrogenesis in a hyaluronic acid scaffold: comparison between chondrocytes and MSC from bone marrow and adipose tissue.** *Knee Surg Sports Traumatol Arthrosc* 2010, **18:**1407–1416.
57. Ho ST, Cool SM, Hui JH, Hutmacher DW: **The influence of fibrin based hydrogels on the chondrogenic differentiation of human bone marrow stromal cells.** *Biomaterials* 2010, **31:**38–47.
58. Patrascu JM, Kruger JP, Boss HG, Ketzmar AK, Freymann U, Sittinger M, Notter M, Endres M, Kaps C: **Polyglycolic acid-hyaluronan scaffolds loaded with bone marrow-derived mesenchymal stem cells show chondrogenic differentiation in vitro and cartilage repair in the rabbit model.** *J Biomed Mater Res B Appl Biomater* 2013, **101:**1310–1320.
59. Akiyama H: **Control of chondrogenesis by the transcription factor Sox9.** *Mod Rheumatol* 2008, **18:**213–219.
60. Dominici M, Le Blanc K, Mueller I, Slaper-Cortenbach I, Marini F, Krause D, Deans R, Keating A, Prockop DJ, Horwitz E: **Minimal criteria for defining multipotent mesenchymal stromal cells. The international society for cellular therapy position statement.** *Cytotherapy* 2006, **8:**315–317.
61. Ning J, Li C, Li H, Chang J: **Bone marrow mesenchymal stem cells differentiate into urothelial cells and the implications for reconstructing urinary bladder mucosa.** *Cytotechnology* 2011, **63:**531–539.
62. Boxall SA, Jones E: **Markers for Characterization of Bone Marrow Multipotential Stromal Cells.** *Stem Cells Intern* 2012, **2012:**975871.
63. Asahara T, Murohara T, Sullivan A, Silver M, van der Zee R, Li T, Witzenbichler B, Schatteman G, Isner JM: **Isolation of putative progenitor endothelial cells for angiogenesis.** *Science* 1997, **275:**964–966.

Modular assembly of designer PUF proteins for specific post-transcriptional regulation of endogenous RNA

Zhanar Abil[1], Carl A Denard[2] and Huimin Zhao[1,2,3*]

Abstract

Background: Due to their modular repeat structure, Pumilio/fem-3 mRNA binding factor (PUF) proteins are promising candidates for designer RNA-binding protein (RBP) engineering. To further facilitate the application of the PUF domain for the sequence-specific RBP engineering, a rapid cloning approach is desirable that would allow efficient introduction of multiple key amino acid mutations in the protein. Here, we report the implementation of the Golden Gate cloning method for an efficient one-step assembly of a designer PUF domain for RNA specificity engineering.

Results: We created a repeat module library that is potentially capable of generating a PUF domain with any desired specificity. PUF domains with multiple repeat modifications for the recognition of altered RNA targets were obtained in a one-step assembly reaction, which was found to be highly efficient. The new PUF variants exhibited high *in vitro* binding efficiencies to cognate RNA sequences, corroborating the applicability of the modular approach for PUF engineering. To demonstrate the application of the PUF domain assembly method for RBP engineering, we fused the PUF domain to a post-transcriptional regulator and observed a sequence-specific reporter and endogenous gene repression in human cell lines.

Conclusions: The Golden Gate based cloning approach thus should allow greater flexibility and speed in implementing the PUF protein scaffold for engineering designer RBPs, and facilitate its use as a tool in basic and applied biology and medicine.

Keywords: Protein engineering, RNA-binding protein, Post-transcriptional regulation, PUF, Pumilio, Tristetraprolin, TTP, Golden Gate

Background

The presence of RNA-binding proteins (RBPs) in RNA biology is ubiquitous. Hundreds to thousands of eukaryotic proteins are estimated to function as RBPs [1] and govern many aspects of RNA biology including translation, turnover, processing, and cellular localization [2-4]. Despite their great diversity in function, only a few types of RNA-binding domains are known, which are combined in different structural arrangements with a variety of functional domains [5]. This modular architecture makes RBPs an attractive tool for studying the vast complexity of eukaryotic transcriptomes as well as manipulating RNA for therapeutic purposes [6,7].

The function of many RBPs can be studied [8-11] by tethering them to a reporter RNA through a well-characterized RNA-binding peptide with a fixed specificity [12]. However, this approach can only be applied to manipulate heterologous RNA because prior tagging of the RNA is required. In order to manipulate endogenous RNA in its native expression conditions, one could envision a designer RBP with an RNA-binding scaffold that could be easily engineered for sequence specificity. To date, only pentatricopeptide repeat [13,14] and Pumilio/fem-3 mRNA binding factor (PUF) [15-18] proteins have been demonstrated to have the potential to be rationally modified for predictable and specific RNA recognition.

* Correspondence: zhao5@illinois.edu
[1]Department of Biochemistry, University of Illinois at Urbana-Champaign, Urbana, IL 61801, USA
[2]Department of Chemical and Biomolecular Engineering, University of Illinois at Urbana-Champaign, Urbana, IL 61801, USA
Full list of author information is available at the end of the article

PUF proteins are eukaryotic RBPs that are involved in post-transcriptional gene regulation [19]. The crystal structure of Pumilio homology domain (PUM-HD), the RNA-binding region of the human Pumilio 1 (PUM1) protein (Figure 1a), reveals 8 structural repeats, each containing ~36 amino acids (aa), and flanking N-and C-terminal regions [20,21]. The structure also suggests that recognition of the target RNA sequence is highly modular since each repeat binds to a single RNA base [15]. The N-terminal repeat (R1) binds to the 3′-nucleotide residue (N8) of the target sequence (Figure 1a and b), while the C-terminal repeat (R8) binds to the 5′-nucleotide residue (N1). Residues at positions 12 and 16 in each repeat directly interact with a Watson-Crick edge of a base, whereas the residue at position 13 is involved in a stacking interaction between two adjacent bases [15]. The structure suggests a recognition "code", where residues at positions 12 and 16 in each repeat contribute to specific recognition of a base, with $N_{12}Q_{16}$ recognizing uracil, $C_{12}Q_{16}$ adenine, and $S_{12}E_{16}$ guanine [15]. The residue combination $S_{12}R_{16}$ was engineered to recognize cytosine [17,18]. By swapping the key residues at these positions, it was shown that designed PUF proteins with altered specificity could be engineered [16,22]. In the past several years, engineered PUF domains were successfully fused to different effector domains for polyadenylation of an endogenous gene or repression of a reporter gene in *Xenopus* [23], cleavage of a mitochondrial-encoded gene

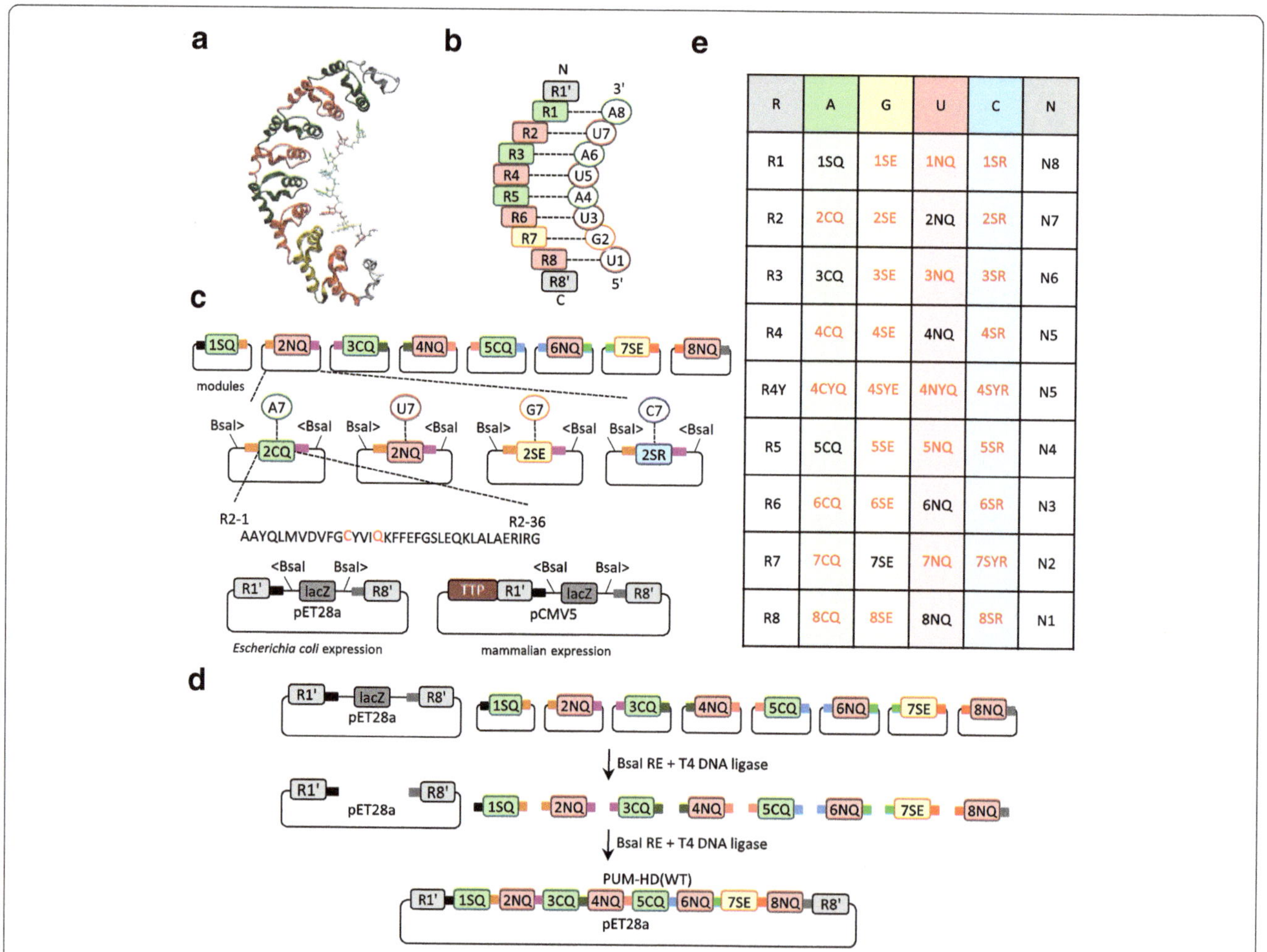

R	A	G	U	C	N
R1	1SQ	1SE	1NQ	1SR	N8
R2	2CQ	2SE	2NQ	2SR	N7
R3	3CQ	3SE	3NQ	3SR	N6
R4	4CQ	4SE	4NQ	4SR	N5
R4Y	4CYQ	4SYE	4NYQ	4SYR	N5
R5	5CQ	5SE	5NQ	5SR	N4
R6	6CQ	6SE	6NQ	6SR	N3
R7	7CQ	7SE	7NQ	7SYR	N2
R8	8CQ	8SE	8NQ	8SR	N1

Figure 1 The GG library and assembly schematic. (a) Crystal structure of PUM-HD bound to RNA, adapted from reference [15], GenBank ID code 1M8Y. **(b)** Schematic of PUM-HD bound to RNA. Filled boxes, PUF modules. Circles, RNA bases. **(c)** Schematic of the main library components: 8 repeat modules, with matching overhangs colored in identical colors; 4 variations of module 2, with corresponding recognition nucleotide indicated above; the aa sequence of module 2, with mutant aa indicated in red; and two receiving vectors. **(d)** The GG assembly schematic. A one-pot reaction that contains 8 modules of choice, a receiving vector, and enzymes allows the creation of 9 unique overhangs. The exact matching of the overhangs results in the predetermined repeat order assembled in the receiving vector. **(e)** Schematic of the GG library. R, module; N, nucleotide; recognized nucleotides indicated in the top row. First and last letters in the module names represent aa residues 12 and 16, in each module, respectively. Middle letter, if present, represents the "stacking" aa 13. Black font, WT modules. Red font, mutant modules. Green, yellow, pink, and blue fillings for modules recognizing A, G, U, and C, respectively.

in human cells [24], splicing regulation of endogenous *Bcl-X* pre-mRNA in human cells [25], and imaging endogenous RNA [26-28]. These advancements demonstrate the growing potential for the RBPs with various functional domains and engineered specificity.

However, PUF-based RBPs are still far from widespread implementation. One of the limitations in engineering PUF domains with novel specificities is the lack of a cloning platform capable of rapid and efficient introduction of multiple mutations in separate repeats simultaneously. In this study, we report the implementation of the Golden Gate (GG) cloning, Type IIS restriction endonuclease-based approach [29] for engineering of PUF-based RBPs. To demonstrate the efficiency of this approach, we first used this cloning method for construction of several mutant PUF domains with novel specificities and assayed their binding affinities. Second, we linked the RNA-binding activity of engineered PUF domains to the translational repression activity of tristetraprolin (TTP), and used the fusion protein in a functional reporter system to assay the PUF domain activity in HeLa cells. Finally, we showed the application of the engineered TTP-PUF fusion proteins for the post-transcriptional regulation of an endogenous gene in HEK293 cells.

Results and discussion

Efficient assembly of a custom-designed PUF domain

The GG cloning method, which is implemented here for the assembly of custom PUF domains, is based on the ability of Type IIS restriction enzymes to cleave outside of their non-palindromic recognition sequence [29], thus creating overhangs unrelated to the recognition sequence. This polarity and flexibility in the overhang sequence allows for a seamless removal of the original restriction site as well as a ligation of multiple fragments in one step.

As a scaffold for the development of the assembly toolkit, we used the human PUM-HD, which consists of the amino acids 828-1176 of the full-length PUM1 [21]. Each of the 8 structural repeats of PUM-HD was cloned individually into a pNEB193-based "intermediate vector" and was used as a separate assembly module (Figure 1c). We designed all 8 modules as well as the *lacZα* gene in the "receiving vector" to be flanked by two BsaI sites in such a way that would allow creation of 9 unique overhangs (Figure 1c). In a one-pot reaction, the 8 modules and the receiving vector can be efficiently cut and religated in a predefined order (Figure 1d).

These 8 modules were further expanded into a library where each module has 4 variations for the recognition of any of the 4 nucleotides (nt), consistent with the PUF recognition "code" [15,17,18]. The variants of the same module position have the same overhangs and the same amino acids as the wild type (WT) module except at positions 12 and 16 (Figure 1c). The introduced mutations are uniform across modules (Figure 1e), except for module 7 for the recognition of cytosine, where the "stacking" residue was also substituted with tyrosine, in accordance with a previous report by Dong *et al.* [18]. Since module $3S_{12}R_{16}$ requires a tyrosine as a "stacking" residue in the adjacent module 4 for *in vivo* activity [18], four additional module 4 variants were created, where the "stacking" residue at position 13 was mutated from histidine to tyrosine.

Finally, we constructed two receiving vectors, pET28-GG-PUF for *Escherichia coli* expression and pCMV-TTP (C147R)-GG-PUF for mammalian expression. They both contain a *lacZα* gene for easy identification of assembled clones using blue-white screening, as well as the flanking N- and C-terminal regions of PUM-HD, which we refer to as R1′ and R8′, respectively. The entire library hence consists of 36 intermediate vectors and 2 receiving vectors. Thus, our library of PUF repeat modules is potentially capable of a one-step assembly of PUF domains with specificity for any RNA sequence of 8 nt, given that they can be expressed in a soluble manner.

In order to test the efficiency of the method, we reassembled the original PUM-HD from the WT modules into both receiving vectors. We analyzed 10 randomly chosen clones from each assembly by restriction digestion and found that all produced the anticipated digestion pattern (Additional file 1: Figure S1a and b). Next, we sequenced one clone from each assembly and found that both sequences were correct, indicating that the assembly process is highly efficient regardless of a receiving vector. Assembly of the consecutive mutant PUF domains was as efficient as the assembly of the WT PUF domain (data not shown). The entire process takes 3 days, with the GG reaction and *E. coli* transformation on day 1, colony picking on day 2, and plasmid purification and digestion-confirmation of the clones on day 3. The procedure is therefore ideal for the rapid introduction of multiple mutations in a PUF domain with high efficiency.

Binding activity of custom-designed PUF domains *in vitro*

With the combinatorial assembly tool in hand, we set to determine if increasing the number of mutations affects the activity or specificity of a PUF domain. We assembled four variant PUF domains with 2, 4, 6, and 8 mutant modules that were named PUF (S2), PUF (S4), PUF (S6), and PUF (S8), respectively (Table 1). To test their *in vitro* activity, we assayed the binding affinity of the proteins to WT and their cognate RNA sequences using a fluorescence polarization assay (Additional file 1: Figure S2b-f). We found that the WT PUF as well as the PUF variants all bound to their cognate RNA sequences with high affinity (Table 1). As predicted, all

Table 1 Mutations, cognate and noncognate RNA oligonucleotide sequences, and binding affinities of PUF (WT), and PUFs S2-S8

Protein	Protein modules	Cognate RNA sequence	K_D, nM	Noncognate RNA sequence	K_D, nM
	Ct 8 – 7 – 6 – 5 – 4 – 3 – 2 – 1 Nt	5′ 1 2 3 4 5 6 7 8 3′		5′ 1 2 3 4 5 6 7 8 3′	
PUF(WT)	NQ-SE-NQ-CQ-NQ-CQ-NQ-SQ	CCAGAAU/UGUAUAUA/UUCG	0.043±0.023	CCAGAAU/**A**GUAUAU**U**/AUUCG	3.56±0.90
PUF (S2)	**CQ**-SE-NQ-CQ-NQ-CQ-NQ-**NQ**	CCAGAAU/**A**GUAUAU**U**/AUUCG	0.76±0.11	CCAGAAU/UGUAUAUA/UUCG	6.29±2.80
PUF (S4)	NQ-SE-**CQ**-**NQ**-**CYQ**-**NQ**-NQ-SQ	CCAGAAU/UG**AUAU**UA/UUCG	0.59±0.10	CCAGAAU/UGUAUAUA/UUCG	5.77±1.33
PUF (S6)	**CQ**-SE-**CQ**-**NQ**-**CYQ**-**NQ**-NQ-**NQ**	CCAGAAU/**A**G**AUAU**U**U**/UUCG	6.05±0.25	CCAGAAU/UGUAUAUA/UUCG	89.6±16.5
PUF (S8)	**CQ-NQ-CQ-NQ-CYQ-NQ-SE-NQ**	CCAGAAU/**AUAUAUGU**/UUCG	2.79±0.69	CCAGAAU/UGUAUAUA/UUCG	ND

Bold, mutant modules and changed RNA bases compared with the WT target sequence. Target RNA sequences are delimitated by forward slashes. Ct, C-terminus, and Nt, N-terminus of the protein. K_D values were determined using nonlinear curve fitting and represent the mean ± SD (n = 3). *ND*, K_D not determined due to little binding at 300 nM PUF (S8) to WT RNA.

of the PUF proteins exhibited lower affinity to noncognate RNA, which contained between 2 and 8 nucleotide mismatches with the cognate RNA (Table 1, Additional file 1: Figure S2b-f). The binding affinities to cognate sequences decreased from PUF (WT) to PUF (S8), indicating that there is a weak negative correlation between the number of mutations and binding affinity to cognate RNA, although the correlation is not strictly linear (Table 1). However, even PUF (S8), which has 8/8 mutant modules, binds specifically and with high affinity to its cognate RNA. These results further corroborate the study by Cheong and Hall, who demonstrated the specificity and modularity of PUF protein target recognition [16].

Engineering and assessment of designer repressor-RBP, TPUF

As an example of PUF-based RBP engineering, we fused the PUM-HD domain to TTP, a well-studied post-transcriptional regulator. TTP binds to AU-rich elements (AREs) in the 3′-untranslated region (UTR) of the target genes and promotes mRNA degradation by recruiting a deadenylase complex [30]. TTP is also known to promote ARE-dependent gene knockdown via translation repression through cooperation with a general translation inhibitor RCK/P54 [31], though the details of this mechanism remain unknown. We reasoned that the RNA-binding activity of TTP was undesirable due to possible interference of TTP towards efficient binding of PUF to PUF-binding sites (PBSs). We therefore introduced the C147R mutation that was shown to abolish the binding of TTP to ARE [32]. TTP (C147R) was fused to the N-terminus of PUM-HD through a $(G_4S)_3G_4$ linker and expressed in HeLa cells for functional analysis (Additional file 1: Figure S3a).

To assay the gene repression activity of TTP-PUF fusion constructs, a dual luciferase assay was implemented. To the 3′ UTR of the firefly luciferase (FL) reporter, 10 PBSs separated by 6-18 nt were cloned (Additional file 1: Figure S3 b). *Renilla* luciferase (RL) lacking any PBSs was co-transfected with FL as a transfection control. FL_{Random}, also lacking any PBSs, was used as a repression control. Values of FL_{PBS}/RL normalized to FL_{Random}/RL were reported as "relative FL/RL activity." We observed that TTP (C147R) alone did not significantly repress the $FL_{PBS\ (WT)}$ activity, whereas PUM-HD (WT) alone repressed $FL_{PBS\ (WT)}$ by 20%. The observed weak activity of PUM-HD RNA-binding motif alone can be explained by previous findings that the Pumilio RNA-binding domain is also a translational regulator that is capable of recruiting deadenylases to the concave surface of repeats 7 and 8 [33]. However, the TTP (C147R)-PUM-HD (WT) fusion construct (hereafter referred to as TPUF (WT)) repressed the $FL_{PBS\ (WT)}$ activity by 80% (Figure 2a), thus demonstrating that the fusion construct exhibits both specific RNA binding and high repression activity.

We also tested the repression levels of FL containing 1, 3, and 5 PBSs in the 3′ UTR (Additional file 1: Figure S4) and observed repression activity of TPUF (WT) ranging between 31-55%. However, to obtain the greatest dynamic range of our reporter assay, we conducted consequent experiments using 10 PBSs. Although using 10 PBSs is in contrast compared with those using 1 PBS in previous assays [23], it is comparable to using 5-6 binding sites in tethering assays of TTP and other ARE-mediated decay activation domains [31,34,35].

To determine if the TPUF (WT) construct functions by promoting degradation of target RNA, we performed real-time PCR (RT-PCR) analysis on the FL reporter, and used RL as the internal control. We did not observe decrease in RNA levels in $FL_{PBS\ (WT)}$ compared with FL_{Random} in the presence of TPUF (WT) (Additional file 1: Figure S5). Several FL RT-PCR primer pairs were used, and consistently no RNA destabilization was observed (data not shown). These results are in accordance with a similar TTP tethering assay [31], where luciferase activity was knocked down despite little RNA destabilization. We suggest that the TPUF constructs function similarly, by promoting translational repression rather than RNA degradation [31].

To further test the design concept, we used our GG toolkit to assemble two previously reported PUF variants

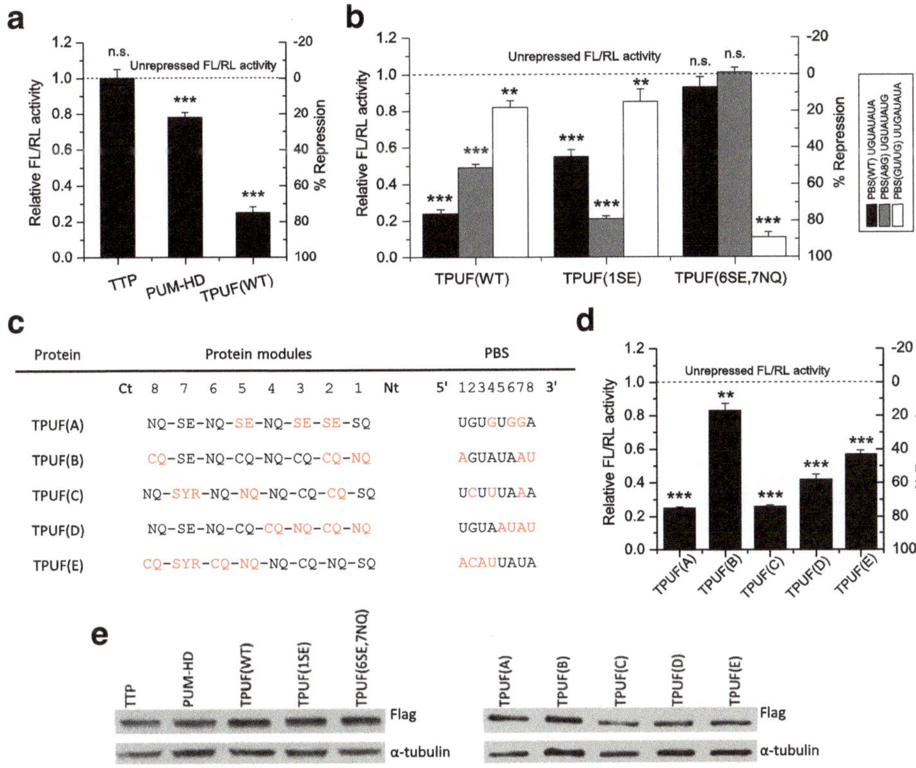

Figure 2 Various TPUF repression activity assessment in HeLa cell line. (a) Dual luciferase assay shows that TPUF (WT), a fusion of TTP (C147R) and PUM-HD (WT), exhibits the greatest down-regulation activity on $FL_{PBS\ (WT)}$ expression, compared with TTP and PUM-HD (WT) alone. Data represented as mean fold change relative to cells transfected with FL_{Random} (dashed line, unrepressed FL/RL activity) ± SD: n.s., not significant, ***P ≤ 0.001 (n = 3, t test). **(b)** Luciferase assay shows predicted specificity of previously reported PUF mutants [16]. TPUF (WT) prefers PBS (WT), TPUF (1SE) prefers PBS (A8G), and TPUF (6SE,7NQ) prefers PBS (GU/UG). Data represent means ± SD: n.s., not significant, **P ≤ 0.01, ***P ≤ 0.001 (n = 3, t test). **(c)** Mutations and PBSs of TPUFs A-E with 3-4 randomly chosen mutant modules. Black, WT PUF modules and corresponding RNA bases. Red, mutant PUF modules and corresponding RNA bases. Ct, C-terminus, Nt, N-terminus of the protein. **(d)** A graph of luciferase activity, where TPUFs A-E repress FLs with cognate PBSs. Data represent means ± SD: **P ≤ 0.01, ***P ≤ 0.001 (n = 3, t test). **(e)** Western blot of effector proteins using anti-Flag antibody shows no major difference in the expression. Anti-α–tubulin antibody was used as a loading control.

[16]. TPUF (1SE) has repeat 1 replaced for recognition of G8 [PBS (A8G)], and TPUF (6SE,7NQ) has repeats 6 and 7 replaced for the recognition of G and U at positions 2 and 3, respectively [PBS (GU/UG)]. TPUF (WT) and the two PUF variants exhibited highest repression activities towards their cognate PBSs, with repression levels of 76%, 79%, and 88%, respectively. Compared with cognate PBSs, we observed a diminished activity towards PBSs with 1 or 2 mismatches, though some cross-reactivity is evident (Figure 2b). These observations are consistent with the cross-reactivity between WT and PUF (1SE) in *in vitro* assays [16] and similar cross-reactivity between WT and other mutant PUF proteins [16,25] that differ by 1-2 repeats. Overall, luciferase repression by the TPUF constructs was sequence-specific, corroborating the validity of the TPUF-reporter system.

In order to further verify the functionality of the cloning method and the TPUF platform, we assembled more TPUF constructs with mutant repeats randomly introduced throughout the PUF domain (Figure 3c), denoted as TPUFs A-E. Out of 5 TPUF variants, only TPUF (B) showed low (17%) repression activity towards a cognate PBS, whereas TPUFs A, C-E showed repression activities ranging from 43% to 75% (Figure 3d). TPUFs with all 8 replaced modules demonstrated poor repression activities (data not shown), indicating that accumulation of mutations in the PUF domain does not always result in active TPUF proteins *in vivo*.

The difference between TPUF activities in a cell line could be dependent on the expression and solubility levels of the fusion proteins. We therefore investigated soluble expression levels of all of the effector proteins mentioned above using Western blotting (Figure 2e). We found that TTP, PUM-HD, and TPUF (WT), as well as mutants TPUF (1SE) and TPUF (6SE,7NQ) were expressed at similar levels, thus excluding the possibility of protein abundance variability distorting the observed repression activities. On the other hand, we noticed some noticeable variability among soluble expression levels of TPUFs (A-E). However, this variability does not

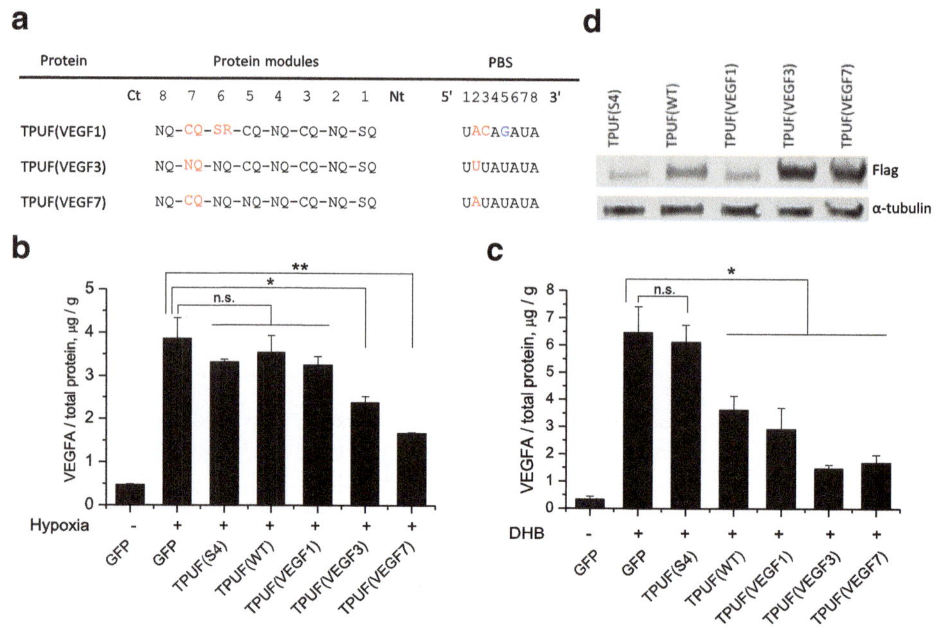

Figure 3 TPUF represses endogenously expressed VEGFA gene in HEK293 cell line. (a) Mutations and binding sequences of TPUFs designed for VEGFA 3′ UTR recognition. Black, WT modules and corresponding RNA bases. Red, mutant modules and corresponding RNA bases. Blue, a mismatch in the recognition sequence. Ct, C-terminus, Nt, N-terminus of the protein. **(b)** The graph demonstrates inhibition of hypoxia-induced VEGFA expression in cells transfected with engineered TPUFs VEGF3 and VEGF7. In hypoxic (+) cultures, VEGFA expression was induced with 500 μM $CoCl_2$ 24 hours after transfection and then cultivated for 24 hours. Secreted VEGFA levels measured by ELISA were normalized to total protein amounts from lysed cells measured by Bradford Assay. Data represented as mean ± SD: n.s., not significant, *$P \le 0.05$, **$P \le 0.01$, (n = 3, t test). **(c)** The graph demonstrates inhibition of DHB-induced VEGFA expression in cells transfected with TPUFs WT, VEGF1, VEGF3, and VEGF7. HEK293 cells with the integrated V24P-GS60 transcriptional activator of endogenous VEGFA promoter were treated with 100 nM DHB 24 hours after TPUF transfection and then cultivated for 24 hours. ELISA and normalization to total protein amounts as in the previous panel. Data represented as mean ± SD: n.s., not significant, *$P \le 0.05$. **(d)** Western blot of effector proteins using anti-Flag antibody shows greater expression of TPUFs VEGF3 and VEGF7 compared with TPUF (WT) and other mutants. Anti-α-tubulin antibody was used as a loading control.

seem to be consistent with the observed TPUF activities. In particular, it does not explain the apparent lower activities of TPUF (B) and TPUF (E).

We conclude that *in vivo* TPUF activities must be dependent on various other factors in addition to protein stability. For example, unequal contributions of different repeat-base interactions to the binding energy of the RNA-protein complexes [16,36] may result in different overall binding affinities to cognate RNA. In addition, binding to noncognate RNA [37] could allow sequestration of the protein to non-target RNA *in vivo*. Finally, sporadic domain interactions in fusion proteins could interfere with RNA binding. We cannot yet predict the contribution of each of these factors on PUF activity *in vivo*, and therefore suggest that the effects of these and other factors on engineered PUF activity have to be systematically investigated.

Endogenous gene regulation using TPUF

As a proof of concept for implementing designer RBP for endogenous gene regulation, we designed TPUFs that bind to the 3′ UTR of human vascular endothelial growth factor-A (*VEGFA*) mRNA. VEGFA is one of the central mediators of angiogenesis, and was shown to be overexpressed in many human tumors. VEGFA is up-regulated in these tumors under hypoxic growth conditions that many tumors create [38]. As such, VEGFA is an attractive target for the development of therapeutics to inhibit pathological angiogenesis. We reasoned that for our TPUF repression assay, elevated VEGFA levels are more therapeutically relevant than physiological lower levels, and therefore employed two different strategies to up-regulate endogenous VEGFA expression. In the first strategy, we incubated HEK293 cells with μM 500 $CoCl_2$, an agent that causes conditions mimicking hypoxia [39], and achieved an 8-fold VEGFA induction compared to cells grown in the absence of the agent (Figure 3b). In the second strategy, we used a HEK293 cell line in which a small molecule-responsive gene switch for VEGFA expression was stably integrated [40]. Upon induction with 4,4′-dihydroxybenzyl (DHB), endogenous VEGFA was up-regulated 19-fold.

For the post-transcriptional down-regulation of VEGFA, we assembled TPUFs VEGF1, VEGF3, and VEGF7 (Figure 3a), which have, respectively, 1, 3, and 7 PBSs in the 3′ UTR of all known transcript variants of

VEGFA gene. PUF domains in the TPUF VEGF3 and VEGF7 constructs each carry one mutant repeat and were previously reported to be active *in vitro* [16]. TPUF (VEGF1) has repeats 6 and 7 replaced for the recognition of 2A and 3C in the PBS. WT repeat $4N_{12}Q_{16}$ was left unchanged for the recognition of 5G (Figure 3a, blue), since this repeat is known to be promiscuous [41]. TPUF (S4) was shown to be active for its cognate RNA using the luciferase assay (Additional file 1: Figure S6); however, it does not have binding sites in the 3′UTR of *VEGFA* mRNA, and therefore was not expected to repress the gene. Fortuitously, PBS (WT) is present 3 times in the 3′ UTR of *VEGFA* mRNA, which allowed us to use TPUF (WT) as a reference for assaying activities of mutant TPUFs.

To assay the repression activity, we transiently transfected recombinant HEK293 cells with equimolar amounts of GFP or one of the TPUF constructs, induced the expression of VEGFA 24 hours after transfection, and assayed secreted levels of VEGFA by ELISA 24 hours after induction. VEGFA levels were normalized to total protein concentration, which we assumed to correlate with cell number. Bradford Assay was used to measure the total protein concentrations from cell lysates. We found that VEGFA amounted to less than 1% of total protein concentration by mass (Figure 3b and c), hence the variations of VEGFA concentrations in the cell would have no significant effect on total protein concentration. In both induction strategies, the VEGFA levels were not significantly affected by TPUF (S4) compared to samples transfected with GFP, as expected (Figure 3b and c). In cells treated with $CoCl_2$, only TPUF (VEGF3) and TPUF (VEGF7) significantly repressed VEGFA expression, which was down-regulated by 38% and 57%, respectively (Figure 3b). On the other hand, in cells induced with DHB, TPUF (WT) caused considerable (44%) repression of VEGFA, whereas TPUFs VEGF1, VEGF3, and VEGF7 knocked down VEGFA levels by 55%, 77%, and 74%, respectively (Figure 3c). We are currently unaware of the reason why cells exposed to hypoxia were more resistant to VEGFA repression by our TPUF constructs compared with cells that expressed the gene switch. We speculate that it could be due to yet undiscovered transcriptional and/or translational gene regulatory response of cells to hypoxia. Increasing the TPUF expression time to 41 hours and decreasing exposure to hypoxia to 7 hours resulted in similar repression levels as in the previous conditions (data not shown), and therefore we exclude the possibility that this resistance is simply due to lower expression of TPUFs at hypoxia. The lower TPUF activity at hypoxia might hence be due to an unknown interference with TTP function. Nevertheless, we observed a considerable sequence-specific down-regulation of VEGFA in cells expressing the VEGFA transcriptional gene switch, and confirmed the efficacy of the TPUF platform in the down-regulation of an endogenous gene in human cells as yet another demonstration of the applicability of PUF-based RBPs at post-transcriptional gene regulation [23,24].

To compare the expression levels of TPUF constructs among each other, we performed a Western blot analysis of the effector proteins under normoxia (Figure 3d). We observed a substantially greater soluble expression of TPUF (VEGF3) and TPUF (VEGF7) which might be the main reason of these constructs' higher overall activity.

These VEGFA repression levels are comparable to those in similar assays conducted with zinc finger transcriptional repressors or small interfering RNA (siRNA). For example, hypoxia-induced VEGFA protein levels in HEK293 cells were knocked down by 74% by an engineered zinc finger transcriptional repressor [42] and VEGFA mRNA was knocked down by 50% by another zinc finger repressor in HEK293 cells grown in normoxia [43]. On the other hand, siRNA knocked down endogenous VEGFA by up to 43% in ID8 cells [44] and up to 71% in hypoxic HEK293 cells [42]. Thus, the TPUF system that we have engineered is as effective as some other existing technologies that are often used to knock down gene expression levels, and can be a powerful alternative.

Based on these findings, we conclude that the implementation of TPUFs for down-regulation of other endogenous genes with high efficacy is possible. However, factors determining tight binding of Pumilio domains to RNA *in vitro*, as well as factors determining high *in vivo* activities are still largely unknown, and a systematic investigation is needed. For example, based on the results reported above, either effective target site recognition or protein stability may play a major role in a protein's functional efficacy *in vivo* in any given mutant. It is difficult, at this stage, to predict which contribution would predominate. *In vitro* RNA binding affinities and *in vivo* activities show only a rough correlation [23,25], and soluble expression levels do not always predict activity differences (Figure 2d and e). Therefore, for engineering of TPUFs with novel specificities or PUM-HD-based proteins with novel functionalities, we recommend building a reporter system suitable for the given protein activity, and directly screening the functional efficacy of the assembled PUF variants.

Conclusions

In this study, we established a toolkit for rapid engineering of designer RBPs that can be used for manipulation of endogenous genes. This approach should allow greater flexibility and speed at creating PUF domains with user-defined specificities and thus facilitate the use of PUF-based designer RBPs as a tool in research and therapeutics. We demonstrated that PUF mutants with as many as 8/8 mutant modules can be cloned with high efficiency and that the resulting proteins retain their specificity and high affinity to their cognate RNA *in vitro*. This result is consistent with the study by Cheong and Hall [16] and confirms the RNA recognition code and modularity of PUF domains. Finally, we were able to demonstrate the implementation of the protein-based post-transcriptional repressor for down-regulation of an endogenous gene. The TPUF platform introduced in this work exhibits modular and sequence-specific recognition and down-regulation of genes. We envision the future development of PUF-based RBPs with various functionalities that could range from endogenous RNA splicing, imaging, and localization to various base modifications and more. The development of rapid assembly tools for PUF specificity engineering, as presented in this work, could play a critical role in facilitating and enhancing these endeavors.

Methods

Materials

All the chemicals and solutions were purchased from Fisher Scientific (Pittsburgh, PA), unless noted otherwise. Oligonucleotides were purchased from Integrated DNA Technologies (Coralville, IA). All the enzymes were purchased from New England Biolabs (Ipswich, MA), unless noted otherwise.

Library creation

The amino acid and DNA sequences of all the modules in our GG cloning library are listed in Additional file 1: Table S1. All the amplification primers used for the creation of the library are listed in Additional file 1: Table S2. The intermediate vector, pChlr-pNEB193, was created by replacing the original *amp* resistance gene in pNEB193 (New England Biolabs) plasmid with the *cam* resistance gene from pACYC (New England Biolabs) plasmid by Gibson Assembly (GA) [45]. Plasmid pTYB3-PUM1-HD [GenBank:D43951] was a gift of Dr. Traci M. Tanaka Hall (Addgene plasmid 17543). The WT GG assembly modules were amplified from pTYB3-PUM1-HD and inserted in the SacI and HindIII sites of the intermediate vector. The amplification primers also contained BsaI sites for subsequent GG cloning. Some of the modules' 5′ and 3′ ends were modified with silent mutations (Additional file 1: Table S1) for the creation of non-overlapping BsaI overhangs. The mutant GG modules were created by GA from the corresponding WT modules.

The receiving vector pET28-GG-PUF was GA-cloned from the following fragments: pET28a (Novagen) digested with NdeI and SalI, flanking PUM-HD repeats R1′ and R8′ amplified from pTYB3-PUM1-HD, and *lacZα* amplified from pNEB193. BsaI sites were introduced next to R1′ and R8′ for GG cloning of PUM-HD. The receiving vector pCMV-TTP (C147R)-GG-PUF was cloned in two steps. First, pCMV-TTP-GG-PUF was created by replacing PUM-HD with a *lacZα* gene flanked by BsaI sites and removal of 3 existing BsaI sites in the pCMV-TTP (WT)-PUM-HD (see the effector plasmids section). Next, pCMV-TTP (C147R)-GG-PUF was GA-cloned by replacing Flag with 3xFlag and mutating the C147R of TTP in the plasmid pCMV-TTP-GG-PUF.

For the availability to the scientific community, we are making arrangements to deposit all the plasmids constituting the PUM-HD repeat library that we have developed here to the Addgene non-profit plasmid repository. These plasmids should be available to researchers within a month of the publication of this manuscript.

Reporter plasmids

pCMV-Fluc plasmid was created by amplification of the firefly luciferase gene from pGL3 plasmid (Promega) and insertion into SacI and KpnI sites of pCMV5 vector (a gift of Dr. David Russell). All the pCMV-Fluc-10xPBS plasmids, as well as pCMV-Fluc-Random were cloned by primer-extension of 6 primers (Additional file 1: Table S3) carrying 10 PBSs and subsequent GA-cloning into PstI and XmaI sites of the pCMV-Fluc plasmid. The 6-18 nt spacers between the 10 PBS in the 3′ UTR of the FL were the same in different FL-PBS sequences, and were designed in such a way to minimize secondary structure formation that would involve these spacers. The pCMV-Fluc-Random was created by replacing all the PBSs in the pCMV-Fluc-10xPBS with 10 different scrambled sequences of 8 nt with approximately 50% GC content. Plasmid pRL-SV40 was a gift of Dr. David J. Shapiro.

Effector plasmids

The effector plasmid pCMV-TTP (WT)-PUM-HD was GA-cloned from the following fragments: 2.2 kb and 2.4 kb pCMV5 fragments, GS-PUM-HD amplified from pTYB3-PUM1-HD, and TTP-GS amplified from cDNA (Open Biosystems catalog number MHS4768-99609440 [GenBank: BC009693.1]). pCMV-TTP (WT) and pCMV-PUM-HD have been assembled from the same vector backbone fragments, as well as TTP-stop or Flag-PUM-HD fragments, respectively (for primers, see Additional file 1: Table S4).

Golden gate assembly of mutant effector plasmids

His-tagged PUF or 3xFlag-tagged TPUF constructs for *E. coli* or mammalian expression were assembled in pET28-GG-PUF or pCMV-TTP (C147R)-GG-PUF, respectively. Receiving vector of choice (50 ng) and 8 modules of choice (75 ng each) were combined with 1 μl T4 DNA ligase and 1 μl BsaI-HF in 10 μl 1× T4 DNA ligase buffer. The reactions were cycled 10 times for 5 min at 37°C and 10 min at 16°C, and a final incubation of 15 min at 37°C. TOP10 *E. coli* cells (Invitrogen) were then transformed with the cloning reactions and plated on LB plates (Cell Media Facility, UIUC) with either *kan* or *amp* selection, and supplemented with 10 μl 0.4 M IPTG (GoldBio) and 40 μl 20 mg/ml Bluo-Gal (Invitrogen) for blue-white screening. All the plasmids for *E. coli* expression were purified using Qiagen Qiaprep Spin Miniprep kit, and plasmids for mammalian expression were purified using Qiagen Plasmid Mini kit.

Protein expression and purification

His-tagged recombinant PUF proteins were expressed in *E. coli* strain BL21 (DE3) (Novagen). The transformed BL21 cultures were grown in 100 ml LB until they reached an OD_{600} of 0.8, induced with 0.4 mM IPTG and expressed at 18°C, 250 RPM overnight. Bacterial pellets were resuspended in lysis buffer (25 mM Tris-HCl pH 7.5, 0.3 M NaCl, 0.5% Triton (Bio-Rad), 5% glycerol (Sigma), 1 mg/ml lysozyme (Sigma), and 0.002 U/μl DNase I) and lysed by sonication. The proteins were purified using Talon Spin Columns (Clontech), according to manufacturer's instructions. The eluted protein was flash-frozen in 25% glycerol in dry ice and stored in aliquots at -80°C.

Fluorescence polarization assay

RNA oligomers were modified with 6-carboxyfluorescein (IDT) at the 5′-end. To determine active protein fractions, we performed saturation assays for PUF proteins against their cognate RNA (a representative saturation curve is shown in Additional file 1: Figure S2a). High concentrations (100 nM) of RNA oligomers in fluorescence anisotropy buffer (20 mM Tris-HCl pH 7.5, 0.5 mM EDTA, 50 mM KCl, 0.1 mg/ml BSA) were mixed with various protein concentrations (determined by Bradford assay), and 200 μl protein-RNA mixtures were assayed (for fluorescence polarization measurements, see below) in black 96-well plates (Corning). The stoichiometric points were used to estimate the active protein fractions, which were determined to be 31% for PUF (WT), 30% for PUF (S2), 30% for PUF (S4), 33% for PUF (S6), and 29% for PUF (S6). Corrected active protein concentrations were used in the subsequent binding curves for the determination of the dissociation constants K_D, where RNA oligomers (250 pM RNA for PUF (WT) and 1 nM RNA for PUF (S2)-PUF (S8)) in the fluorescence anisotropy buffer were mixed with various protein concentrations, and duplicates of 200 μl protein-RNA mixtures were assayed.

Fluorescence polarization measurements were taken on Tecan Infinite 200Pro using excitation and emission wavelengths of 485 nm and 535 nm, respectively. The fluorescence polarization values were converted to fluorescence anisotropy values using Equation 1, where A is anisotropy and P is polarization. The K_D was calculated by curve fitting on Origin 8.5 using Equation 2, where A is observed anisotropy value, A_f is anisotropy of free RNA, A_b is anisotropy of bound RNA, L_T is total ligand (RNA) concentration, and R_T is total receptor (protein) concentration.

$$A = \frac{2P}{3-P} \tag{1}$$

$$A = A_f + (A_b - A_f) * \frac{(L_T + K_D + R_T) - \sqrt{(L_T + K_D + R_T)^2 - 4L_T R_T}}{2L_T} \tag{2}$$

Cell line transfection and dual luciferase assay

Transfection of HeLa cells (ATCC) was performed in triplicates in a 24-well plate format with Fugene-HD transfection reagent (Promega). Transfection mixtures contained 150 ng FL, 2 ng pRL-SV40, and 75 ng TPUF or equimolar amounts of other effector DNA constructs, and empty vector pCMV5 to 500 ng total. Cells were lysed in Passive Lysis Buffer (Promega) 48 h after transfection and FL and RL activities were measured in white 96-well plates (Greiner Bio One) using Dual-Glo Luciferase Assay System (Promega) with measurements taken on Analyst HT microplate reader at the High-Throughput Screening Facility at UIUC.

RT-PCR

Total RNA was isolated from HeLa cells 48 hours after transfection using the RNeasy Mini Kit (Qiagen) following manufacturer's instructions, and DNA was removed from samples with Turbo DNase (Life Technologies). RNA was reverse transcribed into cDNA with ProtoScript First Strand cDNA Synthesis kit (NEB) using the $d(T)_{23}VN$ primer. Reverse transcriptase was omitted in control samples. RT-PCR was performed using Power SYBR Green Master Mix (Life Technologies) with the 7900HT Fast Real-Time PCR System (Applied Biosystems). Reactions were carried out in triplicates in 20 μl reactions with 500 nM of each primer. The primer sequences for FL were 5′-GCGCGGAGGAGTTGTGTT TG and 5′-ATCTTTCCGCCCTTCTTGGC; and for RL 5′-GCAGCATATCTTG AACCATTC and 5′-TTGTA CAACGTCAGGTTTACC. ΔΔCT method was used for RNA level analysis, where FL mRNA levels were normalized to RL mRNA, and $FL_{PBS\ (WT)}$ mRNA levels were normalized to FL_{Random}.

VEGF induction and ELISA assay

For hypoxia-induced VEGFA, HEK293 cells were transfected in a 24-well plate format in triplicates with Fugene-HD. Transfection mixtures contained 500 ng TPUF DNA constructs or 350 ng pmaxGFP (Lonza) and

150 ng pCMV5. The cells were induced 24 h after transfection with 500 μM $CoCl_2$, and the supernatant was collected for assay 24 hours after induction. For gene-switch-induced VEGFA, HEK293 cell line with retrovirally integrated DHB-inducible V24P-GS60 transcription activator was used [40]. The cells were transfected in a 24-well plate format in triplicates as above. The cells were induced 24 h after transfection with 100 nM DHB, in the presence of *pen/strep* (Gibco). The supernatant was collected 24 h after induction and subjected to ELISA. The assay was performed by pre-coating the 96-well clear plate with a goat anti-mouse antibody (Thermo Scientific) at 4°C overnight, and then following the instructions of human VEGF DuoSet kit (R & D Systems). The absorption readings were taken on a SpectraMax 340PC microplate reader. The cell monolayer was saved for Bradford assay.

Bradford assay

The cell monolayers were lysed using RIPA lysis buffer. The protein concentrations were measured in technical duplicates by mixing 4 μl of cell lysate with 295 μl of Coomassie Plus Protein Assay Reagent (Thermo Scientific) in a 96-well clear plate. Quick Start Bovine Serum Albumine Standard Set (Bio-Rad) was used to build a protein standard curve. A_{595} was measured 5 min later using a SpectraMax 340PC microplate reader. The total protein concentrations measured by Bradford assay were used to normalize the VEGF concentrations.

Western blotting

V24P-GS60-integrated 293 cells as well as HeLa cells were transfected in a 6-well plate format with Fugene-HD and 3 μg of effector plasmid. Cells were lysed using RIPA lysis buffer (Santa Cruz Biotech). The proteins were detected using mouse anti-Flag and anti-α-tubulin antibodies (GeneScript) and imaged using SuperSignal West Dura chemiluminescent substrate (Thermo Scientific).

Additional file

Additional file 1: Figure S1. Confirmation through restriction enzyme digestion and gel-electrophoresis of GG assembled plasmids from randomly picked clones. First and last lanes, 1 kb DNA ladder (NEB). (a) KpnI and HindIII digestion of PUF (WT) clones assembled into pET28-GG-PUF receiving vector. 1 kb fragment contains the full length of the assembled PUF domain. (b) SalI and KpnI digestion of PUF (WT) clones assembled into pCMV-TTP-GG-PUF receiving vector. 1 kb fragment contains the assembled PUF domain region. **Figure S2**. Representative fluorescence anisotropy data for RNA binding to various PUF proteins. (a) Representative saturation curve of PUF (S4). (b) Binding curves of PUF (WT) (c) Binding curves of PUF (S2) (d) Binding curves of PUF (S4) (e) Binding curves of PUF (S6) (f) Binding curves of PUF (S8). Black, binding to cognate RNA. Red, binding to noncognate RNA. Each data point is represented by the mean ± SD. K_D values were calculated from nonlinear curve fitting. **Figure S3**. Schematics of the luciferase reporter assay and TPUF platform. (a) Schematic of full-length PUM1, TTP (WT), and TPUF constructs. CCCH, zinc finger domain; GS_L, glycine-serine linker. (b) Schematic of luciferase reporters. Orange boxes, PUF-binding sites. **Figure S4**. Dual luciferase assay showing TPUF (WT) repression of FL with increasing number of PBSs in the 3' UTR of the reporter gene. Data represented as mean fold change relative to cells transfected with FL with no PBS ± SD: **P ≤ 0.01 (n=3, t test). **Figure S5**. Relative levels of FL/RL mRNA, normalized to FL_{Ran}/RL mRNA in the presence of effectors. Fluorescence RT-PCR data were analyzed by $\Delta\Delta C_T$ method. Data represented as mean fold change relative to cells transfected with FL $_{Random}$ (dashed line, unrepressed level) ± SD: n.s., not significant (n=3, t test). **Figure S6**. Dual luciferase assay showing FL reporter repression activity of TPUF (S4). Data represented as mean fold change relative to cells transfected with FL $_{Random}$ ± SD: ***P ≤ 0.001 (n=3, t test). **Table S1**. GG library sequences (a) Aa sequences of WT and mutant modules. Black, WT aa. Red, mutant aa. (b) DNA sequences of WT and mutant modules. Black, WT sequence. Red, mutant nucleotides. **Table S2**. Primer list for GG library creation. **Table S3**. Primer list for FL cloning. **Table S4**. Primer list for effector plasmid cloning.

Abbreviations

PUF: Pumilio/fem-3 mRNA binding factor; RBP: RNA-binding protein; PUM-HD: Pumilio homology domain; PUM1: Pumilio 1; GG: Golden Gate; TTP: tristetraprolin; ARE: AU-rich element; PBS: PUF-binding site; TPUF: TTP (C147R)-PUM-HD; FL: firefly luciferase; RL: *Renilla* luciferase; VEGFA: vascular endothelial growth factor-A; DHB: 4,4'-dihydroxybenzyl; GA: Gibson Assembly.

Competing interests

The authors declare that they have no competing interests.

Authors' contributions

ZA conceived and designed the study, carried out the experiments, analyzed and interpreted the data, and wrote the manuscript. CAD and HZ contributed to the experimental design, analysis and interpretation of the data. All authors have read and approved the final manuscript.

Acknowledgements

We thank Dr. Chen Zhang and Dr. Chengjian Mao (University of Illinois) for discussions, Jing Liang and Ryan E. Cobb for manuscript revision and editing, Ran Chao for automating PBS primer design, Stella Xinzi Wu for her help with cell culture, Dr. Traci M.T. Hall (NIEHS, NIH), Dr. David Shapiro (UIUC), and Dr. David Russel (UTSW) for plasmids. This work was supported by the Centennial Chair Professorship in the Department of Chemical and Biomolecular Engineering at the University of Illinois at Urbana-Champaign.

Author details

[1]Department of Biochemistry, University of Illinois at Urbana-Champaign, Urbana, IL 61801, USA. [2]Department of Chemical and Biomolecular Engineering, University of Illinois at Urbana-Champaign, Urbana, IL 61801, USA. [3]Department of Bioengineering, Department of Chemistry, Center for Biophysics and Computational Biology, and Institute for Genomic Biology, University of Illinois at Urbana-Champaign, Urbana, IL 61801, USA.

References

1. Glisovic T, Bachorik JL, Yong J, Dreyfuss G: **RNA-binding proteins and post-transcriptional gene regulation.** *FEBS Lett* 2008, **582**:1977–1986.
2. Pichon X, Wilson LA, Stoneley M, Bastide A, King HA, Somers J, Willis AE: **RNA binding protein/RNA element interactions and the control of translation.** *Curr Protein Pept Sci* 2012, **13**:294–304.
3. Hoskins AA, Moore MJ: **The spliceosome: a flexible, reversible macromolecular machine.** *Trends Biochem Sci* 2012, **37**:179–188.
4. Jansen RP, Niessing D: **Assembly of mRNA-protein complexes for directional mRNA transport in eukaryotes–an overview.** *Curr Protein Pept Sci* 2012, **13**:284–293.
5. Lunde BM, Moore C, Varani G: **RNA-binding proteins: modular design for efficient function.** *Nat Rev Mol Cell Biol* 2007, **8**:479–490.
6. Filipovska A, Rackham O: **Designer RNA-binding proteins: New tools for manipulating the transcriptome.** *RNA Biol* 2011, **8**:978–983.

7. Mackay JP, Font J, Segal DJ: **The prospects for designer single-stranded RNA-binding proteins.** *Nat Struct Mol Biol* 2011, **18:**256–261.
8. De Gregorio E, Preiss T, Hentze MW: **Translation driven by an eIF4G core domain in vivo.** *EMBO J* 1999, **18:**4865–4874.
9. Coller JM, Gray NK, Wickens MP: **mRNA stabilization by poly (A) binding protein is independent of poly (A) and requires translation.** *Genes Dev* 1998, **12:**3226–3235.
10. Graveley BR, Maniatis T: **Arginine/serine-rich domains of SR proteins can function as activators of pre-mRNA splicing.** *Mol Cell* 1998, **1:**765–771.
11. Long RM, Gu W, Lorimer E, Singer RH, Chartrand P: **She2p is a novel RNA-binding protein that recruits the Myo4p-She3p complex to ASH1 mRNA.** *EMBO J* 2000, **19:**6592–6601.
12. Coller J, Wickens M: **Tethered function assays: an adaptable approach to study RNA regulatory proteins.** *Methods Enzymol* 2007, **429:**299–321.
13. Barkan A, Rojas M, Fujii S, Yap A, Chong YS, Bond CS, Small I: **A combinatorial amino acid code for RNA recognition by pentatricopeptide repeat proteins.** *PLoS Genet* 2012, **8:**e1002910.
14. Yagi Y, Hayashi S, Kobayashi K, Hirayama T, Nakamura T: **Elucidation of the RNA recognition code for pentatricopeptide repeat proteins involved in organelle RNA editing in plants.** *PLoS ONE* 2013, **8:**e57286.
15. Wang X, McLachlan J, Zamore PD, Hall TM: **Modular recognition of RNA by a human pumilio-homology domain.** *Cell* 2002, **110:**501–512.
16. Cheong CG, Hall TM: **Engineering RNA sequence specificity of Pumilio repeats.** *Proc Natl Acad Sci U S A* 2006, **103:**13635–13639.
17. Filipovska A, Razif MF, Nygard KK, Rackham O: **A universal code for RNA recognition by PUF proteins.** *Nat Chem Biol* 2011, **7:**425–427.
18. Dong S, Wang Y, Cassidy-Amstutz C, Lu G, Bigler R, Jezyk MR, Li C, Hall TM, Wang Z: **Specific and modular binding code for cytosine recognition in Pumilio/FBF (PUF) RNA-binding domains.** *J Biol Chem* 2011, **286:**26732–26742.
19. Quenault T, Lithgow T, Traven A: **PUF proteins: repression, activation and mRNA localization.** *Trends Cell Biol* 2011, **21:**104–112.
20. Edwards TA, Pyle SE, Wharton RP, Aggarwal AK: **Structure of Pumilio reveals similarity between RNA and peptide binding motifs.** *Cell* 2001, **105:**281–289.
21. Wang X, Zamore PD, Hall TM: **Crystal structure of a Pumilio homology domain.** *Mol Cell* 2001, **7:**855–865.
22. Opperman L, Hook B, DeFino M, Bernstein DS, Wickens M: **A single spacer nucleotide determines the specificities of two mRNA regulatory proteins.** *Nat Struct Mol Biol* 2005, **12:**945–951.
23. Cooke A, Prigge A, Opperman L, Wickens M: **Targeted translational regulation using the PUF protein family scaffold.** *Proc Natl Acad Sci U S A* 2011, **108:**15870–15875.
24. Choudhury R, Tsai YS, Dominguez D, Wang Y, Wang Z: **Engineering RNA endonucleases with customized sequence specificities.** *Nat Commun* 2012, **3:**1147.
25. Wang Y, Cheong CG, Hall TM, Wang Z: **Engineering splicing factors with designed specificities.** *Nat Methods* 2009, **6:**825–830.
26. Ozawa T, Natori Y, Sato M, Umezawa Y: **Imaging dynamics of endogenous mitochondrial RNA in single living cells.** *Nat Methods* 2007, **4:**413–419.
27. Tilsner J, Linnik O, Christensen NM, Bell K, Roberts IM, Lacomme C, Oparka KJ: **Live-cell imaging of viral RNA genomes using a Pumilio-based reporter.** *Plant J* 2009, **57:**758–770.
28. Yamada T, Yoshimura H, Inaguma A, Ozawa T: **Visualization of nonengineered single mRNAs in living cells using genetically encoded fluorescent probes.** *Anal Chem* 2011, **83:**5708–5714.
29. Engler C, Kandzia R, Marillonnet S: **A one pot, one step, precision cloning method with high throughput capability.** *PLoS ONE* 2008, **3:**e3647.
30. Marchese FP, Aubareda A, Tudor C, Saklatvala J, Clark AR, Dean JL: **MAPKAP kinase 2 blocks tristetraprolin-directed mRNA decay by inhibiting CAF1 deadenylase recruitment.** *J Biol Chem* 2010, **285:**27590–27600.
31. Qi MY, Wang ZZ, Zhang Z, Shao Q, Zeng A, Li XQ, Li WQ, Wang C, Tian FJ, Li Q, *et al*: **AU-rich-element-dependent translation repression requires the cooperation of tristetraprolin and RCK/P54.** *Mol Cell Biol* 2012, **32:**913–928.
32. Lai WS, Carballo E, Strum JR, Kennington EA, Phillips RS, Blackshear PJ: **Evidence that tristetraprolin binds to AU-rich elements and promotes the deadenylation and destabilization of tumor necrosis factor alpha mRNA.** *Mol Cell Biol* 1999, **19:**4311–4323.
33. Wharton RP, Sonoda J, Lee T, Patterson M, Murata Y: **The Pumilio RNA-binding domain is also a translational regulator.** *Mol Cell* 1998, **1:**863–872.
34. Chou CF, Mulky A, Maitra S, Lin WJ, Gherzi R, Kappes J, Chen CY: **Tethering KSRP, a decay-promoting AU-rich element-binding protein, to mRNAs elicits mRNA decay.** *Mol Cell Biol* 2006, **26:**3695–3706.
35. Lykke-Andersen J, Wagner E: **Recruitment and activation of mRNA decay enzymes by two ARE-mediated decay activation domains in the proteins TTP and BRF-1.** *Genes Dev* 2005, **19:**351–361.
36. Zamore PD, Williamson JR, Lehmann R: **The Pumilio protein binds RNA through a conserved domain that defines a new class of RNA-binding proteins.** *RNA* 1997, **3:**1421–1433.
37. Gupta YK, Nair DT, Wharton RP, Aggarwal AK: **Structures of human Pumilio with noncognate RNAs reveal molecular mechanisms for binding promiscuity.** *Structure* 2008, **16:**549–557.
38. Ferrara N, Gerber HP, LeCouter J: **The biology of VEGF and its receptors.** *Nat Med* 2003, **9:**669–676.
39. Ho VT, Bunn HF: **Effects of transition metals on the expression of the erythropoietin gene: further evidence that the oxygen sensor is a heme protein.** *Biochem Biophys Res Commun* 1996, **223:**175–180.
40. Liang J, McLachlan MJ, Zhao H: **Orthogonal control of endogenous gene expression in mammalian cells using synthetic ligands.** *Biotechnol Bioeng* 2013, **110:**1419–1429.
41. Lu G, Hall TM: **Alternate modes of cognate RNA recognition by human PUMILIO proteins.** *Structure* 2011, **19:**361–367.
42. Kwon HS, Shin HC, Kim JS: **Suppression of vascular endothelial growth factor expression at the transcriptional and post-transcriptional levels.** *Nucleic Acids Res* 2005, **33:**e74.
43. Snowden AW, Zhang L, Urnov F, Dent C, Jouvenot Y, Zhong X, Rebar EJ, Jamieson AC, Zhang HS, Tan S, *et al*: **Repression of vascular endothelial growth factor A in glioblastoma cells using engineered zinc finger transcription factors.** *Cancer Res* 2003, **63:**8968–8976.
44. Zhang L, Yang N, Mohamed-Hadley A, Rubin SC, Coukos G: **Vector-based RNAi, a novel tool for isoform-specific knock-down of VEGF and anti-angiogenesis gene therapy of cancer.** *Biochem Biophys Res Commun* 2003, **303:**1169–1178.
45. Gibson DG, Young L, Chuang RY, Venter JC, Hutchison CA 3rd, Smith HO: **Enzymatic assembly of DNA molecules up to several hundred kilobases.** *Nat Methods* 2009, **6:**343–345.

11

Influence of substrate curvature on osteoblast orientation and extracellular matrix deposition

Marcello Pilia[1,2*], Teja Guda[1,2], Stefanie M Shiels[1] and Mark R Appleford[1,2]

Abstract

Background: The effects of microchannel diameter in hydroxyapatite (HAp) substrates on osteoblast behavior were investigated in this study. Microchannels of 100, 250 and 500 μm diameter were created on hydroxyapatite disks. The changes in osteoblast precursor growth, differentiation, extra cellular matrix (ECM) secretion and cell attachment/orientation were investigated as a function of microchannel diameter.

Results: Curvature did not impact cellular differentiation, however organized cellular orientation was achieved within the 100 and 250 μm microchannels (mc) after 6 days compared to the 12 days it took for the 500mc group, while the flat substrate remained disorganized. Moreover, the 100, 250 and 500mc groups expressed a specific shift in orientation of 17.45°, 9.05°, and 22.86° respectively in 24 days. The secreted/mineralized ECM showed the 100 and 250mc groups to have higher modulus *(E)* and hardness (*h*) (E = 42.6GPa; h = 1.6GPa) than human bone (E = 13.4-25.7GPa; h = 0.47-0.74GPa), which was significantly greater than the 500mc and control groups ($p < 0.05$). It was determined that substrate curvature affects the cell orientation, the time required for initial response, and the shift in orientation with time.

Conclusions: These findings demonstrate the ability of osteoblasts to organize and mineralize differentially in microchannels similar to those found in the osteons of compact bone. These investigations could lead to the development of osteon-like scaffolds to support the regeneration of organized bone.

Keywords: Hydroxyapatite, Osteon architecture, Curvature, Microchannels, Extracellular Matrix, Nano-indentation

Introduction

Natural bone achieves much of its mechanical strength through cortical bone, specifically through the organization of its osteons. The structural organization of native bone directly contributes to the mechanical strength of bone tissue, which is critical since load bearing and providing mechanical support are the primary functions of the skeleton. The lamellar rings that surround the central microchannel-like structure of the osteon are formed by the secretion of Type-I Collagen (Col-I) by osteoblasts during osteonal development. While Col-I and other organic molecules make up 70% of total bone composition, the remaining 30% is inorganic, composed of bone minerals, specifically nano-size crystals of hydroxyapatite (HAp) [1]. The bone minerals are responsible for the hardness of bone whereas the organic portion gives skeletal tissue its elasticity [2]. In addition, the secretion of Col-I along different orientations, followed by the deposition of bone minerals, gives cortical bone its high compressive strength and toughness [3]. The compressive strength of cortical bone ranges between 100–230 MPa, whereas trabecular bone ranges between 2–12 MPa [4].

Regeneration of bony defects is a growing problem, as bone fracture and trauma related injuries account for more than 1.3 million procedures in the United States [5,6]. Specifically, critical sized segmental bone defects (SBDs) are a cause of significant concern in orthopedics because they do not heal on their own [7-12]. Autografts are the gold standard approach favored by surgeons to treat SBDs [7,8,10,12,13]. Although this technique is widely used, it is also associated with various drawbacks, including infections and donor site morbidity [8]. In recent years, the advent of bone tissue engineering has brought new ideas, discovery, development and

* Correspondence: m_pilia@hotmail.com
[1]Department of Biomedical Engineering, University of Texas, San Antonio, TX, USA
[2]Joint Program in Biomedical Engineering, University of Texas Health Science Center at San Antonio/University of Texas, San Antonio, TX, USA

characterization of new biomaterials for regenerative purposes. Tissue engineered grafts for SBDs should promote vascularity through highly interconnected pores [14] while maintaining the structural and mechanical integrity similar to native tissue [15]. Current bone tissue engineering approaches include cell-based, scaffold-based, and delivery-based strategies [10]. Even though in recent times there has been an increase in scientific publications aimed at increasing the mechanical properties of biomaterials [16,17] and underlining the importance of mechanotransduction [18-20], mechanical stability remains a weak component of these approaches. Current tissue engineered scaffolds can only match the strength range of trabecular bone [21] making the implants unsuitable for weight-bearing applications without secondary fixation. Transitory metallic plates have become a necessity to be placed in conjunction with the SBD scaffolds [22,23] to divert the load away from the fragile scaffold. A drawback of these plates is stress shielding effects around the defect area, slowing down the healing response and limiting healthy bone regeneration at the scaffold [24]. The two factors that have received the most attention in bone tissue engineering to overcome these short-comings are material composition and architecture of the scaffolds [25]. Various types of degradable scaffolds have been investigated and developed over the years for bone repair using ceramics, polymers, and combinations of the two material types [21]. Synthetic HAp, has been used as a material for bone tissue engineering because of its biocompatibility, bioactivity [26-28] and its compositional resemblance to the inorganic phase of natural bone [29]. Current investigations waiver toward HAp scaffolds that are highly porous and have architectures resembling that of trabecular bone [23]. In contrast, HAp scaffolds composed of microchannel-like structures resembling the architecture of cortical bone remains virtually untested in the literature. When comparing osteoblast attachment and proliferation, the geometry of the substrate to which the cell attaches differs between the filament-like, convex surface of the trabeculae and the tubular-like, concave surface of the osteons. Yet, the diameter of the osteon ring to which the cell attaches changes continuously [30]. Black *et al.* reported the average osteon size to range between 50 and 500 μm in diameter [30]. When the osteoblast attaches to the outermost layer of the osteon (~500 μm in diameter) the cell (20-30 μm in size) is exposed to a nearly flat surface [31]. As bone is deposited, the concentric layers stack and the cells begin to attach to an increasingly curved substrate, eventually forming the innermost osteonal layer (~50 μm diameter). A graphical representation to help understand this concept can be seen in Figure 1. This leads to the hypothesis that as osteoblasts are exposed to substrates of varying curvature, the ECM that is secreted is organized variably from layer to layer. It is critical to first gain understanding of how osteoblasts grow and differentiate within the osteon and determine how substrate curvature affects cell responses. Few studies have been conducted to analyze the effect of substrate curvature on cell attachment, orientation and growth. Studies have focused on the strength of attachment of cytoskeletal filaments on spherical and micropatterned substrates [32-38]. Specifically, Noireaux *et al.* studied the growth of an actin gel around spherical beads to determine cell membrane deformation and cell motility direction [36], Schwartz *et al.* investigated the role of substrate curvature in actin pushing forces [38], Pathak *et al.* investigated focal adhesion development on Y, V, T, and U shape micropatterned substrate [37], Lehnert *et al.* and Balaban *et al.* investigated cell behavior on linear and hole-micropatterned substrates [32,34]; Lam *et al.* created wavy micropatterns out of silicone rubber and attached myoblast precursor cells to determine myotube formation [33], Nikkah *et al.* investigated the change in fibroblast behavior when exposed to 3D silicon microstructures [35], and Holthaus *et al.* investigated the effects of microchannel width in osteoblast orientation [16]. However, limited literature is available to show how the change in substrate curvature diameter affects cellular response, bone ECM secretion or bone marker activation.

The goal of this study is to determine the effect of curved substrates on osteoblast growth, differentiation and organization within the microchannels, as well as ECM secretion, mineralization and hardness as a function of the substrate curvature. To accomplish this, HAp substrates with microchannels of various diameters were built to match the microchannel diameter range of natural osteons. On these patterned substrates, an osteoblast precursor cell was cultured *in vitro* to investigate cell responses to the various curvatures.

Materials and methods

Mold fabrication

Half cylinder molds were fabricated using stainless steel wire (HM Wire International Inc. – Canton, OH) of three different diameters: 110, 320, and 750 μm. The wire was cut in 25 mm sections that were positioned adjacent to each other with no overlap or additional-spacing. Sufficient wires were arranged to form a square 25 × 25 mm. Dental stone cement (Coecal™ Type III Dental Stone – GC America Inc, Alsip, IL) was mixed with distilled water (3:1 w/v ratio) to form molds into which HAp slurry were cast. Each cement mold measured 5 mm in height (Figure 2). Subsequently a hole (ϕ = 10 mm) was drilled in the center for fabrication of the Hap disks. The cement mold was then placed directly above the wire and was held securely in place by vice compression. Details illustrating mold fabrication are found in Figure 2A-C.

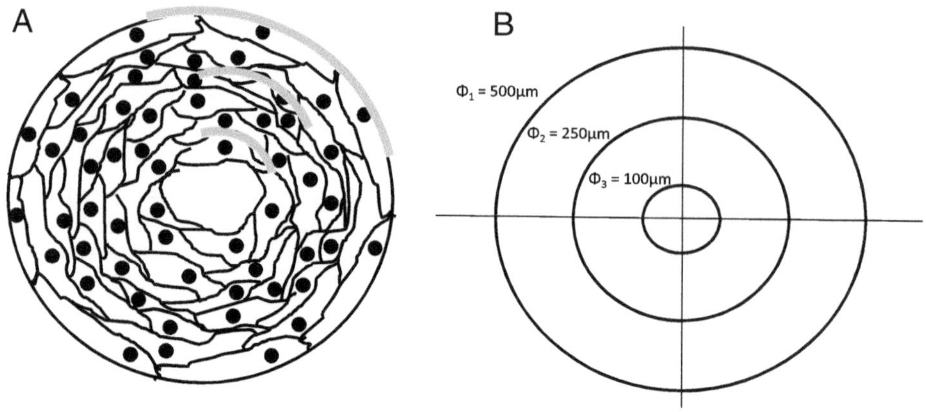

Figure 1 Drawing (A) and representation (B) of the concentric ring structure typical of the osteon. A) The osteonal concentric ring structure ranges from 50 to 500 μm in diameter. As multiple layers of cells are deposited towards the middle of the osteon the osteocytes are exposed to different substrate curvature (gray line). This was simulated by fabricating microchannels of various diameters where HFOb were seeded for up to 24 days. **B)** a graphical representation of the different curvatures the cells are exposed to in all the different substrates. The smaller the diameter of the microchannel, the more curved the substrate becomes.

Fabrication of the HAp disks

The HAp disks were made by solution casting as shown in Figure 2D. A HAp slurry was created by using a previously described method [39] using synthetic nano-size HAp (OssGen, South Korea). Briefly, the binders used to stabilize the slurry structure included 3% high molecular weight polyvinyl alcohol, 1% v/v carboxymethylcellulose, 1% v/v ammonium polyacrylate dispersant, and 3% v/v N, N-dimethylformamide drying agent. The solution was then cast into the constructed molds and sintered in a high temperature furnace (Thermolyne, Dubuque, IA). Before sintering, each disk measured 10 mm diameter and 5 mm height. The sintering process profile contained a ramp increase in temperature of 5°C/min up to 300°C

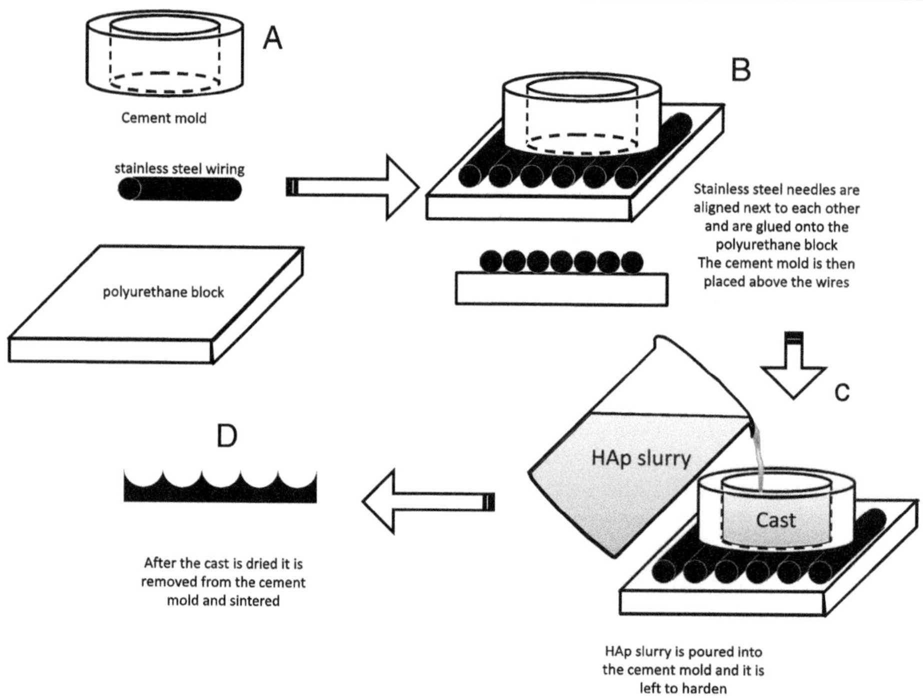

Figure 2 Diagram showing the steps used to create the molds onto which the HAp was cast. (A) First stainless steel wires of different dimensions were cut 25 mm long and were lined up next to each other in the axial direction on a polyurethane block to form the negative template. **(B)** These needles were then adhered with epoxy to the PU block to prevent movement. **(C)** The HAp was then cast into a cement mold (10 mm diameter and 6 mm depth) to form the positive cast. The cast HAp disk was then removed and sintered. **(D)** The resulting pattern was a series of concave curves.

with a hold time of 1 hour, then up at the same rate to 600°C for 1 hour hold, and finally to 1230°C for a 5 hour hold. The sintered disks were then cooled at a rate of 5°C/min until room temperature is achieved.

Characterization of the disks

The resulting disks were analyzed for template geometry and architecture using scanning electron microscopy (SEM) and light microscopy. Laser profilometry (Bruker Contour GT, Tucson, AZ) was used to determine the surface roughness of the disks. First, a disk of each template size was embedded in Pelco® Fast Curing Epoxy Hardener (TedPella Inc, Redding, CA) and a microtome saw (Leica SP1600; Wetzlar, Germany) was used to cut the samples in a direction perpendicular to the template microchannels. Each section cut was roughly 100 μm thick and was polished on both sides using a wet-sander (Struers LaboPol-5, 800 → 1200 grit paper; Cleveland, OH). Each section was analyzed under a microscope (Leica DMIL LED, Wetzlar, Germany), where software (Bioquant Osteo 2010, Nashville, TN) was used to acquire λ (wavelength), a (amplitude), l (arc length), and d_d (disk diameter) (Figure 3). λ is the average distance between two peaks; a is the distance between the peaks and the valleys; l is the length of the arc between two peaks. The surface area (A_e) was calculated using Equations 1–3, where n is the number of channels within each disk and d_e is the diameter of the ellipse. Following characterization, the disks used for cell culture were sterilized by ethylene oxide (AN74i Anprolene gas sterilizer, Andersen Sterilizers, Inc) for a 12 hour cycle.

$$n = \frac{d_d}{\lambda} \tag{1}$$

$$d_e = n \times l \tag{2}$$

$$A_e = \pi \frac{d_d}{2} \frac{d_e}{2} \tag{3}$$

Human fetal osteoblast cell culture

Human Fetal Osteoblasts (HFObs) (Cell Applications, Inc.- San Diego, CA) were used to evaluate bone cell response to the varying curvatures. Even though primary Mesenchymal cells are pluripotent compared to HFObs which are committed to the osteogenic lineage, they were not used because the objective of this study was to investigate the effect of local micro-architecture on the matrix production, organization and maturation of osteoblast-like cells, rather than to investigate the commitment of progenitor cells to an osteoblast-like phenotype. The cells were cultured in growth media containing Dubecco Modifed Eagle Medium (DMEM), 10% Fetal Bovine Serum (FBS), and 1% Penicillin Streptomycin Amphotericin B Solution (PSA) (all purchased from Invitrogen, USA). When cells reached confluence on the cell culture-flask, the HFObs were washed with phosphate buffered saline (PBS) and then 0.25% Trypsin/EDTA was added in osteogenic media (DMEM, 3% FBS, 1% PSA, 10 mM Glycerolphosphate, 50 μg/mL Ascorbic acid and 10nM Dexamethasone). The cells in solution were counted (Z_2 Coulter® Particle Count and Size Analyzer; Beckman Coulter™ - Brea, CA) and seeded on the disks at confluence (55,000 cells/cm^2). Four time points were tested: 6, 12, 18 and 24 days (n = 12). For n = 8 disks, media was collected. Each disk was then washed with PBS, followed by cell permeabilization using 0.1% Triton X-100 in PBS (PBS-T), and after a freeze/thaw cycle the supernatant was collected. The remaining n = 4 disks were stored in 4% formaldehyde for imaging.

In vitro osteoblast tracking on the HAp disks

Cell numbers were measured directly from the cell lysate solution. Specifically, 25 μL of lysate was added to the Quant-iT™ PicoGreen® dsDNA kit (Invitrogen, USA). This assay was performed in black opaque 96 well plates and the fluorescence was assessed using a microplate reader (Biotek Synergy 2 – Winooski, VT). The plate was excited at 485/20 nm, and the emitted light was measured at 528/20 nm. This reading was used to assess the change in cell numbers over time.

In vitro differentiation assays of the osteoblasts on the HAp disks

HFOb differentiation was determined by testing the cell lysates for the protein product of the bone-specific transcription factor runt-related transcription factor 2

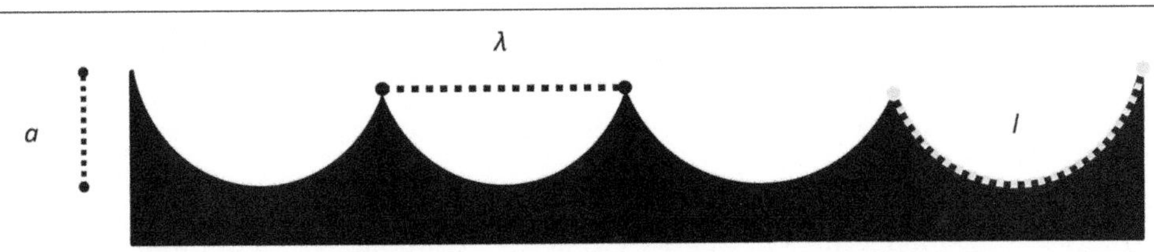

Figure 3 Graphical representation of the longitudinal section of the newly developed HAp disks. During characterization the wavelength (λ), amplitude (*a*) and the arc length (*l*) were measured from ceramic image analysis.

(RUNX2), Alkaline Phosphatase (ALP), Dental Matrix Protein 1 (DMP1), and Osteopontin (OPN) activity. RUNX2 activity was measured from the lysate using a phosphospecific antibody cell-based enzyme linked immunoabsorbent assay (PACE). Specifically, 50 μL of lysate were pipetted into a protein-attachment ready microplate and diluted in PBS-T solution at a ratio of 1:1. After 24 hours the wells were endogenous peroxide quenched in 0.6% H_2O_2, and blocked in 10% fetal bovine serum (FBS). Anti-RUNX2 primary antibody (Cat # 41–1400, Invitrogen) was then added overnight using a concentration of 1 μg/mL, followed by PBS-T washes and a secondary antibody (Cat # 81–6720, Invitrogen) for one hour using a concentration of 0.214 μg/mL. Following 3 PBS washes, Pierce 1-step ultra TMB was added to each well and the reaction was stopped using 2 M H_2SO_4. The plate absorbance was measured operating the same microplate reader used in the DNA analysis. The absorbance was read at 450 nm with reference at 655 nm. Primary to secondary antibody ratio was optimized by identifying highest signal-to-noise ratio (highest signal to noise ratio found using 1ug/mL primary, 0.214 μg/mL secondary with these antibodies). ALP activity was assessed from the cell lysate using an ALP Fluorescence Detection Kit (APF, Sigma-Aldrich). 10 μL of lysate were added in black opaque 96 well plates to fluorescence assay buffer following kit manufacturer instructions. Fluorescence was read after exactly 45 minutes with an excitation of 360 nm, and an emission of 460 nm. DMP1 was detected and quantified using the same PACE technique used for the RUNX2 assay. All of the steps remained the same with the difference that the primary antibody used was anti-DMP1 (code ab76632, Abcam) at a concentration of 2.5 μg/mL, and the secondary antibody used was the same used in the RUNX2 and was used at a concentration of 0.30 μg/mL. As in RUNX2, primary to secondary antibody ratio was optimized for maximum signal-to-noise ratio (strongest signal to noise ratio found using 2.5 μg/mL primary, 0.30 μg/mL secondary with these antibodies). OPN detection from the cell lysate solution was performed using the bone panel Milliplex kit (Millipore, USA). This kit also tested for osteocalcin and osteoprotegerin. Specifically, 10 μL of lysate were used for this test and combined with assay buffer following kit manufacturer instructions.

In vitro Collagen I semi-quantification assays of the HAp disks

A total of 10 fluorescence immunohistochemistry readings for each group were analyzed after testing semi-quantitatively for the presence of Col-I. The disks, previously fixed in 4% formaldehyde, were quenched hydrogen peroxide in 0.6% H_2O_2, blocked in 10% FBS, and a polyclonal anti-Col-I primary antibody (code ab34710, Abcam) was added overnight at a concentration of 10 μg/mL. This step was followed by PBS-T washes and the addition of a FITC linked secondary antibody (Code ab96895, Abcam) also using a concentration of 5 μg/mL. As in RUNX2 quantification, the strongest signal to noise ratio was found using 10 μg/mL primary, 5 μg/mL secondary with these antibodies. The disks were washed 2x in PBS-T and ProLong® Gold Antifade with DAPI was added to the bottom of the plate where the disks were inverted for fluorescent microscopy. A total of 10 intensity readings were obtained from different channels of the disks. Readings were averaged and Col-I was quantified using intensity measurements with reference to control surfaces.

In vitro cell orientation assays of the HAp disks

After staining the disks for Col-I and nuclei, the disks were analyzed under fluorescent microscope (SFL7000, Leica) for both Col-I and DAPI. Twenty random images were taken of each disk inside the artificial microchannels for both collagen and nuclei. The DAPI pictures were used to determine the angle of orientation of the cells with respect to the microchannel direction. This was done assuming that when the nucleus of the cell was elliptical in shape, the long axis of the nucleus matched the long axis of the cytoskeleton of the cell [40]. Cells undergoing mitosis were not measured for alignment. Images were analyzed using Bioquant Osteo system (Bioquant Osteo 2010, Nashville, TN). The angle of the cell was measured in degrees, with 0° being parallel to the microchannel direction, and 1 to 90° and −1 to -90° being of an angle to the right or to the left of the microchannel respectively. A schematic to describe this quantification is shown in Figure 4. When analyzing the flat control disks, bias measurement orientation due to the media-meniscus effect was prevented by taking pictures from five independent regions of interest on each sample. Successively, the blue channel images from the nuclei and the green channel from the Col-I were merged using Adobe Photoshop®.

In vitro mechanical ECM characterization of the HAp disks

Representative disks from each investigated group were tested to determine the mechanical properties of the cell's ECM secretions. The cells on the surface of the disks were gradually dehydrated in ethanol to preserve cellular structure. Specifically, the disks were moved into a 20/80 (vol/vol) ethanol in water for 1 hour, then 40/60 for another hour, 60/40, 80/20, and 100% ethanol two more times. The disks were then epoxied onto a 3 cm diameter metal cylinder using a very thin layer of epoxy glue. Using a commercial nano-indenter (MTS nano-indenter XP, MTS System Corporation, MN), the reduced elastic modulus (E_r) and the microhardness (H)

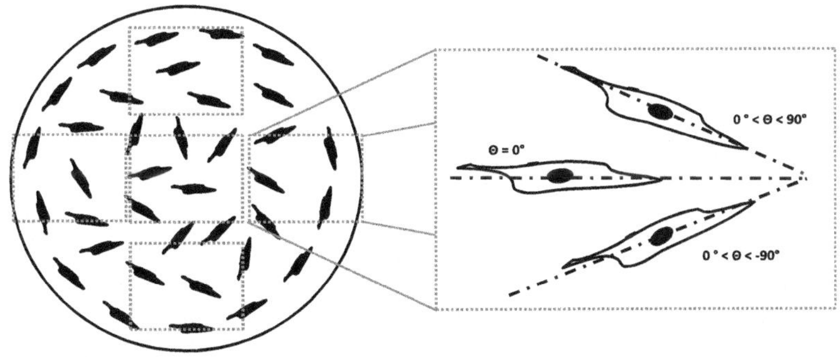

Figure 4 Graphical representation of the technique used to measure the orientation of the cells growing in the flat control HAp disks. Particular attention was paid to avoid bias orientation due to the meniscus effect due to the media on the well plate wall during cell culture. This was accomplished by measuring the cells from top to bottom and from left to right. When the long axis of the cell nuclei was parallel to the direction of the microchannels the angle was set to 0°. All the cells that were angled to the right were measured anywhere between 0 and 90°, whereas the cells angled to the left were measured with a negative number, 0 to -90°.

of the secreted ECM were determined. Before measuring each group the nano-indenter location was calibrated using a 5-indent template. The indenter's microscope (20x objective) was then used to find the peaks and the valleys of the disk substrate. The difference between the channel peaks and the valleys were easily noticed at this magnification. For each experimental group the indents were made in the valleys of the microchannels. This was done because the size and the geometry of the microchannels did not allow for the positioning of the indenter tip in a normal direction to that of the surface, which has a continuously changing angle. Also, we wanted to circumvent the possibility of the nano-indenter tip coming into contact with and sliding down the curved substrate wall. The indents were performed using a Berkovich tip and a standard trapezoidal loading profile with a loading rate of 250 μN/s until a maximum load of 500 μN was applied [41]. E_r and H were determined using a previously described method [42]. Briefly, the Young's modulus for the tested material (E_m) was calculated by assuming that the cells had a Poisson's ratio (ν_m) of 0.3 [41,43,44]. For the diamond indenter the Young's modulus (E_i) was set to 1140 GPa, with a Poisson's ratio (ν_i) of 0.07. Equation 4 was used to calculate E_m [42]:

$$E_m = \frac{1-\nu_m^2}{\frac{1}{E_r} - \frac{1-\nu_i^2}{E_i}} \tag{4}$$

300 random readings were taken from each disk. Hardness instead was calculated using Equation 5:

$$H = \frac{P_{\max}}{A} = \frac{P_{\max}}{F(h_c)} \tag{5}$$

In which P_{max} is the peak load and A is the projected area of contact at peak load evaluated by a function which relates the cross-sectional area of the indenter to the vertical distance from its tip (h_c) [42].

The modulus and hardness readings that fell into the control HAp readings ± standard deviation were discarded because it was assumed that the indent was measuring the actual HAp layer and not the cell ECM.

Statistical analysis

All data is reported as average ± standard error. The statistical test used to analyze the data was a two-way analysis of variance, and Tukey's comparison test was used to determine statistical significance between individual groups. All statistical tests were performed using SigmaPlot® (version 11.0, Systat Software, Inc.). Differences were considered significant at $P < 0.05$. The difference between variances of distribution was evaluated using an F-Test (MedCalc® V 12.2.1.0).

Results

Characterization of the Hap disks

Reproducible disks with longitudinal curved microchannels spanning the entire disk were created using the casting technique. This novel geometry was analyzed under SEM and constant microchannel diameter was seen throughout the four disk groups (Figure 5). Constant crystal size was previously determined with our slurry HAp technique [39]. The measurements for λ, a, and l are reported in Table 1. The unsintered template microchannels measured 120, 320, and 750 μm in diameter, and after the sintering process HAp underwent compaction due to binder loss that resulted in channel sizes of 100, 250 and 500 μm in diameter, the same λ dimensions that match the physiological size range of naturally occurring osteons [30]. For group identification these three groups will be labeled as 100 μm microchannel diameter (100mc), 250mc and 500mc.

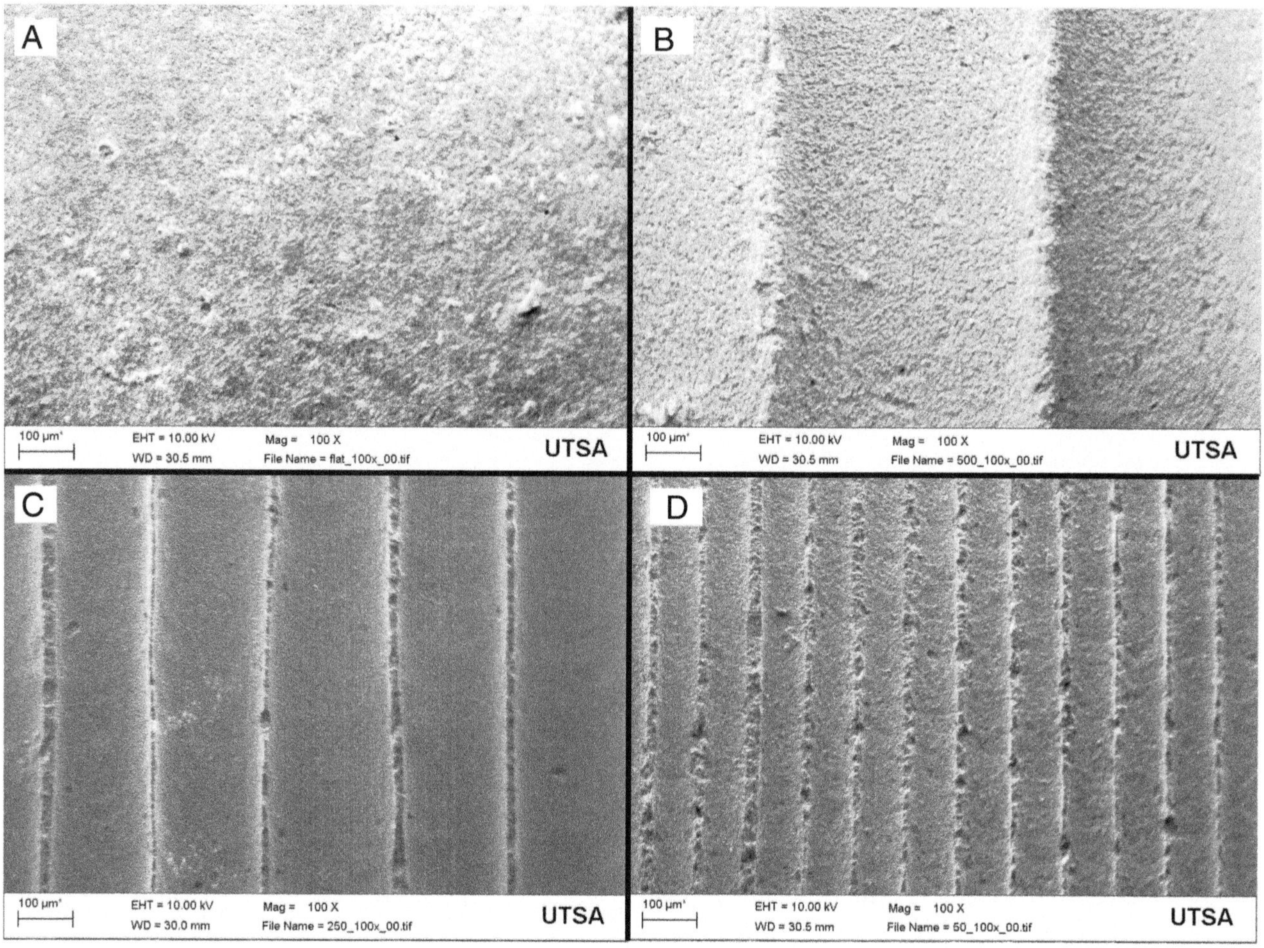

Figure 5 Scanning Electron Microscopy (SEM) pictures of the microchannel templates in each of the three dimensions fabricated: the flat control substrate (A), the 500 μm (B), the 250 μm (C) and the 100 μm template (D). Magnification is 100x, and the scale bar length is 100 μm.

HFOb growth and differentiation in the microchannels

The dsDNA assay demonstrated that throughout the 24 day experiment the cells did not show significant changes in cell number. This is consistent with a mature osteoblast phenotype. This is likely to have occurred because all the cells that were seeded in the disks were signaled to differentiate by the dexamethasone in the osteogenic media. These results are shown in the chart in Figure 6A.

The four differentiation markers analyzed were chosen for being indicative of early through late stage osteoblast differentiation. Of the four markers, RUNX2 is activated at the earliest time [45], Figure 6B. RUNX2 results showed the highest levels at day 6 and then decreased through the remainder of the experiment. The earlier time point shows that the 250mc and the 100mc have the highest trend of RUNX2 expression. By day 12 the 250mc decreased to its lowest level while the control group spiked, demonstrating delayed differentiation. The second chronological differentiation marker is ALP, Figure 6C. The data supports this by showing an ALP spike on day 12, and on day 24 [46]. At this time point, the 500mc and the 250mc were significantly different from the flat control group ($P < 0.001$ and $P < 0.05$

Table 1 Results of morphology characterization of the HAp disks

Template (μm)	λ (μm)	a (μm)	l (μm)	Area (mm^2)
100	92.32 ± 0.57	42.83 ± 0.92	169.55 ± 2.12	79.78 ± 0.87
250	229.92 ± 1.07	111.84 ± 2.24	430.27 ± 5.41	81.25 ± 0.92
500	525.07 ± 2.32	203.18 ± 3.40	880.27 ± 9.00	72.78 ± 0.51

All measurements were taken after thin histological longitudinal sections of the disks were cut, polished, and analyzed using Bioquant Osteo®. All values reported with standard error of the mean (SEM).

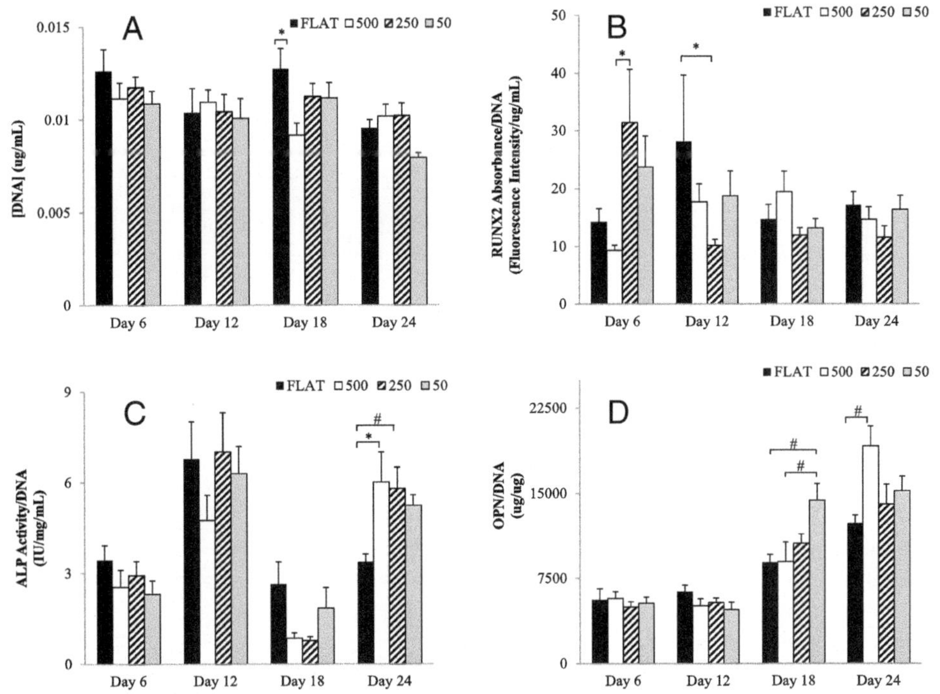

Figure 6 HFOb cell count and differentiation assay for the four groups tested. (A) cell count for the Flat, 500, 250 and 100 mc throughout four time points: 6, 12, 18, and 24 days. No proliferation is seen throughout the four weeks. There is a significant decrease in cells found on the 100 group at the last time point. Also seen are the differentiation markers RUNX2 **(B)**, ALP **(C)**, OPN **(D)**. The RUNX2 groups do not show any significant difference between the groups although we can see a trend in which the 250 μm diameter microchannels have lower levels than the rest of the groups. The ALP activity shows two different peaks at day 12 and at day 24. At the last time point the 500 and 250 μm diameter microchannels showed a significant higher ALP activity than the Flat control group. The OPN levels do show significant difference on day 18 between the 100 μm diameter microchannels and the control group. At day 24 the significant difference was seen between the 500 μm and the Flat. All data shown as mean ± standard error. * $P < 0.05$; ^ $P < 0.01$; # $P < 0.001$.

respectively). OPN is chronologically an even later differentiation marker of osteoblasts, and the results are shown in Figure 6D. This data also correlates with our findings and after being minimal on days 6 and 12, they rise on day 18 to have the highest spike at day 24. In particular, the 100mc had a significantly higher amount of OPN than the flat control group on day 18 ($P < 0.05$) while the 500mc showed a significantly higher level than flat control on day 25 ($P < 0.05$). The last differentiation marker tested was DMP1. These results can be seen in Figure 7A. This marker tests for osteoblast differentiation into osteocytes. The data for this marker shows very slight to no change in time in the expression of this marker. Although possible spikes of DMP1 are seen in the control flat disks at days 12 and 18, and for the 500mc at day 24, none of these changes are significantly different.

In vitro extracellular matrix (ECM) assays of the HAp disks

A total of 10 readings were analyzed in the quantification of Col-I fluorescence in each group. These findings are shown in Figure 7B. Although the data was not significantly different, a trend is seen in which all of the tested groups from day 18 produced greater Col-I than the flat control group. The highest values of Col-I were seen at day 18 and slowly decreased by day 24 to roughly the same levels as day 12. Figure 8 shows collagen staining within the different size microchannels.

In vitro cell organization and orientation assays of the HAp disks

The cell organization and orientation within the microchannels was determined from computational analysis of DAPI-stained cell nuclei. Although the same data was used, cell organization was analyzed comparing the change in orientation between the different mc diameters at each time point, whereas cell orientation was studied by comparing the orientation of each individual investigational group through time. The results showed that the cells growing in the flat control group did not become organized in 24 days. Figure 9 shows the measured frequency distribution of cellular orientation. Narrow frequency curves were characterized by lower standard deviation, and translated to higher cellular

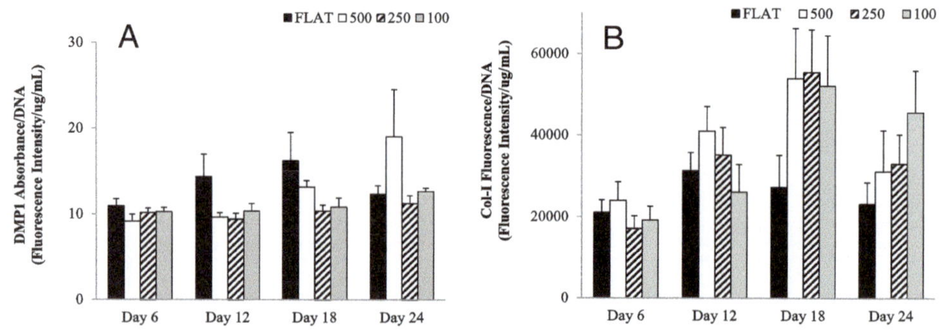

Figure 7 Chart showing HFOb differentiation marker DMP1 (A) and Collagen-I fluorescent assay (B). (A) shows no significance between any of the groups or time points, probably due to the fact that this marker is very late for differentiation and we saw no osteoblasts becoming osteocytes. **(B)** shows Col-I fluorescent assay results that are normalized to DNA. The results show a trend in which the flat control group has less collagen than the rest of the groups. There is no significant difference between the groups at each time point. Other markers tested included osteocalcin and osteoprotegering, but levels detectable levels were too low. All data shown as mean ± standard error.

organization on the substrate. Correspondingly, wider frequency curves were represented by a higher standard deviation, indicating low organization of the cells. At day 6 (Figure 9A), the variances of all the samples were statistically significant from one another ($P < 0.01$), with the 250mc having the lowest variance followed by the 100mc and the unorganized 500mc. At day 12 (Figure 9B) the 100mc displays statistically higher organization than the rest of the groups ($P < 0.05$), followed by the 250mc and the 500mc. The 500mc in fact shows early organization trends that differ from the control ($P < 0.001$). At day 18 (Figure 9C) the most organized groups remain the 100mc and the 250mc ($P < 0.01$). The 500mc continues to become more organized as well. By day 24 (Figure 9D) the 100mc once again shows higher organization than the rest of the groups ($P < 0.001$), followed by the 250mc and 500mc which show no difference from one another. The change

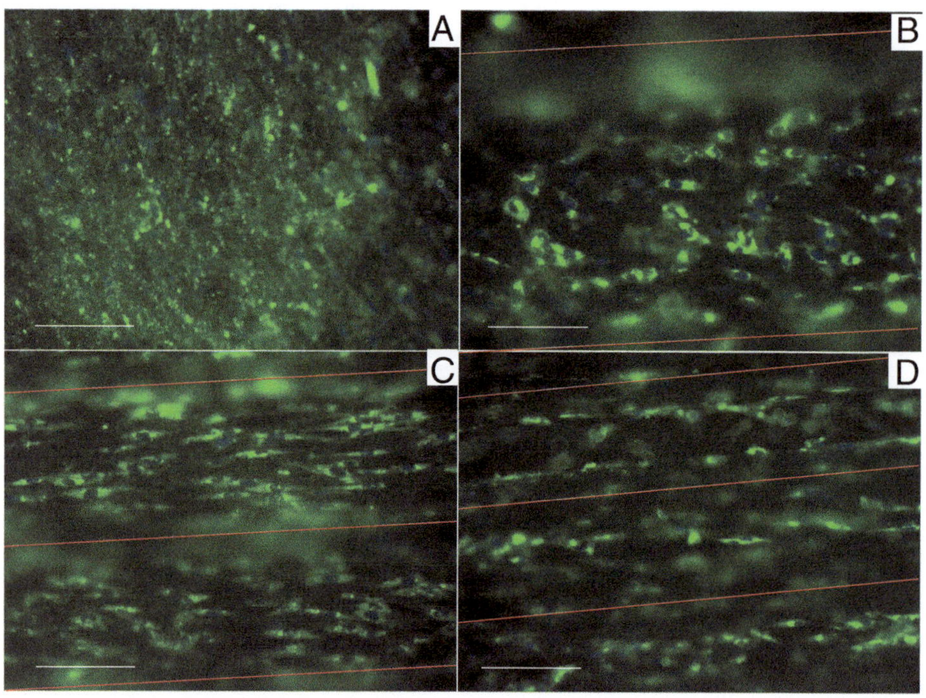

Figure 8 Fluorescent photographs of Col-I stain at day 6 in the experimental groups. The flat control **(A)** does not show any type of organization. The 500mc group **(B)** shows very early signs of organization even though the individual cells have not completely chosen direction. The 250mc group **(C)** shows cellular distribution which is allot more organized than the previous groups. Last, the 100mc group **(D)** shows early signs of differentiation already at day 6. Scale bar corresponds to 100 μm. In each figure, the red lines indicate the direction of the microchannels as compared to the cell animation.

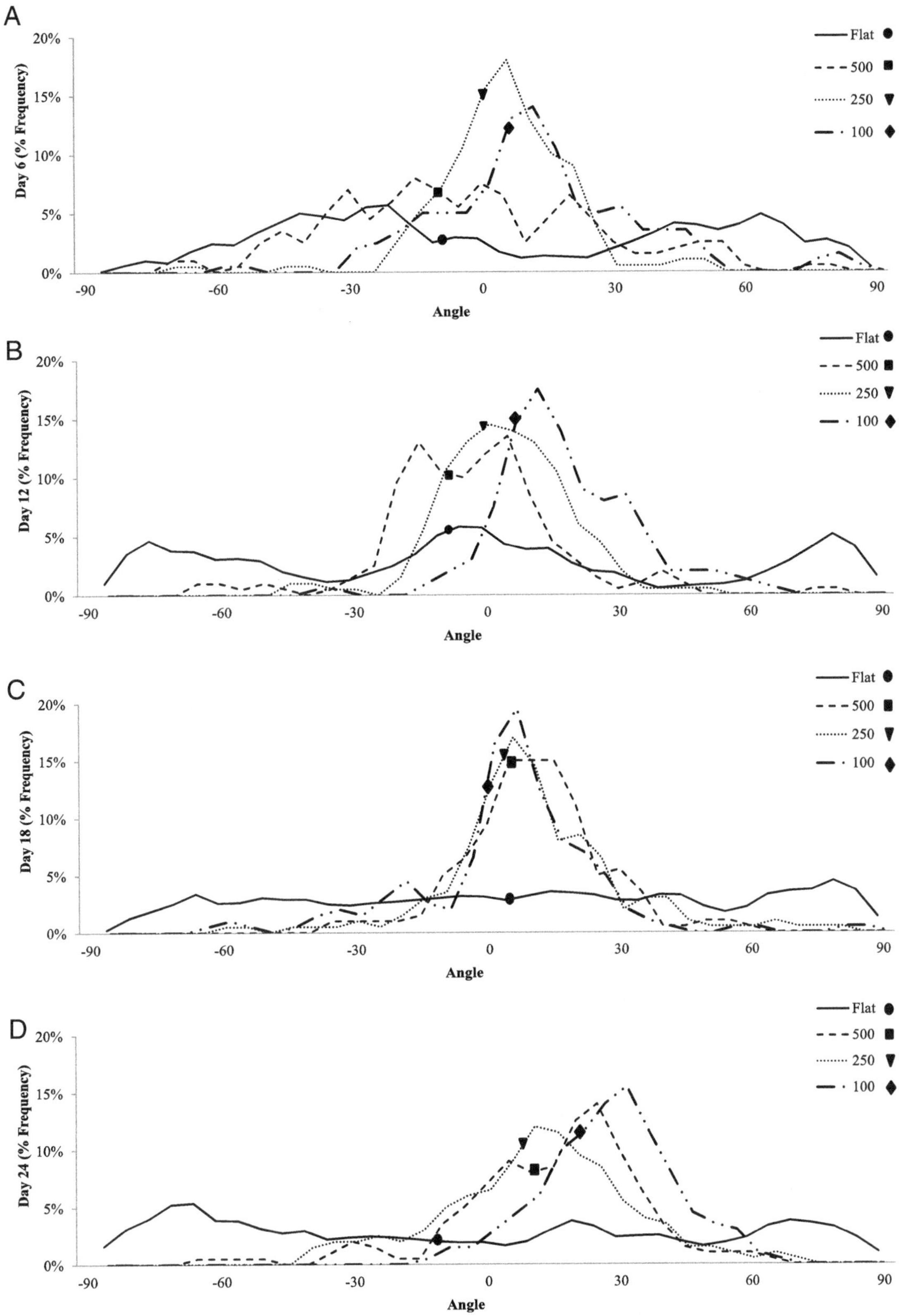

Figure 9 (See legend on next page.)

(See figure on previous page.)

Figure 9 Frequency distribution of the cell angles during the four time points. The dots represent the median angle of orientation for each group. This graph shows the early stages of cell organization following cell attachment and early differentiation. In the control flat group the cells appear to be very disorganized and are scattered in every position. Overall all of the experimental groups showed higher organized than the flat control groups which remained disorganized for the entire 24 days (significantly higher variance#). At day 6 **(A)** the 250mc group shows earlier orientation than the rest of the groups#, followed by the 100mc group. The 500mc group does not quite achieve early organization. By day 12 **(B)** the 100mc group gains the most organization* followed by the 250mc and the 500mc group. The 500mc group reaches organization by this time. At day 18 **(C)** the most organized groups remain the 100mc and the 250mc^. The 500mc becomes even more organized as well. At the very last time point **(D)** we see the 100 group that shows higher organization than the rest of the groups ($P < 0.001$), followed by the 250mc and 500mc which show no difference from one another.* $P < 0.05$; ^ $P < 0.01$; # $P < 0.001$.

in cellular orientation was also evaluated within each group over time from day 6 to day 24. The control group was never able to gain orientation, and therefore the median angle was never analyzed. However the 500mc showed a significant change in median angle orientation with time ($P < 0.05$) from −11.30°, to −7.05°, to 5.40° and finally 11.56°. The 250mc displayed a significant change in orientation from day 6 to day 24 ($P < 0.001$) starting from −0.89° and ending at 8.16°, with the most significant changes happening between days 12 (−1.69°) and 18 (2.69°) ($P < 0.001$). The 100mc showed a very significant change between day 6 and 24, going from 4.43° to 21.88° respectively ($P < 0.05$), exhibiting perhaps the largest shift in cell orientation amongst the groups. The 250mc remained stable and close to the 10° angle. Overall, all the microchannel groups were more organized than the control group which remained disorganized for the entire 24 days.

In vitro mechanical ECM characterization of the HAp disks

Elastic modulus and hardness of the secreted ECM was analyzed using a nano-indentation technique. The experimental samples were compared to pure HAp substrate to determine when the indenter was in contact with cell ECM or in contact with the pure HAp substrate. The results from the experimental groups were all significantly different from the control HAp substrate for both the modulus and the hardness ($P < 0.001$). When analyzing the elastic modulus of the secreted ECM (Figure 10A), the 100mc had the highest value, and was significantly different from all other groups ($P < 0.05$). Even though the 500mc and the 250mc were not significantly different, a trend was observed in which the 500mc had higher modulus than the 250mc group. All of the experimental groups had higher moduli than the flat control group ($P < 0.05$). The same trend was seen in the hardness values (Figure 10B), with all microchannel groups having significantly greater hardness than the flat HAp disk ($P < 0.05$).

Discussion

Multiple materials, surface morphologies, and cell types have been previously investigated to identify the role of surface architecture on tissue regeneration. In fact, micropatterned surfaces have been previously created on different material substrates using a variety of techniques. Primarily these methods were micromachining [47-49], low voltage electron beam lithography [34], standard photolithographic [50-52] and photopolymerization process [53], plasma oxidation [33,54], and single mask fabrication technique [35]. All these techniques can achieve a different array of substrates, but none of them was useful to create the microchannel template on HAp. Other materials used to identify the effects of substrate on cell behavior include silicon wafers [33-35,48], polystyrene [36,52], methacrylate [53], Perspex [51], SU-8 5 photoresist [50], latex [36], and ligand patches [37]. HAp posed as a more challenging material to machine and/or pattern due to the strength and brittleness of its nature. Multiple cell types including tenocytes [50], fibroblasts [32,35,37,48], myoblasts [33,55], neurons [51], spiral ganglion cells [22], retinal endothelial cells [56], epithelial cells [9], astroglial cells [52], HeLa S3 cancerous cell line [36], and Mouse B16F1 melanoma cells [36] have been previously investigated regarding their response to surface architectural cues. However, there have been very few (or no known reports to the authors knowledge) reports on the response of fetal osteoblasts on such curved or micro-patterned substrates. In this study, a mold template/casting technique for HAp was employed. This allowed for precise, repeatable (<5% variation), and consistent (< 2.5% variation) patterned morphologies resembling the natural range of osteon curvature (Table 1). Using this system, we systematically studied the effect of a changing substrate curvature within the range of native osteon diameter on osteoblast response while simulating lamellar organization.

Osteoblast precursor cells such as the ones used in this study are not fully differentiated bone cells, however they are already committed to the osteogenic lineage. These cells undergo three key phases of cell activity: proliferation, differentiation and mineralization [57]. These steps are well defined and can overlap one another [58]. The osteoblast precursor cells seeded on the disks were induced to differentiate into mature

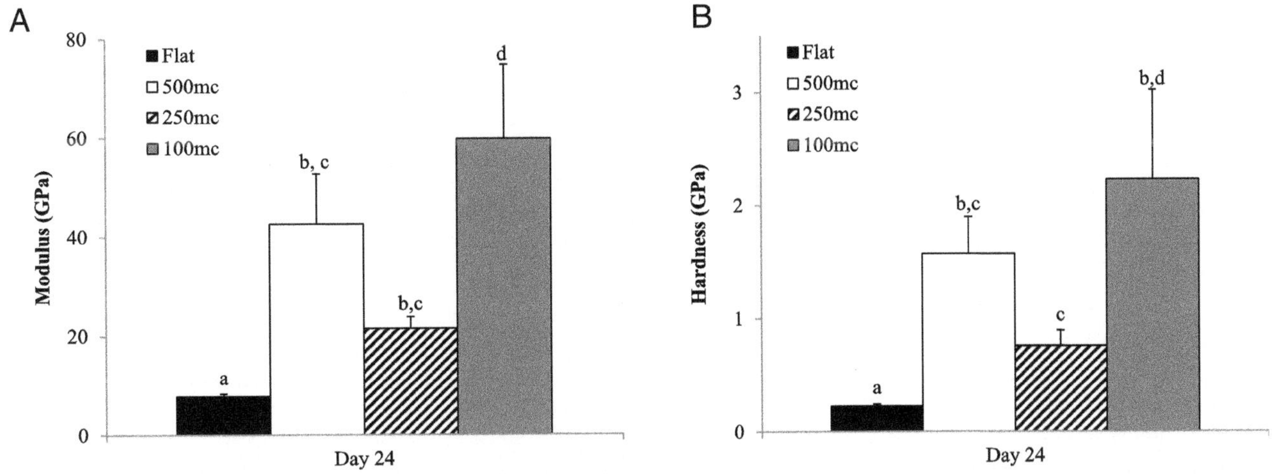

Figure 10 The elastic modulus (A) and the hardness (B) of the secreted ECM at day 24 across different substrates are shown. In both graphs all of the experimental groups were significantly different from the control flat group. The 100mc group had the highest modulus value than the other groups, and also had a higher hardness value than the 500mc group. Thus it can be concluded that the curvature had an effect on ECM secretion and mineralization in the 100mc group that is not seen in the other flat substrates. Letters show statistical significance between groups ($p < 0.05$), where a,b,c,d are used to indicate statistical differences between groups where any two groups unconnected by a common letter are significantly different ($p < 0.05$).

osteoblasts using the glucocorticoid dexamethasone. A common drawback of this steroid is a reduction in the rate of replication of the cells [59]. Thus, as osteoblast precursor cells start to differentiate, proliferation ceases. This is demonstrated with the findings on DNA analysis in the current study which showed no significant change in cell number but rather only differentiation. The assay used to determine cell numbers relies on testing from cell lysate solution. Given both the geometry of the substrate and the time for which the cells were cultured, it is not surprising that the DNA yield was not optimal, translating into high standard error. Another drawback related to use of Dexamethasone is the inhibition of bone formation in vivo due to decreased collagen synthesis [59]. Overall in this experiment the dexamethasone ρpdid not inhibit collagen secretion altogether, since actual Col-I deposition was seen (Figure 8). Our results show a strong differentiation response consistent with the effect of dexamethasone, glucocorticoid receptor induced activation of osteoblast gene expression such as osteocalcin, collagen Iα1 and transforming growth factor-β1. Osteoblasts differentiation can be detected by analyzing specific early and late markers which have been thoroughly studied and characterized [60]. Perhaps the earliest activation factor is RUNX2, followed by ALP (an early-mid marker), OPN, OC, ON [61]. The four week experiment resulted in no changes in osteoblast differentiation/mineralization rates between the four experimental groups, concluding that the curvature associated with different microchannel diameters had little or no effect on HFOb rate of differentiation. The rate of differentiation for all groups in this study was consistent with the literature. RUNX2 peaked early at week 1, followed by ALP, which was activated first at day 12 and then at day 24. OPN, which is activated by RUNX2 [62], did not show activation until day 18, with the highest levels seen at day 24, showing continuous differentiation throughout the study. The expression of DMP1 was limited after 24 days, which does not agree with the study by Mikami *et al.* which showed early expression of DMP1 from mRNA using RT-PCR when using dexamethasone as stimulant [63]. However, the current study did not analyze DMP1 expression from mRNA but rather quantified the actual Human DMP1 protein present in the cell lysate at the different time points. It is difficult to determine whether natural DMP1 expression follows the same *in vitro* trend. Narayanan *et al.* gave an insight into the role DMP1 [64]. According to their study in the very early stages of differentiation, DMP1 serves as a transcription factor that stimulates further differentiation and expression of such markers as RUNX2. Only at the last stage of Ob maturity (near collagen mineralization) is DMP1 secreted by the differentiating osteoblast and triggers mineralization [64]. This contradicts the findings that showed different modulus and hardness of the deposited ECM, suggesting that mineralization is occurring in the curved substrates, whereas there were no differences in the flat substrate. According to the findings in the current study, a noticeable increase in DMP1 should have been expected at four weeks. These contradicting discoveries could be due to the deterioration of the DMP1 after freeze/thaw cycles, or due to the low sensitivity of the DMP1 ELISA. Thus, it is reasonable to conclude from the present study that negligible increases in DMP1 expression observed from day 6 to 24 in all tested groups indicated the absence of phenotypic transition from early osteoblast to late mineralization behavior. DMP1 is also a marker for osteoblast differentiation into osteocyte. The lack of DMP1 could be explained by the cells not reaching mature osteoblast status at 24 days, or by the inhibited proliferation of the cells in culture, preventing layering effects in the microchannels.

When Luan *et al.* tested osteoblast cultures for Col-I secretion at 4, 8, and 12 days they were unable to see any significant differences at each time point [65]. Refitt *et al.* instead investigated the function of silicone in Col-I secretion. The Col-I measurements they described were done indirectly on the amount of carboxy-terminal propeptide of type 1 procollagen liberated into the culture medium [66]. In this study, although a definite trend can be seen in collagen secretion up to day 18 in all tested groups, no significant differences in collagen deposition over time or between groups were seen. However, a change of trend in the Col-I fluorescence was observed at day 24 and was characterized by a decrease in fluorescence intensity throughout all four experimental groups. This change, although not significant, was attributed to early mineralization that contributed to the masking of the Col-I antibody labeling. The control group did not show any changes in Col-I secretion within the study, which is consistent with data reported by Luan *et al.* [65].

Previous studies on cell orientation within micropatterned substrata all agreed that the narrower the channel, the higher the degree of alignment the cell presents, and the deeper the groove, also the higher the degree of alignment [49-53,55]. There are however a few basic differences between the cited research and those performed in the current study. The micropatterning described in the literature is in the range of few micrometers and does not necessarily reflect the concave shape that this study created. In fact Recknor *et al.* and Kapoor *et al.* had squared grooves [50,52], Brunette *et al.* had both vertical and sloped walls [49], and Charest *et al.* had 10 μm holes and 10 μm grooves [55]. Another important difference is that most studies only lasted up to 3 days and the cells were seeded at minimal confluence with the purpose of analyzing individual cell's behavior. The present study, however, lasted 24 days, and the disks were seeded at full confluence with the goal of analyzing osteoblast behavior as a tissue, observing its mineralization and tissue orientation in relation to the substrate. The behavior of the individual cells was not as imperative. This experiment was able to show overall organization in the 100 μm, 250 μm and 500 μm, as well as the change in orientation with time which may be part of the signaling mechanism for the change in lamellar alignment observed in natural alternating osteons. Although initial cell layering was observed in the channel valleys from the beginning, it was not shown to change density over 24 days. A previous study by Kacena et al. showed that gravity plays no effect on fully confluent osteoblast that are not in proliferation mode [67]. No effect of gravity or cell slippage within the microchannels was observed in the current study either.

The nanomechanical analysis demonstrated that by day 24 the ECM modulus and hardness values were much higher in the 100mc than the flat substrate and other microchannel groups. It is hypothesized that the high modulus and hardness values seen were due to the process of mineralization of the ECM by day 24, a process that perhaps was not seen in the flat substrates. The cell line investigated lacked phenotypic transition from early osteoblast to one that has late mineralization behavior. This can be explained by the cells not reaching mature osteoblast status at 24 days, or by the inhibited proliferation of the cells in culture, preventing layering effects in the microchannels.

It can be speculated that the curvature of the substrate indirectly increased the rate of mineralization of the HFObs on the ECM. This mineralization trend was also supported by the ALP activity and the OPN levels, which were significantly higher in the 100mc group than the flat control group. When visually comparing the organization of the cells and their change in orientation with the modulus and the hardness, it was observed that the higher the organization of the cells in the substrate, the higher the modulus and the hardness. As observed above, at day 24 the 100mc displayed the highest cell organization, followed by the 500mc and the 250mc. On the other hand, the flat substrate never demonstrated orientation. As the cells aligned onto the substrate, they were able to organize faster and secrete organized ECM with higher modulus and hardness. After 24 days in cell culture, cells produced their own extracellular matrix that could modify the topography of the substrates. In fact, the rate of change on each of the microchannels groups is probably different since the cells are probably laying down ECM to vary their local micro-environment. This could explain the different temporal response elicited on each microchannel. During nano-indentation testing we measured E and h of the plain HAp disks to determine baseline values ($E = 125.30 \pm 9.52$ and $h = 8.98 \pm 1.25$). This value was considered the internal control to verify that all the indents on the in vitro cultured substrates were measuring actual secreted ECM properties and not the underlying HAp surfaces. Also, HAp's E was 2 fold higher than the 100mc group and 6 fold higher than the control flat groups. The hardness was 4 fold higher than the 100mc group and 9 fold higher than the Flat disk control. The elastic modulus value of the flat substrate (7.8GPa) was slightly lower than what has been previously reported in the literature for dehydrated human bone (12.4GPa) [41,43,68-70]. However, the three experimental groups with different diameter microchannels resulted in a much higher modulus. Specifically, the 100 μm group showed the highest modulus amongst the groups (42.6GPa) and tested above human bone (13.4-25.7GPa) [41]. The hardness values of the flat control disk (0.2GPa) averaged below previously reported values for human bone (0.47-0.74GPa) [41,43,68-70]. The experimental groups with microchannels have much higher hardness than what was previously reported, and once again, the hardness in the 100mc (1.6GPa) is above natural bone values. It is hypothesized that the reason this group tested above cortical bone is because the experimental setup in this study allowed us to measure modulus and hardness directly on the surface of the osteon. When testing natural bone, the values come from an area outside the osteon, and this could contribute to the differences seen here. Potentially due to a largely two dimensional environment, in this study there were no significant changes in differentiation markers, although a significant difference in matrix modulus and hardness was observed, indicating different collagen expression and mineralization between groups occurred. This is something that should be further investigated in future studies, especially in a 3D environment.

The goal of this study was to determine osteoblast response to being cultured on a curved substrate mimicking the native curvature of an osteon. This was investigated through the alignment of cells in the microchannels (organization), the change in orientation with time, the secretion of collagen, and the production of a mineralized extracellular matrix. In this study we showed that the substrate affects physical properties of cellular organization and orientation, as well as the composition and quality of the ECM secreted. Regardless of what substrate the cells were cultured in, they all maintained equal number of cells over time while differentiating at statistical equal rates. Although this model is only the first step in depicting osteoblast behavior within a three dimensional osteon, it is critical since it has allowed us to demonstrate that when cultured in the appropriate substrate curvature and appropriate conditions, osteoblasts can organize as a tissue and change orientation with time, producing a mineralized ECM that is hard and tough. Based on these findings, future studies will focus on recreating three-dimensional scaffolds with longitudinal microchannels that resemble naturally occurring osteons. The scaffolds will then be characterized for their ability to recreate an *in vitro* osteon.

Conclusion

Overall, in this study microchannels of 100, 250 and 500 μm diameter were successfully created on the surface of HAp disk. A 24 day *in vitro* study was performed on these different substrates and demonstrated an early osteogenic phenotype and extensive collagen deposition Moreover, cell alignment within the microchannels were assessed to find that the 100mc and the 250mc groups induced fast orientation and a higher degree of organization, while the 500mc only started showing organization after 12 days and the control flat group did not show alignment towards a specific direction. This study was also able to demonstrate a significant increase in elastic modulus and hardness in the microchannel substrates compared to a flat control, suggesting that curvature plays an important role in the quality of the ECM secreted and mineralized. Understanding how bone cells grow and secrete ECM in these curved substrates is the first step in understanding the mechanism involved in creating artificial osteons and in the long run regenerating organized cortical bone.

Competing interests

The authors declare that they have no competing interests.

Authors' contributions

Study design: MP, MA, SS and TG. Study conduct: MP. Data collection: MP and SS. Data analysis: MP and TG. Data interpretation: MP and MA. Manuscript preparation: MP, TG, MA and SS. Approving final version of manuscript: MP, MA, TG and SS. MP takes responsibility for the integrity of the data analysis. All authors read and approved the final manuscript.

Acknowledgements

Thanks to Dr. Daniel Oh for his help and expertise in the casting of HAp materials, Beth Pollot for her assistance with scanning electron microscopy, and Cameron Taylor for his support analyzing cell orientation. Also a special thanks to the dental research staff at the United States Army Institute of Surgical research at Fort Sam Houston, San Antonio TX for allowing us to use their laser profilometer. This study was supported in part by the UT System South Texas Technology Management (STTM) POC Sparc grant program.

References

1. Murugan R, Ramakrishna S: **Development of nanocomposites for bone grafting.** *Compos Sci Technol* 2005, **65**:2385–2406.
2. Teo WE, Liao S, Chan C, Ramakrishna S: **Fabrication and characterization of hierarchically organized nanoparticle-reinforced nanofibrous composite scaffolds.** *Acta Biomater* 2011, **7**:193–202.
3. Martin BR, Burr DB, Sharkey NA: *Skeletal Tissue Mechanics.* New York: Springer-Verlag; 1998.
4. Carter CB, Norton MG: *Ceramic Materials: Science and Engineering.* New York: Springer; 2007.
5. Chim H, Hutmacher DW, Chou AM, Oliveira AL, Reis RL, Lim TC, Schantz JT: **A comparative analysis of scaffold material modifications for load-bearing applications in bone tissue engineering.** *Int J Oral Maxillofac Surg* 2006, **35**:928–934.
6. Gibson I, Savalani MM, Lam CXF, Olkowski R, Ekaputra AK, Tan KC, Hutmacher DW: **Towards a medium/high load-bearing scaffold fabrication system.** *Tsinghua Sci Technol* 2009, **14**(1):13-–19.
7. Brandoff JF, Silber JS, Vaccaro AR: **Contemporary alternatives to synthetic bone grafts for spine surgery.** *Am J Orthop (Belle Mead NJ)* 2008, **37**:410–414.

8. Buma P, Schreurs W, Verdonschot N: **Skeletal tissue engineering-from in vitro studies to large animal models.** *Biomaterials* 2004, **25:**1487–1495.
9. Cowan CM, Soo C, Ting K, Wu B: **Evolving concepts in bone tissue engineering.** *Curr Top Dev Biol* 2005, **66:**239–285.
10. Drosse I, Volkmer E, Capanna R, De Biase P, Mutschler W, Schieker M: **Tissue engineering for bone defect healing: an update on a multi-component approach.** *Injury* 2008, **39**(Suppl 2):S9–S20.
11. LeGeros RZ: **Properties of osteoconductive biomaterials: calcium phosphates.** *Clin Orthop Relat Res* 2002, **395:**81–98.
12. Mistry AS, Mikos AG: **Tissue engineering strategies for bone regeneration.** *Adv Biochem Eng Biot* 2005, **94:**1–22.
13. Bauer TW, Togawa D: **Bone graft substitutes: towards a more perfect union.** *Orthopedics* 2003, **26:**925–926.
14. Roohani-Esfahani S-I: **The influence hydroxyapatite nanoparticle shape and size on the properties of biphasic calcium phosphate scaffolds coated with hydroxyapatite-PCL composites.** *Biogeosciences* 2010, **31:**5498–5509.
15. Delloye C: **Bone substitutes in 2003: an overview.** *Acta Orthop Belg* 2003, **69:**1–8.
16. Gleeson JP, Plunkett NA, O'Brien FJ: **Addition of hydroxyapatite improves stiffness, interconnectivity and osteogenic potential of a highly porous collagen-based scaffold for bone tissue regeneration.** *Eur Cell Mater* 2010, **20:**218–230.
17. Roshan-Ghias A, Terrier A, Bourban PE, Pioletti DP: **In vivo cyclic loading as a potent stimulatory signal for bone formation inside tissue engineering scaffold.** *Eur Cell Mater* 2010, **19:**41–49.
18. Aguirre A, Gonzalez A, Navarro M, Castano O, Planell JA, Engel E: **Control of microenvironmental cues with a smart biomaterial composite promotes endothelial progenitor cell angiogenesis.** *Eur Cell Mater* 2012, **24:**90–106. discussion 106.
19. Seefried L, Mueller-Deubert S, Schwarz T, Lind T, Mentrup B, Kober M, Docheva D, Liedert A, Kassem M, Ignatius A, *et al*: **A small scale cell culture system to analyze mechanobiology using reporter gene constructs and polyurethane dishes.** *Eur Cell Mater* 2010, **20:**344–355.
20. Torcasio A, van Lenthe GH, Van Oosterwyck H: **The importance of loading frequency, rate and vibration for enhancing bone adaptation and implant osseointegration.** *Eur Cell Mater* 2008, **16:**56–68.
21. Wagoner Johnson AJ, Herschler BA: **A review of the mechanical behavior of CaP and CaP/polymer composites for applications in bone replacement and repair.** *Acta Biomater* 2011, **7:**16–30.
22. Babis GC, Soucacos PN: **Bone scaffolds: the role of mechanical stability and instrumentation.** *Injury* 2005, **36:**S38–S44.
23. Nandi SK, Roy S, Mukherjee P, Kundu B, De DK, Basu D: **Orthopaedic applications of bone graft & graft substitutes: a review.** *Indian J Med Res* 2010, **132:**15–30.
24. Gerich TG, Wilmes P, Nackenhorst U, Gosling T, Ziefle M, Krettek C: **A clinical, radiological and computational analysis of the thrust plate prosthesis in young patients.** *Bull Soc Sci Med Grand Duche Luxemb* 2011, **2:**57–70.
25. Hutmacher DW: **Scaffolds in tissue engineering bone and cartilage.** *Biogeosciences* 2000, **21:**2529–2543.
26. Bhumiratana S, Grayson WL, Castaneda A, Rockwood DN, Gil ES, Kaplan DL, Vunjak-Novakovic G: **Nucleation and growth of mineralized bone matrix on silk-hydroxyapatite composite scaffolds.** *Biogeosciences* 2011, **32:**2812–2820.
27. Wu C, Zhang Y, Zhu Y, Friis T, Xiao Y: **Structure–property relationships of silk-modified mesoporous bioglass scaffolds.** *Biogeosciences* 2010, **31:**3429–3438.
28. Zhou H: **Nanoscale hydroxyapatite particles for bone tissue engineering.** *Acta Biomater* 2011, **7:**2769–2781.
29. Tanase CE, Popa MI, Verestiuc L: **Biomimetic bone scaffolds based on chitosan and calcium phosphates.** *Mater Lett* 2011, **65:**1681–1683.
30. Black J, Mattson R, Korostoff E: **Haversian osteons: size, distribution, internatl structure, and orientation.** *J Biomed Mater Res* 1973, **8:**299–319.
31. Wheeless CR: *Wheeless Textbook of Orthopaedics*, Book Wheeless Textbook of Orthopaedics. City: Editor ed.^eds; 1996.
32. Balaban NQ, Schwarz US, Riveline D, Goichberg P, Tzur G, Sabanay I, Mahalu D, Safran S, Bershadsky A, Addadi L, Geiger B: **Force and focal adhesion assembly: a close relationship studied using elastic micropatterned substrates.** *Nat Cell Biol* 2001, **3:**466–472.
33. Lam MT, Sim S, Zhu X, Takayama S: **The effect of continuous wavy micropatterns on silicone substrates on the alignment of skeletal muscle myoblasts and myotubes.** *Biogeosciences* 2006, **27:**4340–4347.
34. Lehnert D, Wehrle-Haller B, David C, Weiland U, Ballestrem C, Imhof BA, Bastmeyer M: **Cell behaviour on micropatterned substrata: limits of extracellular matrix geometry for spreading and adhesion.** *J Cell Sci* 2004, **117:**41–52.
35. Nikkhah M, Strobl JS, Agah M: **Geometry-dependent behavior of fibroblast cells in three-dimensional silicon microstructures.** *Conf Proc IEEE Eng Med Biol Soc* 2007, **2007:**6078–6081.
36. Noireaux V: **Growing an actin gel on spherical surfaces.** *Biophys J* 2000, **78:**1643–1654.
37. Pathak A, Deshpande VS, McMeeking RM, Evans AG: **The simulation of stress fibre and focal adhesion development in cells on patterned substrates.** *J R Soc Interface* 2008, **5:**507–524.
38. Schwartz IM, Ehrenberg M, Bindschadler M, McGrath JL: **The role of substrate curvature in actin-based pushing forces.** *Curr Biol* 2004, **14:**1094–1098.
39. Appleford MR, Oh S, Cole JA, Carnes DL, Lee M, Bumgardner JD, Haggard WO, Ong JL: **Effects of trabecular calcium phosphate scaffolds on stress signaling in osteoblast precursor cells.** *Biogeosciences* 2007, **28:**2747–2753.
40. Yim EKF, Reano RM, Pang SW, Yee AF, Chen CS, Leong KW: **Nanopattern-induced changes in morphology and motility of smooth muscle cells.** *Biogeosciences* 2005, **26:**5405–5413.
41. Kazakia GJ, Nauman EA, Ebenstein DM, Halloran BP, Keaveny TM: **Effects of in vitro bone formation on the mechanical properties of a trabeculated hydroxyapatite bone substitute.** *J Biomed Mater Res A* 2006, **77A:**688–699.
42. Oliver WCaP GM: **An improved technique for determining hardness and elastic modulus using load and displacement sensing indentation experiments.** *J Mater Res* 1992, **7:**1564–1583.
43. Rho JY, Tsui TY, Pharr GM: **Elastic properties of human cortical and trabecular lamellar bone measured by nanoindentation.** *Biogeosciences* 1997, **18:**1325–1330.
44. Reilly DT, Burstein AH, Frankel VH: **The elastic modulus for bone.** *J Biomech* 1974, **7:**271–275.
45. Zhao Z, Zhao M, Xiao G, Franceschi RT: **Gene transfer of the Runx2 transcription factor enhances osteogenic activity of bone marrow stromal cells in vitro and in vivo.** *Mol Ther* 2005, **12:**247–253.
46. Oste L, Bervoets AR, Behets GJ, Dams G, Marijnissen RL, Geryl H, Lamberts LV, Verberckmoes SC, Van Hoof VO, De Broe ME, D'Haese PC: **Time-evolution and reversibility of strontium-induced osteomalacia in chronic renal failure rats.** *Kidney Int* 2005, **67:**920–930.
47. Brunette DM: **Spreading and orientation of epithelial cells on grooved substrata.** *Exp Cell Res* 1986, **167:**203–217.
48. Brunette DM: **Fibroblasts on micromachined substrata orient hierarchically to grooves of different dimensions.** *Exp Cell Res* 1986, **164:**11–26.
49. Brunette DM, Kenner GS, Gould TRL: **Grooved titanium surfaces orient growth and migration of cells from human gingival explants.** *J Dent Res* 1983, **62:**1045–1048.
50. Kapoor A, Caporali EH, Kenis PJ, Stewart MC: **Microtopographically patterned surfaces promote the alignment of tenocytes and extracellular collagen.** *Acta Biomater* 2010, **6:**2580–2589.
51. Clark P, Connolly P, Curtis AS, Dow JA, Wilkinson CD: **Topographical control of cell behaviour: II. Multiple grooved substrata.** *Dev* 1990, **108:**635–644.
52. Recknor JB, Recknor JC, Sakaguchi DS, Mallapragada SK: **Oriented astroglial cell growth on micropatterned polystyrene substrates.** *Biogeosciences* 2004, **25:**2753–2767.
53. Clarke JC, Tuft BW, Clinger JD, Levine R, Figueroa LS, Allan Guymon C, Hansen MR: **Micropatterned methacrylate polymers direct spiral ganglion neurite and Schwann cell growth.** *Hear Res* 2011, **278:**96–105.
54. Bowden N, Huck WTS, Paul KE, Whitesides GM: **The controlled formation of ordered, sinusoidal structures by plasma oxidation of an elastomeric polymer.** *Appl Phys Lett* 1999, **75:**2557–2559.
55. Charest JL, García AJ, King WP: **Myoblast alignment and differentiation on cell culture substrates with microscale topography and model chemistries.** *Biogeosciences* 2007, **28:**2202–2210.
56. Théry M, Pépin A, Dressaire E, Chen Y, Bornens M: **Cell distribution of stress fibres in response to the geometry of the adhesive environment.** *Cell Motil Cytoskeleton* 2006, **63:**341–355.
57. Aubin JE: **Bone stem cells.** *J Cell Biochem Suppl* 1998, **30–31:**73–82.
58. Chen J, Shi ZD, Ji X, Morales J, Zhang J, Kaur N, Wang S: **Enhanced osteogenesis of human mesenchymal stem cells by periodic heat shock in self-assembling peptide hydrogel.** *Tissue Eng Part A* 2013, **19:**716–728.
59. Chyun YS, Kream BE, Raisz LG: **Cortisol decreases bone formation by inhibiting periosteal cell proliferation.** *Endocrinol* 1984, **114:**477–480.

60. Alcantara EH, Shin M-Y, Sohn H-Y, Park Y-M, Kim T, Lim J-H, Jeong H-J, Kwon S-T, Kwun I-S: **Diosgenin stimulates osteogenic activity by increasing bone matrix protein synthesis and bone-specific transcription factor Runx2 in osteoblastic MC3T3-E1 cells.** *J Nutr Biochem* 2011, **22:**1055–1063.
61. Gadeau AP, Chaulet H, Daret D, Kockx M, Daniel-Lamazière JM, Desgranges C: **Time course of osteopontin, osteocalcin, and osteonectin accumulation and calcification after acute vessel wall injury.** *J Histochem Cytochem* 2001, **49:**79–86.
62. Li S, Kong H, Yao N, Yu Q, Wang P, Lin Y, Wang J, Kuang R, Zhao X, Xu J, *et al*: **The role of runt-related transcription factor 2 (Runx2) in the late stage of odontoblast differentiation and dentin formation.** *Biochem Biophys Res Commun* 2011, **410:**698–704.
63. Mikami Y, Takahashi T, Kato S, Takagi M: **Dexamethasone promotes DMP1 mRNA expression by inhibiting negative regulation of Runx2 in multipotential mesenchymal progenitor, ROB-C26.** *Cell Biol Int* 2008, **32:**239–246.
64. Narayanan K, Ramachandran A, Hao J, He G, Park KW, Cho M, George A: **Dual functional roles of dentin matrix protein 1. Implications in biomineralization and gene transcription by activation of intracellular Ca2+ store.** *J Biol Chem* 2003, **278:**17500–17508.
65. Luan Y, Praul CA, Gay CV: **Confocal imaging and timing of secretion of matrix proteins by osteoblasts derived from avian long bone.** *Comparat Biochem Physiol - Part A: Mol; Integ Physiol* 2000, **126:**213–221.
66. Reffitt DM, Ogston N, Jugdaohsingh R, Cheung HFJ, Evans BAJ, Thompson RPH, Powell JJ, Hampson GN: **Orthosilicic acid stimulates collagen type 1 synthesis and osteoblastic differentiation in human osteoblast-like cells in vitro.** *Bone* 2003, **32:**127–135.
67. Kacena MA, Todd P, Gerstenfeld LC, Landis WJ: **Experiments with osteoblasts cultured under varying orientations with respect to the gravity vector.** *Cytotechnology* 2002, **39:**147–154.
68. Rho J-Y, Roy ME, Tsui TY, Pharr GM: **Elastic properties of microstructural components of human bone tissue as measured by nanoindentation.** *J Biomed Mater Res* 1999, **45:**48–54.
69. Roy ME, Rho J-Y, Tsui TY, Evans ND, Pharr GM: **Mechanical and morphological variation of the human lumbar vertebral cortical and trabecular bone.** *J Biomed Mater Res* 1999, **44:**191–197.
70. Turner CH, Rho J, Takano Y, Tsui TY, Pharr GM: **The elastic properties of trabecular and cortical bone tissues are similar: results from two microscopic measurement techniques.** *J Biomech* 1999, **32:**437–441.

12

Parameter-less approaches for interpreting dynamic cellular response

Kumar Selvarajoo[1,2]

Abstract

Cellular response such as cell signaling is an integral part of information processing in biology. Upon receptor stimulation, numerous intracellular molecules are invoked to trigger the transcription of genes for specific biological purposes, such as growth, differentiation, apoptosis or immune response. How complex are such specialized and sophisticated machinery? Computational modeling is an important tool for investigating dynamic cellular behaviors. Here, I focus on certain types of key signaling pathways that can be interpreted well using simple physical rules based on Boolean logic and linear superposition of response terms. From the examples shown, it is conceivable that for small-scale network modeling, reaction topology, rather than parameter values, is crucial for understanding population-wide cellular behaviors. For large-scale response, non-parametric statistical approaches have proven valuable for revealing emergent properties.

Keywords: Biological networks, Non-parametric, Cell signaling, Immune response, Gene expression

The interpretation of dynamic cellular processes is indispensable for biological research. Especially in the last two decades, there have been tremendous efforts that were aimed at understanding complex biological networks in different cell types to various kinds of stimulations or perturbations, and in disease conditions using systems biology approaches. What sort of models do we need to conceptualize biological networks for interpreting or predicting dynamic responses?

In the early 1900s Victor Henri, Leonor Michaelis and Maud Menten thoroughly investigated enzymatic reactions *in vitro*, and developed the hyperbolic rate equation that we now popularly call the Michaelis-Menten enzyme kinetics. This is a more sophisticated form of mass-action type reaction, considering the saturation of kinetics at higher substrate concentrations instead of ever increasing profile for the latter. Subsequent work on this basic principle led to the extension of the kinetics to represent more complex scenarios, such as multi-substrate ping-pong and ternary-complex mechanisms [1].

As the development of computing power progressed significantly in the 1960s, there have been numerous efforts to model complete biological pathway modules, such as the glycolysis, using enzyme kinetic equations with the aim of estimating parameter values by fitting to steady-state concentration levels of metabolites. However, the more truthful abstraction of enzymatic complexity resulted in a dilemma where increased accuracy required increased knowledge of many parameters that were too difficult to obtain precisely. If parameter values are not accurately determined, the enzymatic reaction models will not be able to recapitulate experimental outcome reasonably well.

Most, if not all, studies adopting *in vitro* experiments determine the parameter values of reaction species for computational modeling from an artificial environment where the species are deliberately purified from its physiologic neighbors. This is because, until today, the *in vivo* kinetic parameters cannot be reliably measured using the current experimental technologies. Notably, there have been various reports that claim the kinetic parameters determined through *in vitro* and *in vivo* experiments can differ by several orders of magnitudes [2]. As a result, when combining these errors into the model, the final predictions could differ by several orders of magnitude. For example, the steady-state concentration of the glycolytic metabolite 3-phosphoglycerate in *Trypanosoma brucei* was under-predicted by an order of 7 [3].

Correspondence: kumar@ttck.keio.ac.jp
[1]Institute for Advanced Biosciences, Keio University, Tsuruoka, Yamagata, Japan
[2]Systems Biology Program, Graduate School of Media and Governance, Keio University, Fujisawa, Japan

The difficulty of accurately determining parameter values led to the development of non-parametric approaches such as the flux-balance analysis (FBA) [4]. Here, only the reaction topologies or stoichiometry of the network and steady-state levels are required to be known. Constraints are introduced by the stoichiometric coefficients in the system for the optimization of certain biological function, such as growth or production of certain compounds. Although, the FBA requires the assumption that metabolite concentrations remain at steady-states for analysis, it has been successfully used to interpret important physiological functions of a living cell. For example, Palsson and colleagues experimentally verified their prediction for the primary carbon source and oxygen uptake rates for maximal cellular growth in *E. coli* [5]. So, why is such simple steady-state method relying on stoichiometry of reactions make useful predictions? (Note that FBA requires the network topology to be largely known, as is the case for metabolic networks. For signaling pathways, where the detailed role of numerous molecules are still incomplete, FBA has limited application).

In a pioneering work on understanding the complex dynamics of bacterial chemotaxis, Leibler and colleagues created a highly simplified two-state mass-action model of *E. coli* chemotactic network [6]. Using the model, and subsequently with experiments [7], they showed that the adaptation precision of bacterial chemotaxis was insensitive to the large variation of its network parameter values. This mechanism, therefore, allows *E. coli* to display robust behavior to a wide range of attractant and repellent concentrations. However, at the same time, other properties, such as adaptation time and steady-state tumbling frequency, were variable to the stimulant concentration. Overall, their work demonstrated that bacterial adaptation property is a consequence of network's connectivity and does not require the precision of parameter values. This work is a milestone paper that indicates complex biological phenomena can be understood using simple models that are not sensitive to parameter values.

The observation of simplicity in what appears to be highly dynamic and complex can have profound benefits in understanding and controlling disease conditions. Our research has focused on cell signaling dynamics of innate immune response and cancer cell survival. Over the last decade, we adopted systems biology approaches to study toll-like receptor (TLR) signaling [8-10], tumor necrosis factor (TNF) signaling [11] and TNF-related apoptosis-inducing ligand (TRAIL) signaling [12], from receptor stimulation through downstream gene expressions, via transcription factor activations.

The strategy was to first create a dynamic computational model based on current known pathways of a signaling process. Next, first-order response (mass-action) equation was used to represent each signaling reaction or process (protein binding, complex formation, ubiquitination, etc.). Subsequently, the model parameters were chosen to fit wildtype experimental dynamics, and compared with mutant cells for reliability of the models and their parameters [13]. When a single model is unable to simulate multiple experimental conditions, the model's topology was allowed to be modified, using *response rules*, in accordance with the law of signaling flux conservation [9-13]. This is simply because we do not yet possess complete knowledge of all signaling reactions or molecules involved.

Notably, for all the complex signaling processes that we have investigated so far, we were successful to predict novel signaling features, such as missing intermediates, crosstalk mechanisms, feedback loops [8,11,12], and identify novel targets for controlling proinflammatory response [11] and cancer apoptosis [12]. All the predictions have been experimentally validated [9,11,14,15]. So why do simple models utilizing first-order response equations sufficient to produce insightful results of a complex system?

Firstly, the main reason for us to utilize first-order terms is due to the experimental observation of deterministic response waves of signal transduction within the period of investigations, usually up to 1-2 h after stimulation. That is, stimulating cell population in a dish with respective ligands resulted, in general, to dynamic activation response of intracellular proteins that followed gradual increase from their initial state to reach peak activation levels and, subsequently, decay to their original state (Figure 1A). Such responses are observed for the first round of response waves of myriad signaling species (Figure 1B). Although the kinetics could vary slightly from sample to sample, the general average response profiles are very well reproducible. In other words, regardless of how complex a signaling topology might be, the species' average dynamic responses followed deterministic formation and depletion waves [13,17,18].

Secondly, it can be shown, theoretically, that no matter how complex or non-linear the signaling system is, the dynamic response can be approximated using first-order terms if the perturbation levels are small. Consider the general form of a complex kinetic equation: $\frac{\partial X}{\partial t} = F(X)$. F can be any non-linear function constituting of reaction and diffusion terms of species X. In engineering, for relative changes and when insufficient information is available, such systems are often carefully linearized using power or Talyor series ($\frac{\partial \delta X}{\partial t} = \frac{\partial F(X)}{\partial X}|_{X_a} \delta X + \frac{\partial F^2(X)}{2!\partial X^2}|_{X_a} \delta X^2 + \ldots$, where $X = X_a$ is the point for linearization). Given a small perturbation, the higher order terms become less significant, leaving only the first-order term as the dominant factor. Note that the linearization techniques are approximate

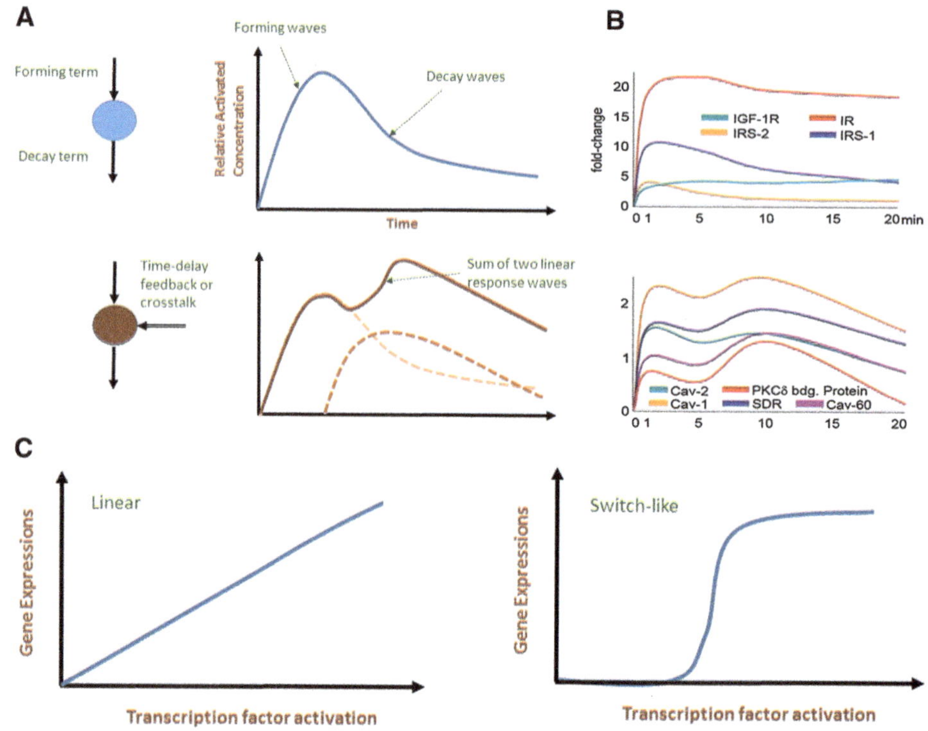

Figure 1 The observation of linear response waves. A) Schematic of activated signaling species, such as protein binding and gene expressions, with respect to time following formation and decay waves. Top panel represents a simple linear cascade with single wave. Bottom panel illustrates two linear waves superposed, as a consequence of an additional time-delay formation term. This may arise from feedback or crosstalk mechanisms. **B)** Quantitative dynamics of key molecules in insulin signaling pathway, showing similar dynamics to schematic in **A)**. Figures adapted from [16]. **C)** Schematic of linear and switch-like relationship between transcription factor concentration and gene expressions.

methods to understand general behaviors and, in many cases, cannot be used to interpret detailed mechanisms of response.

In light of this linear response hypothesis, it is noteworthy to quote the recent findings of two relevant works, that studied the relationships between the transcription factors and gene expressions in TNF-induced [19] and Msn2 overexpressed [20] stress response. Collectively, they found that increasing transcription factor concentration resulted in graded gene expressions that approximately followed a linear relationship (Figure 1C). Although it is known that many transcription factors produce switch-like or digital relationship due to cooperativity in DNA binding, the stress response transcription factors have shown simple graded behavior. This finding may justify that certain key cellular processes, such as the immune response, may be guided by linear response through the signaling cascades. Taken together, it appears that linear response, as a governing principle, is key to invoke precise and optimal response when living cells are faced with immediate threats.

In other studies, even without the need to know graded response, binary (ON/OFF) state approaches have yielded fascinating results in understanding cell signaling. One notable study developed discrete Boolean network modeling to investigate the survival mechanism of cytotoxic T lymphocytes (CTL) in T cell large granular lymphoctye (T-LGL) leukemia [21]. Loughran and colleagues created a T-LGL survival signaling model with 58 nodes, representing molecular species, and 123 edges, representing causal interactions between the species. Using the model, they identified the most significant interactions for activating CTL in disease state compared to normal. Subsequent experiments confirmed their model predictions.

It is conceivable that the arrival to parameter-less approaches may be unrealistic in the realm of complex systems, where non-linear factors and stochastic effects can cause even small variation in perturbations to produce diverse multistable outcomes or oscillatory patterns. Such is the case observed for cell fate decisions where a single fertilized egg can diversify into distinct cell lineages or a bacteria being able to change fate under nutrient-deficient condition [22]. To model such complexity, dynamical systems theory adopting non-linear equations may possibly be used [23]. Also, for understanding self-organizing behaviors such as biological clocks/rhythms, spatial patterns, Hopf bifurcation or other non-linear dynamics, Goodwin, Brusselator, and Lotka–Volterra equations have been widely adopted [24-26]. However, these models require the precision of parameter values and most often reproduce only the general behavior of complex biological responses in one (wildtype) condition.

Another issue to consider is the scale of networks. So far, biological modules or network modeling that have been successfully used consist of molecular species that are relatively small, in the order of tens or a few hundreds. However, the living system invokes response of thousands of species and such large-scale studies probably require different approaches. One common strategy used to tackle large-scale effects is to use statistical techniques that investigate regression or correlation between species and samples, or apply clustering techniques to identify groups of genes with similar temporal or functional behaviors [27,28]. These methods have been instrumental in revealing emergent behaviors, for example, the observation of collective oscillations of numerous cell cycle independent specific metabolic cycle genes in *Saccharomyces cerevisiae* [29,30], and the collective genome-wide expression dynamics, including lowly expressed genes, for innate immune response [31,32] and neutrophil cell differentiation [33,34]. For classifying distinct cancer types for targeted therapy, self-organizing maps on high-dimensional gene expression data have been highly useful [35]. Thus, non-parametric statistical works on high throughput gene expression datasets have been crucial in showing emergent self-organized behaviors in cell populations.

In the future, non-parametric autonomous Boolean circuits, that have been recently shown to generate chaos, with multiple attractor states through time-delayed feedback loops in physical signal propagation [36,37], may also be investigated for biological systems. These could, especially, be valuable for the application of cell signaling related to non-linear cell fate decisions or disease formation.

Acknowledgement

The author thank Tsuruoka city, Yamagata Prefecture for financial support and acknowledge the JSPS Grants-in-Aid for Scientific Research FX132008K3.

References

1. Leskovac V: *Comprehensive enzyme kinetics.* New York: Kluwer Academic/ Plenum Pub; 2003.
2. Gutenkunst RN, Waterfall JJ, Casey FP, Brown KS, Myers CR, Sethna JP: **Universally sloppy parameter sensitivities in systems biology models.** *PLoS Comput Biol* 2007, **3:**1871–1878.
3. Bakker BM, Michels PA, Opperdoes FR, Westerhoff HV: **Glycolysis in bloodstream form Trypanosoma brucei can be understood in terms of the kinetics of the glycolytic enzymes.** *J Biol Chem* 1997, **272:**3207–3215.
4. Edwards JS, Palsson BO: **How will bioinformatics influence metabolic engineering?** *Biotechnol Bioeng* 1998, **58:**162–169.
5. Edwards JS, Ibarra RU, Palsson BO: **In silico predictions of Escherichia coli metabolic capabilities are consistent with experimental data.** *Nat Biotechnol* 2001, **19:**125–130.
6. Barkai N, Leibler S: **Robustness in simple biochemical networks.** *Nature* 1997, **387:**913–917.
7. Alon U, Surette MG, Barkai N, Leibler S: **Robustness in bacterial chemotaxis.** *Nature* 1999, **397:**168–171.
8. Selvarajoo K: **Discovering differential activation machinery of the Toll-like receptor 4 signaling pathways in MyD88 knockouts.** *FEBS Lett* 2006, **580:**1457–1464.
9. Selvarajoo K, Takada Y, Gohda J, Helmy M, Akira S, Tomita M, Tsuchiya M, Inoue J, Matsuo K: **Signaling flux redistribution at toll-like receptor pathway junctions.** *PLoS One* 2008, **3:**e3430.
10. Helmy M, Gohda J, Inoue J, Tomita M, Tsuchiya M, Selvarajoo K: **Predicting novel features of toll-like receptor 3 signaling in macrophages.** *PLoS One* 2009, **4:**e4661.
11. Hayashi K, Piras V, Tabata S, Tomita M, Selvarajoo K: **A Systems Biology Approach to Suppress TNF-induced Proinflammatory Gene Expressions.** *Cell Commun Signal* 2013, **11:**84.
12. Piras V, Hayashi K, Tomita M, Selvarajoo K: **Enhancing apoptosis in TRAIL-resistant cancer cells using fundamental response rules.** *Sci Rep* 2011, **1:**144.
13. Selvarajoo K: *Immuno Systems Biology: A macroscopic approach for immune cell signaling.* New York: Springer; 2013.
14. Kagan JC, Su T, Horng T, Chow A, Akira S, Medzhitov R: **TRAM couples endocytosis of Toll-like receptor 4 to the induction of interferon-beta.** *Nat Immunol* 2008, **9:**361–368.
15. Zanoni I, Ostuni R, Marek LR, Barresi S, Barbalat R, Barton GM, Granucci F, Kagan JC: **CD14 controls the LPS-induced endocytosis of Toll-like receptor 4.** *Cell* 2011, **147:**868–880.
16. Krüger M, Kratchmarova I, Blagoev B, Tseng YH, Kahn CR, Mann M: **Dissection of the insulin signaling pathway via quantitative phosphoproteomics.** *Proc Natl Acad Sci U S A* 2008, **105:**2451–2456.
17. Selvarajoo K, Tomita M, Tsuchiya M: **Can complex cellular processes be governed by simple linear rules?** *J Bioinfor and Comp Biol* 2009, **7:**243–268.
18. Selvarajoo K: **Macroscopic law of conservation revealed in the population dynamics of Toll-like receptor signaling.** *Cell Commun Signal* 2011, **9:**9.
19. Giorgetti L, Siggers T, Tiana G, Caprara G, Notarbartolo S, Corona T, Pasparakis M, Milani P, Bulyk ML, Natoli G: **Noncooperative interactions between transcription factors and clustered DNA binding sites enable graded transcriptional responses to environmental inputs.** *Mol Cell* 2010, **37:**418–428.
20. Stewart-Ornstein J, Nelson C, DeRisi J, Weissman JS, El-Samad H: **Msn2 coordinates a stoichiometric gene expression program.** *Curr Biol* 2013, **23:**2336–2345.
21. Zhang R, Shah MV, Yang J, Nyland SB, Liu X, Yun JK, Albert R, Loughran TP Jr: **Network model of survival signaling in large granular lymphocyte leukemia.** *Proc Natl Acad Sci U S A* 2008, **105:**16308–16313.
22. Selvarajoo K: **Understanding multimodal biological decisions from single cell and population dynamics.** *Wiley Interdiscip Rev Syst Biol Med* 2012, **4:**385–399.
23. Furusawa C, Kaneko K: **A dynamical-systems view of stem cell biology.** *Science* 2012, **338:**215–217.
24. Woller A, Gonze D, Erneux T: **The Goodwin model revisited: Hopf bifurcation, limit-cycle, and periodic entrainment.** *Phys Biol* 2014, **11:**045002.
25. Prigogine I, Lefever R: **Symmetry breaking instabilities in dissipative systems. II.** *J Chem Phys* 1968, **48:**1695.
26. Barrio RA, Romero-Arias JR, Noguez MA, Azpeitia E, Ortiz-Gutiérrez E, Hernández-Hernández V, Cortes-Poza Y, Álvarez-Buylla ER: **Cell patterns emerge from coupled chemical and physical fields with cell proliferation dynamics: the Arabidopsis thaliana root as a study system.** *PLoS Comput Biol* 2013, **9:**e1003026.
27. Gitter A, Lu Y, Bar-Joseph Z: **Computational methods for analyzing dynamic regulatory networks.** *Methods Mol Biol* 2010, **674:**419–441.
28. Chandrasekaran S, Price ND: **Probabilistic integrative modeling of genome-scale metabolic and regulatory networks in Escherichia coli and Mycobacterium tuberculosis.** *Proc Natl Acad Sci U S A* 2010, **107:**17845–17850.
29. Tu BP, Kudlicki A, Rowicka M, McKnight SL: **Logic of the yeast metabolic cycle: temporal compartmentalization of cellular processes.** *Science* 2005, **310:**1152–1158.
30. Tu BP, McKnight SL: **Metabolic cycles as an underlying basis of biological oscillations.** *Nat Rev Mol Cell Biol* 2006, **7:**696–701.
31. Tsuchiya M, Piras V, Choi S, Akira S, Tomita M, *et al*: **Emergent genome-wide control in wildtype and genetically mutated lipopolysaccarides-stimulated macrophages.** *PLoS One* 2009, **4:**e4905.

32. Selvarajoo K, Giuliani A: **Finding self-organization from the dynamic gene expressions of innate immune responses.** *Front Physiol* 2012, **3**:192.
33. Huang S, Eichler G, Bar-Yam Y, Ingber DE: **Cell fates as high-dimensional attractor states of a complex gene regulatory network.** *Phys Rev Lett* 2005, **94**:128701.
34. Tsuchiya M, Piras V, Giuliani A, Tomita M, Selvarajoo K: **Collective dynamics of specific gene ensembles crucial for neutrophil differentiation: the existence of genome vehicles revealed.** *PLoS One* 2010, **5**:e12116.
35. Guo Y, Eichler GS, Feng Y, Ingber DE, Huang S: **Towards a holistic, yet gene-centered analysis of gene expression profiles: a case study of human lung cancers.** *J Biomed Biotechnol* 2006, **2006**:69141.
36. Zhang R, Cavalcante HL DS, Gao Z, Gauthier DJ, Socolar JE, Adams MM, Lathrop DP: **Boolean chaos.** *Phys Rev E Stat Nonlin Soft Matter Phys* 2009, **80**:045202.
37. Cavalcante HL, Gauthier DJ, Socolar JE, Zhang R: **On the origin of chaos in autonomous Boolean networks.** *Philos Trans A Math Phys Eng Sci* 2010, **368**:495–513.

13

Challenges in predicting the evolutionary maintenance of a phage transgene

Matthew Schmerer[1], Ian J Molineux[2,3], Dilara Ally[4], Jabus Tyerman[5], Nicole Cecchini[6] and James J Bull[1,3,6*]

Abstract

Background: In prior work, a phage engineered with a biofilm-degrading enzyme (dispersin B) cleared artificial, short-term biofilms more fully than the phage lacking the enzyme. An unresolved question is whether the transgene will be lost or maintained during phage growth – its loss would limit the utility of the engineering. Broadly supported evolutionary theory suggests that transgenes will be lost through a 'tragedy of the commons' mechanism unless the ecology of growth in biofilms meets specific requirements. We test that theory here.

Results: Functional properties of the transgenic phage were identified. Consistent with the previous study, the dispersin phage was superior to unmodified phage at clearing short term biofilms grown in broth, shown here to be an effect attributable to free enzyme. However, the dispersin phage was only marginally better than control phages on short term biofilms in minimal media and was no better than control phages in clearing long term biofilms. There was little empirical support for the tragedy of the commons framework despite a strong theoretical foundation for its supposed relevance. The framework requires that the transgene imposes an intrinsic cost, yet the transgene was intrinsically neutral or beneficial when expressed from one part of the phage genome. Expressed from a different part of the genome, the transgene did behave as if intrinsically costly, but its maintenance did not benefit from spatially structured growth per se – violating the tragedy framework.

Conclusions: Overall, the transgene was beneficial under many conditions, but no insight to its maintenance was attributable to the established evolutionary framework. The failure likely resides in system details that would be used to parameterize the models. Our study cautions against naive applications of evolutionary theory to synthetic biology, even qualitatively.

Keywords: Phage, Evolution, Biofilm, Tragedy of the commons, Synthetic biology

Background

Biofilms are persistent bacterial structures that arise in many medical and industrial applications. They are usually undesirable and difficult to control because they are recalcitrant to treatments that eliminate planktonic bacteria [1,2]. Bacterial viruses – phages – seem to offer a promising technology for biofilm control because they are predators of bacteria and should in theory amplify and persist at the site of biofilms as long as their prey persists. Yet although phage kill bacteria, they are not necessarily endowed with the means to propagate themselves through or degrade a biofilm's extracellular matrix. Without the ability to penetrate a biofilm, phage may merely graze on planktonic cells exuded by a biofilm without degrading or effecting a long term reduction in bacterial numbers [3].

Lu and Collins [4] proposed using engineered phages to enhance biofilm degradation. Their specific application was the addition of a gene encoding the bacterial enzyme dispersin B to the lytic coliphage T7. Dispersin degrades poly-N-acetyl-glucosamine (PNAG), an essential component of *E. coli* biofilms. The feasibility of their method in short term biofilms was demonstrated with the engineered phage providing better clearing and cell killing than did various control phages lacking the enzyme.

*Correspondence: bull@utexas.edu
[1]Center for Computational Biology and Bioinformatics, The University of Texas at Austin, Austin, TX, USA
[3]Institute for Cellular and Molecular Biology, The University of Texas at Austin, Austin, TX, USA
Full list of author information is available at the end of the article

The purpose of the present study is to explore the evolution of this engineered phage – whether it maintains the transgene as the phage grows on biofilms. Dispersin performs its biofilm-degrading function only after an initial cell has been infected, lysed and released the enzyme into the environment. Because the enzyme performs its function outside the cell, the transgene may be lost through an evolutionary 'tragedy of the commons' in which the gene benefits not only its own genome but the genomes of unrelated phages lacking the gene. A considerable body of theory and empirical work has supported the tragedy of the commons framework at the organismal level [5-12], but it is not known whether and how that framework can be applied to transgenic phages. If the dispersin transgene is evolutionarily stable, it will help alleviate concerns for other engineered phages and provide impetus for their use. If unstable, can a better transgene design extend its longevity? The hope is that the evolutionary theory can provide broad insight to the fate and design of engineered phages across many contexts and system-specific details.

The theory

Basics

The addition of a gene to a genome will often have deleterious effects – a cost, c. We conjecture that a cost is nearly universal for an arbitrary transgene but that the bases of that cost lie in intricate molecular details of the phage life cycle, so the magnitude of the cost will be sensitive to details and be quantitatively unpredictable. A transgene may also provide a benefit (b), and the basis of this benefit will often be known or inferred in advance, as it is the motivation for the engineer. It is straightforward to appreciate that a transgene will be favored only if the benefit outweighs the cost: $b > c$, but understanding how these costs and benefits interact is not always straightforward. For example, in our case, any cost likely operates during the phage intracellular life cycle, whereas benefit comes from modifying the external environment in which the phage grows. The benefit thus varies with environment, but the specific nature of this benefit presents yet an additional problem.

We study a phage genome encoded to produce a transgenic enzyme, dispersin. A transgenic protein of this type is produced within the infected cell, but the expected benefit is to the phage progeny after the bacterium is lysed. After lysis, the enzyme diffuses extracellularly where it can exert its effect on the biofilm exopolysaccharide, releasing bacteria from the biofilm. The benefit is likely one of exposing bacteria to phage that would otherwise be protected by the biofilm matrix. Thus this benefit may well accrue to phages other than the ones encoding the transgene – any phage in the local environment should benefit from the cells exposed by dispersin. Consequently, the transgenic phage may be subject to an evolutionary 'tragedy of the commons' in which it experiences a net benefit of the transgene, but other, non-producing phages experience an even greater benefit because their genomes do not suffer the cost. The tragedy is that, even though the transgene is beneficial, phages lacking it are evolutionarily superior and the transgene is outcompeted to extinction.

The foundation for this problem thus rests on three premises. First, there is likely an intrinsic cost to the transgene of unknown magnitude and perhaps little more than a generic consequence of disrupting a wild-type genome (see Discussion for support). Second, the transgene provides a benefit, one that applies only in some environments. There may be many contexts in which the cost exceeds the benefit, but as the benefit depends on environment, there may also be environments in which the benefit exceeds the cost. Third, the specific nature of the benefit is transferable to other phages, leading to the possible evolutionary demise of the transgene even when its benefit exceeds the cost. Avoiding this sharing of benefits is key to evolutionary success of the transgenic phage, as will be expanded upon below.

A mathematical model

The evolutionary process described above is unintuitive, so we offer a mathematical model. Following several precedents [13,14], the model uses differential equations to describe the dynamics of phage and bacteria in an environment with continuous flow, such as a chemostat, equivalent to a constant death rate. The continuous flow is represented as a constant washout/death rate of all variables. The model assumes mass action, which is most prone to a tragedy of the commons. A more accurate model of the process would include spatial structure (e.g., use partial differential equations), but the main points can be illustrated with the present model without undue emphasis on the mathematics.

The full model (1) has 8 equations that accommodate two bacterial populations and two phage types. One bacterial population exists in a 'free' state and is infected by both types of phages. The other bacteria exist in a refuge that fully escapes infection; the refuge population can be considered a physical region free of phages, or it can equally be regarded as planktonic cells whose surface properties protect them from infection. However, refuge bacteria are released at a low rate into the free state, at which point they are susceptible to infection.

The two phage types consist of (a) a genetically modified (GM) phage and (b) a 'wild' phage. The GM phage produces a diffusible enzyme whose concentration in the environment (E) increases the release rate of refuge bacteria into the free population; the wild phage does not produce the enzyme. The two phages have identical fitness components except for burst sizes (Table 1). In some parameterizations, the GM phage has a lower burst size than the wild phage (17 versus 20), in others, the GM phage has the higher burst size (20 versus 17), in yet others, each phage has a burst size of 20. All other fitness components are the same between both phages.

Parameters and variables are defined in Table 1; bacterial densities are denoted B and R (free and refuge, respective), phage densities as P_G and P_W. Bacterial growth in both the free and refuge populations obeys a logistic function with carrying capacity C that applies to the total density of bacteria, $B+R$. Lysis is modeled as a delay function L minutes after infection, hence a subscript L (e.g., B_L, P_{W_L}) indicates the value of the variable L minutes in the past. The baseline rate at which refuge bacteria are released into the free state is 0.01 (per min), increasing up to 0.2 with E according to the function $h(E)$.

Table 1 Model variables and parameters

Notation	Description	Values
Variables		
B	Density of free bacteria	
R	Density of refuge bacteria	
P_G	Density of GM phage	
P_W	Density of wild phage	
E	Density of enzyme produced by GM phage that increases the release rate of refuge bacteria into the free state rate of both phages	
I_G	Density of bacteria infected with GM phage (before lysis)	
I_W	Density of bacteria infected with wild phage (before lysis)	
Functions of variables		
h(E)	Rate at which refuge bacteria move to free state	
Parameters		
k	Adsorption rate constant	10^{-9}(mL/min)
w	Washout/death rate	0.05
b_G	Burst size of GM phage	17, 20
b_W	Burst size of wild phage	17, 20
L	Lysis time (min)	10
v	Maximum bacterial growth rate	0.1
Z	Enzyme production per lysis	0.1
C	Carrying capacity of environment	5×10^9

In the equations that define this model, a superior dot () indicates a derivative with respect to time:

$$\begin{aligned}
\dot{B} &= v\left(1-\frac{B+R}{C}\right)B - kP_GB - kP_WB - wB + Rh(E)\\
\dot{R} &= v\left(1-\frac{B+R}{C}\right)R - Rh(E)\\
\dot{P}_G &= b_G\, kP_{G_L}B_L\left[e^{-wL}\right] - kP_G\,(B + I_G + I_W) - wP_G\\
\dot{P}_W &= b_W\, kP_{W_L}B_L\left[e^{-wL}\right] - kP_W\,(B + I_G + I_W) - wP_W\\
\dot{I}_G &= -kP_{G_L}B_L\left[e^{-wL}\right] + kP_GB - wI_G\\
\dot{I}_W &= -kP_{W_L}B_L\left[e^{-wL}\right] + kP_WB - wI_W\\
\dot{E} &= ZkP_{G_L}B_L\left[e^{-wL}\right] - wE\\
h(E) &= \frac{0.2}{1+19\,(0.999997^E)}\;.
\end{aligned} \tag{1}$$

Representative short term dynamics are illustrated in Figure 1 and do not require understanding the equations. The top two panels show outcomes for environments with single phages, each with identical parameters except for enzyme production. In each, phage densities start out low and increase in response to the abundance of hosts. In (A), the wild phage cannot infect the refuge population because they cannot free it, so phage and bacteria are both maintained at high density – the refuge population continually generates free bacteria on which the wild phage preys. This equilibrium continues indefinitely. In (B), the GM phage frees refuge bacteria and exposes them to attack; the total bacterial density declines several logs once the phage reach high density, and because hosts are no longer abundant, phage densities then decline as well. If continued longer, the bacteria would rebound, and the phage would again rise to suppress their numbers; this longer time frame is not necessarily relevant to phage treatment applications. Because the GM phage is able to destroy the refuge bacteria, its numbers briefly spike above those of the wild phage (in A), but then the paucity of hosts causes GM phage densities to fall below those of the wild phage. The green curves show the ($\log_{10}$) rate at which refuge bacteria are being released into the free population.

The lower two panels illustrate cultures with both phages together; in each panel, one phage is given a lower burst size than the other (17 for one, 20 for the other). Both lower panels show the same general pattern seen in (B) of phages increasing in abundance initially, eventually overwhelming both bacterial populations, and then declining for lack of abundant hosts. However, of specific interest in (C) and (D) are changes in the *relative* abundances of the two phages (the relative heights of the two phage curves) and the effects of those relative changes

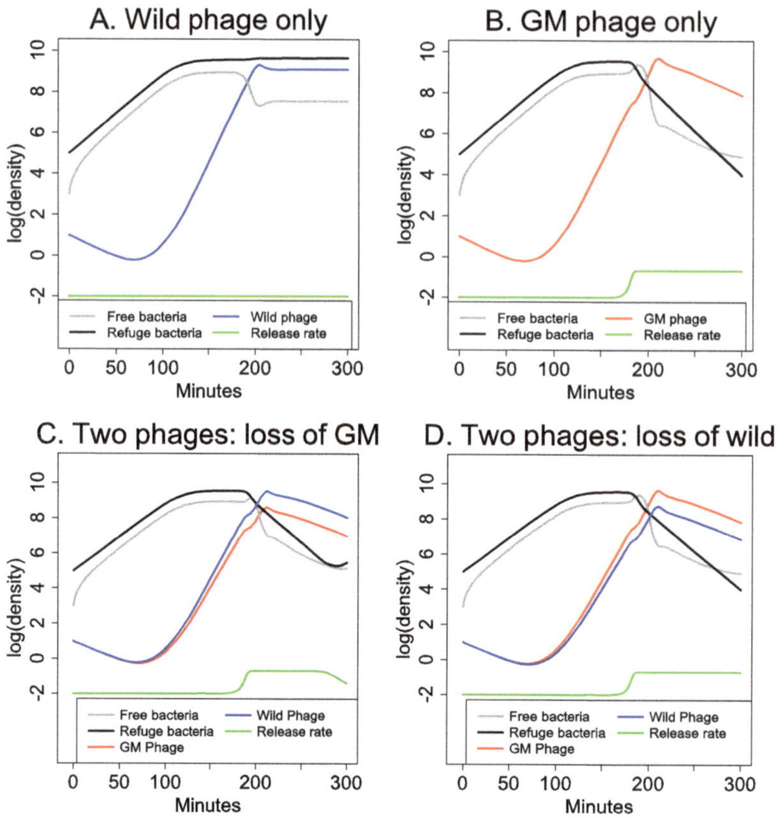

Figure 1 Short term phage-bacterial dynamics and a tragedy of the commons between two phage types. Free bacterial densities are in gray, refuge bacterial densities in black, genetically modified (GM) phage densities in red, and wild phage densities in blue. The rate at which refuge bacteria are released into the free state is in green, affected by enzyme density (not shown). Panels **(A)** and **(B)** represent cultures infected with single phage types. **(A)**: The wild phage type has little effect on bacterial densities because it does not cause release of refuge bacteria; it maintains high levels by feeding on free bacteria released from the refuge. **(B)**: The GM phage produces an enzyme that releases refuge bacteria. At high phage densities, refuge bacteria are released and killed rapidly, bacterial densities drop, and phage densities follow. **(C)**: A tragedy of the commons. Both phages are present, and in the short term, the overall phage and bacterial dynamics follow a similar pattern as when only the GM phage is present, with refuge bacteria depleted (as in B). The wild phage has a slightly higher burst than the GM phage (20 versus 17); although both phages start out equally abundant, the GM phage declines and is eventually lost if the run is continued further, the dynamics returning to the pattern in (A). Loss of the GM phage is the 'tragedy,' since the GM phage releases refuge bacteria. **(D)**: As in (C), but the wild phage is given the lower burst size (17 versus 20) and it is progressively lost. There is no tragedy because GM phage is maintained. When the GM phage is lost (in C), there is an earlier recovery of bacterial densities and earlier return of the release rate to baseline than in (D). The scale of the vertical axis is $\log_{10}$.

on the bacterial dynamics. The two phages start out at 50% frequency in both panels. In (C), the GM phage has the smaller burst and it is progressively lost (dropping to 9% frequency by the end). In (D), the wild phage has the smaller burst, and it is progressively lost (dropping to 10% frequency by the end).

Although the GM phage is responsible for the major impact on bacterial densities (as seen by comparing (A to B), it is progressively lost when the wild phage has the larger burst (in C), which is the tragedy of the commons. The effect of wild phage dominance in (C) is already evident as a reduced impact on and faster rebound of bacterial densities as well as an earlier decline in the release rate of refuge bacteria (compare C to D). If the run in (C) was continued, the system would return to the equilibrium in (A), with a high density of refuge bacteria and only the wild phage present. If the run in (D) was continued, the wild phage would be lost, and both the bacteria and GM phage would cycle between high and low frequencies.

These figures are representative of additional runs (not shown). The details of the dynamics vary (e.g., oscillations dominate some parameter values, densities equilibrate for others), but the GM phage is always progressively lost when it has the lower burst size. Thus the benefit it provides by exposing refuge bacteria does not contribute to its maintenance when both phages are present. It may seem that a tragedy of commons is not relevant when only a single phage type is introduced to a system (e.g., Figure 1B). However, mutations that delete the transgene often quickly arise and convert the population to a mix of types with and without the transgene.

Can the tragedy be avoided?

A widely acknowledged principle for avoiding an evolutionary tragedy of the commons is to impose spatially structured growth [5,7,11,15]. Spatial structure enables an individual and its descendants to be 'selfish' by restricting diffusion of its products just to immediate relatives. The spatial structure must be refreshed often enough (e.g., by dispersal and reestablishment) that selfish mutants do not have time to invade. In theory, an engineered phage introduced into a biofilm might avoid the tragedy if there was adequate structured growth. Some observations of phages in biofilms support structured growth [16], but many dynamical details are unknown, and there may also be a substantial level of phage 'grazing' on planktonic cells produced by biofilms [3].

There is, however, a second layer to the tragedy of the commons that applies to phages: spatially structured growth can work against phages engineered to kill bacteria efficiently and rapidly. In the absence of spatial structure, when phages are mixed in direct competition, phages with higher growth rate have the advantage because they are the fastest to use up all the hosts [10]. In contrast, with spatial structure, phages with low-intermediate growth rates have the advantage because they use their hosts more 'prudently' by allowing host numbers to reach high levels before exploiting them. In the long run, those high bacterial densities yield high levels of phage progeny production – higher progeny levels than achieved by phages that kill quickly. Viewed in this light, the advantage of rapid phage growth when phages compete directly, in the absence of spatial structure, is itself a tragedy of the commons [10].

The benefit to the wild phage of avoiding direct competition with the GM phage can be seen by comparing phage densities in Figure 1A and B. When grown by itself, the GM phage quickly exhausts its hosts and thus realizes only a brief numerical superiority over the wild phage, but outside this window, the wild phage produces more progeny than the GM phage. If spatially structured growth operates so that there is a limited abundance of hosts available at each focus of infection, there may be only a brief window of time in which the GM phage produces more total offspring than the wild phage because it runs out of hosts.

Ironically, therefore, there are two layers to the tragedy of the commons in this problem, and they work against the GM phage both with and without spatial structure. The intrinsic high growth rate of the GM phage would give it the advantage when in direct competition with the wild phage (in the absence of spatial structure) except that it shares its enzyme and thus also boosts the wild phage growth rate - a tragedy that works against the GM phage. Yet the fact that the faster-growing phage wins the direct competition is itself a tragedy, a tragedy that would work in favor of the GM phage if it did not share enzyme. The first tragedy trumps the second, however. Yet when the phages grow with spatial structure, both tragedies are avoided and the wild phage can produce more progeny in the long run because it does not exhaust its hosts quickly.

It is of course possible that spatially structured growth of the phages would conspire to create just the right combination of dynamics to give the advantage to the GM phage. However, even if this were serendipitously achieved in the initial conditions, a further complication is that the degree of spatial structured growth may itself change and decay during phage growth. As phage densities increase and phages diffuse throughout the biofilm, there may be progressively fewer infection foci founded by single phages.

In sum, we can merely conclude at this juncture that the outcome of spatially structured competition is sensitive to subtle and likely unmeasurable details in the growth dynamics. At best, there may be a temporary phase in which the GM phage can avoid a tragedy of the commons, but we are limited to conjecture as to whether a biofilm ever provides the appropriate structure.

A heuristic approach

The value of model (1) is heuristic. With 8 parameters and several functional forms, it cannot be parameterized, and its neglect of explicit spatial structure is also a limitation. There is little alternative, however, as phage dynamics embedded in a biofilm have never been precisely quantitfied. The model nonetheless guides us in a qualitative awareness of the conditions for evolutionary maintenance of a biofilm degrading enzyme:

1. The transgene will be favored if it provides an intrinsic benefit – one that operates in the absence of a biofilm-structured environment (e.g., a larger burst size). This point is obvious, but we also consider intrinsic benefits to be rare outcomes – we can cite no precedent for a transgene that proved beneficial in the absence of a metabolic modification designed to favor the gene, such as drug resistance (see Discussion). However, even if an intrinsic cost proves to be nearly universal, it is conceivable that engineering can be used to minimize it.
2. A gene with an intrinsic cost can be favored if its biofilm degradation exposes cells to attack that would otherwise be protected. This condition is not sufficient however, as there must be spatially structured phage growth that strikes a delicate balance between avoiding enzyme sharing with unrelated phages and killing its hosts too quickly. At

present, the dynamics of phage growth in biofilm environments have too many unknowns for these conditions to be predictable.

On theoretical grounds, the odds oppose the evolutionary maintenance of phages engineered to degrade biofilms.

Results

The ultimate focus of this paper is evolutionary – the maintenance or loss of the dispersin transgene. Yet the selective basis of a transgene product resides in its phenotypic effect. The first part of Results thus addresses the phenotypic effect of dispersin-bearing phages in several environmental contexts. Phenotype assessment is then followed by evolutionary considerations.

1. The dispersin phenotype meets expectations

Lu and Collins [4] used a biofilm-clearing assay of intact phages to infer activity of the dispersin transgene. T7 with dispersin outperformed the parental, empty T7 vector as well as a wild-type T3 control. However, T3 substantially outperformed the empty T7 vector, raising the possibility that the dispersin transgene may have augmented biofilm clearing in ways that did not exploit its enzyme activity. Thus we first explore the nature of dispersin's activity in clearing biofilms: (i) Is biofilm clearing robust to growth conditions? (ii) Does dispersin enzyme alone have the expected effect? Our assays used short-term biofilms broadly similar to those of Lu and Collins – overnight bacterial growth in polystyrene 24-well microtiter plates followed by phage treatment, then assayed with crystal violet (CV) staining. The bacterial strain used was a *csrA* deletion mutant of *E. coli*, which overproduces the PNAG substrate digested by dispersin; the mutant exhibits enhanced biofilm production. Untreated and unstained, these overnight biofilms were evident as thin films (less than 1 mm thick) on the bottom and sides of a well.

Various dispersin phages were tested. The specific notation for a phage is T7$insert_{location}$, where *insert* indicates the set of genes added to the phage genome, and the subscript $_{location}$ indicates the cloning site. 'T7dsp' is used generically for convenience when the specific genome has been made clear from the context.

1.A. Short-term biofilms: transgenic phage superiority across various media

Short-term biofilms were grown as overnight cultures at 37° in 24-well, treated polystyrene microtiter plates, then incubated with phages for 5 h, and finally assayed for clearing with crystal violet (CV, Figure 2). Three types of media were used, full strength broth (LB), dilute broth ($\frac{1}{3}$ LB), and minimal (M9 glucose). With such short-term bacterial growth, a strong effect of media on biofilm mass is expected and was in fact observed: across the 3 types of media, the untreated biofilms grown in full strength broth (LB) yielded the highest CV signal. For different phage treatments within the same media, the lowest CV signal was invariably with T7dsp; heterogeneity within each media was significant when all treatments were included but vanished when the T7dsp treatment was removed (when T7dsp was included, heterogeneity significance levels were $P < 0.0001$ for LB, $P < 0.02$ for $\frac{1}{3}$ LB, and $P < 0.04$ for M9 glucose). However, the CV signal from T7dsp was significantly higher in M9 glucose media than in the other media ($P < 0.0001$). For all 3 media, treatments with phages lacking transgenic dispersin were statistically indistinguishable from phage-free controls in the same media. These results support the main observations of Lu and Collins [4] but they further suggest that the magnitude of effect is media-dependent.

The model requires that the transgene benefit the phage. We have thus far merely conjectured from plausibility that a greater clearing ability translates into greater progeny production. However, Lu and Collins [4] provided assays that supported a link between biofilm clearance and phage

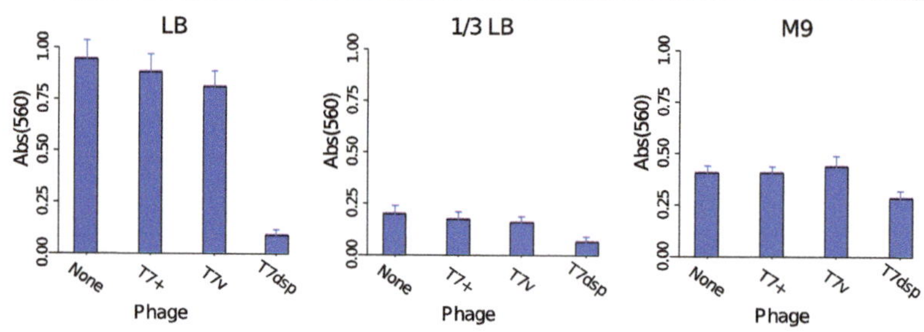

Figure 2 Transgenic phage efficacy on short term biofilms. Each panel represents a different media (full strength LB, $\frac{1}{3}$ LB, and M9 glucose) treated with the same phage stocks (T7+ is wild type T7, T7v is the empty T7 vector, and T7dsp is T7dsp+trx$_{10B}$). The vertical axis is the absorbance (560 nm) of the eluent after crystal violet staining. The bars give mean absorbance with 1 std error. Across each media, there is significant heterogeneity when and only when treatment with T7dsp is included (only marginally so for M9). There is also significant heterogeneity among the T7dsp treatments across the 3 media due to the M9 glucose experiment. Lower signal indicates greater biofilm clearing.

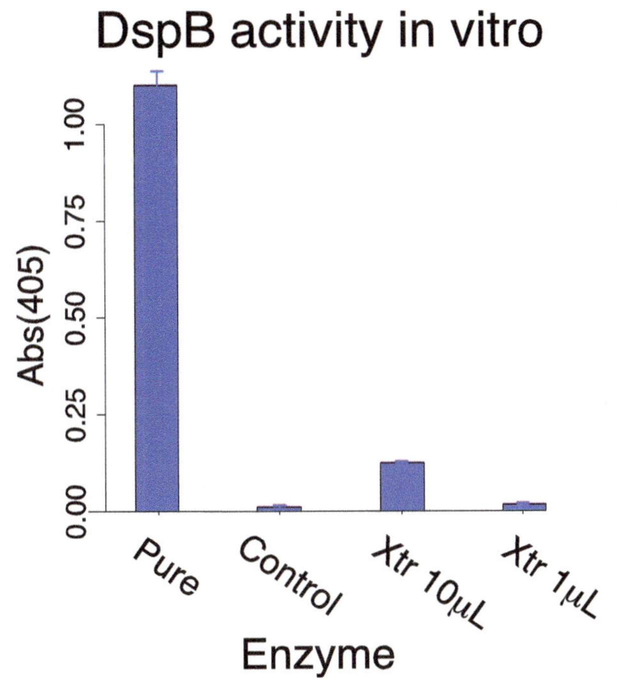

Figure 3 Assays of dispersin activity using a colorimetric substrate. Note that a high absorbance in this assay indicates high enzyme activity (in contrast to the CV assay). The Pure enzyme is a commercial preparation and shows high activity, reflecting its activity as an exoglycosidase on a substrate designed for that activity. The control is from a freeze-thaw lysate of BL21(DE3) cells carrying an empty pET15b plasmid. The bars for enzyme extract (Xtr) are from freeze-thaw lysates of BL21(DE3) cells carrying plasmid pET15b-dspB, with the volume of the extract in parentheses. Bars of 1 std. error are shown (sometimes too small to view).

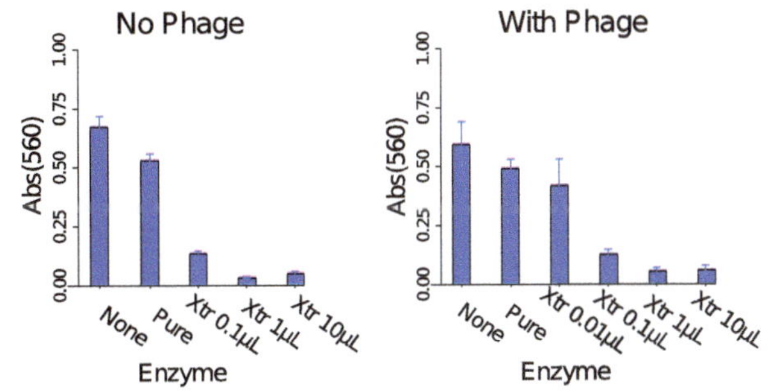

Figure 4 Dispersin enzyme activity on short term biofilms without added phage (left panel) or with T7v (right panel). After 5 hr of enzyme (with or without concurrent phage treatment), the biofilm wells were stained with crystal violet, and absorbance was measured on the eluted wash. A strong effect of enzyme is observed, and phage addition had no significant effect (no pairwise t-tests were significant). The enzyme-free control (None) is leftmost in each panel, followed by the pure, commercial enzyme (Pure), and the 3–4 rightmost histograms are for different volumes of dispersin extract (Xtr), the same as used in Figure 3. The effect of 0.01 μL of extract is shown in the presence of phage primarily to demonstrate that the phage were not having an effect that was masked by enzyme. Error bars indicate 1 std error.

amplification. The dispersin phage consistently showed greater bacterial killing than the control phage (their Figures 3 and 4) as well as a superiority in phage numbers soon after treatment (their Figure 4); the latter effect is visually small because it is plotted on a log scale. Most phage replication occurs on planktonic cells in the media, so the effect of killing biofilm cells is not expected to be easily noted under the conditions used.

1.B. Enzyme alone is sufficient for short-term biofilm degradation

Is the biofilm-clearing effect associated with transgenic dispersin of T7dsp due to enzymatic degradation of the biofilm matrix? Activity of the dispersin protein *per se* was measured directly with two types of assays. One used a colorimetric indicator specific to dispersin [17] and yielded a strong signal of activity with a commercially purified enzyme which is an exodepolymerase (Figure 3). In this assay, phage lysates from T7dsp-infected cells failed to produce a signal above background unless they were first concentrated with ammonium sulfate (not shown). Freeze-thaw lysates of bacteria expressing the dispersin B gene from a plasmid yielded a significant signal, approximately 10-fold less than that of the commercial enzyme (Figure 3).

The second measure of dispersin activity used the short-term biofilm clearing assay (Figure 4). Enzyme was added to wells with or without T7v control phage; clearing was assessed with CV staining at 5h treatment. For enzyme from the freeze-thaw lysate used in Figure 3, a monotonic relationship was observed across dilutions, with a significant effect to at least a 100-fold lower concentration than that needed to produce a significant colorimetric signal. It is noteworthy that the commercial dispersin enzyme was totally ineffective in this assay, perhaps reflecting a different enzymatic activity than dsp B on PNAG. There was no significantly greater clearing of biofilms in which enzyme alone was added versus enzyme plus T7v.

2. Long-term biofilms: phage efficacy independent of dispersin

Biofilms were grown 7 days in silicone tubing at 37° in $\frac{1}{3}$ LB. This combination of time and media concentration enabled a robust biofilm to develop without complications of bacterial growth into the delivery or sampling tubes. In contrast to the short term biofilms, the week-old biofilms were visibly thick (several mm) but also irregular in macroscopic structure. They were also delicate in that their structure shifted if the tubing was physically disturbed.

Pure phage populations or 1:1 mixtures of T7dsp and control phages were added on day 7. After 4–5 days

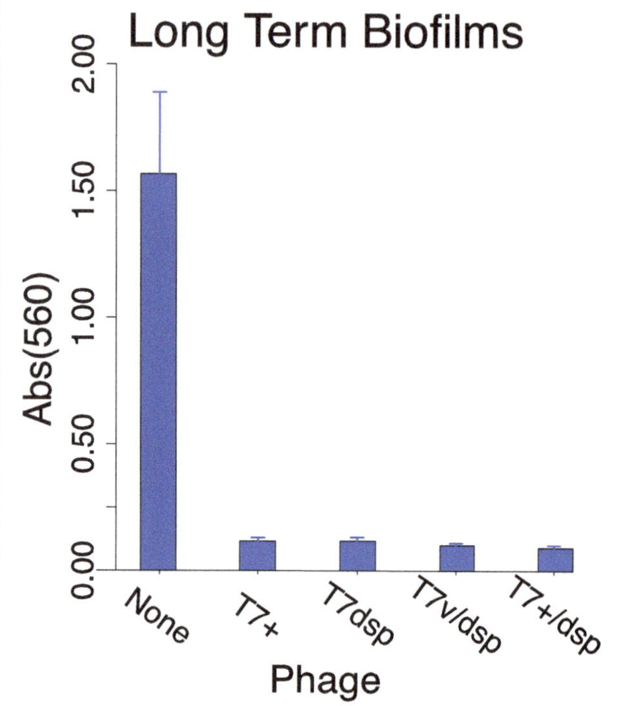

Figure 5 The dispersin transgene provides no benefit over control phages to the clearing of long term biofilms. Biofilm residue was measured with the crystal violet assay in biofilms that were grown 7 days in silicone tubing, then treated 4–5 days with phage (the leftmost bar is a no-phage control). T7+ and T7dsp (T7dsp+trx$_{10B}$) represent single-phage treatments, whereas T7v/T7dsp and T7+/T7dsp are from biofilms treated with equal mixes of control and transgenic phage. No significant heterogeneity is evident across the four phage-treated biofilms. Error bars represent 1 std. error; from left to right, the number of biofilms is 6, 6, 7, 2, and 2.

of phage treatment, a CV assay was performed on the contents of the tubing (Figure 5). In contrast to the superiority of dispersin phages over control phages on short-term biofilms, all phages and phage combinations cleared the biofilm mass substantially, no differences being attributable to the transgenic dispersin. To ensure that dispersin activity had not been lost during these adaptations (which could explain the lack of a difference between dispersin-containing and control phages), dispersin activity of final phage populations was tested in short term biofilm assays; these tests were conducted with eluents of two long term biofilms that had been seeded with 100% T7dsp. Clearance of short term biofilms by these eluents was compatible with maintenance of functional dispersin (not shown). Thus the equivalence of the different phages on long term biofilms was not due to loss of dispersin functionality.

These data clearly present an anomaly: dispersin is essential for clearing short term biofilms but has no significant effect on long term biofilms. It is tempting to attribute the difference to biofilm thickness or gene expression changes with age (e.g., PNAG levels may differ between short term and long term biofilms), but those features are not the only differences between short and long term biofilms. The substrate on which the two biofilms were grown also differed – silicone for the long term biofilms versus treated polystyrene for the short term biofilms. To evaluate whether substrate influenced the dispersin effect, 2 mm thick rings of sterile silicone tubing were added to wells of a polystyrene microtiter plate and a standard short term biofilm was grown; both wells and tubing were then treated with phage in the usual fashion. The polystyrene walls of the microtiter wells were specifically sensitive to T7dsp and were unaffected by control phages; yet by visual inspection the silicone rings in the same wells were equally sensitive to all phages.

The data thus suggest that the properties of a biofilm - and thus its susceptibility to phage infection - is affected by the substrate on which it is initiated. Because substrate-bound cells are only a minor component of a thick, long-term biofilm, it further suggests that the overall architecture of a biofilm is influenced by the nature of binding of that minor component.

Phage titers were monitored daily in the outflow of the silicone tubing in which the long-term biofilm was growing. Regardless of phage identity, titers were maintained near 10^7/mL (Figure 6). Thus despite clearing of the biofilm by the phage as revealed by CV staining, a reservoir of sensitive cells must have persisted to allow phage maintenance. The fact that phage titers were maintained

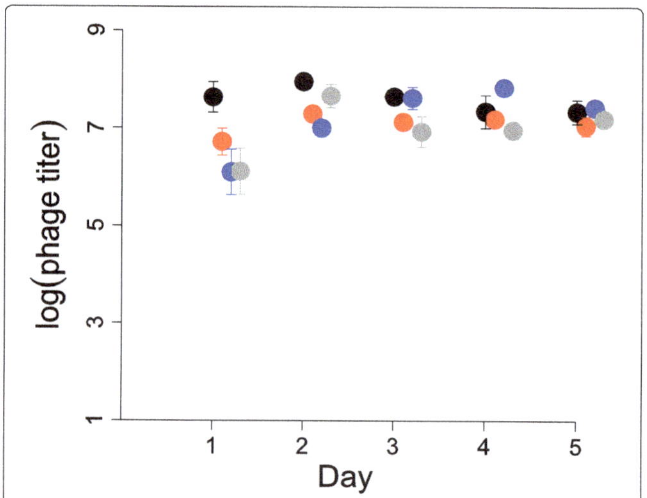

Figure 6 Phage titers in the effluent of long term biofilms (the same biofilms assayed in Figure 5). The horizontal axis gives the number of days of phage treatment. Although titers one day after phage addition show some scatter, all treatments are clustered near 10^7/mL by day 3. Red: T7dsp (T7trx+dsp$_{10B}$), black: T7+, blue: an equal mixture of T7+ and T7dsp, and gray: an equal mixture of T7v and T7dsp. Error bars of ± 1 std. err. are shown. Logs are to base 10.

at the same level across 4 days suggests minimal bacterial evolution to resist phage killing. The maintenance of phage likely reflects residual bacterial 'wall growth' and other refuges that the phage could not access (as noted for glass chambers in [18]), a model further supported by consideration of the biofilm flow rate: a 5 mm length of tubing experiences 90 replacement volumes of media per hour, so a population of purely suspended cells could not have maintained itself against this flow.

3. Evolution

The preceding results testify to the benefit of the dispersin transgene in reducing biofilms under some conditions. A question motivating our study is the evolutionary one of whether and under what conditions dispersin will be maintained. In particular, can the general framework of the tragedy of the commons be applied to understand evolution of the transgene?

3.A. Transgene maintenance on long term biofilms

Our initial assays of evolution were carried out in long term biofilms and revealed that the dispersin transgene was maintained, which we interpreted as indicating spatially structured growth (Table 2). These adaptations included some biofilms inoculated with only T7dsp+trx_{10B} and others inoculated with a 50:50 mix of that and a control phage. The dispersin phage was lost in one biofilm and appeared to decline in two others, but it was maintained at detectable levels after 4 days of growth in 5 of 6 populations. If the dispersin gene imposed a substantial cost on the phage, maintenance of the dispersin gene in these biofilms should indicate avoidance (or reduction) of a tragedy.

Table 2 Phage evolution in long term biofilms[1]

Initial phages[2]	Final dsp frequency[3]	Confidence interval[4]
T7dsp	1.0	(0.94, 1)
T7dsp	1.0	(0.94, 1)
T7dsp, T7v	0.15	(0.07, 0.33)
T7dsp, T7v	0.0	(0, 0.06)
T7dsp, T7+	0.5	(0.24, 0.76)
T7dsp, T7+	0.15	(0.07, 0.33)

[1]The biofilms assayed here are a subset of those shown in Figures 5 and 6.
[2]T7dsp is T7dsp+trx_{10B}. Where two phages were inoculated, the starting frequencies were approximately 0.5 each.
[3]After 96hr phage growth, frequencies were estimated from 8 PCR reactions per sample for each of two primer sets, each reaction comprised of 6 plaque isolates. The PCR product from one primer set indicated presence of the dispersin gene in at least one of the 6 isolates of the reaction; the product from the other set indicated its absence in at least one of the 6 isolates. Frequency was estimated as the maximum likelihood value of the corresponding multinomial term.
[4]The confidence interval was calculated as the Bayesian credible interval using a uniform prior.

3.B. Dispersin cloned in 10B is maintained on planktonic cells, indicating an intrinsic benefit

In the theory section, we suggested that transgene maintenance required either an intrinsic benefit – which seemed unlikely *a priori* – or required the right combination of spatially structured dynamics. In view of the observed maintenance of dispersin in long term biofilms (section 3.A, above), we sought to distinguish these alternatives. The straightforward way to demonstrate an intrinsic fitness benefit is to grow the phage in the absence of PNAG or the absence of spatial structure. If the gene is deleterious, it should be lost – as was observed in previous work with an endosialidase transgene in T7 [15].

Adaptations of various dispersin phages on planktonic IJ1133 (which has normal PNAG expression) revealed that the dispersin transgene was commonly maintained. Many of these outcomes were assessed qualitatively (e.g., the presence of dispersin was indicated by a positive PCR signal in the final culture), but some quantitative experiments were also conducted: (1) a 3 hr adaptation of a recombinant mixture of T7dsp+trx_{10B}/dsp_{10B} on IJ1133 in LB followed by single plaque isolation revealed dispersin in all of 10 plaques tested at both the start and end of the adaptation; (2) a similar test using a T7dsp+trx_{10B}/T7v recombinant mixture detected dispersin in 7 of 10 plaques at the start and in 5 of 10 plaques at the end. These data do not rule out a decline in the fraction of phage genomes that retained the dispersin transgene, but they indicate that any decline is at most modest.

As there is no obvious benefit to degrading extracellular PNAG in planktonic cultures of IJ1133, maintenance of the dispersin transgene evinces either an unexpected benefit to the intracellular phage life cycle or a minimal cost. This is in sharp contrast to transgenic endosialidase [15]. Nevertheless, because dispersin is maintained on planktonic cells, its maintenance in a biofilm cannot be attributed to avoidance of a tragedy of the commons (recall Figure 1D).

3.C. Evolution of dispersin cloned in 3.8: loss on liquid-grown planktonic cells, maintenance in biofilms but also in biofilm-derived planktonic cells

The genomes used in the previous assays harbored dispersin in gene *10B*, in the late or class III region of the phage. The relative costs and benefits of a transgene in different genomic locations are not yet predictable, but as the timing and levels of gene expression in T7 are affected by position in the genome [19], varying effects of insertions at different locations are to be expected. We constructed a T7dsp phage with dispersin inserted into gene *3.8*, a non-essential class II gene of no known function [19]; this phage is denoted here as T7$dsp_{3.8}$ (as T7dsp+$1.2_{3.8}$/$dsp_{3.8}$ in Table 3). One rationale for this choice was that the dispersin gene would be expressed

Table 3 Phage strains

Notation	Genotype	Use	Ref
T7+	Wild-type (Genbank V01146)	Control phage for some adaptations	[20]
T7v	T7 cloning vector, made as T7Select 415b-1 (Novagen) with the left end BclI fragment replaced by that of a phage carrying the C74 deletion (gene *0.3* is intact)	Used as backbone for gene insertions because of the cloning site and the ability to grow on hosts containing type I restriction modification (RM)	$T7_0$ in [15]
T7dsp+trx_{10B}	T7415-C74:trxAdspB. T7v with the *E. coli trxA* gene, T7 ϕ10 promoter and *A. actinomycetemcomitans* CU1000 *dspB* gene recombined in to the multiple cloning site between genes *10* and *11*	Re-engineered version of T7DspB from [4] to grow on hosts with type I RM	This study
T7dsp+trx_{10B}/ dsp_{10B}	T7dsp+trx_{10B} grown on a host with pUC57-T7dspB to remove the *trxA* gene by recombination; not clonally isolated	Used to allow evolution of dispersin free of *trxA*	This study
T7dsp+$1.2_{3.8}$	T7v but most of gene *3.8* replaced with T3 gene *1.2* (enabling growth on F plasmid-bearing strains) and the *dspB* gene. Created by recombination with plasmid pMK-RQ-T7trxAdspB. The multiple cloning site in gene *10B* is unaltered from T7v.	Used to create T7-1.2+$dsp_{3.8}$/$dsp_{3.8}$.	This study
T7dsp+$1.2_{3.8}$/ $dsp_{3.8}$	T7-1.2+$dsp_{3.8}$ recombined with pMKRQ-T7dspB to create variation in the presence/absence of the *1.2* gene. Not purified.	Evolved in a chemostat to study evolution of the dsp gene.	This study
T7$dsp_{3.8}$P	The initial mix of T7-1.2+$dsp_{3.8}$/$dsp_{3.8}$ recombined against the chemostat-evolved population of T7-1.2+$dsp_{3.8}$/$dsp_{3.8}$ in which the dsp gene had been lost.	Multiple recombinations separated dsp from *1.2* and introduced any mutations from a chemostat adaptation that might be beneficial in the background of the phage. This polymorphic population was used to evaluate the effect of selection on the dsp transgene in serial transfer, short term biofilms and long term biofilms.	This study

sooner after infection in T7$dsp_{3.8}$ than in T7dsp_{10B}, providing a greater opportunity for it to have a deleterious effect on phage development. Indeed, a preliminary adaptation on planktonic IJ1133 revealed that the dispersin transgene was completely lost. More thorough adaptations were then conducted with a recombinant mix of that phage.

Starting with a T7$dsp_{3.8}$ shown to have dispersin activity, a polymorphic population was created (denoted T7$dsp_{3.8}$P) in which 26% of the phage carried dispersin [credible interval (0.18, 0.36)]. By starting with a mix of different genotypes at intermediate frequencies – as opposed to a pure type – selection quickly leads to an easily detectable change in frequency of the dispersin transgene [21]. Serial transfer of T7$dsp_{3.8}$P for 10–12 hr (~ 40–50 generations) at 37° on planktonic *csrA* cells in $\frac{1}{3}$ LB broth resulted in apparent loss of the dispersin gene in both of two replicates (undetected in 50 plaques of the final population from one replicate, 40 in the other). The loss of dispersin in liquid serial transfer indicates an intrinsic cost and renders this phage appropriate for testing the role of spatial structure in avoiding a tragedy of the commons. If the biofilm environment is selective for dispersin and if the spatial structure provided by a biofilm avoids a tragedy of the commons, the dispersin gene frequency should be constant or even increase in biofilms.

3.C.1. Long term biofilms. T7$dsp_{3.8}$P was introduced to 7-day biofilms. In principle, long term biofilms enable phage evolution to be studied across days rather than hours. However, if a phage destroys the biofilm environment, it may destroy any spatially-structured selection for maintaining the dispersin gene. Typically, the visibly thick matrix of long term biofilms was cleared within 12–24 hr after phage addition, so sampling was conducted early (5–7 hr) and late (70+ hr). In 5 of 6 biofilms, the dispersin transgene initially rose in frequency over 5–7 hours but then declined by 72–74 hr; in the sixth biofilm, an initial decline was followed by a slight rise (Figure 7). Collectively, the dispersin transgene appears to be favored or maintained early in biofilm attack but not long term.

To distinguish any benefit of structured growth from the physiological state of the bacterial host on maintenance of the dispersin transgene, evolution of dispersin was tested in cells being sloughed off a long-term biofilm. At day 7 of bacteria-only biofilm growth, the effluent from a biofilm was shunted into fresh tubing, which therefore

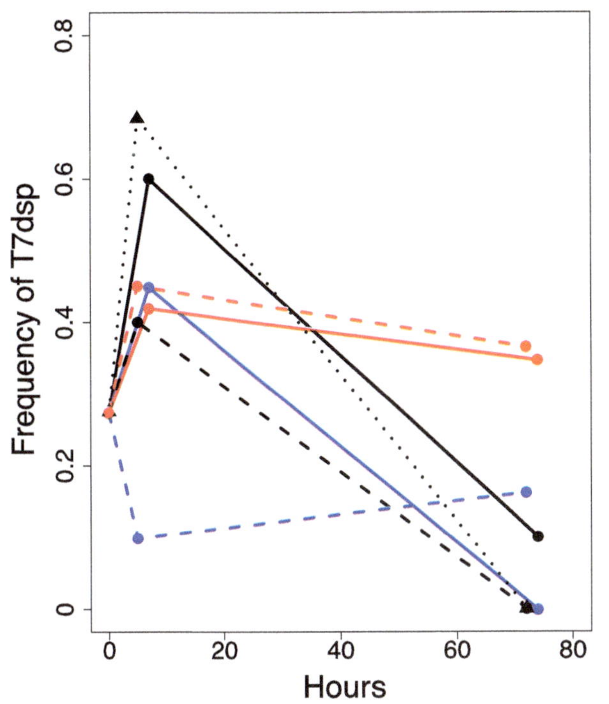

Figure 7 Frequency of T7dsp during growth in long term biofilms. The phage inoculum was T7dsp$_{3.8}$P. Each color-texture line combination represents a different biofilm, each inoculated with the same phage stock; the frequency of T7dsp in that stock is used as the initial frequency for all 7 adaptations. The black dotted line is for cells sloughed from a 7-day biofilm and shunted into a biofilm-free tube two hours before phage addition; those cells should be almost entirely planktonic.

should contain primarily planktonic cells. After 2 hr incubation, phage were added. Evolution of T7dsp shows the same pattern on the shunted cells as on the established biofilms, only more so: a sharp rise in the frequency of the transgene in the short term but loss in the long term (dotted black in Figure 7). This result suggests that the early benefit of T7dsp depends more on the physiological state of the cells than on spatial structure.

3.C.2. Short term biofilms. The dispersin transgene has a dramatic effect on short term biofilms. It was thus of special interest to determine whether the dispersin transgene would also realize a major evolutionary advantage in that treatment. For reasons noted above, it was necessary to use T7dsp$_{3.8}$P for this test, as the transgene appears to be intrinsically beneficial in T7dsp$_{10B}$. To control for as many variables as possible while comparing the effect of planktonic versus biofilm-embedded cells, planktonic cells were obtained from the first rinse of an overnight biofilm well using fresh media and deposited in an empty well where no biofilm was present. (Microscopic observations on washes confirmed that cells were either single or in small clusters of 5–20 cells that should allow free phage access independent of dispersin.) The well containing the original biofilm was replenished with fresh media. Thus one well had a biofilm and the other had only planktonic cells derived from it. T7dsp$_{3.8}$P was added to both wells, and the frequency of the dispersin transgene compared between them after 5 hr of growth.

Surprisingly, there was no significant difference between the planktonic and biofilm environments (Figure 8). Equally surprising, the frequency of T7dsp increased significantly from its initial value in both environments (combining the 6 replicates and using 0.26 as the expected frequency, $\chi^2(1) = 35.7$, P$< 10^{-4}$ for the planktonic cells, $= 9.2$, P< 0.0024 for the biofilm cells). Thus, the profound importance of dispersin in clearing short term biofilms is matched by an apparent selective advantage of the dispersin gene, but any spatial structure of the biofilm has no additional effect.

Discussion

Organisms with engineered genomes are finding ever-increasing applications (e.g., [22-25]). Commonly, those engineered organisms must reproduce themselves to perform their function, whether in a fermenter, patient, or

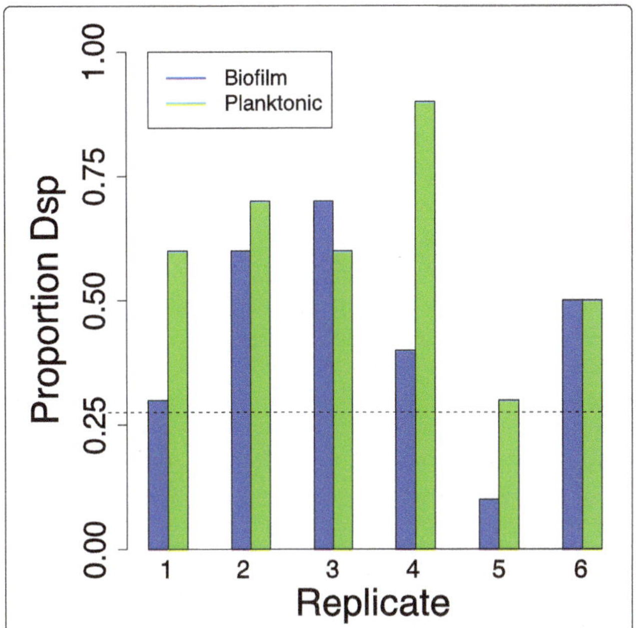

Figure 8 Frequency of phage carrying a dispersin transgene after 5hr growth on cells derived from short term biofilms. Each pair of bars represents an independent, paired replicate: a blue bar is from phage growth on a biofilm, the adjacent green bar is from growth on planktonic cells derived from that biofilm. For each bar, the frequency of dispersin is the fraction of 10 plaque isolates from the final population that carried the gene. The horizontal dotted line is the initial frequency of the dispersin transgene in the T7dsp$_{3.8}$P stock added to the cells.

field. That reproduction virtually ensures their evolution, evolution that may reverse the engineering, improve it, or allow its escape. Being able to predict that evolution will have many benefits that span social impacts to corporate finances. The present study is broadly viewed as a test case of this perspective, predicting the evolution of an engineered genome. Overall, the results challenge the expectation that evolution can be predicted from general principles applied naively.

The effort here used an established transgenic phage system. Lu and Collins [4] engineered a T7 to encode the bacterial gene for dispersin, which degrades an exopolysaccharide component of biofilms (poly-N-acetylglucosamine or PNAG). In support of their design principle, the engineered phage cleared short-term biofilms more completely than did control phages. What was not tested was whether the phage would retain the dispersin gene and activity during growth in biofilms. This latter issue is one of evolutionary stability, and it intersects the domains of ecology and evolution, where there are well established principles that might be expected to govern the fate of engineered phages [5-12].

The crux of the evolutionary problem faced by this phage was thought to reside in a framework known as 'tragedy of the commons'. It was assumed that dispersin would not be beneficial to the intracellular phage life cycle, but that it might benefit the phage population by degrading the biofilm matrix and thereby increasing phage access to cells; data from Lu and Collins supported this latter point. But a benefit to the phage population is not the same as a benefit to the transgenic genome: the dispersin protein is released at lysis as a free protein that diffuses and degrades the matrix in the local environment. Thus any and all phages in this local environment benefit, not just the genome that produced the enzyme. The intrinsic cost of the transgene – incurred only by the genome encoding it – then leads to its evolutionary loss when competing with phages lacking the transgene.

Well established theory suggests that the tragedy might be avoided if the phage is 'selfish', as by growing in a physically structured environment that limits sharing of the transgenic protein. But even this solution requires a delicate balance of dynamical properties. Based on these *a priori* considerations, evolutionary maintenance of the dispersin transgene seemed unlikely but not impossible, but that spatially structured growth was required.

The naive application of the evolutionary framework failed because one of the main assumptions failed: the transgene was not intrinsically deleterious. It is of course possible to offer reasons *post hoc* why an arbitrary transgene might have virtually no cost (e.g., restoring genome length closer to wild-type), but previous work points to nearly ubiquitous deleterious effects of engineered changes in T7 [15,26,27] (including unpublished observations on *trx*A inserts) and even bacteria [28]. The failure is thus one of the approach used here, extrapolating seemingly generic properties of transgene effects observed in other contexts to dispersin. In short, details matter when predicting evolution. This is an unfortunate message, because virtually any application of evolutionary principles to real world systems will require many levels of similarly naive assumptions – implementation will never await knowing all details in advance (e.g., [23,24]).

Aside from the unexpected benefit of dispersin, several other observations reflect on our ability to predict or anticipate behavior of the transgenic phage:

1. The transgenic phage dramatically augmented clearing of short-term biofilms in two different media but had only a modest effect in a third. Adding enzyme alone to short term biofilms could reproduce the effect. The effect of media was not explored further but could be a consequence of varying concentrations of PNAG production relative to other biofilm exopolysaccharides. These observations are broadly compatible with those expected from previous work [4] except for the effect of media.
2. The dispersin transgene offered no benefit in clearing long-term biofilms, the environment of greatest potential utility. The contrast with dispersin's effect in short term biofilms was partly attributable to a different architecture of the two types of biofilms, suggesting that even preliminary laboratory studies on biofilm remediation should be focused on the conditions of end use.
3. During growth in both long- and short-term biofilms, the transgene was beneficial and its frequency increased at least temporarily. Whereas the evolutionary theory dictated that this benefit should stem from or be enhanced by spatial architecture of the biofilm, the benefit merely reflected the specific physiology of cells growing in a biofilm. Bacteria sloughed from the biofilms provided an equal or superior benefit to the transgenic phage. There was thus no insight or outcome attributable to the tragedy of commons framework.

The results indicate that maintenance of a transgene depends both on intrinsic properties of the transgene and on engineering. Although the tragedy framework did not obviously apply to dispersin maintenance, the basis of that model failure is a good one from an engineering perspective: the transgene was intrinsically beneficial. In cases where the transgene is intrinsically costly and a tragedy of the commons is not averted, nature may suggest ways to engineer a solution. Specifically, some naturally-occurring phage depolymerases enhance biofilm degradation [3,29,30]. Those enzymes are typically encoded as virion tailspikes and are required for infection, thus directly benefitting the parent genome; they are not subject to loss through a tragedy of the commons. Engineering enzymes as phage tail components may thus enable maintenance of a transgene when it would otherwise be lost.

Methods

Strains and media

Phage and bacterial strains are described in Tables 3 and 4. All are from the collections of JJB or IJM. LB broth was 10 g NaCl, 10 g Bacto tryptone, 5 g Bacto yeast extract per liter. $\frac{1}{3}$ LB used a third of all ingredients per liter. M9 glucose was 47.8 mM Na_2HPO_4, 22 mM KH_2PO_4, 8.5 mM NaCl, 1.87 mM NH_4Cl, 1 mM $MgSO_4$, 0.1 mM $CaCl_2$ with 0.2% glucose per liter. Plates used LB with 1.5% Bacto agar. Determinations of phage titers used plates overlaid with soft agar (0.7% Bacto agar in LB) containing a suitable density of hosts.

The *E. coli trx*A gene, or alternatively T3 *1.2* was used as a selectable marker in detecting initial recombinants but has no benefit afterward in our design. Consequently, the evolutionary fate of the dispersin B gene may be considered to have been influenced by *trx*A. However, some adaptations used a recombinant mix that deleted *trx*A from the dispersin transgene, and other adaptations were observed to spontaneously delete *trx*A while maintaining dispersin. Thus *trx*A is not expected to affect the evolutionary fate of dispersin. All adaptations of phages in which T3 *1.2* had been used as a selectable marker were recombinant mixtures that included phages carrying dispersin but lacking T3 *1.2*, so again, this marker would not affect selection of the dispersin B gene.

Plasmids

Plasmid pET15b-dspB was the commercial vector pET15b with an insert of bases 1-1086 of the dispersin B (dspB) ORF from *Aggregatibacter actinomycetemcomitans* CU1000 (Genbank AY228551). Although this fragment is not the complete ORF, it has beta-hexosaminidase activity [35]. pUC57-T7dspB consisted of the vector pUC57 with a linear insert in order of (i) T7Select 415b-1 bases 21403-21503 (Novagen), (ii) the $\phi 10$ T7 promoter, (iii) a ribosomal binding sequence, (iv) bases 1-1086 of the *A. actinomycetemcomitans* dspB ORF, and (v) T7Select 415b-1 nts. 21517-21617. pUC57-T7trxAdspB consisted of pUC57-T7dspB with the *E. coli* trxA gene inserted between T7Select 415b-1 nts. 21403-21503 and the $\phi 10$ promoter. pMKRQ-T71.2+dspB consisted of plasmid pMK-RQ (Gene Art) with an insert of T7 bases 11059-11229 (5' of gene *3.8*) followed by T3 gene *1.2*, dspB, and T7 bases 11550-11652 (the 3' end of *3.8*). pMKRQ-T7dspB was pMKRQ-T71.2+dspB without the T3 *1.2* gene.

Recombinant phage construction

Recombinant phages were obtained by first plating T7v on a host containing a plasmid that carried an insert of dspB with a selectable marker (*trx*A or T3 *1.2*) and T7 flanking sequences. Plaques, which invariably contained a mixture of recombinants and T7v, were suspended and replated on an appropriate selection host. These second-step plaques were then usually purified in a subsequent round of plating and the presence of dispersin checked phenotypically and/or by PCR.

The phages at this step carried not only the dispersin gene but a selectable marker that might have imposed a fitness cost. To eliminate the selectable marker, phages were plated on a host carrying a plasmid that would recombine out the selectable marker but leave intact the

Table 4 Bacterial strains

Notation	Genotype:	Use	Ref
IJ1133	*E. coli* K-12 F$^-$ Δ*lacX74 thi* Δ*(mcrC-mrr)102*::Tn10	Host for measuring phage titers and some serial transfers to determine the evolutionary stability of the *dsp B* transgene in the absence of an extracellular PNAG matrix.	[31]
BL21 (DE3)	*E. coli* B Gal$^-$ *dcm ompT hsdS*(r_B- m_B-) [λ(DE3)]	Host for protein expression	[32]
IJ512	*E. coli* K12 Δ*lacX74 supE44 galK2 galT22 mcrA rfbD1 mcrB1 hsdS3* /F′ *lac*	Host for selecting T7 carrying T3 gene *1.2*	
Δ*trxA*	JW5856 (Keio Δ*trxA::kan*)	Host for selection of positive phage recombinants carrying *trxA*	KEIO [33]
Δ*csrA*	*E. coli* K-12 F$^-$ Δ*lacU169* Δ*csrA::kan*	Biofilm host; overproduces PNAG, the substrate for dispersin B	TR1-5 mutant in [34]
Δ*nagZ*	JW1083 (Keio Δ*nagZ::kan*)	Host with low background activity for producing dispersin B protein from phage lysates	KEIO [33]

dsp gene. In some cases, isolates lacking the selectable marker were identified by screening for inability to grow on the ΔtrxA host and then confirmed by PCR. In other cases, the entire polymorphic plaque was used, as selection would be able to separate the selectable marker from dsp.

Dispersin B protein expression and activity assay

BL21 (DE3) cells with pET15b or pET15b-dsp were grown to log phase in 10 mL of LB broth at 37°C. IPTG was added to 1 mM and the cultures grown for 3 hr. Cells were pelleted, resuspended in 0.5 mL of lysis buffer (100 mM NaCl, 20 mM Tris-HCl pH 8.0, 6 mM $MgSO_4$ and 0.5 mg/mL lysozyme), subjected to two rounds of freeze-thaw and cleared by centrifugation. Lysates were stored at 4°C, maintaining activity for at least one month.

Chemical activity of dspB protein was assessed using the colorimetric substrate 4-nitrophenyl N-acetyl-β-glucosaminide (Sigma-Aldrich, St. Louis, MO) at 1 mM [17]. Commercial β-N-acetylglucosaminidase (New England Biolabs, Ipswitch, MA), an exoglycosidase, was used as an enzyme standard. Reactions followed manufacturer instructions using the recommended buffer. Reactions were incubated for 30 min at 37°C and stopped with 100 mM glycine pH 10. Absorbance at 405 nm was measured using the UV-Vis module of the NanoDrop Spectrophotometer (Thermo Scientific, Wilmington, DE).

One preparation of dspB enzyme used lysates of dispersin phages. Δ*nagZ* cells were grown at 37°C with aeration in 50 mL LB broth to a density of approximately 10^8 cells/mL. 1×10^4 phage per mL were added to each culture and the cultures grown until visible lysis (2 hr). Protein from lysates was precipitated with saturated ammonium sulfate, pelleted and resuspended in 0.5 mL of buffer (20 mM Tris-HCl pH 8, 50 mM NaCl, 1 mM EDTA, pH 8). For controls, phage-free lysates of Δ*nagZ* cells were grown, pelleted and lysed by freeze-thaw as described above. Protein from the cleared cell lysate was precipitated using saturated ammonium sulfate, pelleted and resuspended in 0.05 mL buffer.

Short-term biofilms

Aliquots (100 μL) of Δ*csrA* cells were made as 10-fold concentrates of a stationary phase culture, frozen in LB with 20% glycerol. An aliquot was suspended in 12 mL media and 1 mL was added to each well of a 24-well tissue culture plate (Corning with Standard Tissue Culture Treated Surface). Cells were grown overnight in a static 37°C incubator. After 12–16 hr, the media was replaced and supplemented with $10^4 - 10^5$ phage per well or with a lysate containing enzyme. The plate was then incubated 5 hr in a static 37°C incubator. After treatment, wells were rinsed once with water, air dried, stained for 15 min with crystal violet (CV, 0.1% in water), rinsed twice in water, and then air dried. Residual CV was extracted in 95% ethanol and quantified by absorbance at 560 nm using the UV-Vis module of the NanoDrop Spectrophotometer (Thermo Scientific, Wilmington, DE).

Long-term biofilms

Biofilms of Δ*csrA* cells were grown in $\frac{1}{3}$ LB for 7 days in silicone tubing (20 cm, 0.125" I.D. x 0.250" O.D; VWR International LLC, Radnor, PA). An additional length of tubing connected the biofilm chamber to a waste container. Luer-lock stopcocks at each end of the chamber tubing allowed sterile introduction and sampling of phage and cells. Media flow rate was kept at 3 mL/hr. The biofilm chambers were placed on a covered slide warmer set at 37°C. The cell preparation was the same as for short-term biofilms. After inoculation, cells were incubated statically for 1 hr to allow attachment prior to reinstating media flow.

Biofilms were grown for 7 days, then 10^4 phage were added. The biofilm was grown a further 4–5 days. Biofilm effluents were assayed for phage daily. After phage treatment, chambers were rinsed once with water and dried. Crystal violet staining assessed biofilm mass.

Growth of phage for testing loss of transgene

Evolutionary stability of transgene maintenance was carried out on IJ1133 in LB or on Δ*csrA* in $\frac{1}{3}$ LB. Eight lines used 3 hr of serial transfer on IJ1133 following established methods [36,37]. One line was grown 48 hr by continuous culture in a 2-tube chemostat [38].

Statistics of dispersin transgene frequency

Standard PCR was used to measure the presence or absence of the dispersin gene in phage isolates to obtain frequency estimates from a population. Primer pairs consisted of one internal and one flanking dispersin, or both flanking. When the frequency of the target sequence is high or low, nearly all phage will be of the same type, and a large number of single-plaque reactions are required to estimate the frequency. Combining multiple plaques into a single reaction facilitates the detection of uncommon types by reducing the number of reactions needed. However, the statistics of sampling error become somewhat more complicated. We used the following methods:

Suppose the population frequency of phages carrying a transgene is T ($U = 1 - T$ for phages lacking the transgene). Let n phage isolates be combined into a single PCR reaction such that the product reveals either (A) the absence of the transgene in all n isolates (probability

U^n), (B) the presence of the transgene in all n isolates (probability T^n), or (C) the presence of the transgene in some but not all isolates (probability $1 - U^n - T^n$). (Multiple primer sets of the same mixture of plaques may be required to distinguish all 3 outcomes.) If R independent n-plaque reactions are tested, the probability of observing X reactions in which all n isolates lack the transgene, Y reactions in which all isolates carry the transgene, and Z reactions in which at least one plaque of each type occurs is the corresponding multinomial term,

$$M(A,B,C,X,Y,Z) \propto A^X B^Y C^Z \quad ,$$

where $A = U^n$, $B = T^n$, and $C = 1 - U^n - T^n$ and $X + Y + Z = R$ (the combinatorial constant term is omitted). $M(A,B,C,X,Y,Z)$ is easily solved for the maximum likelihood estimate of T, and a 95% Bayesian 'credible' interval obtained by treating the normalized $M(A,B,C,X,Y,Z)$ as a probability density function of T ($0 \leq T \leq 1$) and finding the bounds for the upper and lower 2.5% tails.

Graphics

Figures were drawn in R [39].

Simulations

Differential equations were evaluated numerically in the program Berkeley Madonna (v. 9.0.118 beta) with a step size of 10^{-3} and the Euler method. The numerical output was transferred to R for presentation.

Competing interests

The authors declare that they have no competing interests.

Authors' contributions

MS developed many of the assays reported here and carried out most of them; NC carried out some fitness assays. IJM advised the conduct of several assays and the overall molecular biology dimension of the study and helped write the paper. DA and JT conducted some preliminary adaptations of Tim Lu's T7 dispersin phage at the University of Idaho. JJB designed the study, oversaw its conduct, did the math, and wrote most of the paper. All authors read and approved the final manuscript.

Acknowledgements

We thank L. Forney for advice and instruction on *E. coli* laboratory biofilm growth. Tim Lu sent us his T7 dispersin phage, which got the project started. N. Ramasubbu advised us on a colorimetric dispersin substrate. We thank a reviewer for pointing out that the spatially structured dynamics of phage growth in a biofilm may be dynamic. Supported by NIH GM 57756 and U. Texas Miescher Regents Professorship to JJB. We thank RJ Bull for some sequence analysis of preliminary adaptations (not reported here).

Author details

[1]Center for Computational Biology and Bioinformatics, The University of Texas at Austin, Austin, TX, USA. [2]Department of Molecular Biosciences, The University of Texas at Austin, Austin, TX, USA. [3]Institute for Cellular and Molecular Biology, The University of Texas at Austin, Austin, TX, USA. [4]Bayer Crop Science - Biologics, 1540 Drew Ave, Unit 170, Davis, CA, USA. [5]Total New Energies USA Inc., 5858 Horton Street, Suite 253, Emeryville, CA, USA. [6]Department of Integrative Biology, The University of Texas at Austin, Austin, TX, USA.

References

1. Donlan RM, Costerton JW: **Biofilms: survival mechanisms of clinically relevant microorganisms.** *Clin Microbiol Rev* 2002, **15**(2):167–193.
2. Worthington RJ, Richards JJ, Melander C: **Small molecule control of bacterial biofilms.** *Org Biomol Chem* 2012, **10**(37):7457–7474.
3. Cornelissen A, Ceyssens P, T'Syen J, Van Praet H, Noben J, Shaburova OV, Krylov VN, Volckaert G, Lavigne R: **The T7-related *Pseudomonas putida* phage ϕ15 displays virion-associated biofilm degradation properties.** *PloS One* 2011, **6**(4):e18597.
4. Lu TK, Collins JJ: **Dispersing biofilms with engineered enzymatic bacteriophage.** *PNAS* 2007, **104**(27):11197–11202.
5. West SA, Griffin AS, Gardner A, Diggle SP: **Social evolution theory for microorganisms.** *Nat Rev Microbiol* 2006, **4**(8):597–607.
6. Xavier JB, Foster KR: **Cooperation and conflict in microbial biofilms.** *Proc Natl Acad Sci U S A* 2007, **104**(3):876–881.
7. MacLean RC: **The tragedy of the commons in microbial populations insights from theoretical, comparative and experimental studies.** *Heredity* 2008, **100**(5):471–477.
8. Bull JJ, Harcombe WR: **Population dynamics constrain the cooperative evolution of cross-feeding.** *PloS One* 2009, **4:**e4115.
9. Kummerli R, Griffin AS, West SA, Buckling A, Harrison F: **Viscous medium promotes cooperation in the pathogenic bacterium *Pseudomonas aeruginosa*.** *Proc Biol Sci R Soc* 2009, **276**(1672):3531–3538.
10. Eshelman CM, Vouk R, Stewart JL, Halsne E, Lindsey HA, Schneider S, Gualu M, Dean AM, Kerr B: **Unrestricted migration favours virulent pathogens in experimental metapopulations: evolutionary genetics of a rapacious life history.** *Philos Trans R Soc Lond B Biol Sci* 2010, **365**(1552):2503–2513.
11. Frank SA: **Demography and the tragedy of the commons.** *J Evol Biol* 2010, **23:**32–39.
12. Smith J: **Tragedy of the commons among antibiotic resistance plasmids.** *Evolution* 2012, **66**(4):1269–1274.
13. Levin BR, Stewart FM, Chao L: **Resource-limited growth, competition, and predation - a model and experimental studies with bacteria and bacteriophage.** *Am Nat* 1977, **111**(977):3–24.
14. Levin BR, Bull JJ: **Population and evolutionary dynamics of phage therapy.** *Nat Rev Microbiol* 2004, **2**(2):166–173.
15. Gladstone EG, Molineux IJ, Bull JJ: **Evolutionary principles and synthetic biology: avoiding a molecular tragedy of the commons with an engineered phage.** *J Biol Eng* 2012, **6:**13.
16. Doolittle MM, Cooney JJ, Caldwell DE: **Tracing the interaction of bacteriophage with bacterial biofilms using fluorescent and chromogenic probes.** *J Ind Microbiol* 1996, **16**(6):331–341.
17. Manuel SGA, Ragunath C, Sait HBR, Izano EA, Kaplan JB, Ramasubbu N: **Role of actie-site residues of dispersin B, a biofilm-releasing beta-hexosaminidase from a periodontal pathogen, in substrate hydrolysis.** *FEBS J* 2007, **274**(22):5987–5999.
18. Schrag SJ, Mittler JE: **Host-parasite coexistence: the role of spatial refuges in stabilizing bacteria-phage interactions.** *Amer Natur* 1996, **148**(2):348–377.
19. Molineux I: **The T7 group.** In *The Bacteriophages*. Edited by Calendar R. Oxford: Oxford University Press; 2005:277–301.
20. Dunn JJ, Studier FW: **Complete nucleotide sequence of bacteriophage T7 DNA and the locations of T7 genetic elements.** *J Mol Biol* 1983, **166**(4):477–535. [PMID: 6864790].
21. Rokyta D, Badgett MR, Molineux IJ, Bull JJ: **Experimental genomic evolution: extensive compensation for loss of DNA ligase activity in a virus.** *Mol Biol Evol* 2002, **19**(3):230–238.
22. Burns CC, Campagnoli R, Shaw J, Vincent A, Jorba J, Kew OM: **Genetic inactivation of poliovirus infectivity by increasing the frequencies of CpG, and UpA dinucleotides within and across synonymous capsid region codons.** *J Virol* 2009, **83**(19):9957–9969. [PMID: 19605476].
23. Gould F: **Broadening the application of evolutionarily based genetic pest management.** *Evolution* 2008, **62**(2):500–510.
24. Bull JJ, Turelli M: **Wolbachia versus dengue: evolutionary forecasts.** *Evol Med Public Health* 2013, **2013:**197–207.
25. Paddon CJ, Keasling JD: **Semi-synthetic artemisinin: a model for the use of synthetic biology in pharmaceutical development.** *Nat Rev Microbiol* 2014, **12**(5):355–367.

26. Cecchini N, Schmerer M, Molineux IJ, Springman R, Bull JJ: **Evolutionarily stable attenuation by genome rearrangement in a virus.** *G3 (Bethesda, Md.)* 2013, **3**(8):1389–1397.
27. Springman R, Molineux IJ, Duong C, Bull RJ, Bull JJ: **Evolutionary stability of a refactored phage genome.** *ACS Synthetic Biol* 2012, **1**(9):425–430.
28. Sleight SC, Bartley BA, Lieviant JA, Sauro HM: **Designing and engineering evolutionary robust genetic circuits.** *J Biol Eng* 2010, **4:**12.
29. Hanlon GW, Denyer SP, Olliff CJ, Ibrahim LJ: **Reduction in exopolysaccharide viscosity as an aid to bacteriophage penetration through *Pseudomonas aeruginosa* biofilms.** *Appl Environ Microbiol* 2001, **67**(6):2746–2753.
30. Hughes KA, Sutherland IW, Jones MV: **Biofilm susceptibility to bacteriophage attack: the role of phage-borne polysaccharide depolymerase.** *Microbiology* 1998, **144(Pt 11):**3039–3047.
31. García LR, Molineux IJ: **Rate of translocation of bacteriophage T7, DNA across the membranes of *Escherichia coli*.** *J Bacteriol* 1995, **177**(14):4066–4076.
32. Studier FW, Moffatt BA: **Use of bacteriophage T7 RNA polymerase to direct selective high-level expression of cloned genes.** *J Mol Biol* 1986, **189:**113–130.
33. Baba T, Ara T, Hasegawa M, Takai Y, Okumura Y, Baba M, Datsenko KA, Tomita M, Wanner BL, Mori H: **Construction of *Escherichia coli* K-12 in-frame, single-gene knockout mutants: the Keio collection.** *Mol Syst Biol* 2006, **2:**2006.0008.
34. Romeo T, Gong M, Liu MY, Brun-Zinkernagel AM: **Identification and molecular characterization of csrA, a pleiotropic gene from *Escherichia coli* that affects glycogen biosynthesis, gluconeogenesis, cell size, and surface properties.** *J Bacteriol* 1993, **175**(15):4744–4755.
35. Kaplan JB, Ragunath C, Ramasubbu N, Fine DH: **Detachment of Actinobacillus actinomycetemcomitans biofilm cells by an endogenous beta-hexosaminidase activity.** *J Bacteriol* 2003, **185**(16):4693–4698.
36. Heineman RH, Bull JJ: **Testing optimality with experimental evolution lysis time in a bacteriophage.** *Evolution* 2007, **61**(7):1695–1709.
37. Bull JJ, Heineman RH, Wilke CO: **The phenotype-fitness map in experimental evolution of phages.** *PloS One* 2011, **6**(11):e27796.
38. Bull JJ: **Optimality models of phage life history and parallels in disease evolution.** *J Theor Biol* 2006, **241**(4):928–38.
39. R Development Core Team: *R: A Language and Environment for Statistical Computing*. Vienna: R Foundation for Statistical Computing; 2012. [http://www.R-project.org/]. [ISBN 3-900051-07-0].

Cross-section perimeter is a suitable parameter to describe the effects of different baffle geometries in shaken microtiter plates

Clemens Lattermann[1†], Matthias Funke[2†], Sven Hansen[3], Sylvia Diederichs[1] and Jochen Büchs[1*]

Abstract

Background: Biotechnological screening processes are performed since more than 8 decades in small scale shaken bioreactors like shake flasks or microtiter plates. One of the major issues of such reactors is the sufficient oxygen supply of suspended microorganisms. Oxygen transfer into the bulk liquid can in general be increased by introducing suitable baffles at the reactor wall. However, a comprehensive and systematic characterization of baffled shaken bioreactors has never been carried out so far. Baffles often differ in number, size and shape. The exact geometry of baffles in glass lab ware like shake flasks is very difficult to reproduce from piece to piece due to the hard to control flow behavior of molten glass during manufacturing. Thus, reproducibility of the maximum oxygen transfer capacity in such baffled shake flasks is hardly given.

Results: As a first step to systematically elucidate the general effect of different baffle geometries on shaken bioreactor performance, the maximum oxygen transfer capacity (OTR_{max}) in baffled 48-well microtiter plates as shaken model reactor was characterized. This type of bioreactor made of plastic material was chosen, as the exact geometry of the baffles can be fabricated by highly reproducible laser cutting. As a result, thirty different geometries were investigated regarding their maximum oxygen transfer capacity (OTR_{max}) and liquid distribution during shaking. The relative perimeter of the cross-section area as new fundamental geometric key parameter is introduced. An empirical correlation for the OTR_{max} as function of the relative perimeter, shaking frequency and filling volume is derived. For the first time, this correlation allows a systematic description of the maximum oxygen transfer capacity in baffled microtiter plates.

Conclusions: Calculated and experimentally determined OTR_{max} values agree within ± 30% accuracy. Furthermore, undesired out-of-phase operating conditions can be identified by using the relative perimeter as key parameter. Finally, an optimum well geometry characterized by an increased perimeter of 10% compared to the unbaffled round geometry is identified. This study may also assist to comprehensively describe and optimize the baffles of shake flasks in future.

Keywords: Shaken bioreactors, Maximum oxygen transfer capacity (OTR_{max}), Degree of baffling, Relative perimeter, Out-of-phase phenomena

* Correspondence: Jochen.Buechs@avt.rwth-aachen.de
†Equal contributors
[1]AVT.Biochemical Engineering, RWTH Aachen University, Worringer Weg 1, 52074 Aachen, Germany
Full list of author information is available at the end of the article

Introduction

Shaken small scale bioreactors like shake flasks or microtiter plates are typically used for high throughput screening processes today. In the last years, many efforts have been made to characterize small scale bioreactors by describing and modeling liquid distribution, gas transfer, specific power input as well as mixing (reviewed in [1-7]). These efforts are motivated by the need to understand the yet insufficiently characterized screening systems in more detail. Furthermore, knowledge about important process parameters in an earlier stage of process development is desired [7,8]. While shake flasks are already characterized to some extent, there are several open questions regarding microtiter plates. The small dimensions of these reactor systems result in specific problems like the influence of surface tension or the lack of space for on-line measurement equipment.

For successful screening processes, sufficient oxygen supply is mandatory [7]. In shaken bioreactors higher maximum oxygen transfer capacities can be achieved by reducing the filling volume or increasing the shaking frequency, the vessel diameter or the shaking diameter. However, this approach has clear limits. Another possibility to enhance the maximum oxygen transfer capacity is the use of baffles inside the bioreactor. However, for shake flasks it is known that the reproducibility of oxygen transfer is poor between individual flasks if baffles are introduced [7,9,10]. This is due to the flow characteristics of molten pyrex glass which is very hard to control during the fabrication process of the baffles. As a result, it is nearly impossible to reproduce the exact geometries of the baffles. In most cases this can be proven even by the naked eye. Therefore, some authors even do not recommend the use of baffled shaken bioreactors at all [7,11].

Fabrication of baffles in microtiter plates is much more accurate and reproducible if laser cutting or injection molding is used. A wide variety of different baffled microtiter plates were extensively studied by Funke et al. [12]. An optimal geometry was found, considering different criteria. The best version is meanwhile commercially available. It shows reproducible oxygen transfer properties on an elevated level.

Up to now, no general method to predict the impact of baffles and its geometries on the maximum oxygen transfer capacity of shaken bioreactors is available. This is mainly caused by the problem to obtain a set of different baffled bioreactors with reproducible geometries made from glass. Therefore, in this work we utilized the results of Funke et al. [12] which were generated with the sulfite method for a wide variety of different reproducible geometries of 48-well microtiter plates as shaken model bioreactors. By using 48-well microtiter plates, an influence of surface tension which is quite prominent in 96-well microtiter plates is minimized [4,13]. The absolute cross-section areas were consistently kept constant (112 mm^2) among the different geometries. The relationship between the geometries of the wells and the maximum oxygen transfer capacities were derived and a respective mathematical correlation was developed. This correlation depends on the shaking frequency, the filling volume and the relative perimeter of the cross-section area of the wells as new geometric key parameter. In this equation, the relative perimeter is the only criterion reflecting the degree of baffling. Thereby, all important parameters of baffling like number, size and shape of the baffles are considered in one value. This work is deemed to be the first step in understanding and quantifying the impact of baffles on shaken bioreactor performance. In a future attempt the applicability of the developed approach has to be proven also for baffled shake flasks.

Results and discussion

Influence of baffling on oxygen transfer in microtiter plates

In this study, the maximum oxygen transfer capacity OTR_{max} of thirty different cross-section geometries in 48-well microtiter plates varying in their degree of baffling, as shown in Figure 1, were investigated. In Figure 2 the OTR_{max} values of all geometries are exemplarily shown for a filling volume of 600 μL at increasing shaking frequencies. The OTR_{max} increases with increasing shaking frequency. Furthermore, a clear dependency between the maximum oxygen transfer capacity and the cross-section geometry of the wells can be noticed. In particular, the OTR_{max} values of the baffled geometries shown in Figure 1A (transition edged to round) increases between least pronounced baffling (round) and most pronounced baffling (square) for nearly all shaking frequencies (compare upper diagrams in Figure 2). It can be concluded, that the OTR_{max} systematically depends on the applied baffle geometry. However, until now only an empirical description of the performance of the different baffle designs implemented in this study was available [12]. A mathematical correlation between the degree of baffling and the OTR_{max} could not yet be established. Moreover, the unexpected stagnation or decrease of the OTR_{max} with increasing shaking frequency, as it can be observed for some well geometry in Figure 2 (compare edged star geometry, middle of bottom diagram), could not be explained.

The influence of baffling on the maximum oxygen transfer capacity can be described, if a geometric key parameter is found, which allows an overall description of different cross-section geometries. For the well designs resulting from the transition from edged to round geometry (Figure 1A) an adequate parameter could be the number of edges or the diameter of the edge roundings. With

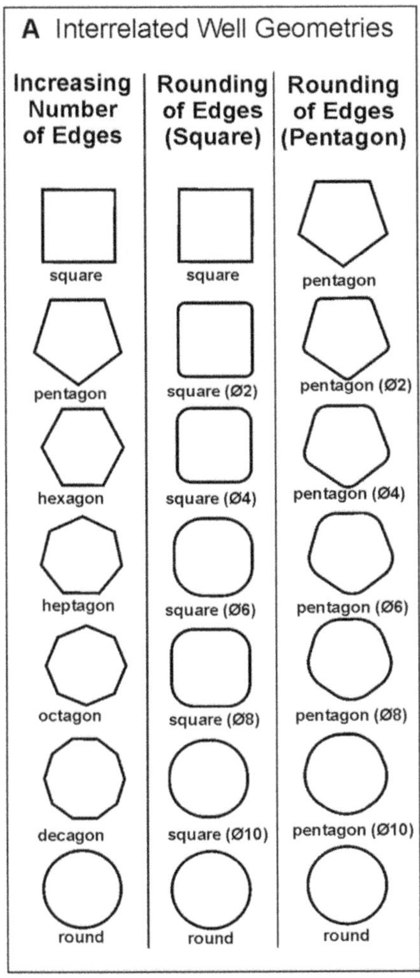

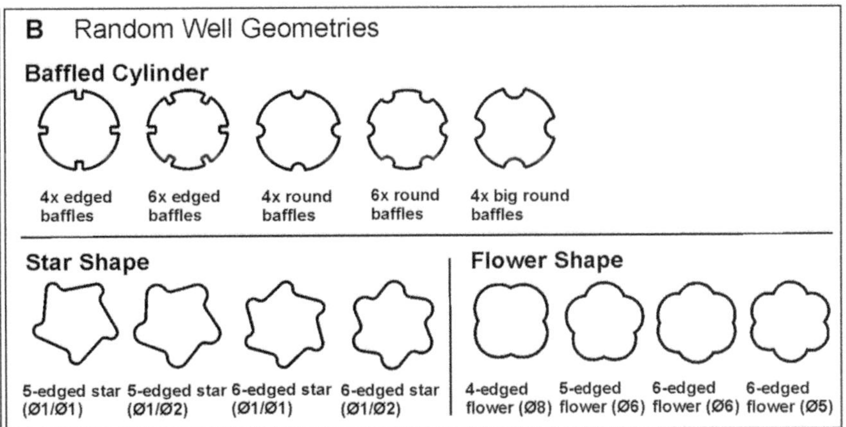

C Sorted Geometries according to their perimeter relative to round geometry

round	1.000	square (Ø4)	1.054
pentagon (Ø10)	1.001	pentagon	1.075
square (Ø10)	1.003	5-edged flower (Ø6)	1.076
pentagon (Ø8)	1.007	square (Ø2)	1.087
square (Ø8)	1.013	6-edged flower (Ø5)	1.120
decagon	1.018	square	1.128
pentagon (Ø6)	1.018	4x rounded baffles	1.135
4-edged flower (Ø8)	1.019	4x big rounded baffles	1.141
octagon	1.027	5-edged star (Ø1/Ø2)	1.179
square (Ø6)	1.030	5-edged star (Ø1/Ø2)	1.185
pentagon (Ø4)	1.032	6x rounded baffles	1.205
heptagon	1.037	5-edged star (Ø1/Ø1)	1.206
hexagon	1.050	6-edged star (Ø1/Ø1)	1.207
6-edged flower (Ø6)	1.050	4x edged baffles	1.283
pentagon (Ø2)	1.051	6x edged baffles	1.430

Figure 1 Designs of MTP well geometries (48-well plate format). (A) 3 different sets of well geometries represent the gradual transition from the most pronounced baffling (square and pentagon geometry, respectively) to the least pronounced geometry (round geometry); **(B)** 3 groups of well designs attained by introducing different types of baffles in round (baffled cylinders) or edged (star and flower shapes) well geometries. All geometries have an equivalent cross-section area of 112 mm^2. The diameters of the edge roundings or the introduced half-circular baffles are given in parentheses. Further design details can be found in Funke et al. (2009) [12]; **(C)** Geometries described in (A) and (B) sorted according to their perimeter relative to the round well geometry.

both parameters, however, it is not possible to adequately correlate all cross-section geometries and measured maximum oxygen transfer capacities (data not shown). The description of the other well geometries shown in Figure 1B is even more difficult with this approach. A more universal key parameter is required. Therefore, various further parameters like the ratio between maximum and minimal diameter have been investigated (data not shown). Finally, the perimeter of the cross-section area of the wells has been identified as the parameter resulting in the best correlation between different cross-section geometries and measured OTR_{max} values in the investigated microtiter plates. For comparison, we normalized the perimeter of the baffled geometry through dividing by the perimeter of the round reference geometry:

$$Peri := \frac{Perimeter_{baffled}}{Perimeter_{round}}. \tag{1}$$

Eq. (1) represents a universal, dimensionless parameter which includes the number as well as the size and the shape of the baffles. For example, the relative perimeter of highly pronounced baffles with a steep slope is greater than that of less pronounced baffles. Also, the number of baffles is considered by using the relative perimeter as correlating parameter. This property characterizes the relative perimeter as a powerful parameter allowing a ranking of investigated geometries as shown in Figure 1C. Furthermore, the relative perimeter enables the pooling of all measurement data for one filling volume (as shown in Figure 2) in one diagram (Figure 3). Thus, the maximum

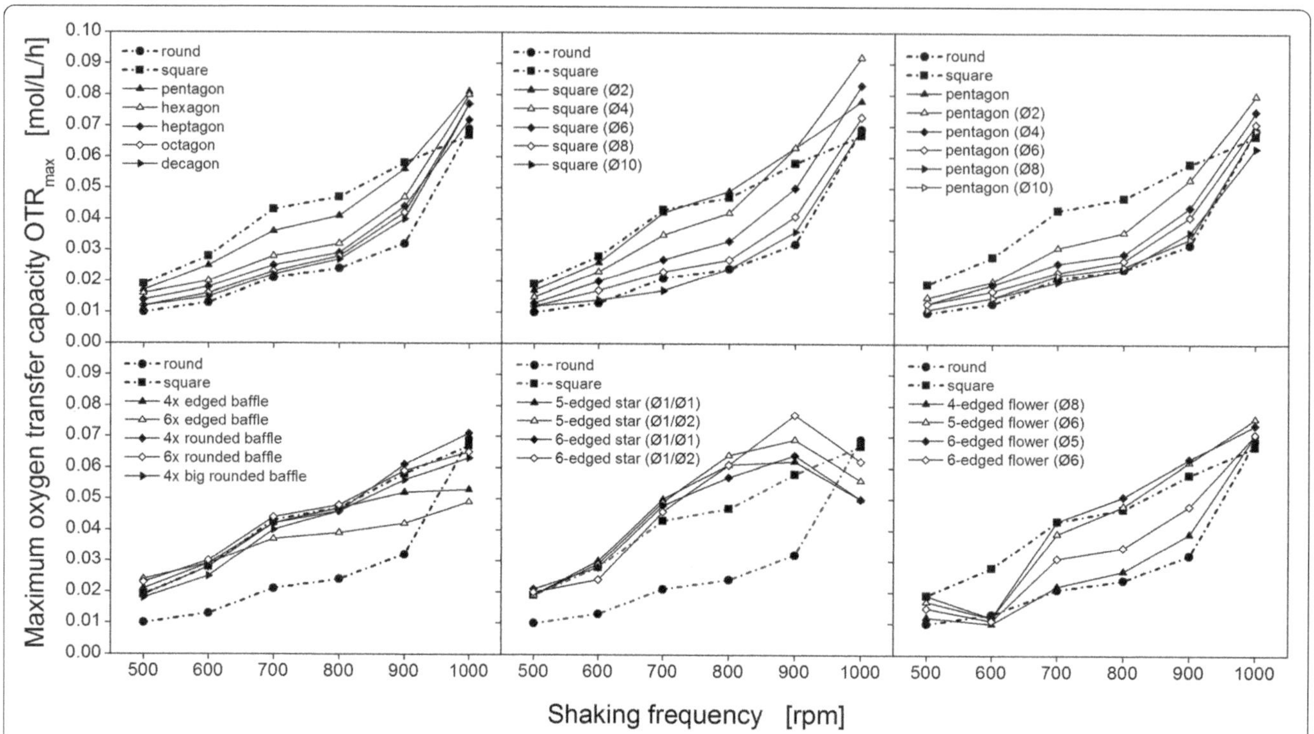

Figure 2 Maximum oxygen transfer capacity (OTR_{max}) obtained by the sulfite system at increasing shaking frequencies. OTR_{max} values are determined for different well geometries with 112 mm^2 cross-section area (48-well plate format). The filling volume is kept constant at 600 μL. Measurements performed with a modified BioLector measurement system: shaking diameter: 3 mm, temperature: 25°C, shaking frequency: 500 – 1000 rpm.

oxygen transfer capacity for all different geometries can easily be compared.

In Figure 3, the OTR_{max} is plotted as a function of the relative perimeter of all investigated geometries to systematically characterize the influence of baffles on the maximum oxygen transfer capacity. The smoothing splines between the measurement points were calculated by applying a modified L-curve criterion. As expected, the maximum oxygen transfer capacity increases with decreasing filling volume, as illustrated in Figure 3. This result agrees with the theory and literature, since the surface-to-volume ratio, and, thus, the oxygen transfer into the bulk liquid increases with decreasing filling volume in surface aerated bioreactors [11]. The measured OTR_{max} value in general increases with increasing shaking frequency. This trend also agrees with literature [14]. Furthermore, the shapes of the OTR_{max} curves over the relative perimeter for filling volumes of 200 - 400 μL equal each other. For example, looking at a specific shaking frequency, the OTR_{max} curves are merely shifted in their absolute height depending on the filling volume. For filling volumes of 500 μL and 600 μL the shape of the OTR_{max} curves seem to be smoother.

The most important result depicted in Figure 3 is the development of the OTR_{max} curves as a function of the relative perimeter. Starting at a relative perimeter of 1.0, up to a relative perimeter of approximately 1.05, a linear increase of the OTR_{max} values can generally be observed. Subsequently, always a maximum is reached at a relative perimeter of about 1.075 to 1.12. For even higher relative perimeters, the OTR_{max} values remain either constant (<700 rpm) or decrease (>700 rpm). Obviously, there is a common limit of the OTR_{max} for all applied shaking frequencies, filling volumes and well geometries. Expressed in other words, increasing the degree of baffling increases the maximum oxygen transfer capacity until an upper limit value is reached. This maximum is generally observed at baffled geometries with a perimeter 10% larger than the one of the unbaffled round geometry. The result is quite astonishing because it reflects the abstract character of the relative perimeter, taking number, size and shape of the baffles indirectly into account. According to Figure 1C, the maximum oxygen transfer capacity at a relative perimeter of 1.1 corresponds to a geometry between square (Ø2) and the 6-edged flower (Ø5). It is remarkable that this result excellently agrees with the qualitative observations made by Funke et al [12]. It can now be quantitatively described by means of the relative perimeter.

Influence of baffling on liquid distribution in microtiter plates

The phenomenon of stagnant or decreasing OTR_{max} values at relative perimeters > 1.1 in Figure 3 has to be

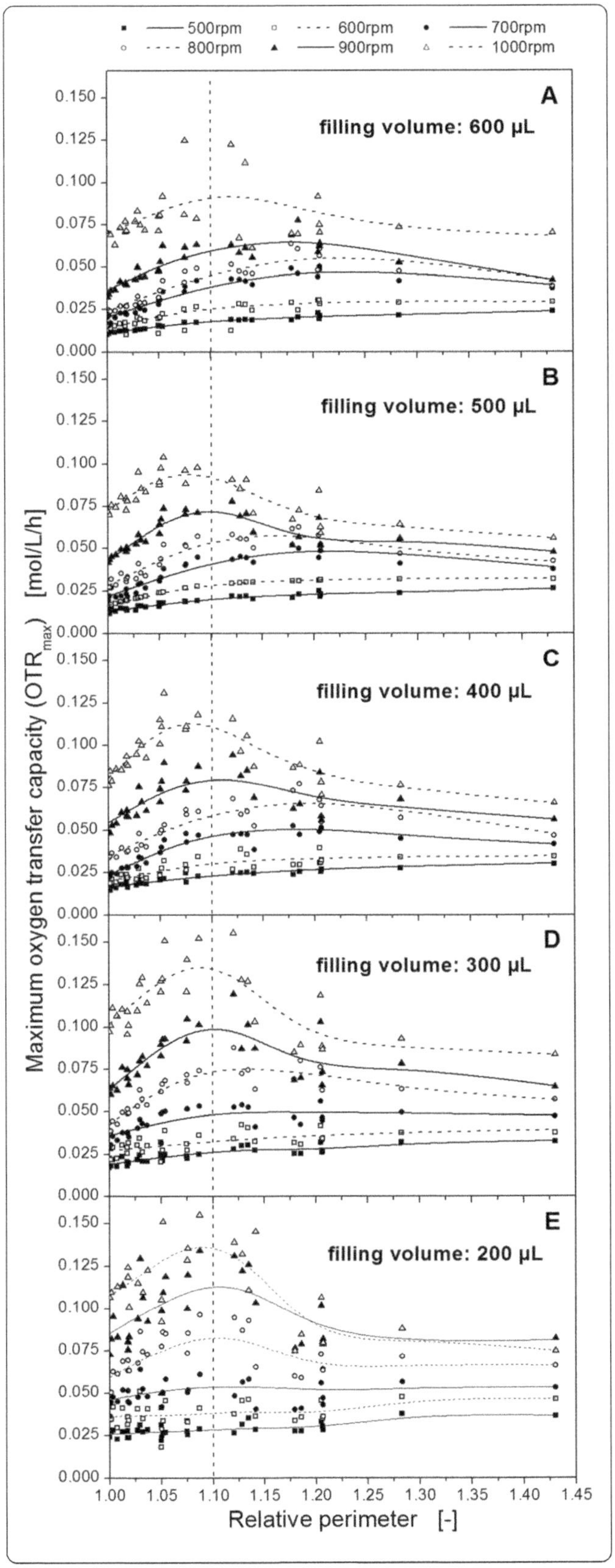

Figure 3 Maximum oxygen transfer capacity (OTR_{max}) as a function of the relative perimeter. The perimeters of the different well geometries are related to the perimeter of the round geometry with the same well cross-section area of 112 mm^2. The dashed line represents the transition from in-phase to out-of-phase operation condition. Measurements performed with a modified BioLector measurement system: shaking diameter: 3 mm, temperature: 25°C, shaking frequency: 500 – 1000 rpm, filling volume: 600 µL **(A)**, 500 µL **(B)**, 400 µL **(C)**, 300 µL **(D)**, 200 µL **(E)**.

evaluated in detail to entirely explain the whole data. In order to understand the mass transfer data, the behaviour of the rotating liquid has to be considered. The bulk liquid in shaken bioreactors typically rotates inside the vessel if in-phase conditions exist. In contrast, at out-of-phase conditions the major part of the liquid remains on the bottom of the vessel and does not move anymore. Simultaneously, the maximum liquid height and the maximum oxygen transfer capacity into the liquid is reduced.

Three different reasons for inducing these unsuitable out-of-phase conditions in shake flasks were identified in literature [15]. One effector is elevated viscosity. In this case, out-of-phase conditions are provoked, if the rotating centrifugal force is not strong enough to overcome the viscous forces [16]. A new Phase-Number Ph was defined and concluded that Ph > 1.26 is the relevant constraint for desired in-phase operation [16]. Another reason for out-of-phase conditions are unsuitable high ratios between the maximum diameter of the shaken bioreactors and the shaking diameter of the applied shaker, as shown for non-baffled large shake flasks by Büchs et al. [15] and for microtiter plate with round geometry by Kensy et al. [17]. The third reason for unsuitable out-of-phase operating conditions in shake flasks is the introduction of baffles which are too in number large in size. Problems may already occure at water-like viscosities. The probability of out-of-phase conditions increases with decreasing shaking diameter [7,15]. The undesired phenomenon appears if the rotating centrifugal force generated by the shaker is not strong enough to overcome the negative impact of the baffles preventing the bulk liquid from rotating in the flask. Although this third reason for out-of-phase conditions resembles the first reason, the above mentioned Phase-Number can not be used to evaluate the impact of baffles.

Considering the decreasing mass transfer at higher degrees of baffling (relative perimeter > 1.1) it can be speculated that out-of-phase conditions exist at relative perimeters higher than 1.1 in the investigated 48-well microtiter plates. Therefore, an experimental examination of the appearance of out-of-phase conditions has been conducted in this study. As mentioned above, the shaken liquid remains at the bottom of the well if out-of-phase conditions are present. This implies that the liquid height at the well center is elevated compared to in-phase conditions. Hence, an optical measurement of the liquid height provides further information about the flow conditions in the microtiter wells and, thus, may help to explain the observations of the mass transfer study.

In Figure 4 the measured liquid heights in microtiter plates are shown for an exemplary filling volume of 500 μL and different relative perimeters and shaking frequencies. For the unbaffled round geometry (relative perimeter = 1.0) the liquid height decreases with increasing shaking frequency (follow data points on the y-axis top down in Figure 4). These measurements agree with the theory that the liquid rises up the well wall the higher the centrifugal forces are and, simultaneously, drains off the well bottom. However, for all elevated perimeters < 1.1, the liquid height at the center of the wells is higher compared to the unbaffled geometry at the same shaking frequencies. In this case, more liquid is located on the well bottom due to the higher degree of baffling and the breakup and restricted movement of the bulk liquid. Although the liquid rotation is reduced, the OTR_{max} still increases at these operating conditions (refer to Figure 3) since the breakup of the bulk liquid also increases its surface area and introduces turbulence into the bulk liquid. For relative perimeters higher than 1.1, the liquid level remains more or less constant at high level. An increased degree of baffling completely prevents the rotation of the liquid. The rotating bulk liquid does not exist anymore and nearly all liquid swashes at the center of the well as illustrated by pictures for shake flasks by Büchs [7]. As a consequence, the maximum oxygen transfer capacity is reduced which can be seen for relative perimeters higher than 1.1 in Figure 3. Finally, these results indicate the transition from in-phase to out-of-phase operating condition beyond the dashed lines in Figure 3 and Figure 4. Thus, the existence of optimal well geometries concerning high maximum oxygen transfer capacity in microtiter plates at a relative perimeter of about 1.1 has been confirmed by applying a completely different measurement method (liquid height) compared to the OTR_{max} measurements.

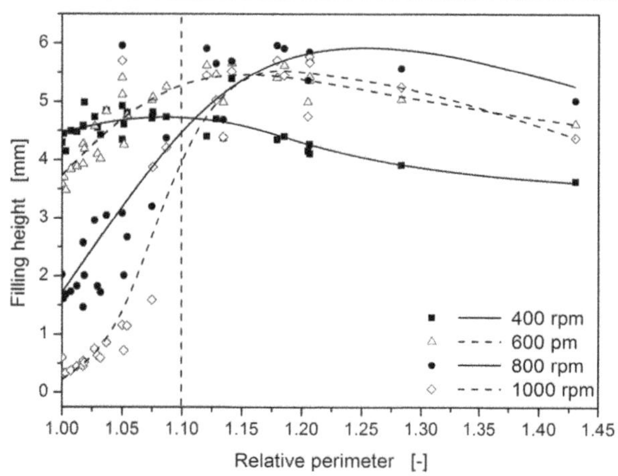

Figure 4 Filling height as a function of the relative perimeter. The liquid height is measured vertically at the center of the well bottom with the modified BioLector measurement system at a total liquid filling volume of 500 μL. The dashed line represents the transition from in-phase to out-of-phase liquid behavior. The perimeters of the different well geometries are related to the perimeter of the round geometry with the same well cross-section area. Shaking diameter: 3 mm, temperature: 25°C, shaking frequency: 400 – 1000 rpm.

The relative perimeter as integral parameter to calculate OTR_{max} in microtiter plates

The aforementioned investigations revealed that the OTR_{max} of the investigated microtiter plates increases linearly with the relative perimeter until a maximal value is reached (Figure 3). For relative perimeters larger than 1.1, a constant or even decreasing trend was noticed. This observation is explained above by prevailing out-of-phase conditions. Therefore, a correlation to calculate OTR_{max} values depending on the relative perimeter, shaking frequency and filling volume can only be derived for relative perimeters smaller than 1.1, where regular in-phase conditions exist. By fitting the experimental results with a least square method, the following equation can be obtained:

$$OTR_{max} = 2.5 \cdot 10^{-7} \cdot Peri^{6.0} \cdot n^{2.37} \cdot V_L^{-0.64} + 3.3 \cdot 10^{-4}. \tag{2}$$

Equation (2) represents a novel equation to pre-calculate the maximum oxygen transfer capacity OTR_{max} in baffled 48-well microtiter plates at in-phase operating condition. The OTR_{max} is calculated dependent on the relative perimeter of the well geometry (Peri), the shaking frequency (n) and the filling volume (V_L). From literature it is known that the shaking diameter also has a strong influence on the oxygen transfer in shaken bioreactors [14]. However, for optical on-line measurements in shaken microtiter plates using the BioLector technique, a constant shaking diameter of 3 mm is established [18]. Therefore, the shaking diameter in this study was kept constant to 3 mm and the influence of the shaking diameter is not considered in Eq. (2). The constant second term in Eq. (2) is explained by the fact that a basal oxygen transfer is accomplished by sole diffusion at static condition (without any convective flow). This effect has originally been discovered by Hermann et al. [13] for 96-well plates and later been verified by Kensy et al. for 48-well plates [17].

With Eq. (2) the OTR_{max} values of the different geometries in Figure 1 were calculated and plotted over the values obtained by the oxygen transfer measurements. Figure 5 illustrates the comparison of the measured and

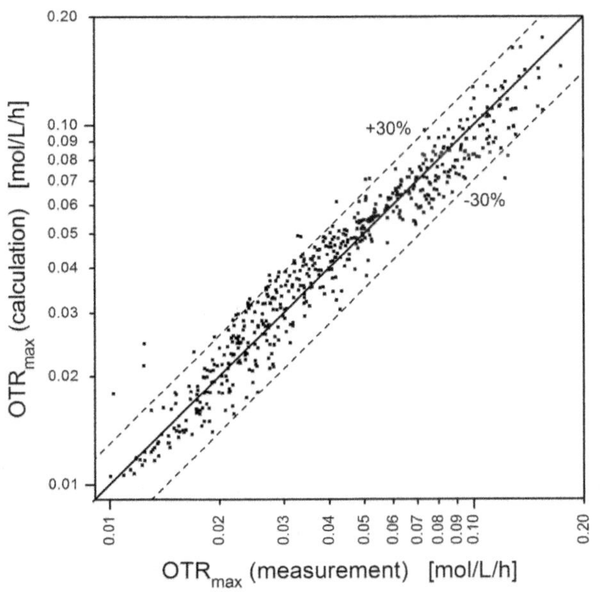

Figure 5 Comparison of calculated and measured OTR_{max} values. Only well geometries with a relative perimeter of < 1.1 are considered. The OTR_{max} values are calculated according Eq. (2). Dashed lines indicate a standard deviation of ± 30%.

calculated OTR_{max} values. It can be seen that all measured data for in-phase operating condition fit to the calculated data within an accuracy of ± 30%. The few outliers which can be seen in Figure 5 do not follow a systematic behavior and cannot be assigned to certain well geometries or operating conditions. Consequently, these outliers are caused by measurement errors in the determination of the OTR_{max} based on optical measurements. Such a correlation can only be derived from a high quantity of experimental data. To reduce the experimental effort, we just considered 48-well microtiter plates in this study. Experimental data from shake flasks are much more time consuming to obtain and less reproducible due to the variability of the baffles, as explained in the introduction. Furthermore, an experimental survey of our results in microtiter plate formats, which differ from the 48-well dimensions, is a topic for further investigations. Thereby, the height of the well geometry will not have a direct influence on the oxygen transfer. It just limits the maximum liquid height in the well and, thus, the maximum shaking frequency. The only defining parameter for different well geometries is already incorporated in our correlation, since the perimeter of the baffled well geometry has been defined as its relative value compared to the perimeter of the round geometry with the same well area (Eq. (1)). By defining the perimeter in that relative way, the obtained results, i.e. the calculation of the OTR_{max} up to a relative perimeter of 1.1, can be transferred to larger well geometries. However, due to changing surface-to-volume ratios, an adaptation of Eq. (2) can be expected, if other microtiter plate formats are used. To complete the discussion, it has to be emphasized that Eq. (2) is restricted so far to a shaking diameter of 3 mm and liquids of water-like viscosities. Furthermore, the influence of surface tension, which is observed in small volume microtiter plates (e.g. 96-well plates), is not taken into account just as the influence of the material surface properties is not considered, which might influence the oxygen transfer as observed in shake flasks [14]. However, from current state of knowledge it is legitimate to assume that the relative perimeter can also be used as key parameter to describe different well formats. Keeping in mind that the investigated baffled geometries vary significantly in number, size and shape, Eq. (2) represents a substantial improvement to characterize the oxygen transfer capacity in small scale shaken bioreactor systems. It is the first time that the maximum oxygen transfer capacities in shaken baffled bioreactors were quantitatively described. With the experience made with 48-well microtiter plates in future work also the effect of baffles in shakes flasks made of glass or plastic material will be investigated.

Conclusion and outlook

In this work the influence of baffling on the maximum oxygen transfer capacity in 48-well microtiter plates has quantitatively been investigated. The relative perimeter of the cross-section area was chosen as geometric key parameter to correlate the maximum oxygen transfer capacity with the shaking frequency and filling volume. In agreement with the qualitative results of Funke et al [12], an optimum maximum oxygen transfer capacity could be found at a relative perimeter of 1.1. Furthermore, optical measurements of the liquid height at the center of the wells were conducted to obtain additional information about the prevailing operating condition. It could be shown that undesired out-of-phase conditions exist if the relative perimeter exceeds values of 1.1. With this investigation, a phenomenological explanation for the measured mass transfer values could be given. Finally, an empirical correlation was derived to calculate the maximum oxygen transfer capacity in baffled microtiter plates depending on the shaking frequency and filling volume. The relative perimeter as geometric key parameter includes the specific properties of the baffles. Therefore, the degree of baffling in microtiter plates can now quantitatively be described for new geometries if liquids with water-like viscosities are used and effects caused by surface tension are negligible. The obtained correlation is limited to microtiter plates shaken at a shaking diameter of 3 mm. Further investigations in other shaken bioreactors like shake flasks or large shaken barrels have to be carried out to extend the correlation to a more universal equation. The influence of specific effects in small scale bioreactors, e.g. caused by surface tension, could also be investigated in future studies.

Materials and methods

Baffled well geometries

A total of thirty different well geometries, varying in number, size and shape of the baffles, were investigated in this work. The baffled geometries shown in Figure 1 have been realized in the wells of 48-well microtiter plates by introducing rectangular or rounded wall structures. The dimensions of the wells were chosen as such that the cross sectional area is always equivalent (112 mm^2) to the round reference geometry.

Prototypes of microtiter plate bodies were fabricated out of a 20 mm thick acrylic glass plate (polymethyl methacrylat, PMMA) with outer dimensions of 128 mm × 85 mm by laser cutting. To seal the bottom, a PMMA plate of 2 mm thickness was glued onto the bottom of the prototype bodies. As illustrated in Figure 1A, 3 different sets of well geometries represent a gradual transition from the most pronounced baffling (square and pentagon geometry, respectively) to the least pronounced baffling (round geometry). The transition was realized in 3 different ways. First, starting at square geometry, the number of edges is continuously increased. Second and third, starting at square and pentagon geometry, the edges are more and more rounded. Moreover, Figure 1B shows 3 groups of well designs attained by introducing different types of baffles in round or edges well geometries. For further details regarding the investigated well geometries refer to [12].

Optical measurement system

The BioLector technique was used to determine the liquid height at the well center of the microtiter plate as well as the maximum oxygen transfer capacity OTR_{max} of the different baffled geometries. This optical measurement system enables non-invasive fluorescence and scattered light measurements in shaken microtiter plates [18]. The shaking process of the microtiter plate is not interrupted and, thus, disturbing influences during measurement are avoided. Thereby, a quasi-continuous measurement in shaken microtiter plates is realized. In this study, a slightly modified BioLector system was used which consists of an orbital shaker (based on Lab-Shaker LS-W, Kühner AG, Basel, Switzerland), a x-y linear drive (Bosch Rexroth AG, Lohr am Main, Germany), a custom-made filter fluorescence spectrometer (PreSens GmbH, Regensburg, Germany) and a computer. The orbital shaker was modified to realize a shaking diameter of 3 mm and shaking frequencies of up to 1000 rpm. A hood was placed above the microtiter plate on the shaker tray. The hood was continuously flushed with humidified air to reduce evaporation. To avoid the influence of a cover foil, no additional cover except the hood was used. For further information regarding the optical measurement device refer to [12].

Characterization of oxygen transfer with a 0.5 M sulfite system

The maximum oxygen transfer capacity OTR_{max} of the different well geometries was characterized by applying a slightly modified sulfite oxidation method originally developed by Hermann et al. [13,17,19]. As described in detail by Funke et al. [12], the pH-indicator of the original recipe [19] was replaced by a $2 \cdot 10^{-8}$ M hydroxy-pyrenetrisulfonic acid (HPTS) solution (Fluka, Buchs, Switzerland), which shows pH dependent fluorescence properties. Corresponding to the mentioned sulfite oxidation method, the time t_{ox} is measured until the oxidation of the 0.5 M sodium sulfite to sulfate is completed. At the end of the reaction the pH value in the test solution drops sharply. The fluorescence of HPTS, excited at 420 nm and measured above 515 nm, decreases with decreasing pH. Thus, the pH drop and, therefore, the reaction time can be exactly monitored by means of optical measurements. For known reaction time (t_{ox}), initial concentration of sulfite ($c_{sulfite}$) and stoichiometric coefficient for oxygen ($\nu_{O2} = 0.5$), the oxygen transfer rate (OTR) from gas to liquid phase can be calculated as follows:

$$OTR = k_L a \cdot \left(c^*_{O_2} - c_L\right) = c_{sulfite} \cdot \nu_{O_2}/t_{ox}. \quad (3)$$

In Eq. (3) the volumetric mass transfer coefficient k_La as well as the oxygen concentrations in the bulk liquid (c_L) and the oxygen concentration at the liquid gas interface (c^*_{O2}) are unknown. According to Weisenberger and Schumpe [20], the oxygen concentration at the liquid gas interface c^*_{O2} can be calculated as the product of oxygen solubility in the liquid and the oxygen partial pressure in the gas phase ($L_{O2} = 8.35 \cdot 10^{-4}$ mol·L^{-1}·bar^{-1} at 25°C, $p_G = 0.2095$ bar). Furthermore, Hermann et al. [13] determined the reaction constant k_1 for the first-order kinetic for the sulfite oxidation out of experiments in stirred tank reactors to $k_1 = 2.385$ h^{-1}. By using a mass balance for oxygen transfer into the bulk liquid of a shaken microtiter plate, the oxygen concentration in the bulk liquid c_L can be determined if the OTR is known [13]:

$$c_L = OTR/k_1. \quad (4)$$

Combining Eq. (3) and Eq. (4), the volumetric mass transfer coefficient k_La can be calculated as follows:

$$k_L a = OTR/(L_{O_2} \cdot p_G - OTR/k_1). \quad (5)$$

If the oxygen concentration in the bulk liquid is close to zero, the OTR becomes maximal. Consequently, the maximum oxygen transfer capacity OTR_{max} can be calculated by the following equation:

$$OTR_{max} = k_L a \cdot c^*_{O_2} = OTR \cdot L_{O_2} \cdot p_G / (L_{O_2} \cdot p_G - OTR/k_1). \quad (6)$$

Using Eq. (6), the maximum oxygen transfer capacity can now be obtained indirectly through measurement of the oxygen transfer rate.

Detection of out-of-phase conditions through liquid height measurements

The liquid height at the well center of the microtiter plate was determined by measuring the fluorescence intensity of a $1.25 \cdot 10^{-6}$ M fluorescein solution (sodium salt, Fluka, Buchs, Switzerland) in 0.2 M sodium phosphate buffer (pH 7) (Roth, Karlsruhe, Germany). The fiber optics of the BioLector prototype was vertically installed underneath the wells. All experiments to determine the liquid height were conducted at a constant filling volume of $V_L = 500$ µL and different shaking frequencies of 400 – 1000 rpm. The fluorescence was excited at 420 nm (bandpassfilter ± 10 nm) and detected above 515 nm (cut-of filter). For calibration, different filling volumes of the fluorescein solution were filled into the microtiter plate prototypes. The fluorescence signal was measured at a shaking frequency of 200 rpm, where no liquid movement occurs [13]. Since the cross-section area for each geometry is constant (112 mm^2), the liquid height can be calculated depending on the filling volume. The fluorescence signal of the fluorescein solution decreases with decreasing liquid height. Thus, a geometry-independent correlation of fluorescence signal and liquid height at the well center is obtained.

Fitting of measurement data by applying the L-curve criterion

For the graphical presentation both the experimental data of the maximum oxygen transfer capacity and of the filling height at the well center were fitted by using smoothing splines. The regularization parameter α of the smoothing spline fit was determined by using the L-curve criterion [21,22]. The L-curve is a plot of the norm of the regularized solution (S) representing the smoothness of the curve (Eq. (7)) versus the norm of the corresponding residual (R) representing the data error (Eq. (8)). A good compromise between both norms, i.e. a suitable regularization parameter can be found at the "corner" of the L-curve plot. This is the point where the curve is closest to the origin.

$$S = \left\| \vec{X} \right\| = \sqrt{(X_{spline,1})^2 + (X_{spline,2})^2 + \ldots + (X_{spline,i})^2} \quad (7)$$

$$R = \left\| \vec{X} \right\| = \sqrt{(X_{spline,1} - X_1)^2 + (X_{spline,2} - X_2)^2 + \ldots + (X_{spline,i} - X_i)^2} \quad (8)$$

A slight modification compared to the literature was used when applying the L-curve criterion. The square of the norm of the corresponding residual (R^2) was doubled in order to obtain a fitting better representing the experimental data. Thus, the suitable regularization parameter α was defined as the radius of a circle around the origin which touches the L-curve at the closest point to the origin. This led to the following minimization problem:

$$min_\alpha \left\| 2 \cdot R^2_\alpha + S^2_\alpha \right\|. \quad (9)$$

Equations (7) to (9) were implemented into an algorithm and solved by applying the MATLAB software package (Version 7.8.0; The Mathworks, MA, USA).

Abbreviations

HPTS: 8-Hydroxypyrene-1,3,6-trisulfonic acid; MTP: Microtiter plate; α: Regularization parameter [-]; c_L: Oxygen concentration in the liquid [mol/L]; c^*_{O2}: Oxygen concentration at the gas-liquid interface [mol/L]; $c_{sulfite}$: Sulfite concentration [g/L]; k_1: First-order kinetic reaction constant [1/h]; k_La: Volumetric mass transfer coefficient [1/h]; L_{O2}: Oxygen solubility in the liquid [mol/L]; n: Shaking frequency [1/min]; OTR: Oxygen transfer rate [mol/L/h]; OTR_{max}: Maximum oxygen transfer capacity [mol/L/h]; p_G: Oxygen partial pressure in the gas phase [bar]; Peri: Normalized perimeter of the baffled geometry [-]; $Perimeter_{baffled}$: Perimeter of the baffled geometry [mm]; $Perimeter_{round}$: Perimeter of the round reference geometry [mm]; Ph: Phase-Number [-]; R: Residual, representing the data error [-]; S: Regularized solution, representing the smoothness of the curve [-]; t_{ox}: Reaction time [s]; V_L: Filling volume [µL]; v_{O2}: Stoichiometric coefficient of oxygen [-]; $X_{spline,i}$: Data point of the spline [-], in this study substituted by calculated data of OTR_{max}; X_i: Data point of measurement data [-], in this study substituted by measured data of OTR_{max}.

Competing interests

The authors declare that they have no competing interests.

Authors' contributions

CL is the author of the final version of this study. MF is responsible for the experimental results. He also drafted the first version of the manuscript. CL, MF and JB masterminded the study and participated in its design, coordination, and drafting and finalizing of the manuscript. SH kindly assisted in fitting the experimental data. SD supported in conducting and evaluating experimental results. All authors read and approved the final manuscript.

Author details

[1]AVT.Biochemical Engineering, RWTH Aachen University, Worringer Weg 1, 52074 Aachen, Germany. [2]Lonza Group Ltd, Münchensteinerstraße 38, 4002 Basel, Switzerland. [3]Evonik Industries AG, Rodenbacher Chaussee 4, 63457 Hanau-Wolfgang, Germany.

References

1. Kirk TV, Szita N: **Oxygen transfer characteristics of miniaturized bioreactor systems.** *Biotechnol Bioeng* 2013, **110:**1005–1019.
2. Klöckner W, Büchs J: **Advances in shaking technologies.** *Trends Biotechnol* 2012, **30:**307–314.
3. Schäpper D, Alam MNHZ, Szita N, Lantz AE, Gernaey KV: **Application of microbioreactors in fermentation process development: a review.** *Anal Bioanal Chem* 2009, **395:**679–695.
4. Duetz WA: **Microtiter plates as mini-bioreactors: miniaturization of fermentation methods.** *Trends Microbiol* 2007, **15:**469–475.
5. Fernandes P, Cabral JMS: **Microlitre/millilitre shaken bioreactors in fermentative and biotransformation processes - a review.** *Biocatal Biotransform* 2006, **24:**237–252.
6. Kumar S, Wittmann C, Heinzle E: **Minibioreactors.** *Biotechnol Lett* 2004, **26:**1–10.
7. Büchs J: **Introduction to advantages and problems of shaken cultures.** *Biochem Eng J* 2001, **7:**91–98.
8. Bareither R, Pollard D: **A review of advanced small-scale parallel bioreactor technology for accelerated process development: current state and future need.** *Biotechnol Progr* 2011, **27:**2–14.
9. Delgado G, Topete M, Galindo E: **Interaction of cultural conditions and end-product distribution in *Bacillus subtilis* grown in shake flasks.** *Appl Microbiol Biotechnol* 1989, **31:**288–292.
10. McDaniel LE, Bailey EG, Zimmerli A: **Effect of oxygen supply rates on growth of *Escherichia coli* - I: studies in unbaffled and baffled shake flasks.** *Appl Microbiol* 1965, **13:**13.
11. Henzler HJ, Schedel M: **Suitability of the shaking flask for oxygen supply to microbial cultures.** *Bioprocess Eng* 1991, **7:**123–131.
12. Funke M, Diederichs S, Kensy F, Müller C, Büchs J: **The baffled microtiter plate: increased oxygen transfer and improved online monitoring in small scale fermentations.** *Biotechnol Bioeng* 2009, **103:**1118–1128.
13. Hermann R, Lehmann M, Büchs J: **Characterization of gas-liquid mass transfer phenomena in microtiter plates.** *Biotechnol Bioeng* 2003, **81:**178–186.
14. Maier U, Büchs J: **Characterisation of the gas-liquid mass transfer in shaking bioreactors.** *Biochem Eng J* 2001, **7:**99–106.
15. Büchs J, Lotter S, Milbradt C: **Out-of-phase operating conditions, a hitherto unknown phenomenon in shaking bioreactors.** *Biochem Eng J* 2001, **7:**135–141.
16. Büchs J, Maier U, Milbradt C, Zoels B: **Power consumption in shaking flasks on rotary shaking machines: II: nondimensional description of specific power consumption and flow regimes in unbaffled flasks at elevated liquid viscosity.** *Biotechnol Bioeng* 2000, **68:**594–601.
17. Kensy F, Zimmermann HF, Knabben I, Anderlei T, Trauthwein H, Dingerdissen U, Büchs J: **Oxygen transfer phenomena in 48-well microtiter plates: determination by optical monitoring of sulfite oxidation and verification by real-time measurement during microbial growth.** *Biotechnol Bioeng* 2005, **89:**698–708.
18. Samorski M, Müller-Newen G, Büchs J: **Quasi-continuous combined scattered light and fluorescence measurements: a novel measurement technique for shaken microtiter plates.** *Biotechnol Bioeng* 2005, **92:**61–68.
19. Hermann R, Walther N, Maier U, Büchs J: **Optical method for the determination of the oxygen-transfer capacity of small bioreactors based on sulfite oxidation.** *Biotechnol Bioeng* 2001, **74:**355–363.
20. Weisenberger S, Schumpe A: **Estimation of gas solubilities in salt solutions at temperatures from 273 K to 363 K.** *Aiche J* 1996, **42:**298–300.
21. Hansen PC, O'Leary DP: **The use of the L-curve in the regularization of discrete ill-posed problems.** *SIAM J Sci Comput* 1993, **14:**1487–1503.
22. Hansen PC: **Analysis of discrete ill-posed problems by means of the L-curve.** *SIAM Rev* 1992, **34:**561–580.

15

Minireactor-based high-throughput temperature profiling for the optimization of microbial and enzymatic processes

Martin Kunze[1], Clemens Lattermann[1], Sylvia Diederichs[1], Wolfgang Kroutil[2] and Jochen Büchs[1*]

Abstract

Background: Bioprocesses depend on a number of different operating parameters and temperature is one of the most important ones. Unfortunately, systems for rapid determination of temperature dependent reaction kinetics are rare. Obviously, there is a need for a high-throughput screening procedure of temperature dependent process behavior. Even though, well equipped micro-bioreactors are a promising approach sufficient temperature control is quite challenging and rather complex.

Results: In this work a unique system is presented combining an optical on-line monitoring device with a customized temperature control unit for 96 well microtiter plates. By exposing microtiter plates to specific temperature profiles, high-throughput temperature optimization for microbial and enzymatic systems in a micro-scale of 200 µL is realized. For single well resolved temperature measurement fluorescence thermometry was used, combining the fluorescent dyes Rhodamin B and Rhodamin 110. The real time monitoring of the microbial and enzymatic reactions provides extensive data output. To evaluate this novel system the temperature optima for *Escherichia coli* and *Kluyveromyces lactis* regarding growth and recombinant protein production were determined. Furthermore, the commercial cellulase mixture Celluclast as a representative for enzymes was investigated applying a fluorescent activity assay.

Conclusion: Microtiter plate-based high-throughput temperature profiling is a convenient tool for characterizing temperature dependent reaction processes. It allows the evaluation of numerous conditions, e.g. microorganisms, enzymes, media, and others, in a short time. The simple temperature control combined with a commercial on-line monitoring device makes it a user friendly system.

Keywords: High-throughput screening, On-line monitoring, Microtiter plate, Optical measurement, Temperature optimum

Background

Bioprocesses, either fermentations or enzymatic catalysis, depend on a number of different operating parameters. Temperature is one of the most important. It is commonly known that processes should be performed under optimal conditions in order to achieve best results, e.g. reaction rates or yields [1]. Thereby, the temperature optimum can be different within one microbial or enzymatic system, either the focus is on growth or product formation [2], on enzyme activity or stability [3,4].

* Correspondence: Jochen.Buechs@avt.rwth-aachen.de
[1]AVT-Chair for Biochemical Engineering, RWTH Aachen University, Worringerweg 1, 52074 Aachen, Germany
Full list of author information is available at the end of the article

Over the last decades the number of potential bioprocesses for the production of valuable products increased and this trend continues. Consequently, more and more microbial and enzymatic systems need to be characterized regarding their temperature optima. Conventional methods are disadvantageous in many ways. Repetitive batch cultivations at varied temperatures, either in shake flasks or bench scale bioreactors, provide only a few data points at the expense of relatively high material and time input. The analysis of temperature-specific activities and kinetic parameters of enzymes is traditionally performed in a spectrophotometric way by use of a temperature-controlled water jacketed single cuvettes [5]. Even though the external thermostat provides constant and accurate temperature control over a broad temperature range, the numbers of samples handled for simultaneous reading in such spectrophotometers are usually very limited.

To face these limitations micro-bioreactors (MBR) became a promising alternative, but sufficient temperature control is quite challenging [6]. The simplest way is to use MBRs, e.g. microtiter plates (MTP), in temperature controlled rooms [7] or incubators [8]. Another option is to link the MBR system to a thermostat system and circulate water through the MBR chamber base [4,9-11]. The major disadvantage of these systems is their limitation in operating parallel reactors at different temperatures. One exceptional system allows the operation of sixteen parallel small scale reactors equipped with ceramic heating jackets but its use requires costly instrumentation and is focused on chemical reactions [12]. The integration of electrical micro-heaters to the bottoms or walls of MBRs allows individual temperature control of parallel MBRs [13-15]. However, for high-throughput application the required hardware and control become exceedingly complex.

Whilst temperature control is rather challenging in MBRs, temperature measurement is relatively simple. There are several reports on the application of temperature dependent resistances [14-16] or thermocouples [10,17]. Also thermal or infrared cameras allow temperature determination in MBRs [18,19]. Furthermore, the so called fluorescence thermometry can be used [11,20-22]. The combination of a temperature dependent fluorophore and a spectrophotometer is a promising method for high-throughput temperature determination, e.g. in MTPs.

In this work, a system is presented combining the on-line monitoring system BioLector [23,24] with a customized temperature control unit for 96 well MTPs, thereby, allowing high-throughput temperature optimization for microbial and enzymatic reaction systems in a micro-scale of 200 μL. On the one hand, the BioLector system allows quasi-continuous measurement of optical signals over time representative for microbial growth and product formation or enzymatic activity. On the other hand, the tempering system, consisting of a special thermostating block connected to two thermostats, creates specific temperature profiles over a MTP. For the single well temperature measurement fluorescence thermometry was used. Combining the fluorescent dyes Rhodamin B (RhB) and Rhodamin 110 (Rh110) a reliable assay was established for application with the BioLector technique [21]. After defining several profiles at different temperature levels, the device was used to characterize microbial and enzymatic systems. The whole temperature profiles and optima for the bacterium *Escherichia coli* and the yeast *Kluyveromyces lactis* regarding growth and recombinant protein production were determined in single shot experiments. Additionally, the commercial cellulase mixture Celluclast as a representative for enzymes was investigated applying a fluorescent activity assay. The results of this temperature high-throughput screening could be used for a mathematical description of the particular temperature dependent behavior of the investigated biological system as an extended Arrhenius model for catalyst activation and deactivation [25].

Material and methods

Temperature control unit

The design of the heating block is based on the work of Rachinskiy et al. [4,9]. It is made of aluminium with tube adapters for the two water circulation systems (Figure 1C). The circulation system for cooling was operated with the FL300 recirculation cooler (Julabo, Seelbach, Germany) allowing temperature set points from −20°C to 40°C. The flow rate was determined to 2.1 L min^{-1}. The heating water circulation worked with the Ecoline E300 thermostat (Lauda, Lauda-Königshofen, Germany) for temperature set points of 25-100°C. A flow rate of 1.1 L min^{-1} was measured. Both circulation systems were operated with tap water. Temperature set points below 5°C and above 95°C were avoided to prevent the water from freezing or boiling.

On-line monitoring system

The optical measurement technique is identical to that described in the work of Samorski et al. [24]. The in-house constructed device is henceforth referred as BioLector. Wavelengths and gain factors for all optical signals are specified in Table 1. For all experiments 96well lumox multiwell plates (Greiner Bio-One GmbH, Frickenhausen, Germany) were used.

Temperature determination

For the temperature measurement via fluorescent dyes a mixture of Rhodamin B (RhB) and Rhodamin 110 (Rh110) was used. Before each experiment a fresh measuring solution was prepared from 100-fold concentrated stock solutions of RhB and Rh110, respectively. For these stock solutions dye powder was dissolved in pure methanol yielding a concentration of 1 g L^{-1} and stored at 4°C. For the measuring solution the two stock solutions were mixed with water and diluted to a final concentration of 10 mg L^{-1}, respectively. 200 μL of the prepared solution were filled into each well of the MTP and shaken at a frequency of 995 rpm at a shaking diameter of 3 mm.

For temperature reference a special in-house constructed PT100 temperature sensor was placed in the arbitrary chosen well A2 of the microtiter plate. The sensor was connected to the serial port of the LAUDA thermostat for data output. Before use the sensor was calibrated with a gauged thermometer.

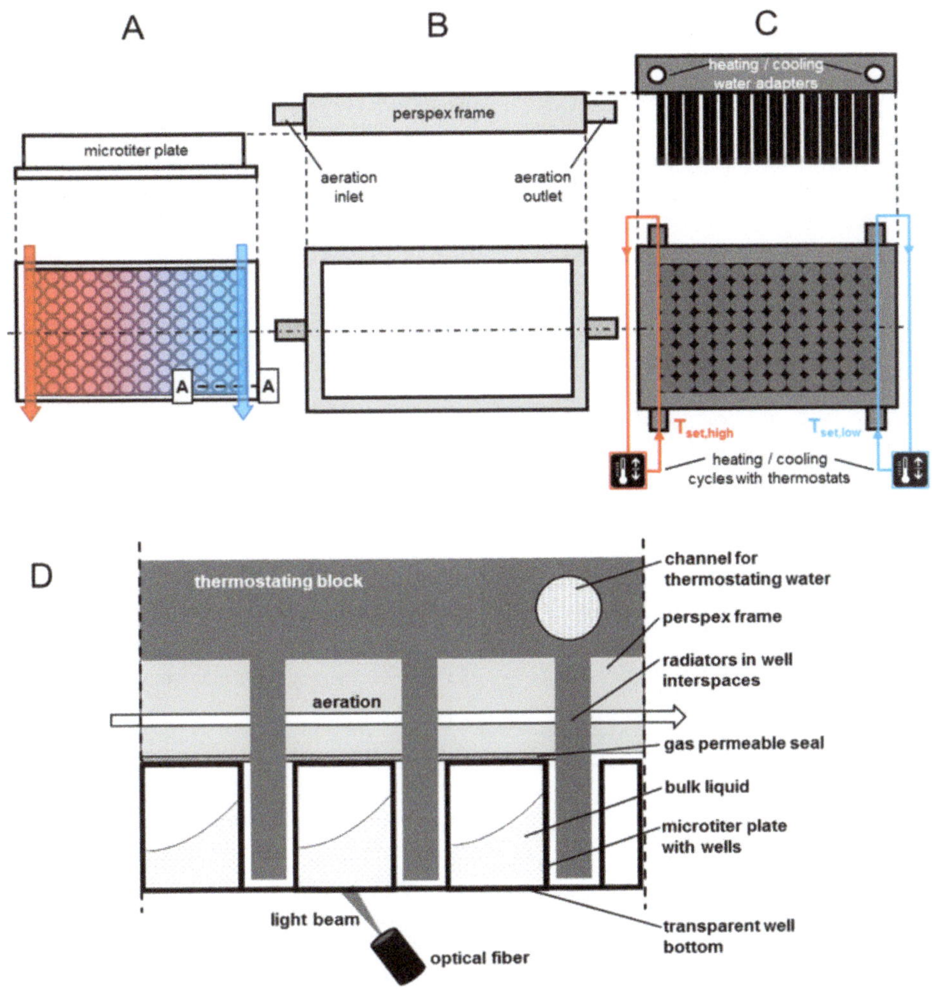

Figure 1 Scheme of the temperature control unit for temperature profiling in microtiter plates (MTP). (A) Side view on the MTP and top view with schematic illustration of thermostating water flows (arrows) and resulting temperature profile. **(B)** Side and top view on the perspex frame for aeration **(C)** Side view on the thermostating block and bottom view with schematic illustration of the heating (red) and cooling (blue) water circulation and their two respective thermostats. $T_{set,high}$ and $T_{set,low}$ stand for thermostat set point temperatures for heating and cooling water, respectively. **(D)** Cut A-A-view (see Figure 1A) on the microtiter plate with attached thermostating block and optical measurement.

For the calibration of the fluorescence signal versus the temperature only the well equipped with the PT100 sensor was monitored in order to accelerate the procedure. For each point five consecutive measuring values were averaged. For temperature profiles twelve consecutive measurement cycles of the whole MTP were performed to get the average for each well. The PT100 sensor was placed as reference in well A2. For each specific temperature profile fresh prepared measuring solution with a new determined calibration curve was used.

Table 1 Optical signals and applied setup for on-line monitoring

Optical signal	λ_{ex} [nm]	λ_{em} [nm]	Gain
Rhodamin B fluorescence	485	589	20
Rhodamin 110 fluorescence	485	520	2
Biomass (scattered light)	620	-	20
FbFP fluorescence	460	490	45
GFP fluorescence	485	510	30
4-methylumbellyferone fluorescence	365	455	25

Microorganisms

The *E. coli* clone expressing FbFP was *E. coli* BL21 (DE3) with the pRhotHi vector and kanamycin resistance. For alcohol dehydrogenase A (ADH-A, from *Rhodococcus ruber*) expression *E. coli* BL21 (DE3) with the pET22b vector and ampicillin resistance was used. For *K. lactis* experiments the strain GG799 with the pKlac1 vector expressing GFP was applied.

Media and cultivation

For *E. coli* pre-cultures terrific broth (TB) medium consisting of 12 g L^{-1} tryptone, 24 g L^{-1} yeast extract,

12.54 g L^{-1} K_2HPO_4, 2.31 g L^{-1} KH_2PO_4, and 5 g L^{-1} glycerol (all ingredients from Roth, Germany) dissolved in water was used. The pH value was 7.2 ± 0.2 without adjustment. For the main cultivation of *E. coli* under non-induced conditions either TB medium or a modified Wilms and Reuss medium (henceforth referred as Wilms-MOPS medium) were used [26,27]. Wilms-MOPS medium consists of 5 g L^{-1} $(NH_4)_2SO_4$, 0.5 g L^{-1} NH_4Cl, 3.0 g L^{-1} K_2HPO_4, 2 g L^{-1} Na_2SO_4, 0.5 g L^{-1} $MgSO_4 \cdot 7H_2O$, 0.01 g L^{-1} thiamine hydrochloride, 20.9 g L^{-1} 3-(N-morpholino)-propanesulfonic acid (MOPS, 0.2 M), 20 g L^{-1} glucose and 1 mL L^{-1} trace element solution. This trace element solution consists of 1.98 g L^{-1} $CaCl_2 \cdot 2H_2O$, 0.54 g L^{-1} $CoCl_2 \cdot 6H_2O$, 0.48 g L^{-1} $CuSO_4 \cdot 5H_2O$, 41.76 g L^{-1} $FeCl_3 \cdot 6H_2O$, 0.3 g L^{-1} $MnSO_4 \cdot H_2O$, 0.54 g L^{-1} $ZnSO_4 \cdot 7H_2O$, 33.39 g L^{-1} Na_2EDTA (Titriplex III). The pH was adjusted with 5 M NaOH to a value of 7. For experiments under induced conditions either the commercially available ready-made complex auto-induction medium Overnight Express Instant TB medium (OnEx, Novagen®/Merck, Darmstadt, Germany) or a modified Wilms-MOPS medium was used. OnEx consists of complex components similar to TB medium and the carbon sources glucose, lactose, and glycerol. For preparing the medium 60 g of the commercial granulate and 12.6 g of glycerol were dissolved in water and filled up to 1 L without pH-adjustment. HPLC analysis of the medium indicated a glucose concentration of 0.5 g/L and a lactose concentration of 2 g/L. For the modification of Wilms-MOPS medium in order to get a mineral auto-induction medium 1 g L^{-1} glucose, 4 g L^{-1} lactose and 5 g L^{-1} glycerol where added instead of 20 g L^{-1} glucose as carbon source. Depending on the clone's resistance, 50 µg mL^{-1} kanamycin or 100 µg mL^{-1} ampicillin were added to the medium from a 1000 fold concentrated stock solution.

For *K. lactis* pre-cultures and cultivation under non-induced conditions yeast extract peptone (YP) [28] medium was used, consisting of 10 g L^{-1} yeast extract, 20 g L^{-1} tryptone and 20 g L^{-1} glucose. For cultivations under induced conditions 20 g L^{-1} galactose instead of glucose served as carbon source and inducer for recombinant protein expression.

For *E. coli* pre-cultivation, 10 mL of TB medium in a 250 ml shake flask were inoculated with 50 µL from a cryoculture, and cultures were grown for 8 h at 350 rpm (shaking diameter 50 mm) and 37°C. *K. lactis* pre-culture conditions were the same aside from being grown in YP medium for 12 h at 30°C.

For all main cultivations the respective medium was inoculated from the pre-culture, resulting in an initial OD_{600} of 0.1. 200 µL of the already inoculated medium was then transferred to each of the wells of the MTP. The plates were sealed with gas-permeable seals (AB-0718, Thermo Scientific, Dreieich, Germany). Subsequently, the pre-tempered thermostating block was mounted atop the MTP and both were fixed on the orbital shaker (Kühner AG, Basel, Switzerland) of the BioLector. The cultivation was performed at a shaking frequency of 995 rpm and a shaking diameter of 3 mm. For aeration pure oxygen was used. For scattered light and fluorescence measurement the initial light intensity (I_0), which is mainly attributed to such factors as the media background or the type of the microtiter plate, was subtracted from the original measured data ($I-I_0$).

Reference shake flask cultures were performed in a Respiration Activity Monitoring System (RAMOS) [29,30] built in-house. Commercial versions of this device are available from Hitec Zang (Herzogenrath, Germany) or Kühner AG (Birsfelden, Switzerland). 10 mL of OnEx medium in a special 250 ml shake flask were inoculated from the pre-culture, resulting in an initial OD_{600} of 0.1. Subsequently, the cultures were shaken at 350 rpm with a shaking diameter of 50 mm at temperatures of 22-37°C with aeration by pressurized air. Cells for ADH-A analysis were harvested in the stationary phase, indicated by the OTR on-line signal of the RAMOS device.

Cellulase experiments

For hydrolysis experiments the substrate 4-methylumbelliferyl-β-D-cellobioside (4MUC) was used in combination with the commercial cellulase mix Celluclast 1.5 L (Novozymes, Bagsvaerd, Denmark). A 0.75 mM stock solution of 4MUC in 0.1 M acetate buffer (pH = 4.8) was prepared. The enzyme stock solution contained 5 g L^{-1} of Celluclast crude extract in 0.1 M acetate buffer (pH = 4.8). Before starting the experiments, 180 µL of the 4MUC stock solution were filled in each well of a 96 well MTP and preheated for 30 min applying the respective temperature profile. Subsequently, 20 µL of the enzyme stock solution were added to each MTP well resulting in final concentrations of 0.6 mM 4MUC and 1 g L^{-1} Celluclast. Since no aeration was necessary, the plate was sealed with a non-permeable foil (AB-0580, Thermo Scientific, Dreieich, Germany). The further procedure was identical to the cultivation experiments in MTPs besides that 4-methylumbellyferone (4MU) fluorescence was the only measuring signal (Table 1). To calibrate the fluorescence signal, 200 µL of solutions with varied 4MU concentrations were filled in a 96 well MTP and their fluorescence intensity was measured using the BioLector. The calibration measuring points are average values from three independent measurements.

Offline analysis

OD_{600} was determined via a Genesys 20 photometer (Thermo Scientific, Dreieich, Germany) in 1.5 mL micro cuvettes (PS, Plastibrand, Roth, Karlsruhe, Germany).

For values higher than 0.5 the samples were appropriately diluted with 0.9% [wt/vol] NaCl solution.

The volumetric activity of the produced alcohol dehydrogenase A (ADH-A) was determined at 30°C by following the oxidation of NADH at a wavelength of 340 nm in 96-well microtiter plates (F-profile, Roth, Germany) using a Synergy-4 Multi-Mode Microplate Reader (BioTek Instruments, Germany). For cell disruption of *E. coli* expressing ADH-A, the cell pellet of 500 μL culture broth was suspended in 100 μL BugBuster Protein Extraction Reagent (Novagen, Merck, Germany) adding 1000 U/mL lysozyme (Roth, Germany) and 25 U/mL DNaseI (AppliChem, Germany). Cell disruption was continued according to the manufacturers' specifications obtaining the soluble fraction with dissolved ADH-A. 200 μL reaction mixture (including enzyme solution) were prepared for measurement of ADH-A activity and contained 50 mM Tris buffer (pH 8, RT), 100 mM 2,5-hexanedione, and 0.5 mM NADH (biomol, Germany). By addition of the enzyme solution the reactions were initiated, thereby, applying appropriate enzyme solution amounts to ensure linear decreases of NADH absorbance over 1 min at least. One unit (U) was defined as the amount of enzyme converting 1 μmol cofactor per min.

Results and discussion

Development of the instrumentation

The here used measurement setup is a modification of an earlier developed system, the so called Enzyme Test Bench [4,9]. It combines an optical on-line monitoring system for cultivations and reactions in microtiter plates with a special system for temperature regulation.

The temperature control unit is depicted in Figure 1. It is similar to the heat transfer block of the Enzyme Test Bench for oxygen consuming reactions described by Rachinskiy et al. [9]. During experiments the MTP (side and top views in Figure 1A) is covered by a perspex frame (side and top views in Figure 1B) with an inlet and an outlet for aeration. The thermostating block (side and top views in Figure 1C) is mounted on top. It includes radiators which fit exactly within the well interspaces. Thus, the wells are jacketed by these radiators providing efficient heat transfer between the thermostating block and the MTP wells. The heat transfer properties of the system were specified before [4]. The whole system itself was placed in a temperature controlled environment. To minimize the evaporation, a gas permeable seal is fixed between the MTP and the perspex frame. In this foil a pattern in the radiators' shape is cut. Additionally, the aeration gas is passed through five bubble columns to saturate it with water vapor. A sectional view (A-A-cut in Figure 1A) of the MTP attached with the perspex frame and thermostating block is shown in Figure 1D.

The original thermostating block described by Rachinskiy et al. was designed to ensure a homogeneous temperature distribution over the microtiter plate [9]. Contrary to that, in this work the aim was to create a temperature gradient. For this purpose, the thermostating block has two water circulation systems instead of one, each with a separate thermostat (Figure 1C). If operated at different temperatures a certain temperature profile can be imposed onto the MTP just by heat conductance through the aluminium block (Figure 1A).

The advantage of the thermostating block mounted on top of the MTP, compared to systems working e.g. with electrical micro-heaters (9, 23, 34, 36) is that the transparent MTP bottom is still available for optical measurements. The applied on-line monitoring system (indicated by the optical fiber and light beam in Figure 1D) is similar to the BioLector technique presented before [24]. Thereby, the equipped fluorescence spectrometer allows measurements at different wavelengths. With this setup, microbial growth (via scattered light) or the formation of various fluorescent components is continuously followed in microtiter plates without interruption of the shaking process.

Optical temperature determination in MTPs

To characterize the behavior of microbial or enzymatic systems at different temperatures it is necessary to know the specific temperature in each single well. Equipping each well with a temperature sensor requires a very high degree of instrumentation. Just by using a temperature dependent fluorophore measurements can easily be done with the optical on-line monitoring system. In this work a combination of the fluorescent dyes Rhodamin B (RhB) and Rhodamin 110 (Rh110) was applied, where RhB is the temperature sensitive compound, whereas Rh110 acts as a reference. This measuring principle was described before [21]. In Figure 2A the fluorescence intensity of both dyes depending on the temperature is shown. Therefore, the thermostating block was tempered only by one thermostat to ensure a constant temperature distribution over the whole MTP. Thermostat set point temperatures from 5-95°C were adjusted. After each temperature shift (dotted vertical lines) the experimental conditions remained unchanged until both fluorescence signals showed constant values. The RhB signal decreases in a step like manner since its fluorescence intensity is decreasing with increasing temperature. After each temperature shift the fluorescence signal drops sharply before it remains constant when the temperature reaches its equilibrium. On the contrary, the fluorescence intensity of Rh110 is almost constant over the whole time. The slight decrease is due to a bleaching effect which is known to happen to RhB as well. To obtain a reliable measuring signal, the ratio of both fluorescence intensities was calculated [21].

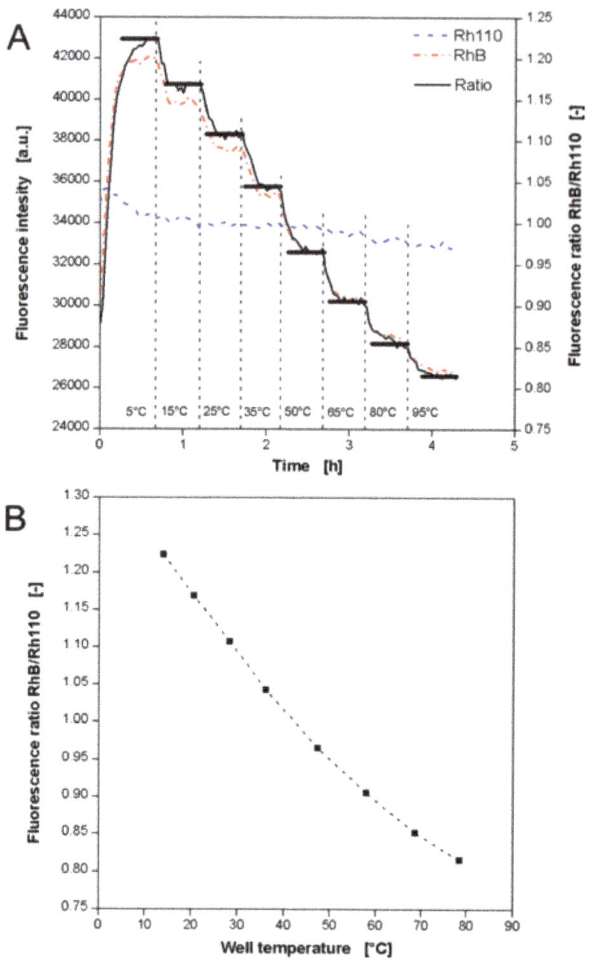

Figure 2 Optical temperature measurement in MTPs applying a fluorescent assay with the dyes Rhodamin B and Rhodamin 110. (A) Progress of the fluorescence signals of Rhodamin B (RhB) and Rhodamin 110 (Rh110) and their calculated ratio in a single well with varied thermostat set point temperature. Contrary to Figure 1C heating and cooling circulation systems were operated with one thermostat. **(B)** Calibration curve of the well temperature (measured via PT100 thermometer) vs. the fluorescence ratio RhB/Rh110. Experimental conditions: 96well MTP, $V_L = 200$ µL, n = 995 rpm, $d_0 = 3$ mm, RT = 37°C.

It must be considered that the thermostat set point temperatures in Figure 2A and the actual temperatures in the MTP's wells are not identical since heat may be lost to the environment. For this reason, one well was equipped with an in-house constructed PT100 temperature sensor. In this way, the corresponding well temperatures for various RhB/Rh110 ratios were determined. In Figure 2B the resulting calibration curve is depicted. It is described by a polynomial equation of second degree applying MS Excel. The fluorescence ratios are average values of five measurements in one well. The maximum relative standard deviation was 0.4%. For further investigation of the measuring accuracy the heating block temperature, as well as the room temperature, was adjusted to 37°C. In this way, a constant temperature of 37°C in each well could be assumed. The regarding measurement of all 96 wells revealed an average value of 37°C with a maximum of 38.7°C and a minimum of 35.6°C. The standard deviation was 0.76 K. The reason for the deviation from well to well cannot be explained completely. Slight deviations in the properties of the transparent microtiter plate bottom are possible which may influence the optical signals. A systematic position effect could be excluded during the experiments.

Temperature profiles

The optical temperature measuring method was then used to characterize the temperature distribution over microtiter plates at varied set point temperatures of the heating and cooling thermostat ($T_{set,high}$, $T_{set,low}$) and at varied room temperature (RT). In Figure 3A the exemplary temperature profile for $T_{set,low}$ of 5°C, $T_{set,high}$ of 50°C and a RT of 30°C is depicted. As expected, the profile shows a clear gradient from the warm right side (column 1) to the cold left side (column 12). The highest measured well temperatures were found in the wells A-D of column 1 (T_{max} = 36.2-36.9°C), the lowest values in the wells A-F of column 12 (T_{min} = 20.4-20.8°C). The rest of the wells cover the whole temperature range between T_{min} and T_{max}. Even though not every MTP row shows exactly the same temperature gradient, a certain repetitive trend can be observed.

By varying the thermostat set temperatures the profile can be shifted to higher or lower temperature ranges. A decrease of $T_{set,high}$ to 40°C results in a temperature range of 17.7-30.3°C, showing that the change of only one set point temperature has influence on both, T_{min} and T_{max} (comp. Figure 3A and B). This is not surprising since the temperature gradient in the thermostating block itself is strongly dependent on the two set point temperatures of the heating and cooling cycle.

The room temperature has a strong influence to the temperature profile, too. Its increase from 30°C to 37°C resulted in a higher temperature range (comp. Figure 3A and C). T_{min} increased from 20.4 to 28.0°C, T_{max} from 36.9 to 39.9°C. This effect is mainly attributed to the MTP's bottom which is thermally completely exposed to the environment. Hence, there is a large heat exchange area. This cannot be avoided, e.g. by insulation, because the transparent bottom needs to be accessible for the optical measurement. On the other hand, this effect might be beneficial by making the RT an additional temperature control parameter.

Figure 3D and E show further profiles at higher temperature levels. By setting the thermostat temperatures to 5 and 60°C, respectively, well temperatures of 30.3-47.3°C can be realized (Figure 3D). To characterize

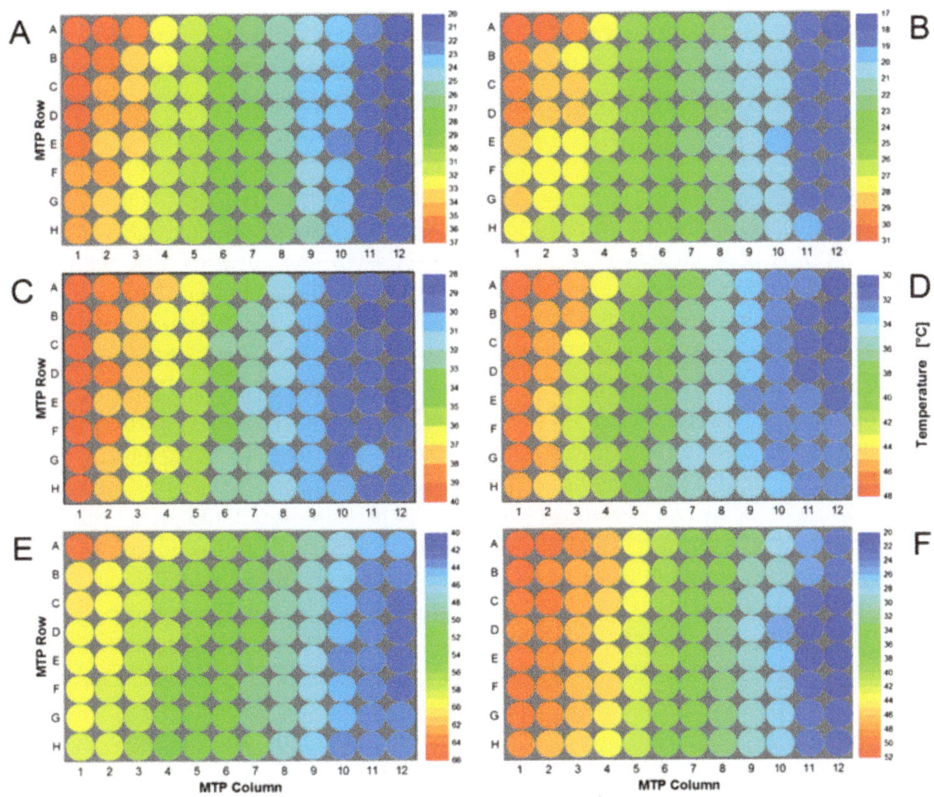

Figure 3 Exemplary temperature profiles in MTPs at varied room temperature and thermostat set point temperatures. **(A)** RT = 30°C, $T_{set,low}$ = 5°C, $T_{set,high}$ = 50°C, n = 995 rpm. **(B)** RT = 30°C, $T_{set,low}$ = 5°C, $T_{set,high}$ = 40°C, n = 995 rpm. **(C)** RT = 37°C, $T_{set,low}$ = 5°C, $T_{set,high}$ = 50°C, n = 995 rpm. **(D)** RT = 37°C, $T_{set,low}$ = 5°C, $T_{set,high}$ = 60°C, n = 995 rpm. **(E)** RT = 37°C, $T_{set,low}$ = 10°C, $T_{set,high}$ = 95°C, n = 995 rpm. **(F)** RT = 37°C, $T_{set,low}$ = 5°C, $T_{set,high}$ = 60°C, n = 0 rpm. Note: Temperature scale differs from A to F for better visualization of temperature profiles. Experimental conditions: 96well MTP, V_L = 200 µL, d_0 = 3 mm.

even thermophilic microbial or enzymatic systems a $T_{set,high}$ of 95°C was chosen (Figure 3E). In this way a profile with T_{max} of 64.5°C was achieved. T_{min} was 42.1°C. It must be considered that under these conditions the cryostat was not able anymore to ensure a set point temperature of 5°C since the cooling water heated up to much while passing through the thermostating block. To work under defined conditions $T_{set,low}$ was increased to 10°C.

In all determined profiles it is obvious that T_{min} and T_{max} differ strongly from $T_{set,low}$ and $T_{set,high}$, respectively. This is due to heat losses in the temperature control system. Some heat transfer might occur from the tubes of the circulation system. Measurements show that this loss is rather low, e.g. for the profile in Figure 3C with $T_{set,low}$ of 5°C and $T_{set,high}$ of 50° the temperatures before entering the block were 5.2°C and 49.3°C, respectively. This proves that the tube insulation worked properly. On the contrary, the current version of the thermostating block has no special insulation to the environment. It can be assumed that the exposed aluminum surface is mainly responsible for the observed temperature differences. The non-insulated MTP bottom is responsible for additional heat dissipation as discussed above. The results in Figure 3F support this argument. Compared to the profile in Figure 3D the MTP was not shaken during the measurement. Interestingly, the differences between the set point temperatures $T_{set,low}$ and $T_{set,high}$ and the regarding minimal and maximal temperatures T_{min} and T_{max} are smaller. This results in a broader well temperature range of 22.2-51.3°C (comp. Figure 3D, 30.3-47.3°C). The phenomenon can be explained by a ventilation effect caused by shaking of the MTP. While an insulating air layer may be generated below a non-shaken MTP decreasing the environment's influence, this layer may be lost when the plate starts moving. Nonetheless, for most processes sufficient shaking is absolutely necessary due to mixing and mass transfer requirements.

For all profiles shown in Figure 3 a slight temperature gradient over the MTP rows (A-H) was observed apart from the intended one over the MTP columns. This is not surprising since the heating water enters the thermostating block closest to well A1, such as the cooling water enters closest to well A12. While passing through the block the heating water gets colder and the cooling water warms up. For the profile in Figure 3C the heating water's temperature decreased by 1.2 K from entry to exit, whereas the cooling water got 0.5 K warmer. Of course these values will differ from one profile to another.

Temperature dependence of microbial systems

In the following experiments the information from the temperature profiles (Figure 3) was used to investigate several microbial systems in order to find optimal conditions for biomass formation or recombinant protein expression. Microbial growth (via scattered light) and the formation of fluorescent proteins was followed on-line applying the BioLector technique. To ensure that really the temperature is the limiting factor during the cultivation it was necessary to exclude other limitations. Previous studies showed that the applied media allow non-limiting growth [27,28,31]. A critical factor is the oxygen supply. For the applied shaking conditions (200 μL filling volume, 995 rpm shaking frequency, 3 mm shaking diameter) and aeration with pressurized air a maximum oxygen transfer capacity of approx. 0.03 mol L^{-1} h^{-1} was determined (data not shown). This value differs among media depending on their oxygen solubility and diffusion coefficient [32-34]. However, *E. coli* grown in rich medium at 37°C may require maximum oxygen transfer capacities up to 0.1 mol L^{-1} h^{-1} (data not shown). To realize such high transfer rates the aeration was shifted from pressurized air to pure oxygen. In this way, almost the 5-fold maximum oxygen transfer capacity can be achieved and cultivations without oxygen limitation are warranted for all applied media and microorganisms.

An *E. coli* strain expressing a fluorescent model protein (FbFP) as product was investigated (Figure 4). Non-induced (MTP row A,C,E,G) and induced (MTP row B, D, F, H) cultivations were performed in parallel in one MTP. The temperature profile was essentially identical to Figure 3D. The results of 12 (out of 48) exemplary non-induced cultures in TB medium are shown in Figure 4A. All cultures begin immediately with their exponential growth without any *lag* phase. After 1 h the curves start spreading indicating different growth rates at different temperatures. The lowest rate is observed at the lowest temperature of 30.3°C. Higher temperatures lead to increased growth rates indicated by steeper curves. The maximum growth rate occurs at temperatures of 41.4-45.3°C. A further increase retarded the microbial growth again. Due to varied growth rate the time point for reaching the stationary phase differs as well. At 41.4°C the culture became stationary after 4.5 h, whereas it needed twice as long at 30.3°C. Comparing the final scattered light intensities, it can be seen that slightly more biomass was formed at lower temperatures. This might be explained by a higher energy demand for cell maintenance at higher temperatures which withdraws metabolic resources from growth [35]. It was already found that organisms growing at temperatures above their optimal growth temperature show lower cell yields. Thereby, it was postulated that biosynthetic reactions at high temperatures do not keep pace with catabolic reactions [36].

Exemplary curves for growth and product formation under induced conditions are depicted in Figure 4B and C, respectively. The auto-induction medium OnEx was used which works as follows: glucose is the preferred carbon source and represses recombinant protein expression to ensure undisturbed initial growth. After glucose depletion lactose is taken up and acts as the inducer of the expression system. Glycerol is an additional energy source. The scattered light curves reflect the medium's working principle (Figure 4B) which was studied before in detail [37]. For all cultures a short *lag* phase (~0.5 h) is followed by a small exponential increase in the scattered light signal after 1.5 h. In this time glucose is consumed allowing undisturbed growth without product formation. As observed for the non-induced cultivations (Figure 4A) the temperature has no influence on the initial behavior of the cells. After the first exponential growth phase the scattered light signal remains at a constant level indicating no further growth. The reason for this is the induction by lactose after glucose depletion. The recombinant protein expression causes a metabolic burden to the cells and inhibits the microbial growth. After a certain time the biomass signal starts increasing again resulting in a second exponential growth phase. From earlier studies it is known that *E. coli* cultures can recover from the metabolic burden after lactose depletion [37]. The duration of the inhibition phase differs over the curves. It is shortest at 40.9-42.7°C and extends at higher and lower temperatures. This might be explained with an accelerated lactose consumption at the temperatures which are most beneficial for the host organism. Consequently, the cells can recover earlier and grow further on the residual glycerol in the medium. For temperatures higher than 38°C the scattered light curves show an unexpected decrease after their maximum value. The reason for this is not yet clear. Morphological changes of the cells when entering the stationary phase might be one explanation [38]. Another could be related to inclusion body formation. It is known that higher temperatures promote protein precipitation in cells [39]. Additionally, there are reports about the influence of inclusion bodies to the cell's light scatter [40,41]. Further flow cytometric investigations should be performed to finally clarify this issue. The corresponding product formation was followed by fluorescence measurement (Figure 4C). After 3 h all cultures started producing the fluorescent protein FbFP and, subsequently, showed a continuous increase of the fluorescence signal.

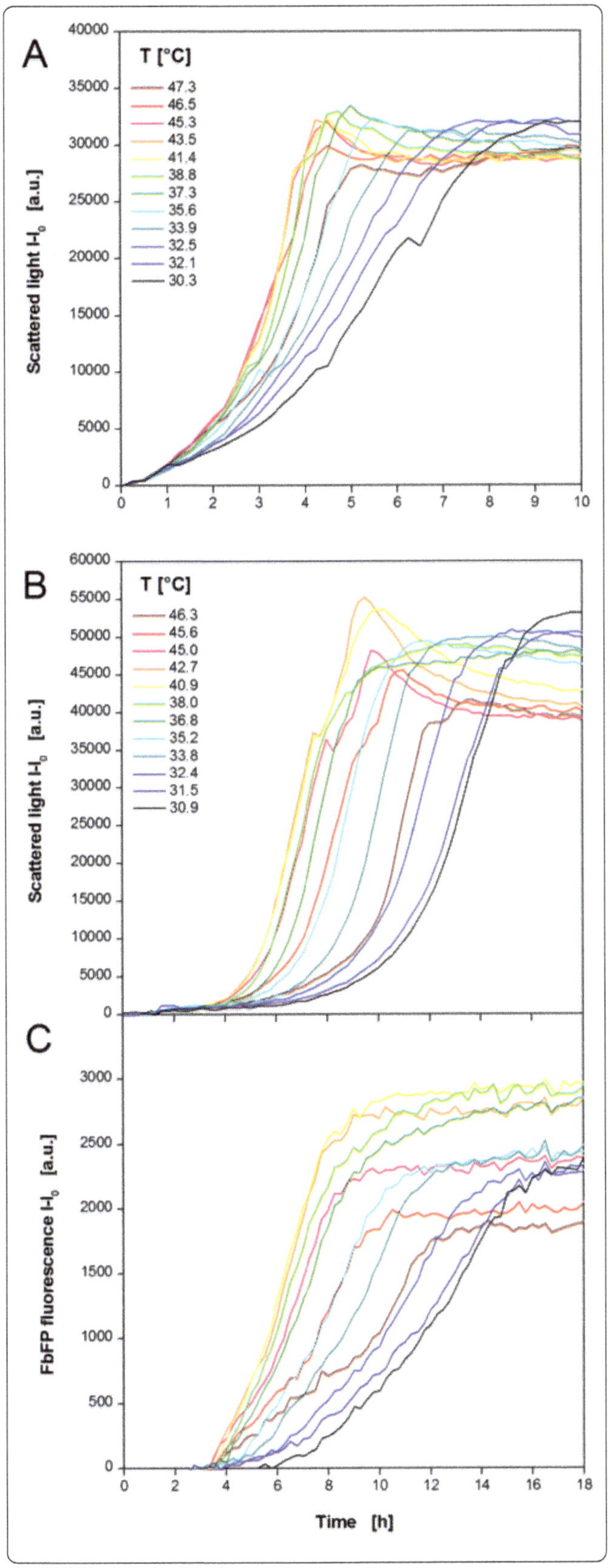

Figure 4 Cultivation of *E.coli* BL21 expressing the fluorescent protein FbFP applying a temperature profile in a MTP. (A) Cultivation and online monitoring of microbial growth (via scattered light) in complex TB medium without induction. **(B)** Cultivation and online monitoring of microbial growth and **(C)** fluorescent protein formation in complex auto-induction OnEx. Culture conditions: 96well MTP, $V_L = 200$ μL, $n = 995$ rpm, $d_0 = 3$ mm, aeration with 100% oxygen. Temperature profile: RT = 37°C, $T_{set,low} = 5$°C, $T_{set,high} = 60$°C (comp. Figure 3D). Data of 12 (from 48) exemplary wells.

The curve's slope depends on the temperature with the strongest increase at 40.9-42.7°C indicating the highest production rate. As a consequence, the maximum FbFP fluorescence is reached in the shortest time after 9 h. The slowest culture is found at 30.9°C reaching its maximum fluorescence after 17–18 h. These times correlate very well with the biomass signal (Figure 4B) which indicates that product formation stops with entering the stationary phase. Additionally, not only the production rate differs among the temperatures, but the maximum product concentration, too. The highest level is observed again at 40.9°C revealing the optimal temperature for FbFP production combining the maximum production rate and product concentration.

As a further microbial system the yeast *K. lactis* expressing GFP as product was investigated (Figure 5). Again, non-induced (MTP row A, C, E, G) and induced (MTP row B, D, F, H) cultivations were performed in parallel in one MTP. The temperature profile was essentially identical to Figure 3A. The results of 12 exemplary non-induced cultures in YP medium containing glucose as carbon source are shown in Figure 5A. Contrary to *E. coli*, the cultures show a temperature dependent *lag* phase. The minimum *lag* time of 2 h was observed at 27.2-33.8°C. During the subsequent exponential growth phase also the highest growth rates occur at those temperatures. Shifts to higher or lower temperatures lead to extended lag phases and reduced growth rates. As already observed for *E. coli*, the time point for reaching the stationary phase differs from 9–18 h due to the temperature dependent *lag* phase and growth rate. Again, the final scattered light intensities are higher at lower temperatures as already discussed for *E. coli*.

Contrary to *E. coli*, the growth behavior of *K. lactis* under induced conditions is not much different from that under non-induced conditions (comp. Figure 5A and B). The substitution of glucose by galactose as carbon source and inducer affects the microbial growth insignificantly. The *lag* phase, the growth rate, and the final biomass concentration show the same temperature dependent behavior as observed for the non-induced *K. lactis* cultures. The product formation shows a slightly different trend (Figure 5C). All GFP fluorescence signals

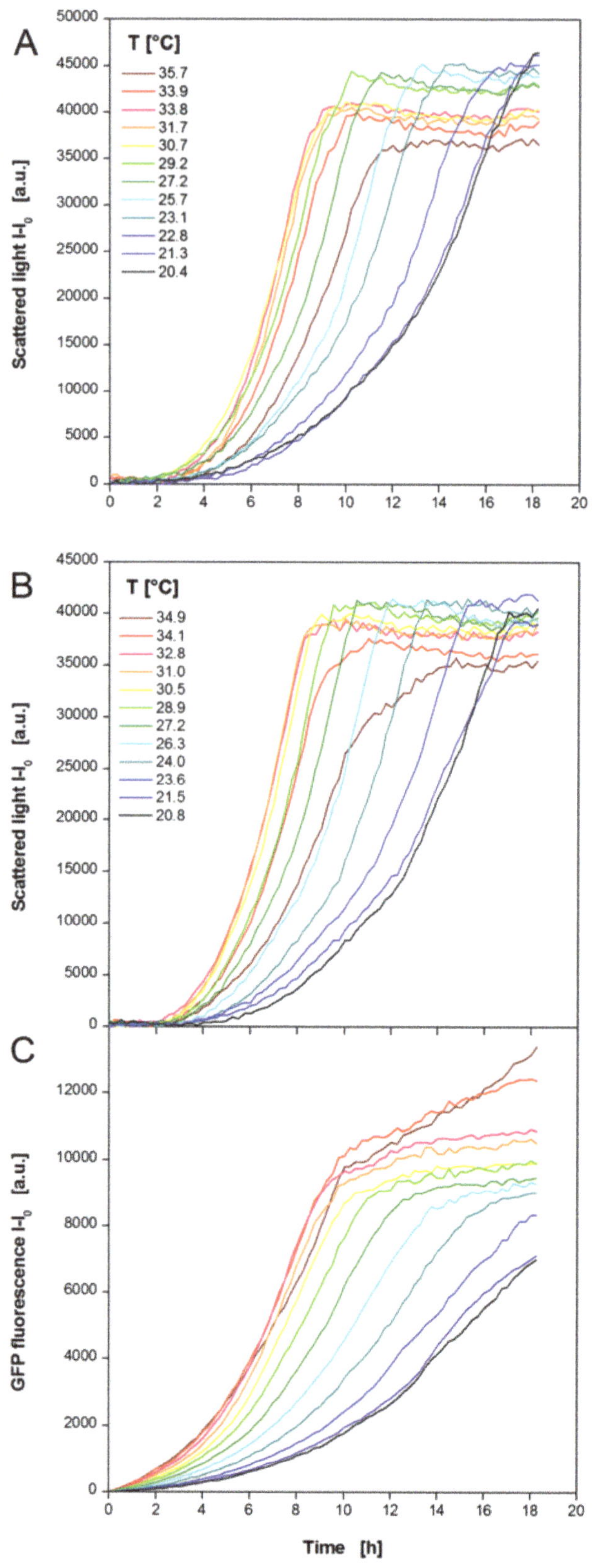

Figure 5 Cultivation of *K. lactis* GG799 expressing the fluorescent protein GFP applying a temperature profile in a MTP. (A) Cultivation and online monitoring of microbial growth (via scattered light) without induction in complex YP medium containing 20 g/L glucose. **(B)** Cultivation and online monitoring of microbial growth and **(C)** fluorescent protein formation in complex YP medium containing 20 g/L galactose as substrate and inducer. Culture conditions: 96well MTP, V_L = 200 μL, n = 995 rpm, d_0 = 3 mm, aeration with 100% oxygen. Temperature profile: RT = 30°C, $T_{set,low}$ = 5°C, $T_{set,high}$ = 50°C (comp. Figure 3A). Data of 12 (from 48) exemplary wells.

increase continuously from the beginning. The strongest increase indicating the highest production rate is observed at 31.0-34.9°C meaning that the GFP expression rather than microbial growth favors higher temperatures. Also the final product concentration increases with rising temperature. The reason for the further increase of the GFP fluorescence in stationary phase observed for 34.1 and 34.9°C is yet unclear. A concentration effect due to increased evaporation cannot explain this phenomenon. The determination of the filling volume after the cultivation revealed relatively low volume decreases of 1.1, 4.5, and 7% at 20.8, 31.0, and 34.9°C, respectively. For this reason, evaporation was not taken into consideration. Higher temperatures may provoke cell lysis in the stationary phase. As a result free GFP may induce brighter fluorescence in the medium without the barrier of the yeast's cell wall and membrane. Further studies, e.g. by applying flow cytometry, could help to clarify this effect.

For cultivation media it is known that temperature changes may result in pH changes as well, hence, two process parameters are unintendedly varied at once. This might lead to distorted results. The buffers used within this work show rather low changes within the applied temperature profiles with a maximum pH change of 0.2 (0.013 K^{-1}). But for more sensitive systems this effect should be considered by adapting the initial pH to the regarding temperature.

From the before described data sets (Figures 4 and 5) it is already possible to get an idea of the temperature optima for microbial growth and product formation in yeast and bacterial cells. But the high throughput of the MTP yields sufficient data for a more detailed characterization. For biomass formation the maximum growth rate under non-induced conditions was chosen as temperature dependent parameter (Figure 6A). As expected, the bacterial and the yeast system show a different behavior. *E. coli* in rich TB medium has a maximum growth rate of 0.41 h^{-1} at 43°C. The highest growth rate determined for *K. lactis* in rich YP medium is 0.16 h^{-1} at 32-33°C. In addition, the growth of *E. coli* in the mineral Wilms-MOPS medium was investigated. Compared to the growth in rich TB medium, a higher maximum growth rate of 0.67 h^{-1} was determined but

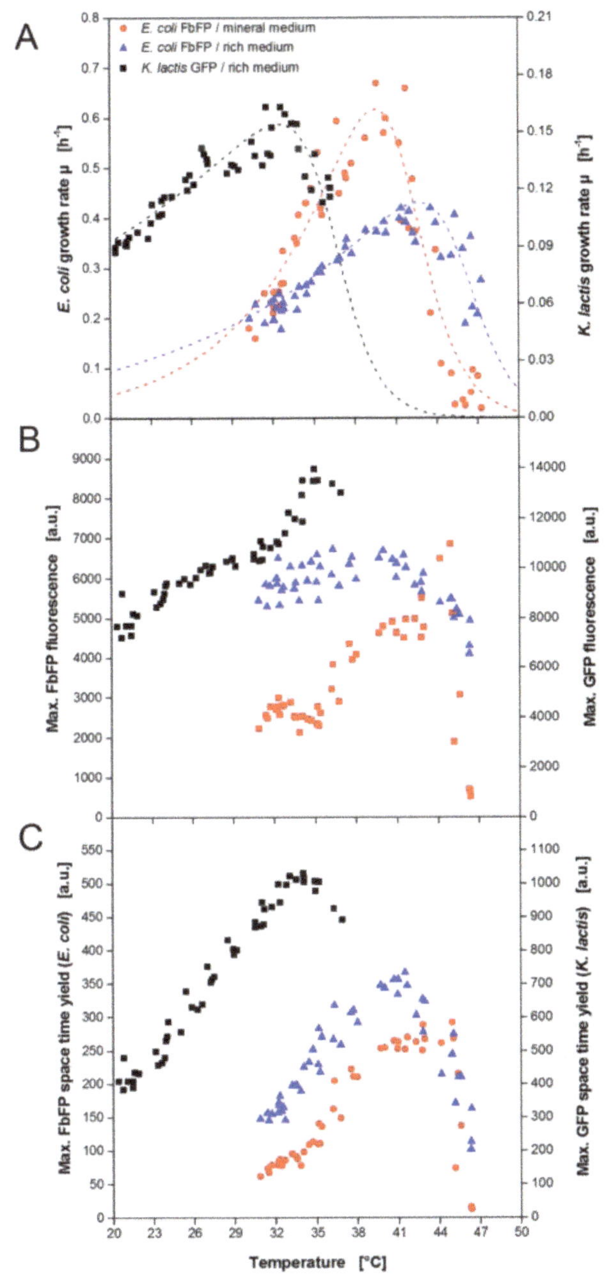

Figure 6 Determination of the optimal temperature for microbial growth and product formation of *E.coli* and *K. lactis* resulting from temperature profile experiments in MTPs. (A) Temperature dependent growth rate of *E.coli* (in rich and mineral medium) and *K. lactis* (in rich medium) under non-induced conditions. Temperature dependent maximum product formation **(B)** and maximum STY **(C)** of *E.coli* producing FbFP (in rich and mineral medium) and *K. lactis* producing GFP (in rich medium) under induced conditions. Dotted lines in A indicate Arrhenius fits due to Eq. 1.

at a lower temperature of 39.5°C. The observed values are in good agreement with earlier studies about *E. coli* [42-45] and *Kluyveromyces* strains [35,46,47]. For the product formation the maximum product fluorescence (indicator for product concentration) and the maximum space time yield were exemplary chosen as characteristic values (Figure 6B and C). Again differences between the microbial systems and cultivation media occur. *K. lactis* has the maximum product fluorescence and highest STY in the same range of 34-35°C which is slightly higher than the temperature for optimal growth. *E. coli* expressing FbFP in the rich auto-induction medium OnEx shows constant high fluorescence values over a relatively broad range from 35-42°C before it drops sharply at higher temperatures. The regarding maximum STY was determined at 41°C. For FbFP production in the mineral auto-induction medium the highest values for fluorescence and STY are observed at 44°C and 40-45°C, respectively.

An overview of all determined temperature optima is given in Table 2. The optima for growth and product formation are not in agreement for some of the investigated expression systems. Interestingly, even the medium composition has an influence. Compared to the standard cultivation temperatures of 30 and 37°C for *K. lactis* and *E. coli*, respectively, the determined optima differ to some extent. Consequently, an individual temperature screening is recommended guided by the desired aim of the process.

It is noticed that the measuring points in Figure 6 show some scattering indicating certain measurement inaccuracies. The fluorescence thermometry method has a standard deviation of ±0.76 K (see chapter Temperature profiles) which might be one explanation. Furthermore, it should be considered that the temperature profiles were determined in separate experiments without the microbial reaction system. It was assumed that no significant changes are encountered during cultivations, thereby, neglecting the possibility of produced reaction heat. In future studies, it should be investigated how fluorescence thermometry can be used directly in the reaction system in order to realize temperature real time monitoring. Nonetheless, the high number of measuring points compensates these disadvantages and gives a clear trend.

A typical procedure to describe the temperature dependent behavior of microorganisms and (bio-)catalysts is the Arrhenius plot. An extended version of the Arrhenius equation was used for mathematical modeling

Table 2 Temperature optima for growth and recombinant protein expression of *E. coli* (in rich and mineral medium) and *K. lactis* (in rich medium) under non-induced (ni) and auto-induced (ai) conditions

Expression system	Temperature optimum		
	Growth	Product concentration	Product STY
E. coli (rich medium)	42°C^{ni}, 43°C^{ai}	35-42°C^{ai}	41°C^{ai}
E. coli (mineral medium)	40.5°C^{ni}, 41.5°C^{ai}	43-44°C^{ai}	40-45°C^{ai}
K. lactis (rich medium)	32°C^{ni,ai}	35°C^{ai}	34°C^{ai}

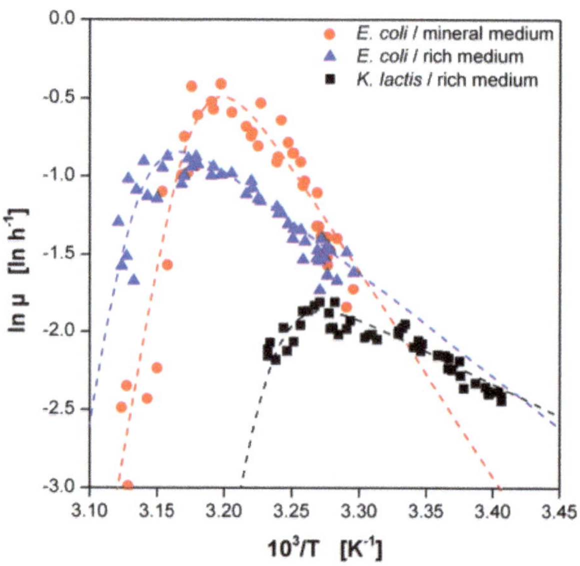

Figure 7 Characterization of the temperature dependent growth behavior of *E.coli* and *K. lactis* applying the Arrhenius plot. Temperature dependent growth rates of *E.coli* (in rich and mineral medium) and *K. lactis* (in rich medium) under non-induced conditions (comp. Figure 6A). Dotted lines indicate Arrhenius fits according to Eq. 1.

(Eq. 1). Thereby, the numerator is the classical Arrhenius equation with the numerical constant A and the activation energy E_g describing the typical increase of the growth rate or catalyst activity with increasing temperatures. Assuming that proteins are temperature-denatured and inactivated by a reversible chemical reaction with free energy change (ΔG_d), the nominator describes the behavior beyond the temperature optimum due to the Hougen-Watson expression for catalyst activity [25]. In Figure 7 the data from Figure 6A for the microbial growth of *E. coli* and *K. lactis* was plotted logarithmically over the reciprocal absolute temperature.

$$v_{max} = \frac{A\,exp(-E_g/RT)}{1 + B\,exp(-\Delta G_d/RT)} \quad (1)$$

By solving Eq. 1 in MS Excel, the parameters for the different investigated microbial systems were determined (Table 3). The model fit for *E. coli* in rich TB medium reveals an activation energy of 55 kJ mol^{-1} which is in very good agreement with earlier reports [42-44]. Interestingly, *E. coli* in the mineral Wilms-MOPS medium shows a much higher value compared to that in TB medium. *K. lactis* shows a lower value of 34.5 kJ mol^{-1}. Literature values for *Kluyveromyces* strains range from 36.6 to 85.2 kJ mol^{-1} [35,46,47]. In contrast to the activation, the deactivation behavior indicated by ΔG_d is similar for all organisms and media.

As an additional expression system *E. coli* producing a recombinant alcohol dehydrogenase A from *Rhodococcus ruber* (ADH-A) was investigated regarding its temperature behavior. In order to look at a broader temperature range, two experimental sets were performed applying a low and a high temperature profile according to Figure 3B and D, respectively. In this way, temperatures of 18.1-46.3°C were realized. To avoid excess evaporation at higher temperatures the cultivation was aborted after 20 h when all cultures had entered the stationary phase. The low temperature profile was applied for 27 h. The biomass formation was followed on-line by scattered light measurement (Figure 8A). As discussed before, the typical growth behavior of *E. coli* in the auto-induction medium OnEx is observed at temperatures of 27.6-39.7°C. After a temperature dependent *lag* phase of 2–4.5 h a first exponential growth is observed. The subsequent growth inhibition indicated by decreasing slopes of the scattered light curves after 4–7 h was caused by the metabolic burden of recombinant ADH-A expression. As seen for *E. coli* expressing FbFP this growth inhibited production phase takes longer at lower temperatures, e.g. 2 h at 39.7°C and 4 h at 27.6°C. When cells recover from that metabolic load a second exponential increase occurs before the stationary phase is reached. At temperatures below 27.6°C the microbial growth becomes more linear so that product formation is not the prior growth inhibitor, but the temperature. At temperatures higher than 39.7°C almost no growth is detected.

Indicator for the product formation was the volumetric ADH-A activity at the end of the cultivation (Figure 8B). An on-line signal was not available and sampling during the cultivation was not possible. Furthermore, the cell suspensions of three wells had to be pooled to have sufficient volume for the enzyme activity assay. Thereby, the information about

Table 3 Arrhenius model parameters in Eq. 1 for the yeast *K. lactis*, the bacterium *E. coli* and the cellulase preparation Celluclast

Parameter		*E. coli* (TB medium)	*E. coli* (Wilms-MOPS)	*K. lactis* (YP medium)	Celluclast
E_g	[kJ mol^{-1}]	55	110	34.5	29
ΔG_d	[kJ mol^{-1}]	549	548	539	579
A	[h^{-1}]	$6 \cdot 10^8$	$1.85 \cdot 10^{18}$	$1.3 \cdot 10^5$	$1.2 \cdot 10^4$
B		$7 \cdot 10^{89}$	$9 \cdot 10^{90}$	$9 \cdot 10^{90}$	$8 \cdot 10^{90}$

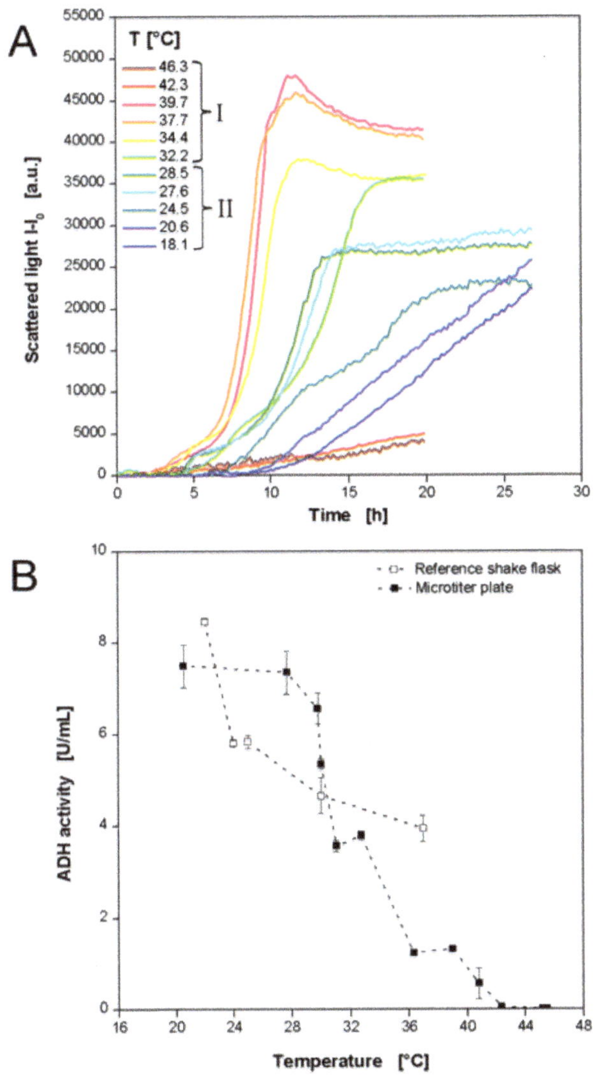

Figure 8 Cultivations of *E.coli* BL21 expressing the recombinant enzyme ADH-A applying 2 temperature profiles in MTPs. (A) Cultivation and online monitoring of microbial growth (via scattered light) in complex auto-induction medium OnEx. **(B)** Temperature dependent volumetric activity at the end of the cultivation in the MTP. Reference values from additional cultivations in shake flasks at different temperatures. Culture conditions: 96well MTP, $V_L = 200$ µL, n = 995 rpm, $d_0 = 3$ mm, aeration with 100% oxygen. Temperature profile I for T = 32.2-46.3°C: RT = 37°C, $T_{set,low} = 5$°C, $T_{set,high} = 60$°C (comp. Figure 3D); Temperature profile II for T = 18.1-28,5°C: RT = 30°C, $T_{set,low} = 5$°C, $T_{set,high} = 40$°C (comp. Figure 3B). Reference cultivation: 250 ml shake flask, $V_L = 10$ mL, n = 350 rpm, $d_0 = 50$ mm, aeration with air. Online data of 10 exemplary wells.

the respective well temperature was beneficial for combining wells of similar temperature. The results of the MTP experiments show a clear trend of increasing volumetric enzyme activities with decreasing temperatures. Maximum values of 7.4 U mL^{-1} were measured at 20-28°C. At the highest temperatures of 42.3 and 45.5°C no product activity was detected. This correlates very well with the on-line biomass signal where almost no growth occurred in this temperature range (comp. Figure 8A). Taking growth and product formation results together into consideration, a certain trend becomes obvious. Higher temperatures (if not too high) favor bacterial growth but reduce product formation, whereas, at low temperatures the behavior is vice versa. The reference values from the shake flask cultivations do not completely reflect the MTP results. Nevertheless, a similar trend is present with increasing volumetric ADH-A activities at lower temperatures. The maximum activity of 8.5 U mL^{-1} was determined at 22°C.

All these investigations of different microorganisms, expression systems and culture media show how the presented high-throughput temperature profiling helps to find optimal conditions for fermentations. The on-line monitoring feature allows a much more detailed insight to the microbial processes at varied temperature than classical end point analysis, e.g. kinetic parameters such as growth and production rates or space time yields.

Temperature dependence of cellulases

As a candidate for the optimization of enzymatic reactions the commercial cellulase cocktail Celluclast was chosen since it is often used for biomass degradation. The Celluclast cocktail contains a mixture of several cellulases from the fungus *Trichoderma reesei* [48]. In order to follow the enzyme reaction on-line, a substrate, namely 4-methylumbelliferyl-β-D-cellobioside (4MUC), was used. It releases the fluorescent dye 4-methylumbelliferone (4MU) when hydrolyzed by cellulases which can be easily detected with a fluorescence spectrometer. Consequently, the 4MUC assay is commonly used for the high-throughput screening of cellulolytic enzymes [49]. Compared to the microbial systems described before, a higher temperature optimum was expected for the cellulases. Consequently, a profile was chosen providing a higher temperature range of 42-65°C (comp. Figure 3E).

Twelve exemplary curves resulting from the 4MUC hydrolysis by Cellucalst are depicted in Figure 9A, whereby, the fluorescence intensity indicates the formation of the product 4MU. For all conditions the typical course of enzymatic reactions is observed with a strong increase of the product concentration in the beginning which runs into saturation after a certain time when no substrate is available anymore. The reaction is strongly temperature dependent. The highest reaction rates are observed at 53.2-56.2°C. Consequently, these curves run into saturation already after approx. 3 h. Furthermore, the highest final product concentrations occur at these temperatures. In the temperature ranges above and below reduced reaction rates and extended reaction times for complete substrate conversion are observed. Pure substrate showed a constant signal close to 0 a.u.,

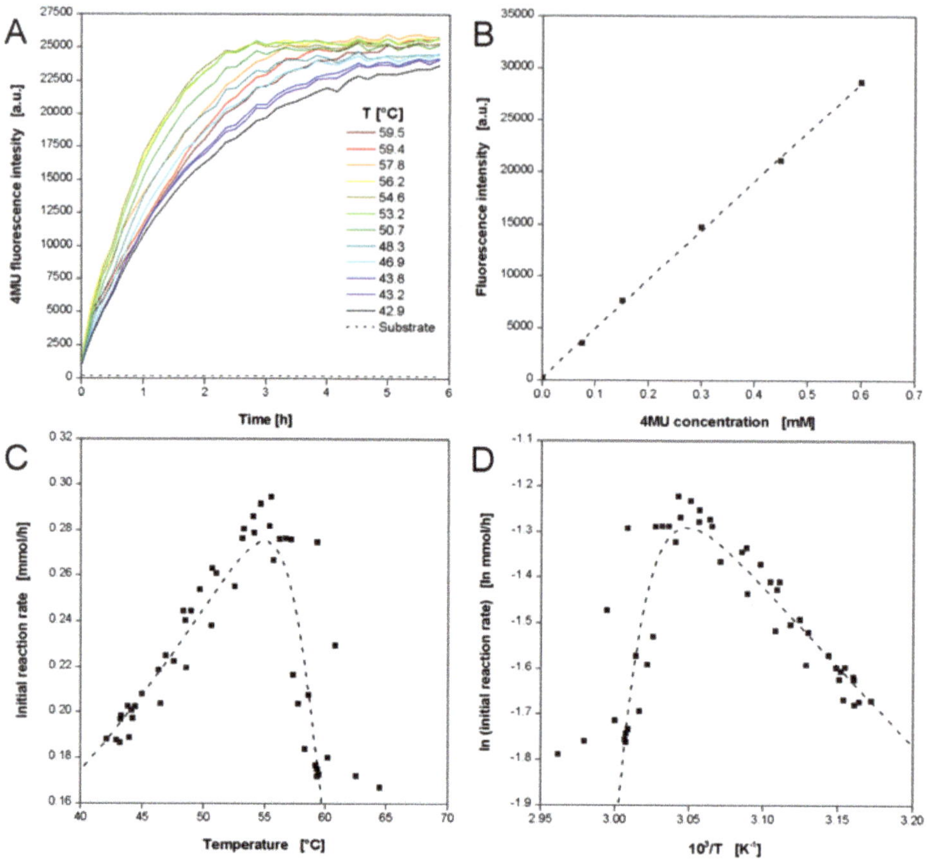

Figure 9 Enzymatic hydrolysis of 4-methylumbelliferyl-ß-D-cellobioside (4MUC) with cellulase from *T. reseei* (Celluclast) applying a temperature profile in a MTP. (A) Progresses of 4-methyl-umbelliferyl (4MU) fluorescence (λ_{ex} = 365 nm, λ_{em} = 455 nm) during the hydrolysis of 4MUC at different temperatures. Online data of 12 (from 48) exemplary wells. **(B)** Calibration curve for fluorescence intensity at varied 4MU concentration. **(C)** Temperature dependent initial reaction rate and **(D)** resulting Arrhenius plot of Celluclast. Dotted lines indicate Arrhenius fit. Experimental conditions: 96well MTP, V_L = 200 µL, n = 995 rpm, d_0 = 3 mm, 1 g/L Celluclast and 0.6 mM 4MUC in 0.1 M acetate buffer, pH = 4.8. Temperature profile: RT = 37°C, $T_{set,low}$ = 10°C, $T_{set,high}$ = 95°C (comp. Figure 3E).

thereby, proving that no substrate reacted in the absence of enzyme. In order to quantify the cellulolytic reaction, the 4MU fluorescence signal was calibrated assuming that each cleaving of a 4MUC molecule releases one 4MU molecule (Figure 9B). In this way reaction rates as well as final product concentrations can be calculated.

To evaluate the temperature dependent behavior, the initial reaction rate was chosen as indicator and plotted over the respective temperature (Figure 9C). A maximum initial rate of 0.295 mmol h^{-1} was determined at 55°C. At the lowest checked temperature of 42°C an initial reaction rate of 0.19 mmol h^{-1} was found, close to 0.17 mmol h^{-1} at the highest temperature. A complete enzyme deactivation was not achieved. The optimum temperature is in good agreement with other reports about cellulases [3,50,51]. As for microbial systems, the corresponding Arrhenius plot was plotted from these results (Figure 9D). By fitting the curve in MS Excel, the parameters in Eq. 1 for Celluclast were determined (Table 2). It must be considered, that Celluclast is no single cellulase but a mixture of endoglucanases (EG), cellobiohydrolases (CBH) and β-glucosidases (BG). In literature E_g values of 20–32 KJ mol^{-1} for EGs [52-54], 18–45 KJ mol^{-1} for CBHs [3,55], and 18–59 KJ mol^{-1} for BGs [55,56] are reported. The here reported value of 29 KJ mol^{-1} is within these ranges.

It can be stated that the presented high-throughput temperature profiling is a useful tool for enzyme characterization. The technique can easily be applied to other enzyme classes as long as fluorescent assays are available. If necessary, the temperature range can be shifted to either higher or lower levels. Sufficient aeration can be assured as discussed for microbial cultivations. With all these presented features the system allows the investigation of a great variety of biocatalysts: psychro-, meso- or thermophilic; oxygen dependent or not.

Conclusion

Running bioprocesses at their temperature optima is essential for their economical operation. Unfortunately, systems for rapid determination of temperature dependent behavior are rare. The presented high-throughput screening system for temperature optimization faces this challenge. A novel temperature control system with a customized thermostating block and two thermostats can generate individual temperature profiles in MTPs with relatively simple instrumentation. Three controlling parameters were identified: $T_{set,high}$, $T_{set,low}$ and RT. For temperature measurement a fluorescent assay with two Rhodamin dyes was established for the use in MTPs which allows the convenient determination of temperature profiles in short time. Applying the high-throughput screening system BioLector which is already commercially available, important process parameters, e.g. biomass and product formation, are monitored on-line. The high throughput of the MTP provides sufficient data output for a detailed characterization of temperature dependent behavior. Consequently, the combination of the thermostating block, temperature determination via fluorescence thermometry, and an optical on-line monitoring system provides extensive information about temperature dependent process behavior, e.g. concentrations of biomass or products, growth or reaction rates, space time yields, and others. Representative investigations with microbial and enzymatic systems proved the system's general applicability for various purposes. Thereby, the repetition of the temperature profiles in the MTP rows allows the investigation of several systems in parallel.

During the experiments some potential for improvement became apparent, considering heat dissipation to the surroundings, temperature dependent pH changes, evaporation effects, and sufficient oxygen supply. All these issues can be addressed in future work to achieve most precise results.

The MBR-based high-throughput temperature profiling is a convenient tool for rapid characterization of temperature dependent reaction processes. It allows the fast investigation of numerous conditions, e.g. microorganisms, enzymes, media, and others, in a short time. The simple temperature control combined with a commercial on-line monitoring device makes it a user friendly system.

Abbreviations

4MU: 4-methylumbellyferone; 4MUC: 4-methylumbellyferyl-cellobioside; A: Numerical constant [h^{-1}]; ADH-A: Alcohol dehydrogenase A from *Rhodococcus ruber*, B: Numerical constant [-]; BG: β-glucosidase; CBH: Cellobiohydrolase; E_g: Activation energy [kJ mol^{-1}]; EG: Endoglucanase; FbFP: Flavinmononucleotide-based fluorescent protein; GFP: Green fluorescent protein; I: Measured signal intensity [a.u.]; I_0: Initial signal intensity [a.u.]; MBR: Micro-bioreactor; MTP: Microtiter plate; OD_{600}: Optical density at 600 nm [-]; OnEx: Overnight express Instant TB medium; OTR: Oxygen transfer rate [mol L^{-1} h^{-1}]; R: Ideal gas constant ~8.314 [J mol^{-1} K^{-1}]; RAMOS: Respiration activity monitoring system; Rh110: Rhodamin 110; RhB: Rhodamin B; RT: Room temperature [°C]; STY: Space time yield [a.u. mL^{-1} h^{-1}]; T: Absolute temperature [K]; T_{max}: Maximum temperature within a temperature profile [°C]; T_{min}: Minimum temperature within a temperature profile [°C]; $T_{set,high}$: Set point temperature of heating thermostat [°C]; $T_{set,low}$: Set point temperature of cooling thermostat [°C]; TB: Terrific broth medium; YP: Yeast peptone medium; ΔG_d: Deactivation energy change [kJ mol^{-1}]; μ: Growth rate [h^{-1}]; μ_{max}: Maximum growth rate [h^{-1}]; v_{max}: Maximum reaction rate [h-1].

Competing interests

The authors declare that they have no competing interests.

Authors' contributions

MK made the conceptual design of the study and the experimental setup and methods, performed experiments and prepared the manuscript. CL established the fluorescence thermometry assay in MTPs. SD performed ADH-A expression experiments. WK developed the ADH-A expression system in *E. coli* and provided it for this work. JB assisted with study's conception, data interpretation and manuscript preparation. All authors read and approved the final manuscript.

Acknowledgements

This work was performed as part of the Cluster of Excellence "Tailor-Made Fuels from Biomass", which is funded by the Excellence Initiative by the German federal and state governments to promote science and research at German universities. The authors thank Ulrich Commandeur and Thomas Schmidt from the Institute for Molecular Biotechnology of the RWTH Aachen University (Germany) as well as Thomas Drepper and Janko Potzkei from the Institute for Molecular Enzyme Technology at the Heinrich-Heine-University Düsseldorf (Germany) for providing the applied microorganisms. The students Lina Tubes and Konrad Smiglak from RWTH Aachen University and Ana Cláudia Gonçalves da Cruz from the University of Minho (Portugal) are acknowledged for experimental support.

Author details

[1]AVT-Chair for Biochemical Engineering, RWTH Aachen University, Worringerweg 1, 52074 Aachen, Germany. [2]Department of Chemistry (Organic and Bioorganic Chemistry), University of Graz, Heinrichstraße 28/II, 8010 Graz, Austria.

References

1. Shuler ML, Kargi F: **Bioprocess engineering : basic concepts.** 2nd edition. Upper Saddle River, NJ: Prentice Hall; 2002.
2. Dragosits M, Frascotti G, Bernard-Granger L, Vazquez F, Giuliani M, Baumann K, Rodriguez-Carmona E, Tokkanen J, Parrilli E, Wiebe MG, Kunert R, Maurer M, Gasser B, Sauer M, Branduardi P, Pakula T, Saloheimo M, Penttila M, Ferrer P, Tutino ML, Villaverde A, Porro D, Mattanovich D: **Influence of growth temperature on the production of antibody fab fragments in different microbes: A host comparative analysis.** *Biotechnol Prog* 2011, **27**:38–46.
3. Pardo AG, Forchiassin F: **Influence of temperature and pH on cellulase activity and stability in Nectria catalinensis.** *Rev Argent Microbiol* 1999, **31**:31–35.
4. Rachinskiy K, Schultze H, Boy M, Bornscheuer U, Büchs J: **"Enzyme Test Bench," a high-throughput enzyme characterization technique including the long-term stability.** *Biotechnol Bioeng* 2009, **103**:305–322.
5. Zakhartsev MV, Portner HO, Blust R: **Environmentally low-temperature kinetic and thermodynamic study of lactate dehydrogenase from Atlantic cod (G-morhua) using a 96-well microplate technique.** *Anal Biochem* 2004, **330**:10–20.
6. Schäpper D, Alam MN, Szita N, Eliasson Lantz A, Gernaey KV: **Application of microbioreactors in fermentation process development: a review.** *Anal Bioanal Chem* 2009, **395**:679–695.
7. Vervliet-Scheebaum M, Ritzenthaler R, Normann J, Wagner E: **Short-term effects of benzalkonium chloride and atrazine on Elodea canadensis using a miniaturised microbioreactor system for an online monitoring of physiologic parameters.** *Ecotoxicol Environ Saf* 2008, **69**:254–262.
8. van Leeuwen M, Heijnen JJ, Gardeniers H, Oudshoorn A, Noorman H, Visser J, van der Wielen LAM, van Gulik WM: **A system for accurate on-line measurement of total gas consumption or production rates in microbioreactors.** *Chem Eng Sci* 2009, **64**:455–458.

9. Rachinskiy K, Kunze M, Graf C, Schultze H, Boy M, Büchs J: **Extension and Application of the "Enzyme Test Bench" for Oxygen Consuming Enzyme Reactions.** *Biotechnol Bioeng* 2014, **111:**244–253.
10. Zanzotto A, Szita N, Boccazzi P, Lessard P, Sinskey AJ, Jensen KF: **Membrane-aerated microbioreactor for high-throughput bioprocessing.** *Biotechnol Bioeng* 2004, **87:**243–254.
11. Zhang ZY, Grattan KTV, Palmer AW: **Fiber optic thermometry based on fluorescence lifetimes of Cr-3+ doped materials.** *Second Int Symp Meas Technol Intelligent Instrum, Pts 1 and 2* 1993, **2101:**476–482.
12. Hoogenboom R, Fijten MWM, Brandli C, Schroer J, Schubert US: **Automated parallel temperature optimization and determination of activation energy for the living cationic polymerization of 2-ethyl-2-oxazoline.** *Macromol Rapid Commun* 2003, **24:**98–103.
13. Krommenhoek EE, van Leeuwen M, Gardeniers H, van Gulik WM, van den Berg A, Li X, Ottens M, van der Wielen LA, Heijnen JJ: **Lab-scale fermentation tests of microchip with integrated electrochemical sensors for pH, temperature, dissolved oxygen and viable biomass concentration.** *Biotechnol Bioeng* 2008, **99:**884–892.
14. Lee HLT, Boccazzi P, Ram RJ, Sinskey AJ: **Microbioreactor arrays with integrated mixers and fluid injectors for high-throughput experimentation with pH and dissolved oxygen control.** *Lab Chip* 2006, **6:**1229–1235.
15. Maharbiz MM, Holtz WJ, Howe RT, Keasling JD: **Microbioreactor arrays with parametric control for high-throughput experimentation.** *Biotechnol Bioeng* 2004, **85:**376–381.
16. Li X, van der Steen G, van Dedem GWK, van der Wielen LAM, van Leeuwen M, van Gulik WM, Heijnen JJ, Krommenhoek EE, Gardeniers JGE, van den Berg A, Ottens M: **Improving mixing in microbioreactors.** *Chem Eng Sci* 2008, **63:**3036–3046.
17. Boccazzi P, Zhang Z, Kurosawa K, Szita N, Bhattacharya S, Jensen KF, Sinskey AJ: **Differential gene expression profiles and real-time measurements of growth parameters in Saccharomyces cerevisiae grown in microliter-scale bioreactors equipped with internal stirring.** *Biotechnol Prog* 2006, **22:**710–717.
18. Liu LY, Peng SL, Niu XZ, Wen WJ: **Microheaters fabricated from a conducting composite.** *Appl Phys Lett* 2006, **89.**
19. Yamamoto T, Nojima T, Fujii T: **PDMS-glass hybrid microreactor array with embedded temperature control device. Application to cell-free protein synthesis.** *Lab Chip* 2002, **2:**197–202.
20. Fernicola VC, Zhang ZY, Grattan KTV: **Investigation of Cr-fluorescence in olivine crystals with potential in fibre optic thermometry.** *Sensors and Their Applications Viii* 1997, **:**367–372.
21. Sakakibara J, Adrian RJ: **Whole field measurement of temperature in water using two-color laser induced fluorescence.** *Exp Fluids* 1999, **26:**7–15.
22. Sun T, Zhang ZY, Grattan KTV, Palmer AW, Collins SF: **Temperature dependence of the fluorescence lifetime in Pr3+:ZBLAN glass for fiber optic thermometry.** *Rev Sci Instrum* 1997, **68:**3447–3451.
23. Kensy F, Zang E, Faulhammer C, Tan RK, Buchs J: **Validation of a high-throughput fermentation system based on online monitoring of biomass and fluorescence in continuously shaken microtiter plates.** *Microb Cell Factories* 2009, **8:**31.
24. Samorski M, Müller-Newen G, Büchs J: **Quasi-continuous combined scattered light and fluorescence measurements: A novel measurement technique for shaken microtiter plates.** *Biotechnol Bioeng* 2005, **92:**61–68.
25. Roels JA: *Energetics and kinetics in biotechnology. Amsterdam.* New York: Elsevier Biomedical Press; 1983.
26. Scheidle M, Klinger J, Buchs J: **Combination of on-line pH and oxygen transfer rate measurement in shake flasks by fiber optical technique and Respiration Activity Monitoring System (RAMOS).** *Sensors* 2007, **7:**3472–3480.
27. Wilms B, Hauck A, Reuss M, Syldatk C, Mattes R, Siemann M, Altenbuchner J: **High-cell-density fermentation for production of L-N-carbamoylase using an expression system based on the Escherichia coli rhaBAD promoter.** *Biotechnol Bioeng* 2001, **73:**95–103.
28. Hahn-Hagerdal B, Karhumaa K, Larsson CU, Gorwa-Grauslund M, Gorgens J, van Zyl WH: **Role of cultivation media in the development of yeast strains for large scale industrial use.** *Microb Cell Factories* 2005, **4:**31.
29. Anderlei T, Büchs J: **Device for sterile online measurement of the oxygen transfer rate in shaking flasks.** *Biochem Eng J* 2001, **7:**157–162.
30. Anderlei T, Zang W, Papaspyrou M, Büchs J: **Online respiration activity measurement (OTR, CTR, RQ) in shake flasks.** *Biochem Eng J* 2004, **17:**187–194.
31. Losen M, Frolich B, Pohl M, Buchs J: **Effect of oxygen limitation and medium composition on Escherichia coli fermentation in shake-flask cultures.** *Biotechnol Prog* 2004, **20:**1062–1068.
32. Duetz WA, Witholt B: **Effectiveness of orbital shaking for the aeration of suspended bacterial cultures in square-deepwell microtiter plates.** *Biochem Eng J* 2001, **7:**113–115.
33. Kensy F, Zimmermann HF, Knabben I, Anderlei T, Trauthwein H, Dingerdissen U, Buchs J: **Oxygen transfer phenomena in 48-well microtiter plates: Determination by optical monitoring of sulfite oxidation and verification by real-time measurement during microbial growth.** *Biotechnol Bioeng* 2005, **89:**698–708.
34. Maier U, Losen M, Buchs J: **Advances in understanding and modeling the gas–liquid mass transfer in shake flasks.** *Biochem Eng J* 2004, **17:**155–167.
35. Rajoka MI, Khan S, Latif F, Shahid R: **Influence of carbon and nitrogen sources and temperature on hyperproduction of a thermotolerant beta-glucosidase from synthetic medium by Kluyveromyces marxianus.** *Appl Biochem Biotechnol* 2004, **117:**75–92.
36. Senez JC: **Some considerations on energetics of bacterial growth.** *Bacteriol Rev* 1962, **26:**95.
37. Kunze M, Huber R, Gutjahr C, Mullner S, Buchs J: **Predictive tool for recombinant protein production in Escherichia coli shake-flask cultures using an on-line monitoring system.** *Biotechnol Prog* 2012, **28:**103–113.
38. Kunze M, Roth S, Gartz E, Buchs J: **Pitfalls in optical on-line monitoring for high-throughput screening of microbial systems.** *Microb Cell Fact* 2014, **13:**53.
39. Chalmers JJ, Kim E, Telford JN, Wong EY, Tacon WC, Shuler ML, Wilson DB: **Effects of temperature on Escherichia coli overproducing beta-lactamase or human epidermal growth factor.** *Appl Environ Microbiol* 1990, **56:**104–111.
40. Hedhammar M, Stenvall M, Lonneborg R, Nord O, Sjolin O, Brismar H, Uhlen M, Ottosson J, Hober S: **A novel flow cytometry-based method for analysis of expression levels in Escherichia coli, giving information about precipitated and soluble protein.** *J Biotechnol* 2005, **119:**133–146.
41. Lavergne-Mazeau F, Maftah A, Cenatiempo Y, Julien R: **Linear correlation between bacterial overexpression of recombinant peptides and cell light scatter.** *Appl Environ Microbiol* 1996, **62:**3042–3046.
42. Herendeen SL, Vanbogelen RA, Neidhardt FC: **Levels of major proteins of Escherichia coli during growth at different temperatures.** *J Bacteriol* 1979, **139:**185–194.
43. Johnson FH, Lewin I: **The growth rate of E. coli in relation to temperature, quinine and coenzyme.** *J Cell Comp Physiol* 1946, **28:**47–75.
44. Mohr PW, Krawiec S: **Temperature characteristics and Arrhenius plots for nominal psychrophiles, mesophiles and thermophiles.** *J Gen Microbiol* 1980, **121:**311–317.
45. Ratkowsky DA, Olley J, Mcmeekin TA, Ball A: **Relationship between temperature and growth-rate of bacterial cultures.** *J Bacteriol* 1982, **149:**1–5.
46. Rajoka MI, Khan S, Shahid R: **Kinetics and regulation studies of the production of beta-galactosidase from Kluyveromyces marxianus grown on different substrates.** *Food Technol Biotechnol* 2003, **41:**315–320.
47. Toyoda T, Ohtaguchi K: **Effect of temperature on D-arabitol production from lactose by Kluyveromyces lactis.** *J Indust Microbiol Biotechnol* 2011, **38:**1179–1185.
48. Jäger G, Wu ZJ, Garschhammer K, Engel P, Klement T, Rinaldi R, Spiess AC, Büchs J: **Practical screening of purified cellobiohydrolases and endoglucanases with alpha-cellulose and specification of hydrodynamics.** *Biotech Biofuels* 2010, **3:**18.
49. Ko KC, Han Y, Cheong DE, Choi JH, Song JJ: **Strategy for screening metagenomic resources for exocellulase activity using a robotic, high-throughput screening system.** *J Microbiol Methods* 2013, **94:**311–316.
50. Karlsson J, Siika-aho M, Tenkanen M, Tjerneld F: **Enzymatic properties of the low molecular mass endoglucanases Cel12A (EG III) and Cel45A (EG V) of Trichoderma reesei.** *J Biotechnol* 2002, **99:**63–78.
51. Krishna SH, Rao KCS, Babu JS, Reddy DS: **Studies on the production and application of cellulase from Trichoderma reesei QM-9414.** *Bioprocess Eng* 2000, **22:**467–470.
52. Busto MD, Ortega N, Perez-Mateos M: **Characterization of microbial endo-beta-glucanase immobilized in alginate beads.** *Acta Biotechnol* 1998, **18:**189–200.
53. Saqib AA, Farooq A, Iqbal M, Hassan JU, Hayat U, Baig S: **A thermostable crude endoglucanase produced by Aspergillus fumigatus in a novel solid state fermentation process using isolated free water.** *Enzyme Res* 2012, **2012:**196853.
54. Xiao ZZ, Wang P, Qu YB, Gao PJ, Wang TH: **Cold adaptation of a mesophilic cellulase, EG III from Trichoderma reesei, by directed evolution.** *Sci China Series C-Life Sci* 2002, **45:**337–343.

55. Steinweg JM, Jagadamma S, Frerichs J, Mayes MA: **Activation energy of extracellular enzymes in soils from different biomes.** *PLoS One* 2013, **8:**e59943.
56. Siddiqui KS, Saqib AAN, Rashid MH, Rajoka MI: **Carboxyl group modification significantly altered the kinetic properties of purified carboxymethylcellulase from Aspergillus niger.** *Enzym Microb Technol* 2000, **27:**467–474.

Effect of substrate stiffness on early human embryonic stem cell differentiation

Nikolai Eroshenko[1,2†], Rukmani Ramachandran[1†], Vamsi K Yadavalli[1] and Raj R Rao[1*]

Abstract

Background: The pluripotency and self renewing properties of human embryonic stem cells (hESC) make them a valuable tool in the fields of developmental biology, pharmacology and regenerative medicine. Therefore, there exists immense interest in devising strategies for hESC propagation and differentiation. Methods involving simulation of the native stem cell microenvironment, both chemical and physical, have received a lot of attention in recent years. Equally important is evidence that cells can also sense the mechanical properties of their microenvironment. In this study, we test the hypothesis that hESCs accept mechanical cues for differentiation from the substrate by culturing them on flexible polydimethylsiloxane (PDMS) of varying stiffness.

Results: PDMS substrates were prepared using available commercial formulations and characterized for stiffness, surface properties and efficiency of cell attachment and proliferation. Across different substrate stiffness, cell numbers, cell attachment and cell surface area were found to be similar. Expression of pluripotency markers decreased with increased time in culture across all PDMS substrates of varying stiffness. Analysis of gene expression of differentiation markers indicates that the differentiation process becomes less stochastic with longer culture times.

Conclusions: We evaluated the utility of PDMS substrates for stem cell propagation and substrate mediated differentiation. The stiffness affected gene expression of pluripotent and differentiation markers with results indicating that these substrate systems could potentially be used to direct hESC fate towards early mesodermal lineages. This study suggests that coupled with soluble factors, PDMS substrates could potentially be useful in generating defined populations of differentiated cells.

Keywords: Stem cells, Biomaterials, Extracellular matrix, Differentiation, Stiffness

Background

Human embryonic stem cells (hESCs) are characterized by their ability to self renew and to differentiate into any diploid human cell type. This property makes them a valuable tool for studying the basic biology of lineage specification, and for applications in fields such as pharmacology and tissue engineering [1]. The control of stem cell fate has chiefly been attributed to genetic specifications and cellular response to signals from the surrounding niche in the form of chemical, mechanical and matrix factors [2]. Multiple studies have shown that soluble factors such as fibroblast growth factors (FGFs), bone morphogenetic proteins (BMPs) and Wnts can regulate stem cell behaviour [3-6] and have been used as chemical cues in methodologies to generate clinically relevant cell populations. Mammalian cells also generate and are exposed to forces *in vivo* and *in vitro*; and these forces can influence stem cell fates by modulating cell shape, cytoskeletal structure and interaction with the extra cellular matrix (ECM) [7,8]. Simulating in-vivo mechanical deformations by imposing substrate strains has been shown to influence stem cell differentiation and such responses to mechanical loading depend not only on the type of stem cell but also on the state of differentiation and the type of strain applied [9-11]. Such studies indicate that even the mechanical properties of the culture system may be modified and tailored to generate a desired cell population. Despite these numerous

* Correspondence: rrao@vcu.edu
†Equal contributors
1Department of Chemical and Life Science Engineering, Virginia Commonwealth University, Richmond, VA, USA
Full list of author information is available at the end of the article

efforts, the creation of efficient, reliable, and scalable differentiation protocols has remained largely elusive.

Recent data suggesting that the extracellular matrix influences stem cell fate has led to interest in research involving control of stem cell fate by directing ECM geometry/ topography, mechanical properties, transmission of mechanical and biophysical factors to the cell, and the control of cell geometry [12,13]. In most cases, individual stem cells do not survive in suspension and their adhesion to a matrix is therefore essential for viability. The ECM as a major niche element provides not only a scaffold for cellular support, migration and proliferation, but also acts as the surrounding microenvironment that influences the cellular fate decision by presenting physical and chemical cues as well as binding soluble factors [12]. In vitro, the substrate acts as the primary ECM component, and while a feeder layer of inactivated mouse embryonic fibroblasts (MEF) has been the traditional gold standard, polymeric materials have also been investigated for their ability to support hESC propagation. We have previously reviewed the potential of bio-inspired polymers in determining human stem cell fate, which not only possess the advantage of being a xeno-free culture system but can also be tailored to very specific needs [14].

Early proof of the influence of ECM stiffness on stem cell differentiation was provided by qualitative studies involving mouse mammary epithelial cells that showed increased differentiation when grown on soft gel collagen substrates as opposed to Tissue Culture Plastic [15]. ECM control of stem cell fate by regulating growth factor diffusion has been demonstrated by artificially tethering a growth factor to a substrate, which increased survival of human mesenchymal stem cells (MSCs) [16]. In additional studies, the ECM was also found to be a more potent differentiation cue for MSCs than standard induction cocktails [17]. Tissue-level elasticity has been shown to be able to determine lineage and phenotype commitment in naïve MSCs. Later studies showed that human MSCs could be kept quiescent by growing them on polyacrylamide substrates that mimicked the properties of marrow while preserving their multilineage potential [18]. When NIH/3 T3 cells were cultured on polydimethylsiloxane (PDMS) substrates patterned with varying stiffness, the cells accumulated preferentially on the stiffer regions of the substrates with differential remodelling of ECM on stiff vs. compliant areas, which led to the suggestion that migration, and not proliferation, was responsible [19]. In seminal studies, Engler and colleagues showed that matrices whose elasticities were comparable to brain tissue ("soft matrices") were neurogenic and stiffer and rigid matrices (with elasticities comparable to muscle and bone tissue, respectively) were respectively myogenic and osteogenic [8]. Substrate compliance was also demonstrated to positively influence survival and functionality of mouse ESC- derived hepatocyte like cells [20].

In this study, we aimed to understand the role of substrate stiffness in two dimensional hESC culture and hoped to devise a PDMS based culture system for directing hESC differentiation. Specifically, we chose to focus on how modulating the mechanical properties of the substrate affected cell density and cell shape, along with maintaining pluripotency and the possibility of lineage specification. hESC (BGO1v) were cultured on commercial Polydimethylsiloxane of varying stiffness both in the presence and absence of basic FGF. Presence of pluripotent cells in culture was determined using Alkaline Phosphatase Assay assay and qPCR was performed at various time points to assess pluripotency and differentiation.

Results

Synthesis and Characterization of PDMS Substrates

Cell culture substrates with varying stiffness were prepared from PDMS by varying the base to cross linking agent ratio from 10:1, 20:1, 40:1. In the following discussion, the various substrates will be referred to by their base: crosslinker ratio e.g. PDMS 10:1. Tangent moduli were calculated from the tensile testing data (Figure 1) and ranged between 0.078 MPa to 1.167 MPa (Table 1). The data for 10:1 PDMS show that the polymer has at least two distinct tangent moduli: lower stiffness was observed at low strain levels than at higher strains. The transition was not observed in 20:1 and 40:1 PDMS, although it is likely that this was because the samples failed before they reached the strain level at which the transition occurs. Surface roughness of the prepared cell culture substrates as determined by tapping mode in AFM was shown to lie between 0.8 nm and 1.0 nm over a 20 μm^2 area (Figure 2) with no major surface features present. Contact angle measurements (Table 2) indicated that surface hydrophilicity prior to fibronectin treatment decreased with decreasing stiffness of the substrate.

Cellular attachment, proliferation and morphology

Cells grown on PDMS substrates initially proliferated at a rate similar to cells grown on mouse embryonic fibroblasts and fibronectin coated Tissue culture polystyrene (TCPS), which were used as controls. However, after seven days, there were more cells on the various PDMS substrates compared to the TCPS controls, indicating that PDMS could be used as a platform for hESC propagation and differentiation (data not shown). In order to determine that this difference in proliferation was a result of the effect of the substrates on the cells and was not merely a consequence of viable cell attachment, cell densities were measured after 12 hours. The results (Figure 3) indicate that in the early stages, even in the absence of cell- cell contact, cellular attachment was

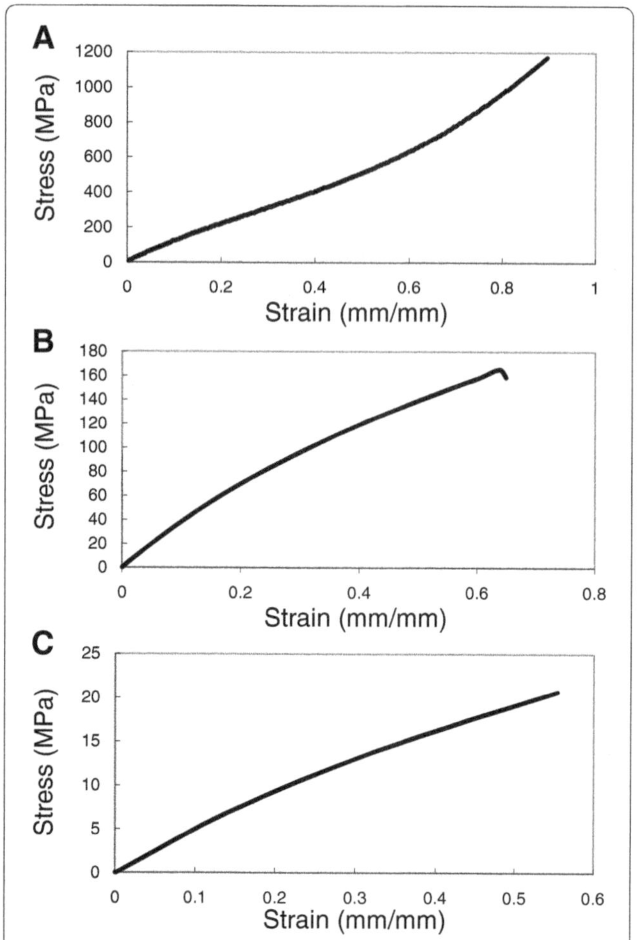

Figure 1 Testing mechanical properties of the three PDMS formulations used in cell culture studies. Stress Strain profiles of **A**) 10:1 PDMS; **B**) 20:1 PDMS and **C**) 40:1 PDMS substrates.

comparable across substrates. Cell shape (also independent of cell-cell contact) and cell surface area was also comparable across substrates at this early time point, with no statistically significant differences with increase in substrate stiffness (Figure 4).

Self renewal and lineage specification

Cells collected from substrates after 4 and 7 days in culture were re-plated onto inactivated MEF layers and cultured in complete hESC medium for 4 days and stained for AP activity. We found that a subpopulation of pluripotent cells remained among the cells cultured on all substrates at Day 4 and at later points (cells collected from substrates on Day 7), while some differentiated cells could be found within most cells stained for AP (Figure 5). These results suggest that these differentiated cells could be early progenitors capable of taking on normal hESC morphology and interacting with cells in a way conducive to proper colony formation.

Table 1 Tangent moduli of PDMS substrates

Substrate	Tangent Modulus (MPa)
PDMS 10:1	1.167 ± 0.088MPa
PDMS 20:1	0.397 ± 0.019MPa
PDMS 40:1	0.078 ± 0.008MPa

Moduli calculated from stress- strain curves. Error reported indicate the standard variance from the mean (n = 3).

Gene expression analysis across all substrates at various time points in culture showed that expression of pluripotency markers tended to decrease over time in the absence of basic FGF, as was anticipated. Differential gene expression Expression Index (EI) analysis (Table 3) indicates that across all time points, cells on the various PDMS substrates were more differentiated than those on TCPS. Among the various PDMS substrates, early differentiation appears to follow a stochastic process, with EI values falling sharply between Day 4 and Day 7.

Across the various PDMS substrates, pluripotency markers decrease over time, with the exception of Nanog in cells cultured on PDMS 20:1(Figure 6A, B, C). Nanog expression peaked at Day 10 on PDMS 20:1 and was significantly higher in cells grown on this substrate than the other two at this time point. Expression of the three representative differentiation markers tested (NeuroD, IGF2, AFP)increased on almost all PDMS substrates from Day 4 to Day 7, with change in NeuroD expression being the most, followed by that of AFP. Few exceptions included IGF2 expression on PDMS 40:1 and AFP expression on PDMS 40:1. On all PDMS substrates, NeuroD expression peaks at Day 7 but by Day 10 falls to slightly lower levels, while still remaining higher than Day 4 (Figure 6D). NeuroD expression on the softest substrate was significantly higher than the other two at this early time point. IGF2 and AFP expression show a consistent increase with time, but were the highest on PDMS 20:1 (Figure 6E, F).

Discussion

We have shown here that substrate stiffness affects cellular spreading, proliferation and gene expression of hESCs. PDMS was chosen as the substrate because it is easy to handle, inexpensive, does not swell in contact with water and can be micropatterned using techniques such as soft lithography. Surface treatments with UV radiation or ethanol, such as those used here; also do not affect material properties [21]. Other properties of PDMS, such as thermal stability, transparency and chemical inertness, make it particularly useful in bioengineering, in spite of potential batch to batch variabilities [22, 23]. We were able to fabricate substrates with stiffness varying from 0.078 to 1.167 MPa, a range similar to that reported by others [24,25]. We generated even stiffer substrates by using a base: crosslinker ratio of 5:1 (PDMS 5:1) and performed initial characterization studies but decided to focus on softer substrates for cell

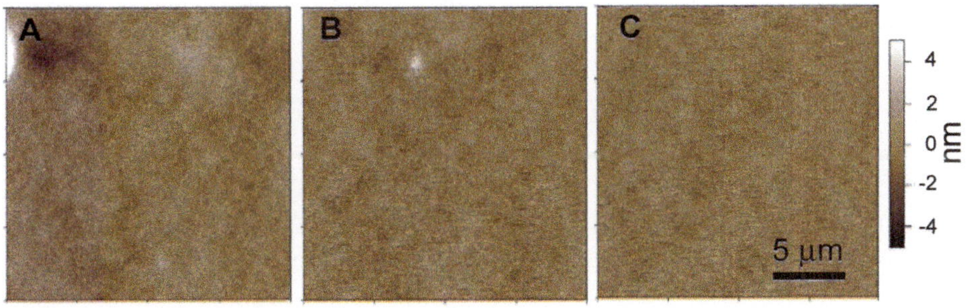

Figure 2 Surface roughness measured by AFM. Height (Topography) of the three PDMS samples. A: PDMS 5: 1 RMS roughness: 1.0 nm over 20 micron area **B**: PDMS 10: 1 RMS roughness: 0.8 nm over 20 micron area **C**: PDMS 20: 1 RMS roughness: 0.8 nm over 20 micron area.

culture and gene expression analysis. However, even within the given range of stiffness tested, substrate mediated biological effects were observed. Although proliferation increased on substrates, cell spreading did not increase with increased stiffness. These results are in direct contrast with previously reported studies conducted with terminally differentiated cells [26,27]. Our results tend to indicate that hESCs might react differently to substrate stiffness when compared with terminally differentiated cells like fibroblasts or endothelial cells [26,27]. These differences in cellular behaviour merits further studies in determining how cells in their undifferentiated state respond to substrate stimuli. Cells on stiffer substrates tend to exert larger traction forces [28] and it may be hypothesised that substrates affect cell development by affecting cellular migration and movement via the dynamics and size of adhesion sites. Substrate compliance are a major factor in cell culture studies and the probable molecular responses to these substrates have been discussed in detail in literature [28,29].

Our data indicate that at early time points, stiffness mediated differentiation follows a rather stochastic process, but certain trends begin to appear with longer duration in culture (Table 3). By Day 10, the mesodermal marker IGF2 was found to be the most highly expressed gene, with its expression being the highest on PDMS 20:1, one of the stiffer substrates. The dynamics of differentiation across different substrates reveals interesting trends. Specifically, the PDMS 20:1 demonstrates stronger dynamics in AFP and IGF2 expression (Figure 6E, F), while other conditions do not show such an obvious trend in differentiation. One possible hypothesis for the mechanism of this increased AFP and IGF2 expressions correlated with stiffness is that stiffer substrates provide an environment that more closely mimics those experienced by migrating mesodermal and endodermal cells in the early embryo. Our data also indicates that there is greater increase in IGF2 (mesoderm) expression (Figure 6E), when compared with AFP (endoderm) expression (Figure 6F), indicating a greater propensity for increased mesodermal differentiation on PDMS 20:1 substrate. At this juncture, it is also interesting to note that MSCs - which originate from the mesoderm - respond highly to substrate stiffness based cues for lineage specification. Recent studies involving hESCs cultured in three dimensional polymeric substrates with a broad range of elasticities also indicated that as stiffness increased, mesodermal differentiation was favoured over endodermal lineages [30].

While the molecular mechanisms linking substrate stiffness and hESC differentiation still remain to be explored, we can speculate this to be either a direct effect of mechanical properties of the substrate on cellular differentiation events, or an indirect effect related to the

Table 2 Contact angles prior to surface treatment

Substrate	Contact angle (degrees)
PDMS 5:1	93 ± 1.5
PDMS 10:1	99 ± 1.0
PDMS 20:1	103 ± 0.5

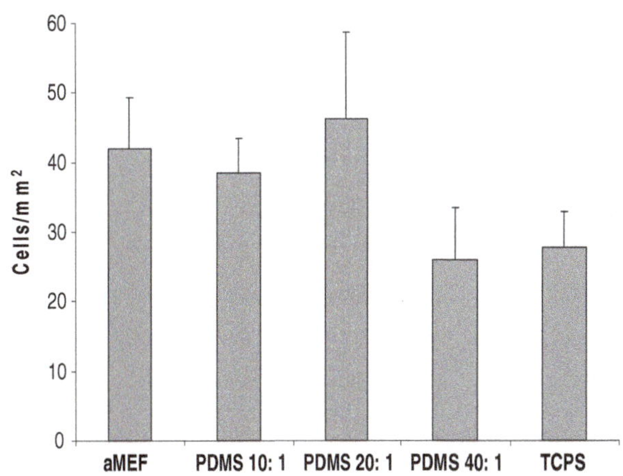

Figure 3 Cell Density across substrates, independent of cell-cell contact. Cell density after 12 hours was used as an indicator for degree of cell attachment across various substrates. Error bars represent standard deviation.

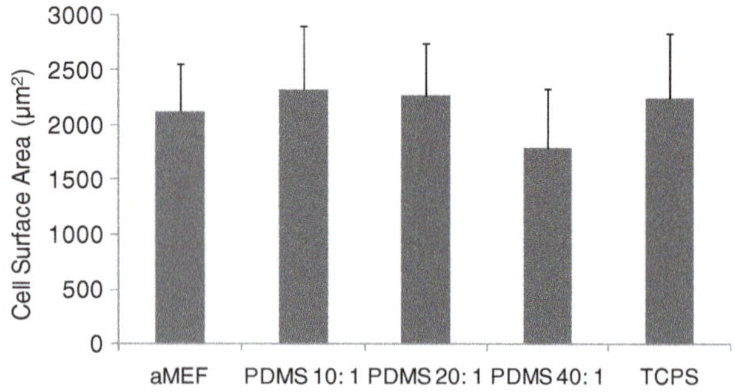

Figure 4 Cell Morphology across substrates, independent of cell-cell contact. Cell surface area after 12 hours on various substrates, with 20 cells measured per condition from triplicate samples. Area of cells grown on PDMS 40:1 was significantly lower than that of the other substrates. Error bars represent standard deviation.

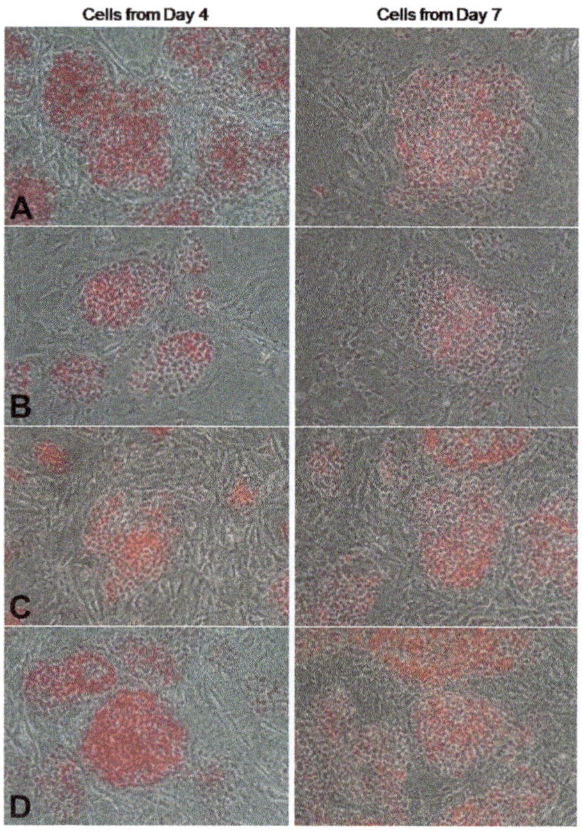

Figure 5 Alkaline Phosphatase Activity for cells collected from various substrates. Cells collected from different substrates after 4 and 7 days in culture were re-plated on feeders in the presence of complete hESC medium for four days, following which they were stained for AP activity. **A**: Cells from 10: 1 PDMS, **B**: Cells from 20: 1 PDMS, **C**: Cells from TCPS, **D**: Cells from MEF.

Table 3 Expression indices across substrates

Substrate	Day 4	Day 7	Day 10
PDMS 10:1	23.13 ± 15.82	1.94 ± 0.52	2.02 ± 1.36
PDMS 20:1	55.86 ± 9.79	1.98 ± 0.17	0.35 ± 0.08
PDMS 40:1	0.36 ± 0.09	2.36 ± 0.41	0.11 ± 0.02
TCPS	83.67 ± 38.20	6.44 ± 3.74	7.22 ± 3.69

Expression Index calculated using Equation 1, errors reported are the standard error of the mean (n = 3).

changes in cell spreading and migration. Thus the stiffness mediated increase in cell attachment might be mimicking the environment of migrating mesoderm cells, thereby supporting the growth and differentiation of more adhesive cells. We also need to take into account that altering substrate rigidity also affects the overall chemical composition of the substrate which could affect differentiation. However, in previous studies on polyacrylamide gels of different formulations but similar stiffness, cell morphology remained similar [26].

One caveat to keep in mind is that not all mechanosensitive cell types respond similarly to changes in substrate stiffness [29]. Multiple tissues may have similar elasticities, and cell types respond differently to mechanical signals, in a manner somewhat similar to what they experience in their native tissue [27].

Conclusions

In this study, we have demonstrated that fibronectin-coated PDMS substrates are capable of supporting hESC attachment and proliferation. We have shown here that substrate stiffness affects proliferation, while cell spreading and cell attachment remains comparable across the range of stiffness tested. With increased time in culture, differentiation increased and gene expression associated with mesodermal differentiation was upregulated as stiffness increased from soft to stiff, suggesting that the substrate is an important variable that needs to be carefully considered in developing protocols for stem cell propagation and differentiation. Future studies will focus on whether such a system can be used to bring about terminal differentiation of hESCs towards defined cell types of the mesodermal lineage.

Materials and methods

ESC culture

A variant hESC line (BG01v) was cultured on MEF feeders that have been inactivated with mitomycin-C. Cells were cultured in hESC medium, which consisted of Dulbecco's modified Eagle's medium (DMEM)/F12 medium (Gibco, Cat. No. 11320–033) supplemented with 20% knockout serum replacement (KSR; Gibco, Cat No. 10828–028), 1 mM L-glutamine (Gibco, Cat. No. 25030–081), 0.1 mM minimal essential medium

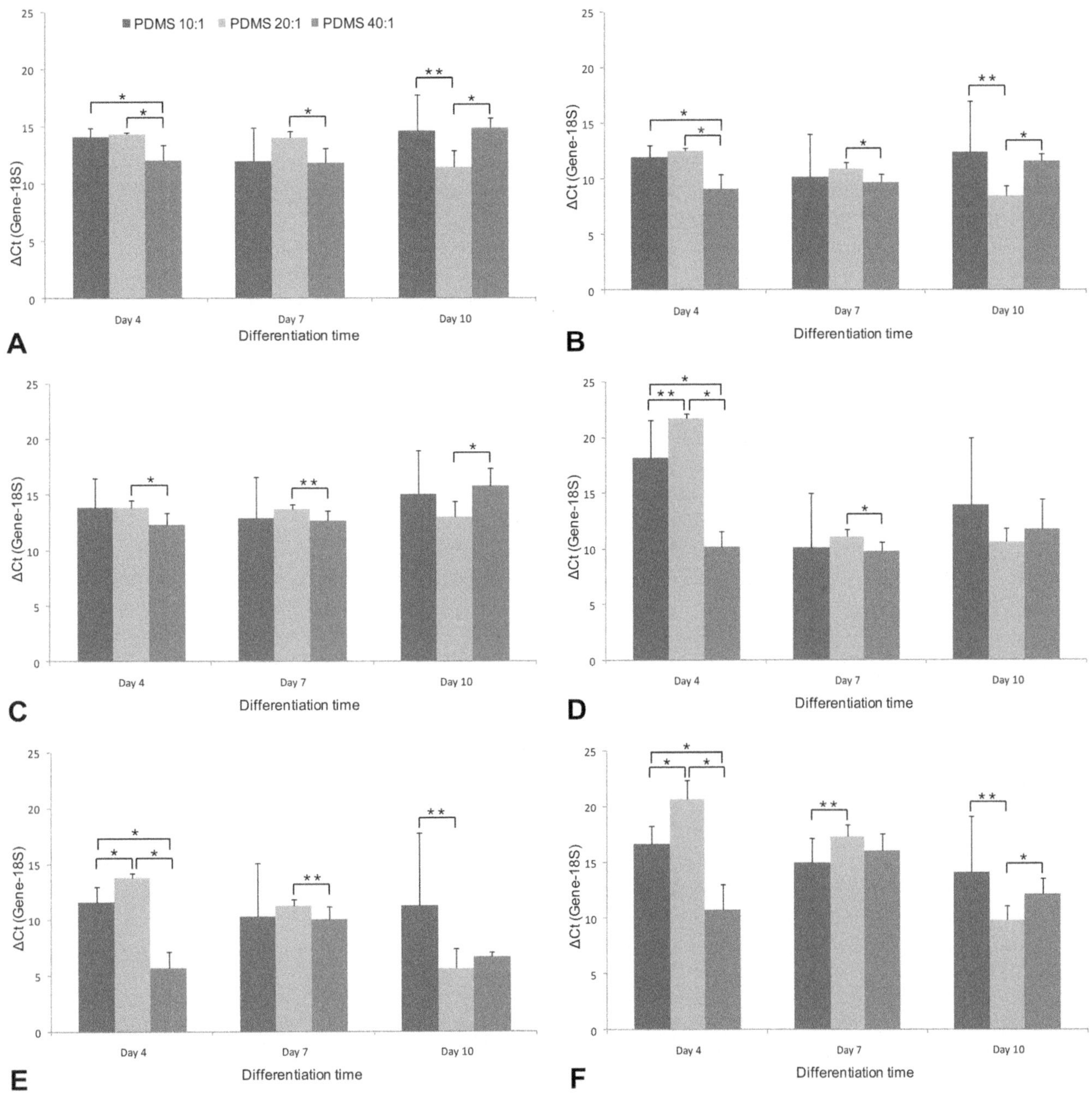

Figure 6 Comparison of individual gene expression across substrates. Genes analyzed included **A**: Oct 4 (Pluripotent) **B**: Nanog (Pluripotent) **C**: Sox 2 (Pluripotent) **D**: Neuro **D** (Ectoderm) **E**: IGF2 (Mesoderm) **F**: AFP (Endoderm). (* indicates $p < 0.01$ and ** indicates $p < 0.05$).

nonessential amino acids (Gibco, Cat. No. 11140–050), 50 U/ml penicillin and 50 μg/ml streptomycin (Gibco, Cat. No. 15070–063), 4 ng/ml basic fibroblast growth factor (Sigma, Cat. No. F-0291), and 0.1mM β-mercaptoethanol (Sigma, Cat. No. M7522). Cells were passaged every 3–5 - days by treatment for 2–3 minutes with 1 mg/ml collagenase type IV (Gibco, Cat. No. 17104–019) in ES cell medium, followed by a 40 second treatment with 0.05% trypsin. Cells were plated at 50,000 cells/35 mm plate and grown at 37°C and 5% CO_2.

Substrate synthesis and characterization

The PDMS substrate was prepared from the commercially available Sylgard 184 silicone elastomer kit (Dow Corning, Midland, MI) by mixing the base and the curing agent in varying ratios. Specifically, PDMS with base: crosslinker w/w ratio 10:1, 20:1, and 40:1 were prepared. The pre-polymer mixtures were mixed thoroughly for at least 5 minutes, degassed, and poured into 35 mm polystyrene tissue culture Petri dishes. PDMS was then cured for at least 60 hours at 22-33°C.

Samples were stored at room temperature in a vacuum desiccator.

Tensile testing was done to characterize the bulk mechanical properties of the substrate. Specifically, 1 mm dog-bone shaped strips were subjected to a tensile load at a strain rate of 10 mm/min and the test was conducted to failure. The elastic modulus was determined manually by calculating the slope of the stress strain curve within linear limits.

Topographical and phase images were taken on an MFP-3D AFM (Asylum Research, Santa Barbara, CA). Images were obtained in non-contact (AC/tapping) mode and height, amplitude and phase images were taken using a silicon cantilever (AC-240 TS, Olympus Instruments) at a scan speed 1 Hz at 512 pixels/line. The scan size was 20 μm × 20 μm.

Surface wetting properties of the various substrates were evaluated by measuring the static water contact angles via the sessile drop method using a Ramé-Hart Goniometer/Tensiometer (Model 500) equipped with a special optical system and a CCD camera and the image was analyzed using DROPImage Advanced for contact angle determination.

Cell culture on PDMS substrates

Before conducting cell culture experiments, PDMS substrates were sterilized by treating them with ethanol under ultraviolet light for 1 hour, followed by a second round of UV exposure for another 30 minutes. To promote cell attachment to the various substrates studied, plates coated with various formulations of PDMS and the polystyrene plates were treated with 2.5 mL of 10 μg/mL of human plasma fibronectin (Chemicon, Cat. No. FC010) overnight at 37°C. Substrates were then washed twice with PBS and cells were seeded at 50,000 cells per 35 mm plate. To promote differentiation, cells were grown in differentiation medium (hESC medium without bFGF). Cells grown on inactivated MEF feeders in hESC medium were used as controls. Medium was changed on the second day and on every following day. On the 4th, 7th, and 10th day cells were collected from the plates by treating them with collagenase and trypsin as described above. Cells that remained attached following enzymatic passaging were collected using a rubber cell scraper. Cell counts were performed using a hemocytometer. The collected cells were snap frozen in liquid nitrogen and stored at −80°C for subsequent gene expression analysis. All experiments were performed with three replicates per condition.

Alkaline phosphatase activity

Some of the collected cells were also seeded onto plates with inactivated MEF feeders and propagated in hESC medium for 4 days after which an Alkaline Phosphatase Substrate Kit (Vector Labs, Cat. No. VC-SK-5100-KI01) was used to assay for alkaline phosphatase (AP) activity.

Cell surface area calculations

To study how substrates affect morphology independent of cell-cell contact, cells were plated at a density of 10,000 cells per 35 mm plate and grown in hESC growth medium. Cells seeded at the same density on acellularized MEF layers were used as controls. After 12 hours, images of cells from different experimental conditions were captured using a Nikon Eclipse TS100 inverted microscope and a Nikon Coolpix 5000 digital camera. Area was calculated using ImageJ software by manually outlining the cell perimeter with each area measurement performed twice. Cell density was calculated by manually counting the number of cells in the field of vision and extrapolating to the cell density in each of the 35 mm dishes. Each condition was performed with three replicate plates, and images of multiple cells were captured from each plate. All cells that were in contact with other cells were excluded from the analysis.

Gene expression analysis

For gene expression analysis, samples were prepared by isolating total RNA using TRIZOL (Invitrogen, Cat. No. 10296–010, Carlsbad, CA) according to manufacturer's instructions. Briefly, cell pellets were treated with TRIZOL and chloroform, RNA from the aqueous phase was precipitated in isopropyl alcohol, washed with 75% ethanol, and dissolved in water. RNA was quantified using a UV–vis Spectrophotometer (Biomate 3, Thermo Scientific, Waltham, MA). cDNA was reverse-transcribed from 1 μg of total RNA according to manufacturer's protocols using the High Capacity cDNA Reverse Transcription Kit (Applied Biosystems, Cat. No. 4368814 Foster City, CA). The reactions were incubated for 10 minutes at 25°C and for 120 minutes at 37°C.

Expression of pluripotent and differentiation genes was analyzed using quantitative real time RT-PCR using Taqman primers (Applied Biosystems, Foster City, CA); performed in an ABI HT7900 system (Applied Biosystems, Foster City, CA) and the data were acquired using sequence detection system software (SDS v2.2.1, Applied Biosystems, Foster City, CA). The original replicates ($n = 3$) for each condition were tested in duplicate, and all failed reactions (termed “undetermined” by the software) were excluded from the analysis. ΔCt values were obtained by normalizing the Ct values against the endogenous 18S ribosomal RNA. Data analysis for differential expression between the different samples was conducted in triplicate and Student's t-test was conducted to ascertain the significance of differential expression.

Expression Index was used to detect the relative differentiation state of cells and was based on average Ct values from triplicate measurements and used the mathematical model described in Noaksson et al. [31]

wherein EI is given by the equation

$$EI = K_{RS}\frac{\sqrt[m]{(1+E_{gene1m})^{Ctgene1m}.(1+E_{gene2m})^{Ctgene2m}...(1+E_{genem})^{Ctgenem}}}{\sqrt[n]{(1+E_{gene1n})^{Ctgene1n}.(1+E_{gene2n})^{Ctgene2n}...(1+E_{genen})^{Ctgenen}}}$$

where E is the PCR efficiency, Ct is the threshold cycle, m and n are the number of genes that are upregulated and downregulated, respectively, upon hESC differentiation. K_{RS} is the relative sensitivity constant (accounts for the differences in fragment lengths of templates) was not determined since it does not affect the relative comparison of samples.

Competing interests
The authors declare that they have no competing interests. There is no conflict of interest in the reporting of this data by any author.

Acknowledgements
We would gratefully like to thank Dr. Jennifer Wayne for granting access to the mechanical testing systems in her laboratory and for help with analyzing mechanical testing data. Funding for this work was provided in part by grant EEC-0234104 from the NSF/NIH Bioinformatics and Bioengineering Summer Institute Program and NSF-CAREER 074556 (RRR). Additional funding was provided by the VCU Honors Summer Undergraduate Research Program (NE).

Authors' contributions
NE fabricated the substrates and performed cell culture and characterization experiments. RR performed cell culture experiments, gene expression analysis and drafted the manuscript. VY performed the AFM and contact angle measurements. RRR is the principal investigator and was responsible for all elements of this research. All authors have read and approved the final manuscript. No writing assistance was used in the production of this manuscript.

Author details
[1]Department of Chemical and Life Science Engineering, Virginia Commonwealth University, Richmond, VA, USA. [2]Current Address: School of Engineering and Applied Sciences, Harvard University, Boston, MA, USA.

References
1. Wobus AM, Guan K, Yang HT, Boheler KR: **Embryonic stem cells as a model to study cardiac, skeletal muscle, and vascular smooth muscle cell differentiation.** *Methods Mol Biol (Clifton, NJ)* 2002, **185**:127–156.
2. Peerani R, Rao BM, Bauwens C, Yin T, Wood GA, Nagy A, Kumacheva E, Zandstra PW: **Niche-mediated control of human embryonic stem cell self-renewal and differentiation.** *EMBO J* 2007, **26**:4744–4755.
3. Xu R-H, Peck RM, Li DS, Feng X, Ludwig T, Thomson JA: **Basic FGF and suppression of BMP signaling sustain undifferentiated proliferation of human ES cells.** *Nat Meth* 2005, **2**:185–190.
4. Chadwick K, Wang L, Li L, Menendez P, Murdoch B, Rouleau A, Bhatia M: **Cytokines and BMP-4 promote hematopoietic differentiation of human embryonic stem cells.** *Blood* 2003, **102**:906–915.
5. Gerrard L, Rodgers L, Cui W: **Differentiation of Human Embryonic Stem Cells to Neural Lineages in Adherent Culture by Blocking Bone Morphogenetic Protein Signaling.** *Stem Cells* 2005, **23**:1234–1241.
6. Dravid G, Ye Z, Hammond H, Chen G, Pyle A, Donovan P, Yu X, Cheng L: **Defining the Role of Wnt/?-Catenin Signaling in the Survival, Proliferation, and Self-Renewal of Human Embryonic Stem Cells.** *Stem Cells* 2005, **23**:1489–1501.
7. McBeath R, Pirone DM, Nelson CM, Bhadriraju K, Chen CS: **Cell shape, cytoskeletal tension, and RhoA regulate stem cell lineage commitment.** *Dev Cell* 2004, **6**:483–495.
8. Engler AJ, Sen S, Sweeney HL, Discher DE: **Matrix elasticity directs stem cell lineage specification.** *Cell* 2006, **126**:677–689.
9. Shimizu N, Yamamoto K, Obi S, Kumagaya S, Masumura T, Shimano Y, Naruse K, Yamashita JK, Igarashi T, Ando J: **Cyclic strain induces mouse embryonic stem cell differentiation into vascular smooth muscle cells by activating PDGF receptor β.** *J Appl Physiol* 2008, **104**:766–772.
10. Kurpinski K, Chu J, Hashi C, Li S: **Anisotropic mechanosensing by mesenchymal stem cells.** *Proc Natl Acad Sci* 2006, **103**:16095–16100.
11. Saha S, Ji L, de Pablo JJ, Palecek SP: **Inhibition of human embryonic stem cell differentiation by mechanical strain.** *J Cell Physiol* 2006, **206**:126–137.
12. Stevens MM, George JH: **Exploring and Engineering the Cell Surface Interface.** *Science* 2005, **310**:1135–1138.
13. Takito J, Al-Awqati Q: **Conversion of ES cells to columnar epithelia by hensin and to squamous epithelia by laminin.** *J Cell Biol* 2004, **166**:1093–1102.
14. Abraham S, Eroshenko N, Rao RR: **Role of bioinspired polymers in determination of pluripotent stem cell fate.** *Regenerative Med* 2009, **4**:561–578.
15. Emerman JT, Burwen SJ, Pitelka DR: **Substrate properties influencing ultrastructural differentiation of mammary epithelial cells in culture.** *Tissue Cell* 1979, **11**:109–119.
16. Fan VH, Tamama K, Au A, Littrell R, Richardson LB, Wright JW, Wells A, Griffith LG: **Tethered epidermal growth factor provides a survival advantage to mesenchymal stem cells.** *Stem cells (Dayton, Ohio)* 2007, **25**:1241–1251.
17. Discher DE, Mooney DJ, Zandstra PW: **Growth factors, matrices, and forces combine and control stem cells.** *Science (New York, NY)* 2009, **324**:1673–1677.
18. Winer JP, Janmey PA, McCormick ME, Funaki M: **Bone marrow-derived human mesenchymal stem cells become quiescent on soft substrates but remain responsive to chemical or mechanical stimuli.** *Tissue Eng Part A* 2009, **15**:147–154.
19. Gray DS, Tien J, Chen CS: **Repositioning of cells by mechanotaxis on surfaces with micropatterned Young's modulus.** *J Biomed Mat Res Part A* 2003, **66**:605–614.
20. Li L, Sharma N, Chippada U, Jiang X, Schloss R, Yarmush ML, Langrana NA: **Functional modulation of ES-derived hepatocyte lineage cells via substrate compliance alteration.** *Ann Biomed Eng* 2008, **36**:865–876.
21. Mata A, Hsu L, Capito R, Aparicio C, Henrikson K, Stupp SI: **Micropatterning of bioactive self-assembling gels.** In *Book Micropatterning of bioactive self-assembling gels.* Edited by. The Royal Society of Chemistry:1228.
22. Levental I: **Soft biological materials and their impact on cell function.** *Soft Matter* 2007, **3**:299.
23. Hengsberger S, Kulik A, Zysset P: **Nanoindentation discriminates the elastic properties of individual human bone lamellae under dry and physiological conditions.** *Bone* 2002, **30**:178–184.
24. Tzvetkova Chevolleau T: **The motility of normal and cancer cells in response to the combined influence of the substrate rigidity and anisotropic microstructure.** *Biomaterials* 2008, **29**:1541.
25. Goffin JM, Pittet P, Csucs G, Lussi JW, Meister J-J, Hinz B: **Focal adhesion size controls tension-dependent recruitment of α-smooth muscle actin to stress fibers.** *J Cell Biol* 2006, **172**:259–268.
26. Pelham RJ: **Cell locomotion and focal adhesions are regulated by substrate flexibility.** *Proc Natl Acad Sci U S A* 1997, **94**:13661.
27. Yeung T, Georges PC, Flanagan LA, Marg B, Ortiz M, Funaki M, Zahir N, Ming W, Weaver V, Janmey PA: **Effects of substrate stiffness on cell morphology, cytoskeletal structure, and adhesion.** *Cell Motil Cytoskeleton* 2005, **60**:24–34.
28. Discher DE, Janmey P, Wang Y-l: **Tissue cells feel and respond to the stiffness of their substrate.** *Science* 2005, **310**:1139–1143.
29. Georges PC, Janmey PA: **Cell type-specific response to growth on soft materials.** *J Appl Physiol* 2005, **98**:1547–1553.
30. Zoldan J, Karagiannis ED, Lee CY, Anderson DG, Langer R, Levenberg S: **The influence of scaffold elasticity on germ layer specification of human embryonic stem cells.** *Biomaterials* 2011, **32**:9612–9621.
31. Noaksson K, Zoric N, Zeng X, Rao MS, Hyllner J, Semb H, Kubista M, Sartipy P: **Monitoring differentiation of human embryonic stem cells using real-time PCR.** *Stem Cells* 2005, **23**:1460–1467.

17

A survey of enabling technologies in synthetic biology

Linda J Kahl* and Drew Endy

Abstract

Background: Realizing constructive applications of synthetic biology requires continued development of enabling technologies as well as policies and practices to ensure these technologies remain accessible for research. Broadly defined, enabling technologies for synthetic biology include any reagent or method that, alone or in combination with associated technologies, provides the means to generate any new research tool or application. Because applications of synthetic biology likely will embody multiple patented inventions, it will be important to create structures for managing intellectual property rights that best promote continued innovation. Monitoring the enabling technologies of synthetic biology will facilitate the systematic investigation of property rights coupled to these technologies and help shape policies and practices that impact the use, regulation, patenting, and licensing of these technologies.

Results: We conducted a survey among a self-identifying community of practitioners engaged in synthetic biology research to obtain their opinions and experiences with technologies that support the engineering of biological systems. Technologies widely used and considered enabling by survey participants included public and private registries of biological parts, standard methods for physical assembly of DNA constructs, genomic databases, software tools for search, alignment, analysis, and editing of DNA sequences, and commercial services for DNA synthesis and sequencing. Standards and methods supporting measurement, functional composition, and data exchange were less widely used though still considered enabling by a subset of survey participants.

Conclusions: The set of enabling technologies compiled from this survey provide insight into the many and varied technologies that support innovation in synthetic biology. Many of these technologies are widely accessible for use, either by virtue of being in the public domain or through legal tools such as non-exclusive licensing. Access to some patent protected technologies is less clear and use of these technologies may be subject to restrictions imposed by material transfer agreements or other contract terms. We expect the technologies considered enabling for synthetic biology to change as the field advances. By monitoring the enabling technologies of synthetic biology and addressing the policies and practices that impact their development and use, our hope is that the field will be better able to realize its full potential.

Keywords: Synthetic biology, Biological engineering, Enabling technologies, Survey, Intellectual property rights, Licensing, Regulation

* Correspondence: linda.sciscript@gmail.com
Bioengineering Department, Stanford University, Y2E2 Room 269C, 473 Via Ortega, Stanford, CA 94305-4201, USA

Background

Synthetic biology is an emerging, interdisciplinary field that aims to make the design, construction, and optimization of biological systems easier and more reliable. Advances in synthetic biology will deepen our understanding of how biological systems work, and should enable faster and cheaper development of useful medicines, chemicals and materials, new means for information processing and data storage, and sources of food and energy that could help promote human health and preserve the environment. Although the field of synthetic biology is relatively young, there have already been promising advances in engineering microorganisms to produce important drugs [1,2], exploring strategies for biofuel production [3,4], and designing and DNA-based information storage [5,6] and genetically-encoded communications and processing systems [7,8].

As is true for many emerging fields of research, realizing the full potential for constructive applications of synthetic biology will require not only the continued development of enabling technologies but also the implementation of policies and practices to ensure that these technologies remain accessible to those working in basic and applied research. The enabling technologies for synthetic biology can be defined, broadly, as any reagent or method that, alone or in combination with associated technologies, provides the means to generate any new research tool or application in synthetic biology. Because the field of synthetic biology spans a wide range of disciplines – from engineering and biology to mathematics and computer science – the technologies considered "enabling" by synthetic biology researchers may be expected to cover a broad range, depending on the focus and nature of the research. Monitoring the enabling technologies of synthetic biology is an important step towards understanding the needs, abilities and accomplishments of this diverse research community.

Here, we conducted a survey among a self-identifying community of practitioners engaged in synthetic biology research to obtain their opinions and experiences with technologies that support the engineering of biological systems. The aim of this first study was to define a set of enabling technologies for the field of synthetic biology, with a focus on technologies used in research laboratories in both academia and industry. Our goal was to gather information about the technologies considered enabling by practitioners in the field so that we and others might better evaluate the landscape of synthetic biology and explore policies and practices that best promote continued innovation. For example, it is likely that useful applications of synthetic biology will embody multiple patented inventions and monitoring the enabling technologies of synthetic biology will facilitate the systematic investigation of the intellectual property rights coupled to those technologies. Investors and funders of synthetic biology research also may find this information useful in guiding funding decisions and establishing policies for the patenting and licensing of enabling technologies. Information gained from monitoring the enabling technologies of synthetic biology also may be useful in identifying technology trends, and could help government agencies and non-governmental organizations in crafting policy frameworks to address the safety and security concerns raised by synthetic biology research.

Results

Demographic data

During the period the survey was active, from August 31, 2012 to January 30, 2013 a total of 160 responses were received. Six responses were excluded because they did not contain answers to any of the substantive questions on technology use. Seventeen responses were from participants who answered "no" to Survey Question 11 that asked survey participants to indicate whether they considered themselves to be a synthetic biologist or to be engaged in basic or applied synthetic biology research or development. Responses from these 17 survey participants were analyzed separately, and the remaining 137 responses were used for most analyses.

Responses originated from the United States (121 responses, 88%) and ten other countries (16 responses, 12%). The distribution of responses from outside the United States was Australia (1), Canada (1), Germany (2), Israel (1), Italy (1), Japan (3), Mexico (1), Sweden (1), and the United Kingdom (5). Responses were received from researchers working exclusively in a non-commercial organization (n = 91, 66%), exclusively in a commercial organization (n = 39, 28%) and in both commercial and non-commercial organizations (n = 7, 5%). Among survey respondents working exclusively in a non-commercial organization, most indicated that they worked in a college or university (n = 64), research institution (n = 9), government laboratory (n = 1), were affiliated with both a college/university and research institution (n = 12) or were independent (n = 5). Among survey respondents working exclusively in a commercial organization, most were from small companies of fewer than 50 employees (n = 21) and the rest were from companies of more than 1000 employees (n = 11), fewer than 1000 employees (n = 3), and fewer than 250 employees (n = 4). Of the 7 survey respondents that worked in both commercial and non-commercial organizations, all worked in a small company of fewer than 50 employees as well as a college/university or research institution.

Experience with the iGEM competition

Survey Question 2 asked participants to provide information about their experience as student or non-student participants of the International Genetically Engineered Machines (iGEM) competition (http://igem.org), a synthetic biology competition aimed at undergraduate students, high school students, and entrepreneurs. Of the 136 survey participants that responded to this question, 112 (82%) indicated that they did not participate in the iGEM competition as a student, 21 (15%) indicated that they previously participated in the iGEM competition as a student, and 3 (2%) indicated that they were a student currently participating in the iGEM competition.

In addition, 58 survey participants provided free-text responses describing their experience as non-student participants of the iGEM competition. Of these, 49 stated that they had advised or mentored iGEM teams or served as judges for the iGEM competition, 6 supported or sponsored iGEM teams, and 3 stated that they had other experience with the iGEM competition, including participating in organizing the software division, assisting in evaluation of the iGEM program, and reading iGEM research reports Thirteen of the 58 survey participants who provided free-text responses were also former students of the iGEM competition, while 45 had no prior experience with iGEM as students.

Use of publicly available registries

Survey Question 3 asked participants to indicate whether they used publicly available registries to obtain natural or engineered biological materials or information. This question allowed survey participants to select from a list of publicly available registries and to write in any additional registries of which they were aware. The listed registries included publicly available collections of information (including DNA sequences) or tangible materials that could be used for synthetic biology research, including plasmids encoding specific biological functions, DNA-binding proteins, microorganisms and cell lines (hereinafter referred to as biological parts). Specifically, the publicly available registries initially listed included the Registry of Standard Biological Parts supporting the iGEM competition (iGEM Registry), the American Type Culture Collection (ATCC), Addgene, the Coli Genetic Stock Center (CGSC), the Synthetic Biology Engineering Resource Center (SynBERC) Registry, the Joint BioEnergy Institute Public Registry (JBEI-ICE Public), the European *Saccharomyces cerevisiae* Archive for Functional Analysis (EUROSCARF), the Agricultural Research Service NRRL collection (ARS/NRRL), the BIOFAB: International Open Facility Advancing Biotechnology (BIOFAB), the Dana-Farber/Harvard Cancer Center (DF/HCC) PlasmID Repository, the DNASU Plasmid Repository (DNASU), the Belgian Coordinated Collections of Micro-organisms (BCCM), and the Leibniz-Institut DSMZ - German Collection of Microorganisms and Cell Cultures (DSMZ).

The majority of synthetic biology researchers in academia (n = 88) used biological parts from or contributed parts to the iGEM Registry (n = 60, 68%) and the ATCC (n = 53, 60%), and many used or contributed to Addgene (n = 42, 48%) (Figure 1A). Other publicly available registries that were widely used among academic researchers included the CGSC (n = 18, 20%), the SynBERC Registry (n = 13, 15%), JBEI-ICE Public (n = 11, 12%) and EUROSCARF (n = 8, 9%). Fewer researchers in academia reported use or contribution of parts to the ARS/NRRL (n = 5), the BIOFAB

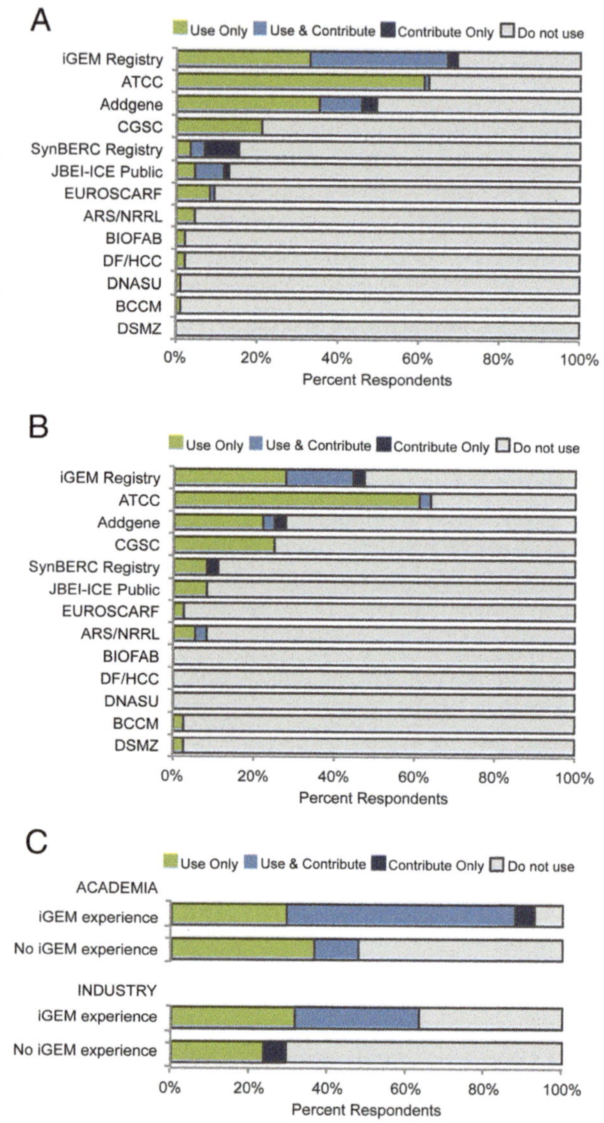

Figure 1 Publicly available registries of natural or engineered biological materials or information. (**A**) Percentage of synthetic biology researchers in academia that use biological parts from, or contribute parts to, publicly available registries. (**B**) Percentage of synthetic biology researchers in industry that use biological parts from, or contribute parts to, publicly available registries. (**C**) Impact of iGEM experience on use of the iGEM Registry by synthetic biology researchers in academia and industry.

(n = 2), the DF/HCC (n = 2), the DNASU (n = 2), the DSMZ (n = 1), and the BCCM (n = 1). Additional registries identified by academic researchers included the CyanoBase-Kazusa Genome Resources (http://www.ncbi.nlm.nih.gov/pmc/articles/PMC2808859) (n = 1), and the various yeast collections distributed by Invitrogen (n = 1).

Synthetic biology researchers in industry (n = 39) reported different usage rates for publicly available registries (Figure 1B). The majority of industry researchers used or contributed parts to the ATCC (n = 26, 67%), and many used or contributed parts to the iGEM Registry (n = 18, 46%). Other publicly available registries used by industry researchers included Addgene (n = 12, 31%), the CGSC (n = 9, 23%), the SynBERC Registry (n = 4, 10%), the JBEI-ICE Public (n = 3, 8%) and the ARS/NRRL (n = 4, 10%). Fewer industry researchers reported use or contribution of parts to the EUROSCARF (n = 2), the BCCM (n = 1), and the DSMZ (n = 2). Additional registries identified by industry researchers included the Keio/ASKA collection of *E. coli* strains (http://www.shigen.nig.ac.jp/ecoli/strain/top/top.jsp) (n = 1) and the pZ series expression vectors developed by Lutz and Bujard [9] (http://www.expressys.com) (n = 1).

Among the publicly available registries included in this survey, only the iGEM Registry showed a statistically significant difference in usage rates between researchers in academia and industry ($p = 0.03$). This difference was explored further by examining the impact of prior iGEM experience on use of the iGEM Registry (Figure 1C). Among synthetic biology researchers in academia, those having experience with the iGEM competition either as a student or non-student participant (e.g., advisors, judges, sponsors, etc.) were significantly more likely to contribute parts to or use parts from the iGEM Registry as compared to academic researchers without iGEM experience (93% and 46%, respectively, $p < 0.00001$). Similarly, industry researchers having experience with the iGEM competition were significantly more likely to use or contribute parts to the iGEM Registry than industry researchers lacking iGEM experience (65% and 26%, respectively, $p = 0.02$). Academic researchers having iGEM experience also were significantly more likely than industry researchers with iGEM experience to use or contribute to the iGEM Registry (93% and 65%, respectively, $p < 0.01$). No significant difference was observed in use of the iGEM Registry between academic and industry researchers without iGEM experience (46% and 26%, $p = 0.17$).

Use of private registries

Survey Question 4 asked participants to indicate whether the laboratories or organizations in which they worked maintained a private registry of biological parts, and whether and how these parts were made available to others. Most synthetic biology researchers in academia reported that the laboratory or organization in which they worked maintained its own registry of biological parts (55/88, 62%) (Figure 2). Of these, the vast majority of academic researchers made these materials available to others outside their own laboratory (53/55, 96%). A significantly greater proportion of academic researchers sent parts directly to others (40/53, 75%) as compared to those that distributed parts through a publicly available registry (15/53, 28%) ($p < 0.00001$).

Similarly, most synthetic biology researchers in industry reported that the laboratory or organization in which they worked maintained its own registry of biological parts (21/39, 54%). However, fewer than half of industry researchers made these parts available to others (9/21, 43%). A significantly greater proportion of industry researchers also sent materials directly to others (6/9, 67%) as opposed to distributing parts through a publicly available registry (1/9, 11%) ($p = 0.05$).

A comparison of the use and distribution rates between academic and industry researchers revealed no statistically significant difference in the likelihood of maintaining a private registry of biological parts ($p = 0.43$). However, researchers in academia were significantly more likely to make parts available to others than researchers in industry ($p < 0.000001$).

Favorite or most useful biological parts

A total of 43 participants responded to Survey Question 5, which was an open question asking respondents to list their favorite or most useful biological parts (Table 1). Biological parts that survey participants identified in more general terms included the Anderson promoter library (n = 4), in-house promoters (n = 2), three-color

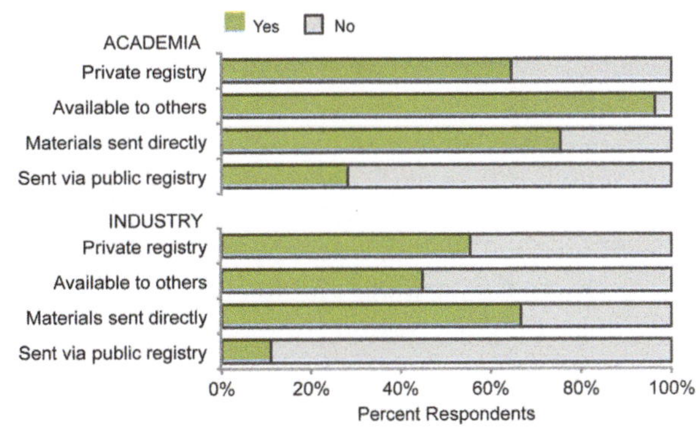

Figure 2 Private registries of natural or engineered biological materials or information. Percentage of synthetic biology researchers in academia and industry that maintain a private registry of biological parts, make these materials available to others, send materials to others directly, and send materials via a publicly available registry.

Table 1 Favorite or most useful biological parts from publicly available registries

ID	Category	N	Description
Addgene			
25712:pAKTaq	Plasmid	1	Bacterial expression vector encoding the DNA polymerase from *Thermus aquaticus*
Coli Genetic Stock Center			
CGSC #12119	Chassis	1	*E. coli* strain BW27783 bearing 9 known mutations
iGEM Registry			
BBa_J23100 series	Regulatory	8	BBa_J23100 through BBa_J23119 is a family of constitutive promoter parts that can be used to tune the expression level of constitutively expressed parts
BBa_B0034	RBS	4	RBS based on Elowitz & Liebler repressilator
BBa_B0015	Terminator	1	Double terminator including BBa_B0010 and BBa_B0012
BBa_C0062	Coding	1	luxR repressor/activator
BBa_E2050	Coding	1	derivative of mRFP1, yeast-optimized
BBa_F2620	Signaling	1	A signaling device whereby the input is $3OC_6HSL$ and the output is PoPS from a LuxR-regulated operator
BBa_I15010	Coding	1	Chimeric Cph1 light receptor/EnvZ protein
BBa_I744210	Generator	1	TetR regulated LuxN-Tsr Chimeric Receptor B
BBa_J04450	Reporter	1	RFP coding device
BBa_J15001	RBS	1	strong synthetic *E. coli* RBS with SacI site
BBa_J153000	Plasmid Backbone	1	broad-host-range shuttle vector pPMQAK1 that provides ampicillin and kanamycin/neomycin resistance
BBa_J176005	Protein Domain	1	Codon optimized mCherry red fluorescent protein
BBa_J176006	Coding	1	Mammalian venus fluorescent protein
BBa_J176022	Protein Domain	1	Human codon-optimized AmCyan1 from pAmCyan1-C1
BBa_J33207	Reporter	1	lac promoter and lacZ
BBa_J61009	Plasmid	1	pAC-LuxGFP that places GFP under the wildtype Vibrio lux device
BBa_J64032	Device	1	pCASP SPI-1 Secretion Circuit
BBa_J85226	Composite	1	Kanamycin resistance (KanR)_off version of J85224
BBa_J176027	Regulatory	1	Constitutive cytomegalovirus promoter
BBa_J176122	Plasmid Backbone	1	pcDNA3.1 plus puromycin resistance
BBa_K566002	Regulatory	1	Biphasic switch
BBa_P1010	Generator	1	ccdB cell death gene
BBa_R0040	Regulatory	1	TetR repressible promoter

scaffold for monitoring gene expression (n = 1), dose-dependent promoters (n = 1), BIOFAB promoters (n = 1), fluorescent proteins (n = 9), colored proteins made at DNA2.0 (n = 1), all the working reporters in the iGEM Registry (n = 1), BioBrick vectors (n = 2), pRS vector series (n = 1), pZ vectors (n = 1), vectors (n = 1), high copy plasmid backbones from the iGEM Registry (n = 1), bicistronic design parts (n = 1), quorum sensing parts (n = 1), terminator variants (n = 1), and TetR (n = 2).

Use of physical assembly standards and methods

Survey Question 6 asked participants to indicate their current and past use of physical assembly standards and methods. This question allowed survey participants to select from a list of physical assembly methods and to write in any additional assembly methods that they used. Of the 134 survey participants that answered this question, most indicated that they currently use or previously have used the Gibson assembly method (48% current, 22% past) and *de novo* DNA synthesis (50% current, 18% past) (Figure 3). The original BioBrick standard (18% current, 31% past) and Gateway cloning (15% current, 26% past) were selected by a significant number of survey participants, although most indicated past use of these specific methods. Survey participants indicated lower overall usage rates for other physical assembly

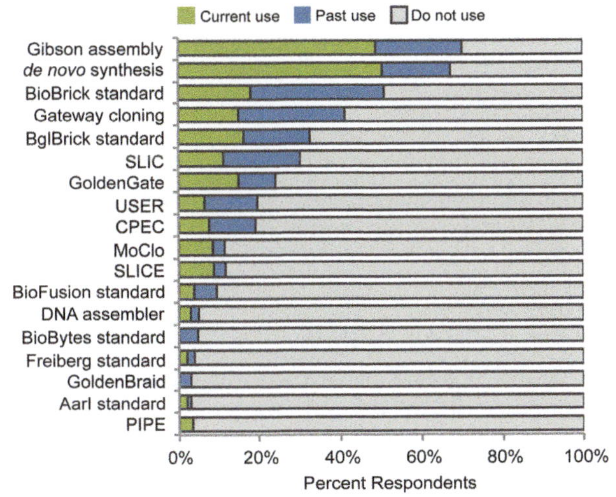

Figure 3 Physical assembly standards and methods. Current and past use of physical assembly standards and methods by synthetic biology researchers.

methods, including the BglBrick standard (16% current, 16% past), Sequence and Ligase Independent Cloning (SLIC) (14% current, 21% past), GoldenGate (16% current, 10% past), and others.

In addition to the physical assembly methods listed in Question 6, survey participants providing free-text responses to this question (n = 33) identified additional assembly methods that they used in the course of synthetic biology research (Table 2).

Table 2 Additional physical assembly methods identified by survey participants

Physical Assembly Method	N
Commercial/Proprietary	
Proprietary method, not specified	3
GeneArt Seamless Assembly	2
GeneArt High Order Assembly	1
Clontech In-Fusion HD cloning kit	1
Ginkgo assembly method	1
Invitrogen TOPO cloning	1
Non-Commercial	
Conventional PCR [10,11]	10
Yeast *in vivo* recombinational cloning [12]	5
Home-brew method, not specified	4
Restriction-site Associated DNA (RAD) assembly [13]	2
Anderson 2 antibiotic (2ab) assembly [14]	1
Inverse PCR [15]	1
Splicing by Overlap Extension (SOE) [16]	1
In development	
A new enzymatic, scarless synthesis and assembly technology	1
Extensions of BioBytes assembly standard [17]	1

Use of measurement, functional composition and data exchange standards and tools

A total of 120 and 127 survey participants responded to Survey Questions 7 and 8, respectively, which asked respondents to indicate their current or past use of measurement tools, functional composition standards, and data exchange tools. These questions allowed survey participants to select from a list of tools and to write in any additional measurement, functional composition and data exchange tools that they used. Usage rates for measurement standards, functional composition standards, and data exchange standards were relatively low (Figure 4). Specifically, the number of survey respondents reporting current and past use were: Relative Promoter Unit (RPU) (7% current, 14% past), Polymerase Per Second (PoPS) (2% current, 14% past), Relative Mammalian Promoter Unit (RMPU) (0 current, 2% past), Expression Operating Unit (EOU) (4% current, 3% past), Synthetic Biology Open Language (SBOL) (18% current, 10% past), SBOL visual (SBOLv) (16% current, 3% past), JBEI-ICE repository platform (8% current, 2% past), electronic datasheets (7% current, 7% past), and visual datasheets (4% current, 4% past).

Additional measurement tools identified by survey participants included fluorescence reporter protein measurement (n = 4), Miller assay (n = 2), beta-galactosidase assay (n = 2), cell auto-fluorescence (n = 1), dual-luciferase reporter measurement (n = 1), comprehensive metabolite measurement (n = 1), comprehensive proteome (n = 1), specific mRNA or protein measurements (n = 1), RNAseq (n = 1), and qPCR (n = 1). Additional data exchange tools included custom laboratory management information systems (n = 2), JERM (n = 1), RightField (n = 1), Systems Biology Markup Language (SMBL) (n = 1), GoogleDocs (n = 1), and the GenBank file format (n = 1).

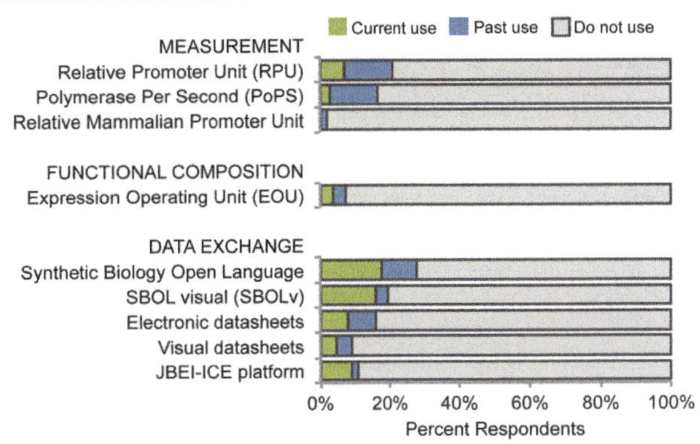

Figure 4 Standards and methods for measurement, functional composition, and data exchange. Current and past use of standards and methods for measurement, functional composition, and data exchange by synthetic biology researchers.

Use of additional tools, reagents and methods

Survey Question 9 asked participants to indicate their current and past use of additional tools, reagents, and methods. This question allowed survey participants to select from a list and to write in any other tool reagent or method that they considered enabling for the field of synthetic biology. Of the survey participants that responded to this question, the vast majority indicated that they used GenBank as their preferred genomic database (84% current, 12% past) and employed tools for search (84% current, 15% past), alignment (78% current, 16% past), and analysis (68% current, 13% past) of DNA sequences (Figure 5). Most survey participants used commercial DNA synthesis services for short oligos (86% current, 11% past) and gene-size fragments (71% current, 12% past), while fewer researchers used in-house DNA synthesis for short oligos (14% current, 15% past) and gene-size fragments (26% current, 44% past). The vast majority of survey participants indicated that they used established culture techniques, as well as other established technologies such as the polymerase chain reaction (PCR), green fluorescent protein (GFP), and non-GFP reporter molecules, while fewer researchers used newer techniques such as directed evolution (e.g., MAGE) (22% current, 10% past).

In addition to the tools, reagents and methods that were listed, survey participants providing free text responses (n = 23) identified additional technologies that they considered enabling for synthetic biology (Table 3).

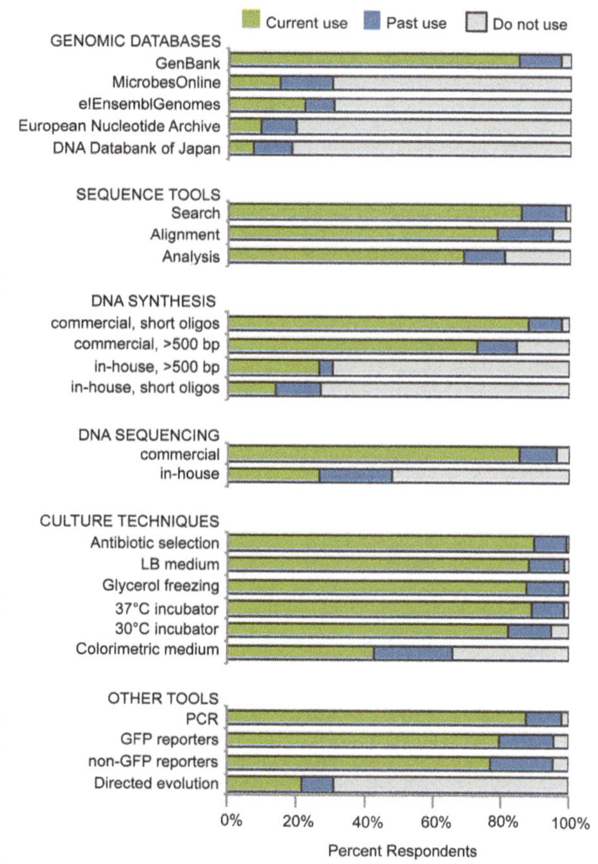

Figure 5 Additional tools, reagents and methods. Current and past use of genomic databases, sequence tools, DNA synthesis tools, DNA sequencing tools, culture techniques, and other tools, reagents, and methods by synthetic biology researchers.

Use of software tools

A total of 133 survey participants responded to Survey Question 10, which asked participants to indicate their current and past use of software tools. This question allowed survey participants to select software tools that were listed and to write in any additional software tools they used. The highest rates of current use were reported for ApE (41% current, 24% past), Primer 3 (33% current, 24% past), Mfold (34% current, 21% past), and the Ribosome Binding Site (RBS) Calculator (28% current, 28% past) (Figure 6). Vector NTI (17% current, 51% past), GeneDesigner (22% current, 27% past), and Mathematica (11% current, 35% past) were selected by a significant number of survey participants, although most indicated past use of these software tools. Lower but increasing-over-time usage rates were reported for the j5 DNA Assembly (16% current, 9% past), Genome Compiler (11% current, 6% past), and GenoCAD (10% current, 6% past) software tools. In addition to the software tools that were listed, survey participants providing free text responses (n = 36) identified additional software tools used in the course of their synthetic biology research (Table 4).

Technology choices and self-identification as a synthetic biologist

Survey Question 11 asked participants whether they considered themselves to be a synthetic biologist or to be engaged in basic or applied synthetic biology research or development. Because this question was introduced on Day 6 of the survey, not all participants were able to respond to this question. A total of 58 survey participants had submitted responses prior to Day 6 and because all of these participants were students or postdoctoral fellows in the Endy or Smolke labs or researchers affiliated with SynBERC they were considered to be synthetic biology researchers for the purposes of the survey. Of the 96 survey participants that submitted responses on Day 6 and later, 79 answered "yes" and 17 answered "no" to this question. Responses from the 79 participants that affirmatively self-identified as synthetic biologists and the 58 participants that responded prior to Day 6 were grouped together for most analyses. Responses from the 17 participants that did not self-identify as synthetic biologists were analyzed separately.

Table 3 Additional technologies considered enabling for synthetic biology by survey participants

Tool, Reagent or Method	Description (URL or reference)	N
BioCyc	a collection of 1962 pathway and genome databases (http://biocyc.org)	1
Bioinformatics	the application of computational techniques to analyze the information associated with biomolecules on a large-scale (http://www.ncbi.nlm.nih.gov/About/primer/bioinformatics.html)	1
CAGE: Conjugative Assembly Genome Engineering	a technology that permits the hierarchical consolidation of modified genomic regions [18]	1
High Throughput Computing	the ability to run many copies of software at the same time across many different computers, reviewed in [19]	1
in vitro screens	tests for biological activity such as metal binding screens, electron uptake, and other enzymatic activity	2
IonTorrent	an approach to DNA sequencing that enables a direct connection between chemical and digital information and aims to place DNA sequencing within the reach of any laboratory or clinic [20]	1
EcoCyc	a database for *Escherichia coli* K-12 MG1655 (http://ecocyc.org)	1
Flow Cytometry	a technology that uses the principles of light scattering, light excitation, and emission of fluorochrome molecules to generate specific multi-parameter data from particles and cells in the size range of 0.5 μm to 40 μm diameter (http://crl.berkeley.edu/flow_cytometry_basic.html)	3
KEGG: Kyoto Encyclopedia of Genes and Genomes	a database resource for understanding high-level functions and utilities of the biological system, such as the cell, the organism and the ecosystem, from genomic and molecular-level information (http://www.genome.jp/kegg)	1
Mass spectrometry	a technology for targeted protein quantification, reviewed in [21]	2
MetaCyc	a database of nonredundant, experimentally elucidated metabolic pathways (http://metacyc.org)	1
Molecular biology technologies, generally	includes methods and reagents for creating competent cells, nucleic acid transfer, digestion, primer extension, ligation, assembly of DNA molecules, etc.	9
OptForce	an algorithm that identifies all possible metabolic interventions that lead to the overproduction of a biochemical of interest [22]	1
PDB: Protein DataBank	an information portal to biological macromolecular structures (http://www.rcsb.org/pdb/home/home.do)	1
Protein purification technologies	methods for purifying a protein of interest efficiently, reviewed in [23]	1
Recombineering	an *in vivo* method of genetic engineering applicable to chromosomal and episomal replicons in *E. coli* [24]	1
Robotic automation	use of robots for repetitive laboratory tasks such as pick and place, liquid and solid additions, heating, cooling, mixing, shaking, etc.	2
Single cell microscopy	a technology that enables visualization of gene expression with exquisite spatial and temporal sensitivity, reviewed in [25]	1
Standards, needed	includes standards for calibrating and sharing data from plate readers, standards for test, measurement and characterization, standards for documentation and sharing of biological modules, for example see Arkin, 2008 [26] and Endy, 2005 [27]	3
SOLiD	a next generation sequencing technology that allows identification of hundreds of millions of short RNAs in a sample in a single run [28]	1
Transcription Activator-Like (TAL) effector technology	a technology that allows proteins to be designed to specifically target and bind to a desired sequence of DNA [29]	1
UniProt: Universal Protein Resource	a collaboration between the European Bioinformatics Institute (EBI), the SIB Swiss Institute of Bioinformatics and the Protein Information Resource (PIR) that aims to provide a comprehensive resource for protein sequence and annotation data (http://www.uniprot.org)	1
Yeast *in vivo* recombination	methods for assembling large DNA constructs in the yeast *Saccharomyces cerevisiae*, for example see Gibson et al., 2008 [30] and Jaschke et al., 2012 [31]	1

Of the 17 participants that did not self-identify as synthetic biologists, 12 were from the US, 2 were from the UK, 1 was from Austria, 1 was from Norway, and 1 was from Portugal. Eleven participants worked exclusively in non-commercial organizations – 5 in a college or university, 2 in a research institution, 1 in both a college/university and research institution, and 3 were independent. The remaining 6 worked exclusively in commercial organizations – 3 in a company with more than 1,000 employees, 2 in a company of fewer than 50 employees, and 1 in a company of fewer than 250 employees. Twelve of the 17 participants had no experience with the iGEM competition, 1 had been a sponsor for iGEM teams, and 1 collaborated with an iGEM team as a DIY biologist.

Among the 17 participants who did not self-identify as synthetic biologists, two indicated that they used parts from the iGEM Registry (one was a former iGEM

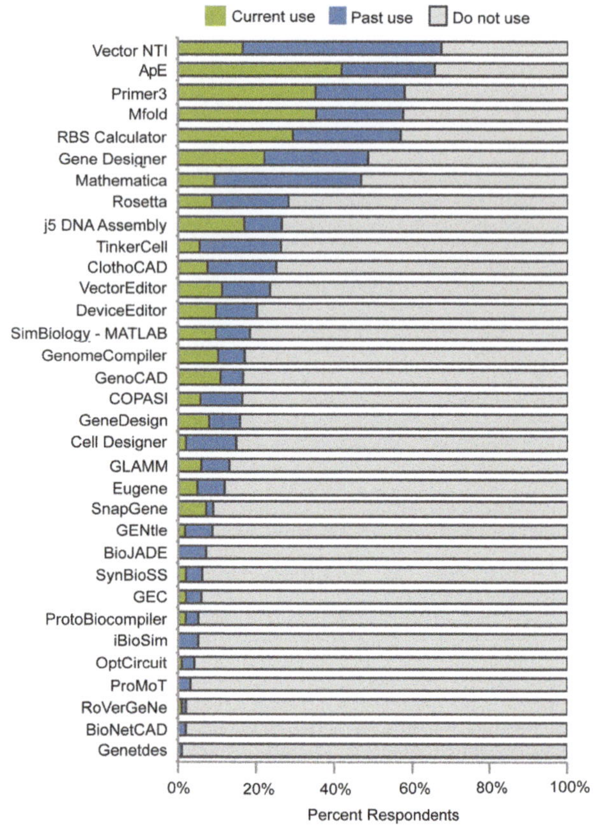

Figure 6 Software tools. Current and past use of software tools by synthetic biology researchers.

student, the other had no iGEM experience). In addition, some participants used parts from other publicly available registries, including the ATCC (n = 5), Addgene (n = 3), the ARS/NRRL (n = 2), the BIOFAB (n = 1), and the CGSC (n = 1). With regards to private registries, 4 participants (3 from academia, 2 from industry) indicated that the laboratories or organizations in which they worked maintained a private registry of biological parts. Of these, 3 made parts available to others (all from academia) – 3 by sending parts directly, and 1 by also distributing parts through a publicly available registry. None of these 17 participants reported current use of standards and methods for measurement, functional composition, or data exchange, although past use of the RPU (n = 1), electronic datasheets (n = 1), and visual datasheets (n = 1) was noted. Both current and past use was reported for other technologies covered in this survey, including physical assembly methods, software tools, and other tools, reagents, and methods (Table 5).

Discussion

The emerging field of synthetic biology has captured the interest and energy of researchers from a variety of disciplines – biology, chemistry, computer science, engineering, and more. Our survey tapped the experiences and perspectives of this diverse research community in order to glean initial insights into the technologies that are considered enabling for the field by its practitioners. The results of the survey offer a first snapshot view of the technologies previously and now in use by synthetic biology researchers, and give a sense of the many and varied technologies that support work in synthetic biology.

One of our objectives in conducting this survey was to establish a set of technologies considered enabling for the field of synthetic biology, so that we and others might systematically investigate the intellectual property rights coupled to these technologies. For example, one overall consideration regarding enabling technologies and property rights is whether or not these technologies are accessible for use – either by virtue of being in the public domain or through legal tools such as non-exclusive licensing – to researchers in academic, government, and commercial organizations. The extent to which innovation in synthetic biology, and biotechnology more generally, may be impeded by broad foundational patents that cannot be licensed or patent thickets remains unclear [32-36]. Identifying the technologies to which wide, unrestricted access is needed to promote continued innovation in synthetic biology is an important step towards understanding the impact of patenting and licensing practices on access to the enabling technologies underlying this field.

Consistent with the postulate that past scientific achievements lay the foundation for future innovation, the results of the survey showed that many of the technologies that enable research in synthetic biology are well established and in the public domain. For many of these earlier technologies patent protection was either not sought or, even if patent protected, sufficient time has lapsed for the technologies to enter the public domain. For example, the vast majority of survey respondents reported use of bacterial cell culture technologies such as LB medium or glycerol freezing (Figure 5), yet these technologies were published in the scientific literature as early as the 1950's [37-39] and are squarely in the public domain. Similarly, the vast majority of survey respondents reported use of PCR technology, yet elements of PCR technology have entered the public domain or will do so shortly. Specifically, foundational patents covering amplification methods (e.g., US 4,683,195 and EP 0 200 362 B), thermal cycling instruments (e.g., US 5,038,852 and EP 0 395 736 B), and thermostable DNA polymerases (e.g., US 4,889,818 and EP 0 258 017 B) have now expired. Although patents continue to be filed on improvements to PCR technologies, many subsequent patents such as those covering thermostable polymerases with enhanced activities (e.g.

Table 4 Additional software tools used for synthetic biology research by survey participants

Software tool	Description (URL)	N
ABySS: Assembly By Short Sequences	a *de novo*, parallel, paired-end sequence assembler that is designed for short reads (http://www.bcgsc.ca/platform/bioinfo/software/abyss)	1
AlignDNA	a pairwise DNA alignment tool (http://www.geneinfinity.org/sms/sms_aligndna.html)	1
BLAST: Basic Local Alignment Search Tool	a program that finds regions of local similarity between biological sequences (http://blast.ncbi.nlm.nih.gov)	7
CentroidFold	a program that predicts an RNA secondary structure from an RNA sequence (http://www.ncrna.org/centroidfold)	1
CLC Genomics Workbench	a comprehensive and user-friendly analysis package for analyzing, comparing, and visualizing next generation sequencing data (http://www.clcbio.com/products/clc-genomics-workbench)	5
CodonCode Aligner	a program for sequence assembly, contig editing, and mutation detection (http://www.codoncode.com/aligner)	1
Cytoscape	an open source software platform for visualizing and integrating complex networks (http://www.cytoscape.org)	1
FastPCR	a program for PCR primer design (http://en.bio-soft.net/pcr/FastPCR.html)	1
Gene Construction Kit	a program for plasmid mapping (http://www.textco.com/gene-construction-kit.php)	1
Geneious	a program for handling and managing bioinformatics data (http://www.geneious.com)	7
Gibthon Ligation Calculator	a software tool for calculating reactant concentrations for DNA ligation (http://django.gibthon.org/tools/ligcalc)	1
JWS online	a tool for simulation of kinetic models from a curated model database (http://jjj.biochem.sun.ac.za)	1
Lasergene-DNAStar	comprehensive software for DNA and protein sequence analysis, contig assembly and sequence project management (http://www.dnastar.com)	7
Mascot	a search engine which uses mass spectrometry data to identify proteins from primary sequence databases (http://www.matrixscience.com/search_intro.html)	1
Mauve	a system for efficiently constructing multiple genome alignments in the presence of large-scale evolutionary events such as rearrangement and inversion (http://gel.ahabs.wisc.edu/mauve)	1
Merlin	a M.A.G.E. optimization tool developed by the Cross-disciplinary Integration of Design Automation Research group at Boston University (http://cidar1.bu.edu:8080)	1
OligoAnalyzer	software for comprehensive oligonucleotide analysis (http://www.idtdna.com/analyzer/Applications/OligoAnalyzer)	1
ORF Finder: Open Reading Frame Finder	a graphical analysis tool which finds all open reading frames of a selectable minimum size in a user's sequence or in a sequence already in the database (http://www.ncbi.nlm.nih.gov/projects/gorf)	1
PaR-PaR	software that allows researchers to use liquid-handling robots effectively (http://prpr.jbei.org)	1
Pigeon	Synthetic Biology Open Language picture generator (http://cidar1.bu.edu:5801/pigeon1.php)	1
PlasMapper	software that automatically generates and annotates plasmid maps using only the plasmid DNA sequence as input (http://wishart.biology.ualberta.ca/PlasMapper)	1
PyMOL	a user-sponsored molecular visualization system for rendering and animating 3D molecular structures on an open-source foundation (http://pymol.org)	1
RNAstructure	a complete package for RNA and DNA secondary structure prediction and analysis (http://rna.urmc.rochester.edu/RNAstructure.html)	1
Serial Cloner	freeware with an intuitive interface that assists in DNA cloning, sequence analysis and visualization (http://serialbasics.free.fr/Serial_Cloner.html)	3
Sequencher	DNA sequencing software (http://genecodes.com/sequencher-features)	1
SSC: Stochastic Simulation Compiler	a tool for creating exact stochastic simulations of biochemical reaction networks (http://web.mit.edu/irc/ssc)	1
SWISS-PDB viewer (aka DeepView)	an application that provides a user friendly interface allowing analysis of several proteins at the same time (http://spdbv.vital-it.ch)	1
Synbiota	a platform of collaborative services to design, store, post, organize, access, or share information (https://mozillalabs.com/en-US/synbiota)	2
Velvet	a set of algorithms to manipulate de Bruijn graphs for genomic sequence assembly (http://www.ebi.ac.uk/~zerbino/velvet)	1

US 5,436,149 and US 5,556,772) have either expired or are due to expire within the next year.

Other more recently developed technologies used by synthetic biology researchers are currently patent protected or have patent applications pending, yet are accessible through non-exclusive licensing. For example, the majority of survey respondents reported current or past use of the Gibson assembly method [40] for *in vitro*

Table 5 Technologies used by survey participants that did not self-identify as synthetic biologists (n = 17)

Tool, reagent or method	Current use	Past use	Tool, reagent or method	Current use	Past use
Physical assembly methods			Genomic database		
Gateway recombinatorial cloning	4	1	GenBank	10	3
de novo DNA synthesis	3	2	E!EnsemblGenomes	7	3
Gibson assembly	3	1	MicrobesOnline	3	1
Conventional restriction site-based cloning	2	0	European Nucleotide Archive	2	2
CPEC	1	2	DNA Databank of Japan	2	0
SLIC	1	1	Sequence tools		
PIPE	1	1	Search (e.g., BLAST)	12	3
USER	1	1	Alignment (e.g., ClustalW2)	10	1
InFusion cloning	1	0	Analysis (e.g., OligoCalc)	7	2
RAD assembly	1	0	Software tools		
Transfer PCR	1	0	ApE	6	1
Yeast *in vivo* cloning	1	0	Rosetta	4	1
BioBrick assembly standard	0	1	Vector NTI	3	5
Measurement, functional composition, data exchange			Primer3	3	3
RPU	0	1	Mathematica	2	5
Electronic datasheets	0	1	Mfold	1	2
Visual datasheets	0	1	Gene Designer	1	2
DNA synthesis			Blast	1	1
Commercial, short oligos	9	4	J5 DNA Assembly	1	1
Commercial, gene size (>500bp)	4	3	Vector Editor	1	1
In-house, gene-size(>500 bp)	3	2	GenoCAD	1	1
In-house, short oligos	1	4	Cell Designer	1	0
DNA sequencing			ClothoCAD	1	0
Commercial	8	3	iBioSim	1	0
In-house	2	4	Gene Design	1	0
Culture technique			ProtoBiocompiler	1	0
LB broth or agar	10	3	SimBiology	1	0
37°C incubator	10	3	SnapGene	1	0
Antibiotic selection	9	4	TinkerCell	1	0
Glycerol freezing	8	3	GenomeCompiler	0	1
30°C incubator	7	3	GEntle	0	1
Colorimetric medium	4	2	GLAMM	0	1
Other tools, reagents and methods			COPASI	0	1
PCR	9	5	DeviceEditor	0	1
GFP reporters	7	4	Lasegene-DNA Star	0	1
Non-GFP reporters	6	4	RBS Calculator	0	1
Directed evolution	3	2			

physical assembly of DNA constructs (Figure 3). Several granted patents and pending patent applications are relevant to this method, including US Patents 7,723,077 and 7,776,532, US Applications 2010/0184187 and 2010/0311126, and International Application PCT/US2006/031214. Although the exclusive period for these patents is expected to extend through at least 2026, access to the Gibson assembly method has been made available through a non-exclusive licensing agreement between Synthetic Genomics, Inc. and New England BioLabs, Inc. [41]. As such, components may be purchased from New England BioLabs, Inc., albeit with significant restrictions that limit use to "internal research purposes for the sole benefit of the purchaser only" [42]. Those desiring additional rights to the Gibson assembly method, for example to manufacture commercial products, must contact Synthetic Genomics, Inc. directly.

Still other technologies that survey respondents consider enabling for the field of synthetic biology are heavily patent protected and the ability of researchers to access these technologies through licensing is less clear. For example, the majority of survey respondents indicated current or past use of GFP or non-GFP reporter molecules (Figure 5). Fluorescent proteins are commonly used as genetically encoded reporter molecules that enable researchers to observe the activity of particular genetic elements and biomolecules inside live cells or tissues. One of the foundational patents covering uses of GFP (US 5,491,084) is due to expire in September 2013. However, the exclusivity periods for other foundational patents on GFP and its uses are expected to continue for a number of years (e.g., US 5,741,668, US 6,146,826, EP 0 759 170 B1). Furthermore, there are literally hundreds, if not thousands, of issued patents covering variants of GFP and their uses. For example, a search of CAMBIA's Patent Lens (http://www.patentlens.net) for the term "green fluorescent protein" in the same claim [i.e., Expert Search of (green near/2 fluorescent) and (fluorescent near/2 protein) and (green near/2 protein) in claims] yielded 770 granted US patents and 256 granted European patents. Similar searches for yellow fluorescent protein, red fluorescent protein, and blue fluorescent protein yielded over 400 patents granted in the US and Europe. The large number of patents covering variants of GFP and their uses presents a considerable challenge to synthetic biology researchers who wish to use fluorescent reporters in creating standards for characterizing biological parts and devices, such as the Relative Promoter Unit (discussed below). Some relief to navigating this thicket of patents may be available through negotiating a license agreement with Life Technologies or GE Healthcare. For example, the ATCC has recently announced that they have secured a license agreement that enables them to distribute GFP-containing biological materials to non-commercial and government researchers [43]. However, for-profit customers must have a separate license with Life Technologies or GE Healthcare to obtain and use these materials.

In addition to the observations noted above, the survey results indicate several trends for the use of technologies by synthetic biology researchers. The majority of survey respondents reported that they used biological parts from or contributed parts to publicly available registries (Figure 1) as well as private registries maintained within individual laboratories (Figure 2). Among the most widely used publicly available registries were the iGEM Registry, the ATCC, and Addgene. These registries, as well as many of the others listed in the survey, provide researchers with tangible materials (e.g., cultures, plasmids, and other reagents) as well as information relevant to the material (e.g., source, nucleic acid sequence, performance specifications). From a technical perspective, public registries are useful only to the degree that the biological materials contained within are reliable and accurately described. Towards that end, several publicly available registries have undertaken steps to curate the parts received and to verify nucleic acid sequence information. From an intellectual property perspective, access to materials from public registries is limited not only by considerations of patent protection, but also by the terms of material transfer agreements or other contracts that may govern the transfer of tangible materials. Registries of biological parts such as the SynBERC registry, the JBEI-ICE Public, and the BIOFAB currently provide information only. To the extent that the genetically-encoded materials described in these registries may be readily synthesized from the sequence information provided, use of these materials is limited primarily by considerations of patent protection. Although the SynBERC and JBEI-ICE Public registries currently indicate whether the material is "encumbered" or "not encumbered," no other information is provided to assist researchers wishing to use these materials with identifying relevant patents. As for private registries, many survey respondents indicated that they shared materials with others and that they distributed materials by direct transfer as well as through publicly available registries. The types of materials maintained in private registries and the terms for their transfer were not queried in this survey, though considerations of patent protection and possibly additional contract terms would be relevant to use of materials from private registries as well.

Not all of the technologies queried in this survey were used by the majority of survey participants. Unlike the high usage rates reported for physical assembly standards and methods (Figure 3, Table 2), relatively few survey participants indicated current or past use of standards and methods for measurement, functional composition, and data exchange (Figure 4). One possibility that could account for such relatively low usage rates is that some types of standards and methods have been introduced only recently. For example, the EOU has been presented at meetings as early as 2010 [44], but an initial formal description and example applications of the EOU have only recently been published [45,46]. Another possibility that could account for the relatively low usage rate is that some types of standards and methods require tools that are not readily licensed by industry. For example, the RPU requires measurement of fluorescent reporter molecules [47] and it may be necessary to work through the patent thicket surrounding uses of

GFP before widespread adoption of such standards is possible. Finally, it may also be that some technologies identified in this survey are useful to only a subset of survey respondents. The synthetic biology research community is comprised of diverse group of individuals with varied interests and goals. As such, the technologies that are considered vitally important for the work of some researchers may not be used at all by others.

To the extent that one or more patented technologies could eventually become widely adopted as a standard in the field, it may be advisable for the synthetic biology research community to consider creating more formal organizational structures and to articulate policies and best practices for the disclosure and licensing of patented technologies. Although standardization in synthetic biology is still at a relatively young stage [48,49], there are indications that the patent landscape is becoming quite complex [50]. To mitigate potential difficulties in the development and implementation of standards using patented technologies, the field of synthetic biology could benefit from the lessons learned by the information and communications technologies (ICT) industry, where a multitude of patented technologies have been incorporated into standards [51]. There, the creation of standards development organizations with formal policies requiring fair, reasonable and non-discriminatory (F/RAND) licensing terms have helped to alleviate some of the problems the ICT industry has faced in incorporating patented technologies into standards [52]. Because the creation of a standards development organization is not a trivial undertaking and could potentially raise antitrust concerns, it will be important to work with counsel and abide by the recommendations of governmental agencies [53,54].

Several limitations should be taken into account when interpreting the results from this survey. First, this survey sampled only a fraction of the global synthetic biology research community and included responses from 137 individuals, mostly from universities or research institutes within the United States. It is difficult to estimate the number of individuals conducting research in synthetic biology, though one study identified nearly 3,000 authors of scientific publications who were working on or writing about synthetic biology [55]. Given the relatively small sample size, the survey results may be subject to sampling bias (i.e., the demographics of the survey respondents may not accurately reflect the demographics of the synthetic biology research community) as well as potential reporting biases (i.e., the voluntary reporting of this survey necessarily excludes those who chose not to volunteer responses). A larger sampling of the synthetic biology research community, with greater representation of researchers outside of the United States as well as researchers working in industry, might provide more balanced insight into the technologies considered enabling for synthetic biology. Second, although in-person interviews were conducted to identify relevant technologies and create questions for the survey, the majority of responses were obtained through an online format. Without an interactive format, such as in-person or telephone interviews, to clarify any ambiguities in the wording of survey questions, respondents may have misunderstood some of the questions posed. Third, the survey questions focused on the technologies actually used by the synthetic biology research community and did not explore the potential reasons underlying why certain technologies were not used. Additional questions that directly query whether intellectual property rights covering certain technologies represented a selection barrier against the use of those technologies would also be informative. Finally, the results of this survey reflect the experiences of synthetic biology researchers at only one point in time. Re-administration of surveys such as the one developed here could provide a more complete view of the technologies that are considered enabling for synthetic biology as the field develops over time.

Beyond facilitating the systematic investigation of property rights, monitoring the enabling technologies of synthetic biology could also help inform governmental and non-governmental organizations in crafting policy frameworks to address the safety and security concerns raised by innovation in this field. Access to concrete data on the technologies used by those working in basic and applied synthetic biology research can be vital for making changes to existing policies as well as for creating new options for governance. For example, advances in DNA synthesis technology and the resulting commercial availability of larger synthetic DNA constructs [56] have led to a shift from research conducted with recombinant DNA to research conducted with synthetic nucleic acid molecules. This shift in the technologies used for research, in turn, has prompted the U.S. Department of Health and Human Services to issue amended guidelines for research involving recombinant or synthetic nucleic acid molecules [57] and to develop recommendations for a framework for synthetic nucleic acid screening [58]. Similarly, monitoring the enabling technologies of synthetic biology could help alert policy makers and stakeholders to advances in technology that may exert a comparable impact on innovation and research practices in this field.

Conclusion

The survey results presented here provide insight into the enabling technologies of synthetic biology. As innovation in this field continues to advance we expect that the reagents, methods, and tools considered enabling for synthetic biology will change. The policies and practices of government, funding, and community organizations that impact the regulation, patenting, and licensing of these technologies are also subject to change. Because research in synthetic biology is conducted across multiple institutions in many countries, it will be important to adopt policies and practices that promote cross-institutional and transnational exchange of ideas, data, and technology. Moreover, because it is likely that useful applications of synthetic biology will embody multiple patented inventions, it will be important to create structures for managing intellectual property rights that will promote access to the technologies used to comprise and create commercially available products. By monitoring the enabling technologies of synthetic biology and advancing policies and best practices for the patenting, licensing and regulation of these technologies, our hope is that the field will be better able to reach its full potential to promote human health and preserve the environment.

Methods

Survey design, distribution and analysis

We designed a web-based survey soliciting responses on technologies that could be considered enabling by practitioners engaged in synthetic biology research. For purposes of the survey, enabling technologies were defined as tools, reagents, and methods that, alone or in combination with associated technologies, provide the means to generate any new research tool or application in synthetic biology. Technologies included in the survey were compiled through review of the scientific literature and personal interviews with synthetic biology researchers from both academia and industry. Researchers working in the field of synthetic biology were located by means of personal references, professional networking, and referrals from synthetic biology organizations. For this first survey, we focused on technologies used in research 1laboratories for the engineering of biological systems. Given its seminal importance in promoting a sense of community [59], we also examined the role of the iGEM competition in fostering the adoption of certain technologies by synthetic biology researchers. Technologies associated with safety and security were not within the scope of this survey, nor were other potentially enabling resources such as professional societies or technology roadmaps.

We first piloted the survey by sending the questionnaire to researchers working in the laboratories of Drs. Christina Smolke and Drew Endy in the Bioengineering Department at Stanford University, and made adjustments based on initial responses. We then sent the survey to members of the Synthetic Biology Engineering Resource Center (SynBERC) and made further adjustments based on responses received by Days 6 and 19. On Day 19 and at various times thereafter, a link to the survey was forwarded to additional researchers working in the field of synthetic biology by the BioBricks Foundation (BBF), a public benefit organization that represents the public interest in the field of synthetic biology (http://biobricks.org), the ERASynBio, a program for the development and coordination of synthetic biology in the European Research Area (http://www.erasynbio.eu), the iGEM Foundation, a public benefit organization that organizes the iGEM competition (http://igem.org), the organizers of SynBioBeta, an industry conference for synthetic biology startup companies (http://synbiobeta.com), and individuals working in community biolabs.

The survey was available via an interactive website (http://www.surveymonkey.com) or as a Word document directly from the authors, and the responses reported were collected from August 31, 2012 through January 30, 2013. Instructions provided at the beginning of the survey encouraged respondents to answer the questions based on their own experience and perspective. A PDF file of the survey questions is available as Additional file 1.

Questionnaire data were exported into Microsoft Excel for analysis and only valid responses were evaluated (i.e., only responses to the specific questions were included in each analysis). Analyses that involved comparison of researchers in academia and industry included only responses from survey participants that worked exclusively in a non-commercial or commercial setting. Survey participants were considered to be working in a non-commercial setting if they indicated that they worked in a college or university, research institute, government laboratory, or were independent (e.g., citizen scientist, amateur biologist, etc.). Survey participants were considered to be working in a commercial setting if they indicated that they worked in a for-profit company of any size. Statistical significance was evaluated using Fisher's exact test over binary contingency tables [60].

Registries of natural or engineered biological materials or information

Eleven publicly available registries were listed in Question 3 throughout the duration of the survey: the iGEM Registry (http://partsregistry.org), the JBEI-ICE Public (https://public-registry.jbei.org), the SynBERC Registry (https://registry.synberc.org), Addgene (http://www.addgene.org), the DNASU Plasmid Repository (http://dnasu.asu.edu/DNASU/Home.jsp), the DF/HCC PlasmID Repository (http://plasmid.med.harvard.edu/PLASMID), the ATCC (http://www.atcc.org), the CGSC (http://cgsc.biology.yale.edu), the EUROSCARF (http://web.uni-frankfurt.de/fb15/mikro/euroscarf/index.html), the Félix d'Hérelle Reference Center for Bacterial Viruses (http://www.phage.ulaval.ca/no_cache/en/accueil), and the ARS/NRRL culture collection (http://nrrl.ncaur.usda.gov). This group of publicly available registries includes both registries that distribute tangible materials (e.g., cell lines, plasmids, DNA binding proteins) and registries that serve solely as repositories of information (e.g., DNA sequences, plasmid construction, performance specifications). Two additional publicly available registries were added to the list for Question 3 after the collection of 85 responses, based on free text responses – the BCCM (http://bccm.belspo.be/index.php) and the DSMZ (http://www.dsmz.de).

Physical assembly standards and methods

Sixteen physical assembly standards and methods were listed in Question 6 throughout the duration of the survey: the original BioBrick assembly standard (BBF RFC 10) [61], the BglBrick assembly standard (BBF RFC 21) [62], the BioFusion standard (BBF RFC 23) [63], Freiberg standard (BBF RFC 25) [64], the AarI cloning standard (BBF RFC 28) [65], the BioBytes assembly standard (BBF RFC 47) [17], Circular Polymerase Extension Cloning (CPEC) [66], DNA assembler [67], Gateway recombinatorial cloning [68], Gibson assembly [40], GoldenBraid assembly [69], GoldenGate shuffling [70], Modular Cloning (MoClo) [71], Seamless Ligation Cloning Extract (SLICE) [72], Sequence and Ligase Independent Cloning (SLIC) [73], and *de novo* DNA synthesis [74]. Two additional physical assembly methods were added to the list in Question 6 after the collection of 52 responses based on free text responses – Polymerase Incomplete Primer Extension (PIPE) [75] and Uracil Specific Excision Reagent (USER) [76].

Tools for measurement, functional composition and data exchange

Tools supporting functional composition of genetic objects, measurement of intracellular molecular activities, and data exchange were listed in Survey Questions 7 and 8. Measurement tools included Polymerase Per Second (PoPS) [77], relative promoter unit (RPU) [47], relative mammalian promoter unit (RMPU) [78], functional composition tools included the expression operating unit (EOU) [45], and data exchange tools included Synthetic Biology Open Language (SBOL) [79], SBOL Visual (SBOLv) [80,81], and visual or electronic data sheets for biological parts and devices [82,83].

Additional tools, methods and reagents

Additional tools, reagents, and methods that survey participants could consider enabling were listed in Survey Question 9. Genomic databases included the DNA Databank of Japan (http://www.ddbj.nig.ac.jp), European Nucleotide Archive (http://www.ebi.ac.uk/ena), GenBank (http://www.ncbi.nlm.nih.gov/genbank), e!EnsemblGenomes (http://www.ensemblgenomes.org), and MicrobesOnline (http://www.microbesonline.org). Tools for searching, alignment, and analysis of DNA sequences such as BLAST [84], ClustalW2 [85], and OligoCalc [86], respectively, as well as commercial and in-house methods for DNA synthesis [74] and DNA sequencing [87] were also included. Long-established cell culture technologies included antibiotic selection, temperature selection, lysogeny broth (a.k.a., Luria-Bertani medium or LB medium) [37,88], colorimetric media [89], and glycerol freezing of bacterial strains [38,39]. More recently established tools such as PCR [10,11], fluorescent reporter molecules [90], and directed evolution [91] were also included.

Software tools

Thirty software tools, many of which have been recently reviewed [92], were listed in Question 10 throughout the duration of the survey: ApE (http://biologylabs.utah.edu/jorgensen/wayned/ape), BioJADE (http://web.mit.edu/jagoler/www/biojade), BioNetCAD (http://www.sysdiag.cnrs.fr/BioNetCAD), Cell Designer (http://celldesigner.org), ClothoCAD (http://www.clothocad.org), COPASI (http://www.copasi.org), DeviceEditor (replaced by AutoBioCAD; http://j5.jbei.org/index.php/Main_Page), Eugene (http://eugenecad.org), GEC (http://research.microsoft.com/en-us/projects/gec), Gene Designer (https://www.dna20.com/genedesigner2), GeneDesign (http://www.genedesign.org), GenoCAD (http://www.genocad.org), Genetdes (http://jaramillolab.issb.genopole.fr/display/sbsite/Download), GLAMM (http://glamm.lbl.gov), iBioSim (http://www.async.ece.utah.edu/iBioSim), j5 DNA Assembly Design Automation Software (http://j5.jbei.org/index.php/Main_Page), Mathematica (http://www.wolfram.com/mathematica), Mfold (http://mfold.rna.albany.edu/?q=mfold), OptCircuit (http://maranas.che.psu.edu/research_circuits.htm), Primer3 (http://simgene.com/Primer3), ProMoT (http://www.mpi-magdeburg.mpg.de/projects/promot), ProtoBiocompiler (http://proto.bbn.com/Proto/Proto.html), RBS Calculator (https://salis.psu.edu/software), Rosetta (http://www.rosettacommons.org), RoVerGeNe (http://iasi.bu.edu/~batt/rovergene/rovergene.htm), SimBiology - MATLAB (http://www.mathworks.com/products/simbiology), SynBioSS (http://www.synbioss.org), TinkerCell (http://www.tinkercell.com), VectorEditor (http://j5.jbei.org/index.php/Main_Page), and Vector NTI (http://www.invitrogen.com/site/us/en/home/Products-and-Services/Applications/Cloning/vector-nti-software.html). Three additional software tools were added to the list for Question 10 after the collection of 51 responses based on free text responses – GenomeCompiler (http://www.genomecompiler.com), GENtle (http://gentle.magnusmanske.de), and SnapGene (http://www.snapgene.com).

Additional file

Additional file 1: PDF file of the survey questions.

Abbreviations

ARS/NRRL: Agricultural research service NRRL culture collection; ATCC: American type culture collection; BBF: BioBricks foundation; BCCM: Belgian coordinated collections of micro-organisms; BIOFAB: BIOFAB international open facility advancing biotechnology; CGSC: Coli genetic stock center; DF/HCC: Dana-farber/harvard cancer center; DSMZ: Leibniz-Institut DSMZ - German collection of microorganisms and cell cultures; EOU: Expression operating unit; EUROSCARF: European *Saccharomyces cerevisiae* archive for functional analysis; GFP: Green fluorescent protein; iGEM: International genetically engineered machines; JBEI-ICE Public: Joint BioEnergy institute public registry; LB medium: Luria-bertani medium; MoClo: Modular cloning; PIPE: Polymerase incomplete primer extension; PoPS: Polymerase per second; RBS: Ribosome binding site; RMPU: Relative mammalian promoter unit; RPU: Relative promoter unit; SLIC: Sequence and ligase independent cloning; SLICE: Seamless ligation cloning extract; SynBERC: Synthetic biology engineering resource center; SBOL: Synthetic biology open language; SBOLv: SBOL visual; SMBL: Systems biology markup language; USER: Uracil specific excision reagent.

Competing interests

DE is a co-founder and director of a commercial DNA assembly company (Gen9, Inc.). No other competing interests have been declared by the authors.

Authors' contributions

LJK and DE conceived the study. LJK designed the survey, analyzed the responses, and wrote the manuscript. DE provided critical feedback and suggestions during the development and analysis of the survey. Both authors reviewed and approved the final manuscript.

Acknowledgements

We thank all of the participants that responded to the survey questions, members of the Endy and Smolke labs for contributing to the development of the survey, and Kevin Costa at the SynBERC, Holly Million at the BBF, Andy Boyce at the BBSRC and the ERASynBio, Meagan Lizarazo and Randy Rettberg at the iGEM Foundation, John Cumbers and Lisa Comeau at the SynBioBeta conference, Mackenzie Cowell at DIYbio, and Antony Evans at Glowing Plant for their efforts in promoting the survey. We also thank Paul Jaschke and Monica Ortiz for sharing their technological expertise, Nicolas Kahl for insight on motivating survey participants, Markus Sommer and Eddie Moler for advice on statistical analysis, and Megan Palmer for helpful discussions throughout the project. Funding for this work was provided by the NSF-sponsored Synthetic Biological Engineering Research Center (SynBERC) and Stanford University.

References

1. Ro DK, Paradise EM, Ouellet M, Fisher KJ, Newman KL, *et al*: **Production of the antimalarial drug precursor artemisinic acid in engineered yeast.** *Nature* 2006, **440**:940–943.
2. Ajikumar PK, Xiao WH, Tyo KE, Wang Y, Simeon F, *et al:* **Isoprenoid pathway optimization for Taxol precursor overproduction in Escherichia coli.** *Science* 2010, **330**:70–74.
3. Colin VL, Rodriguez A, Cristobal HA: **The role of synthetic biology in the design of microbial cell factories for biofuel production.** *J Biomed Biotechnol* 2011, **2011**:601834.
4. Georgianna DR, Mayfield SP: **Exploiting diversity and synthetic biology for the production of algal biofuels.** *Nature* 2012, **488**:329–335.
5. Bonnet J, Subsoontorn P, Endy D: **Rewritable digital data storage in live cells via engineered control of recombination directionality.** *Proc Natl Acad Sci U S A* 2012, **109**:8884–8889.
6. Church GM, Gao Y, Kosuri S: **Next-generation digital information storage in DNA.** *Science* 2012, **337**:1628.
7. Ortiz ME, Endy D: **Engineered cell-cell communication via DNA messaging.** *J Biol Eng* 2012, **6**:16.
8. Bonnet J, Yin P, Ortiz ME, Subsoontorn P, Endy D: **Amplifying genetic logic gates.** *Science* 2013, **340**:599–603.
9. Lutz R, Bujard H: **Independent and tight regulation of transcriptional units in Escherichia coli via the LacR/O, the TetR/O and AraC/I1-I2 regulatory elements.** *Nucleic Acids Res* 1997, **25**:1203–1210.
10. Saiki RK, Gelfand DH, Stoffel S, Scharf SJ, Higuchi R, *et al:* **Primer-directed enzymatic amplification of DNA with a thermostable DNA polymerase.** *Science* 1988, **239**:487–491.
11. Saiki RK, Scharf S, Faloona F, Mullis KB, Horn GT, *et al*: **Enzymatic amplification of beta-globin genomic sequences and restriction site analysis for diagnosis of sickle cell anemia.** *Science* 1985, **230**:1350–1354.
12. Oldenburg KR, Vo KT, Michaelis S, Paddon C: **Recombination-mediated PCR-directed plasmid construction in vivo in yeast.** *Nucleic Acids Res* 1997, **25**:451–452.
13. Etter PD, Preston JL, Bassham S, Cresko WA, Johnson EA: **Local de novo assembly of RAD paired-end contigs using short sequencing reads.** *PLoS One* 2011, **6**:e18561.
14. Leguia M, Brophy J, Densmore D, Anderson JC: **Automated assembly of standard biological parts.** *Methods Enzymol* 2011, **498**:363–397.
15. Ochman H, Gerber AS, Hartl DL: **Genetic applications of an inverse polymerase chain reaction.** *Genetics* 1988, **120**:621–623.
16. Horton RM, Hunt HD, Ho SN, Pullen JK, Pease LR: **Engineering hybrid genes without the use of restriction enzymes: gene splicing by overlap extension.** *Gene* 1989, **77**:61–68.
17. Lloyd D, Robinson K, Garside E, Fedor J, Ridgway D, *et al: BBF RFC 47: BioBytes Assembly Standard.* 2009. http://dspace.mit.edu/handle/1721.1/49518.
18. Isaacs FJ, Carr PA, Wang HH, Lajoie MJ, Sterling B, *et al*: **Precise manipulation of chromosomes in vivo enables genome-wide codon replacement.** *Science* 2011, **333**:348–353.
19. Morgan M, Grimshaw A: **High-throughput computing in the sciences.** *Methods Enzymol* 2009, **467**:197–227.
20. Rothberg JM, Hinz W, Rearick TM, Schultz J, Mileski W, *et al*: **An integrated semiconductor device enabling non-optical genome sequencing.** *Nature* 2011, **475**:348–352.
21. Pan S, Aebersold R, Chen R, Rush J, Goodlett DR, *et al*: **Mass spectrometry based targeted protein quantification: methods and applications.** *J Proteome Res* 2009, **8**:787–797.
22. Ranganathan S, Suthers PF, Maranas CD: **OptForce: an optimization procedure for identifying all genetic manipulations leading to targeted overproductions.** *PLoS Comput Biol* 2010, **6**:e1000744.
23. Walls D, McGrath R, Loughran ST: **A digest of protein purification.** *Methods Mol Biol* 2011, **681**:3–23.
24. Sharan SK, Thomason LC, Kuznetsov SG, Court DL: **Recombineering: a homologous recombination-based method of genetic engineering.** *Nat Protoc* 2009, **4**:206–223.
25. Larson DR: **What do expression dynamics tell us about the mechanism of transcription?** *Curr Opin Genet Dev* 2011, **21**:591–599.
26. Arkin A: **Setting the standard in synthetic biology.** *Nat Biotechnol* 2008, **26**:771–774.
27. Endy D: **Foundations for engineering biology.** *Nature* 2005, **438**:449–453.
28. Guo L, Liang T, Lu Z: **A comprehensive study of multiple mapping and feature selection for correction strategy in the analysis of small RNAs from SOLiD sequencing.** *Biosystems* 2011, **104**:87–93.
29. Boch J, Scholze H, Schornack S, Landgraf A, Hahn S, *et al*: **Breaking the code of DNA binding specificity of TAL-type III effectors.** *Science* 2009, **326**:1509–1512.
30. Gibson DG, Benders GA, Andrews-Pfannkoch C, Denisova EA, Baden-Tillson H, *et al*: **Complete chemical synthesis, assembly, and cloning of a Mycoplasma genitalium genome.** *Science* 2008, **319**:1215–1220.
31. Jaschke PR, Lieberman EK, Rodriguez J, Sierra A, Endy D: **A fully decompressed synthetic bacteriophage oX174 genome assembled and archived in yeast.** *Virology* 2012, **434**:278–284.
32. Rai A, Boyle J: **Synthetic biology: caught between property rights, the public domain, and the commons.** *PLoS Biol* 2007, **5**:e58.
33. Calvert J: **Ownership and sharing in synthetic biology: a 'diverse ecology' of the open and the proprietary.** *BioSocieties* 2012, **7**:169–187.
34. Oye KA, Wellhausen R: **The intellectual commons and property in synthetic biology.** In *Synthetic Biology: The Technoscience and Its Societal Consequences.* Edited by Schmidt M, Kelle A, Ganguli-Mitra A, de Vriend H. Heidelberg, Germany: Springer; 2009:121–140.

35. Rutz B: **Synthetic biology and patents. A European perspective.** *EMBO Rep* 2009, **10**(Suppl 1):S14–S17.
36. Henkel J, Maurer SM: **Parts, property and sharing.** *Nat Biotechnol* 2009, **27**:1095–1098.
37. Bertani G: **Studies on lysogenesis. I. The mode of phage liberation by lysogenic Escherichia coli.** *J Bacteriol* 1951, **62**:293–300.
38. Hollander DH, Nell EE: **Improved preservation of Treponema pallidum and other bacteria by freezing with glycerol.** *Appl Microbiol* 1954, **2**:164–170.
39. Howard DH: **The preservation of bacteria by freezing in glycerol broth.** *J Bacteriol* 1956, **71**:625.
40. Gibson DG, Young L, Chuang RY, Venter JC, Hutchison CA 3rd, *et al*: **Enzymatic assembly of DNA molecules up to several hundred kilobases.** *Nat Methods* 2009, **6**:343–345.
41. *Synthetic Genomics, Inc. Press Release: February 7, 2012. Synthetic Genomics, Inc. Announces Agreement with New England Biolabs to Launch Gibson Assembly™ Master Mix Product for Synthetic and Molecular Biology Applications.* Available: http://www.syntheticgenomics.com/media/press/020712.html. Accessed 11.20.2012.
42. *New England BioLabs, Inc., Gibson Assembly™ Master Mix, Limited Use Label License.* Available: http://www.neb.com/nebecomm/products_intl/productE2611.asp. Accessed 11.20.2012.
43. *ATCC Press Release: June 18, 2012. ATCC Announces Agreement with Life Technologies Corporation to Sell and Distribute Green Fluorescent Protein (GFP)-based Materials.* Available: http://www.prweb.com/releases/2012/6/prweb9612892.htm Accessed 11.29.2012.
44. Mutalik VK: *Expression Operating Unit: Origins and Current Status*, BIOFAB Community Meeting. 2010. June 19, 2010. http://www.biofab.org/sites/default/files/BIOFAB_CommunityMeeting1_Vivek.pdf.
45. Mutalik VK, Guimaraes JC, Cambray G, Lam C, Christoffersen MJ, *et al*: **Precise and reliable gene expression via standard transcription and translation initiation elements.** *Nat Methods* 2013, **10**:354–360.
46. Mutalik VK, Guimaraes JC, Cambray G, Mai QA, Christoffersen MJ, *et al*: **Quantitative estimation of activity and quality for collections of functional genetic elements.** *Nat Methods* 2013, **10**:347–353.
47. Kelly JR, Rubin AJ, Davis JH, Ajo-Franklin CM, Cumbers J, *et al*: **Measuring the activity of BioBrick promoters using an in vivo reference standard.** *J Biol Eng* 2009, **3**:4.
48. Rai A: **Unstandard standarization: the case of biology.** *Communications of the ACM* 2010, **53**:37–39.
49. Torrance AW, Kahl LJ: **Bringing standards to life: synthetic biology standards and intellectual property.** *Santa Clara Computer & High Tech L J.* 2013. in press.
50. Kumar S, Rai A: **Synthetic biology: the intellectual property puzzle.** *Texas Law Review* 2007, **85**:1745–1768.
51. Bekkers R, West J: **The limits to IPR standardization policies as evidenced by strategic patenting in UTMS.** *Telecommunications Policy* 2009, **33**:80–97.
52. Bekkers R, Iverson E, Blind K: **Emerging ways to address the reemerging conflict between patenting and technological standardization.** *Indus Corp Change* 2012, **21**:901–931.
53. U.S. Dep't of Justice, Fed Trade Comm'n: *Antitrust Enforcement and Intellectual Property Rights: Promoting Innovation and Competition.* 2007. http://www.usdoj.gov/atr/public/hearings/ip/222655.pdf.
54. U.S. Dep't of Justice, U.S. Patent and Trademark Office: *Policy Statement on Remedies for Standards-Essential Patents Subject to Voluntary F/Rand Commitments.* 2013. http://www.justice.gov/atr/public/guidelines/290994.pdf.
55. Oldham P, Hall S, Burton G: **Synthetic biology: mapping the scientific landscape.** *PLoS One* 2012, **7**:e34368.
56. Carlson R: **The changing economics of DNA synthesis.** *Nat Biotechnol* 2009, **27**:1091–1094.
57. U.S. Department of Health and Human Services, National Institutes of Health: *NIH Guidelines for Research Involving Recombinant DNA Molecules, revised October 2011.* available at http://oba.od.nih.gov/rdna/nih_guidelines_oba.html.
58. U.S. Department of Health and Human Services, National Institutes of Health: *Screening Framework Guidance for Synthetic Double-Stranded DNA Providers, October 2010.* available at http://www.phe.gov/Preparedness/legal/guidance/syndna/Documents/syndna-guidance.pdf.
59. Smolke CD: **Building outside of the box: iGEM and the BioBricks foundation.** *Nat Biotechnol* 2009, **27**:1099–1102.
60. Carlson JM, Heckerman D, Shani G: *Estimating false discovery rates for contingency tables. Microsoft Research Tech Report MSR-TR-2009-53.* 2009. available at http://research.microsoft.com/apps/pubs/?id=80571.
61. Knight T: *Idempotent Vector Design for Standard Assembly of Biobricks. MIT Synthetic Biology Working Group Technical Report.* 2003. http://dspace.mit.edu/handle/1721.1/21168.
62. Anderson JC, Dueber JE, Leguia M, Wu GC, Goler JA, *et al*: **BglBricks: A flexible standard for biological part assembly.** *J Biol Eng* 2010, **4**:1.
63. Phillips IE, Silver PA: *BBF RFC 23: A New BioBrick Assembly Strategy Designed for Facile Protein Engineering.* 2006. http://dspace.mit.edu/handle/1721.1/32535.
64. Muller KM, Arndt KM, iGEM 2007 Team Freiburg, Grunberg R: *BBF RFC 25: Fusion Protein (Freiburg) BioBrick assembly standard.* 2009. http://dspace.mit.edu/handle/1721.1/45140.
65. Peisajovich SG, Horwitz A, Hoeller O, Rhau B, Lim WA: *BBF RFC 28: A method for combinatorial multi-part assembly based on the Type IIs restriction enzyme AarI.* 2009. http://dspace.mit.edu/handle/1721.1/46721.
66. Quan J, Tian J: **Circular polymerase extension cloning of complex gene libraries and pathways.** *PLoS One* 2009, **4**:e6441.
67. Shao Z, Zhao H, Zhao H: **DNA assembler, an in vivo genetic method for rapid construction of biochemical pathways.** *Nucleic Acids Res* 2009, **37**:e16.
68. Hartley JL, Temple GF, Brasch MA: **DNA cloning using in vitro site-specific recombination.** *Genome Res* 2000, **10**:1788–1795.
69. Sarrion-Perdigones A, Falconi EE, Zandalinas SI, Juarez P, Fernandez-del -Carmen A, *et al*: **GoldenBraid: an iterative cloning system for standardized assembly of reusable genetic modules.** *PLoS One* 2011, **6**:e21622.
70. Engler C, Kandzia R, Marillonnet S: **A one pot, one step, precision cloning method with high throughput capability.** *PLoS One* 2008, **3**:e3647.
71. Weber E, Engler C, Gruetzner R, Werner S, Marillonnet S: **A modular cloning system for standardized assembly of multigene constructs.** *PLoS One* 2011, **6**:e16765.
72. Zhang Y, Werling U, Edelmann W: **SLiCE: a novel bacterial cell extract-based DNA cloning method.** *Nucleic Acids Res* 2012, **40**:e55.
73. Li MZ, Elledge SJ: **Harnessing homologous recombination in vitro to generate recombinant DNA via SLIC.** *Nat Methods* 2007, **4**:251–256.
74. Carr PA, Church GM: **Genome engineering.** *Nat Biotechnol* 2009, **27**:1151–1162.
75. Klock HE, Lesley SA: **The polymerase incomplete primer extension (PIPE) method applied to high-throughput cloning and site-directed mutagenesis.** *Methods Mol Biol* 2009, **498**:91–103.
76. Bitinaite J, Rubino M, Varma KH, Schildkraut I, Vaisvila R, *et al*: **USER friendly DNA engineering and cloning method by uracil excision.** *Nucleic Acids Res* 2007, **35**:1992–2002.
77. Endy D, Deese I, MIT Synthetic Biology Working Group, and illustrated by Wadey C: *Adventures in Synthetic Biology. A comic book.* 2005. available online at http://dspace-test.mit.edu/handle/1721.1/46337.
78. Velten L, Iwamoto N, Hiller C, Uckelmann H, Zhu C, *et al*: *BBF RFC 41: Units for Promoter Measurement in Mammalian Cells.* 2009. http://dspace.mit.edu/handle/1721.1/49501.
79. Galdzicki M, Wilson ML, Rodriguez CA, Pocock MR, Oberortner E, *et al*: *BBF RFC 87: Synthetic Biology Open Language (SBOL) Version 1.1.0.* 2012. http://dspace.mit.edu/handle/1721.1/73909.
80. Johnson J, Galdzicki M, Sauro HM: *BBF RFC 68: Standard for the Electronic Distribution of SBOLv Diagrams.* 2010. http://dspace.mit.edu/handle/1721.1/60086.
81. Quinn JP, Beal J, Bhatia S, Cai P, Chen J, *et al*: *BBF RFC 93: Synthetic Biology Open Language Visual (SBOL Visual), version 1.0.0.* 2013. http://dspace.mit.edu/handle/1721.1/78249.
82. Canton B, Labno A, Endy D: **Refinement and standardization of synthetic biological parts and devices.** *Nat Biotechnol* 2008, **26**:787–793.
83. Lee TS, Krupa RA, Zhang F, Hajimorad M, Holtz WJ, *et al*: **BglBrick vectors and datasheets: a synthetic biology platform for gene expression.** *J Biol Eng* 2011, **5**:12.
84. Altschul SF, Madden TL, Schaffer AA, Zhang J, Zhang Z, *et al*: **Gapped BLAST and PSI-BLAST: a new generation of protein database search programs.** *Nucleic Acids Res* 1997, **25**:3389–3402.
85. Larkin MA, Blackshields G, Brown NP, Chenna R, McGettigan PA, *et al*: **Clustal W and clustal X version 2.0.** *Bioinformatics* 2007, **23**:2947–2948.
86. Kibbe WA: **OligoCalc: an online oligonucleotide properties calculator.** *Nucleic Acids Res* 2007, **35**:W43–W46.
87. Stranneheim H, Lundeberg J: **Stepping stones in DNA sequencing.** *Biotechnol J* 2012, **7**:1063–1073.
88. Bertani G: **Lysogeny at mid-twentieth century: P1, P2, and other experimental systems.** *J Bacteriol* 2004, **186**:595–600.
89. Kocagoz T, Altin S, Turkyilmaz O, Tas I, Karaduman P, *et al*: **Efficiency of the TK culture system in the diagnosis of tuberculosis.** *Diagn Microbiol Infect Dis* 2012, **72**:350–357.

90. Chalfie M, Tu Y, Euskirchen G, Ward WW, Prasher DC: **Green fluorescent protein as a marker for gene expression.** *Science* 1994, **263**:802–805.
91. Wang HH, Isaacs FJ, Carr PA, Sun ZZ, Xu G, *et al*: **Programming cells by multiplex genome engineering and accelerated evolution.** *Nature* 2009, **460**:894–898.
92. Slusarczyk AL, Lin A, Weiss R: **Foundations for the design and implementation of synthetic genetic circuits.** *Nat Rev Genet* 2012, **13**:406–420.

18

Endothelial cells derived from embryonic stem cells respond to cues from topographical surface patterns

Rachel Hatano[1], Kevin Mercurio[2], Jesus Isaac Luna[2], Drew E Glaser[2], Valerie J Leppert[2,3] and Kara E McCloskey[2,3*]

Abstract

The generation of micro- and nano-topography similar to those found in the extra cellular matrix of three-dimensional tissues is one technique used to recapitulate the cell-tissue physiology found in the native tissues. Despite the fact that ample studies have been conducted on the physiological significance of endothelial cells alignment parallel to shear stress, as this is the normal physiologic arrangement for healthy arterial EC, very few studies have examined the use of topographical signals to initiate endothelial cell alignment. Here, we have examined the ability for our mouse embryonic stem cell-derived endothelial cells (ESC-EC) to align on various microchip topographical systems. Briefly, we generated metal molds with 'wrinkled' topography using 1) 15 nm and 2) 30 nm of gold coating on the pre-strained polystryene (PS) sheets. After thermal-induced shrinkage of the PS sheets, polydimethylsiloxane (PDMS) microchips were then generated from the wrinkled molds. Using similar Shrink™-based technology, 3) larger selectively crazed acetone-etched lines in the PS sheets, and 4) fully crazed acetone-treated PS sheets of stochastic topographical morphology were also generated. The 15 nm and 30 nm gold coating generated 'wrinkles' of uniaxial anisotropic channels at nano-scaled widths while the crazing generated micron-sized channels. The ESC-EC were able to respond and align on the 320 nm, 510 nm, and the acetone-etched 10.5 μm channels, but not on the fully 'crazed' topographies. Moreover, the ESC-EC aligned most robustly on the wrinkles, and preferentially to ridge edges on the 10.5 μm-sized channels. The ability to robustly align EC on topographical surfaces enables a variety of controlled physiological studies of EC-EC and EC-ECM contact guidance, as well as having potential applications for the rapid endothelialization of stents and vascular grafts.

Keywords: Endothelial cells, Alignment, Contact guidance, Topography, Embryonic stem cells, Differentiation

Introduction

It is well documented that many cells respond to topographical surface features by changing their proliferation, adhesion, migration and/or cell orientation [1]. This interesting phenomenon is referred to as 'contact guidance'. The physiological significance of controlling cell shape for enhancing cell-tissue function is important in a wide variety of cell types including: neurons [2-4], skeletal muscle [5], cardiac muscle [6-16], and even corneal and lens epithelial cells [17,18]. However, despite the fact that ample studies have been conducted on the physiological significance and atheroprotective benefits of endothelial cell (EC) alignment in response to shear stress (reviewed in [19]), few have incorporated the use of topographical signals to initiate EC alignment and its potential downstream physiological consequences for atheroprotection and EC repopulation after injury.

Vascular EC adhere to their underlying extracellular matrix, the basement membrane (BM), which provides a variety of biophysical and biochemical cues shown to regulate EC behavior [20,21]. The BM has many unique features, including a complex three-dimensional topography that, in turn, can influence endothelial cell function [22]. Studies also suggest that the biomechanical properties, including mechanical strength [23] and compliance

* Correspondence: kmccloskey@ucmerced.edu
[2]Graduate Program in Biological Engineering and Small-scale Technologies, University of California, Merced, CA, USA
[3]School of Engineering, University of California, P.O. Box 2039, Merced, CA 95344, USA
Full list of author information is available at the end of the article

[21], of the endothelial substratum influence endothelial cell behaviors. Quantitative analysis of the topographic features of aorta, carotid, saphenous, and inferior vena cava vessels isolated from the rhesus macaque indicate that vascular BM is a complex meshwork of pores and fibers mostly in the nanoscale range, but do exhibit significant differences in BM thickness and pore diameters between the different vessel tissue locations [24]. The fiber widths (i.e. ridge diameters) of the different BM isolated from a rhesus monkey remained between 25–30 nm [24] with pore sizes between 49–63 nm [25].

Mimicking the ECM topography for facilitating cell alignment has been accomplished using a variety of microfabrication approaches including: microcontact printing, abrasion, photolithography, hot embossing, electrospinning, and laser ablation [6,10,11,13,15,16], and nanofabrication approaches including: e-beam lithography andnanoimprint lithography (reviewed in [26]). Because most of these approaches are very time consuming and expensive, we used an ultra-rapid, tunable, and inexpensive non-photolithographic fabrication method to create cell culture platforms with controllable nano- and micro- scale topographical cues. The alignment grooves are created by leveraging the mismatch in stiffness between a pre-strained polymer sheet and an overlying thin metal film [27]. The wavelength of the wrinkles is tunable based upon the thickness of the metal layer on the surface of the pre-strained sheet prior to thermal exposure. When the plastic sheet retracts upon heating, the stiffer metal film buckles proportional to the thickness of the metal coating [27], causing anisotropic 'wrinkles' for alignment of various cell phenotypes, including cardiac cells [12]. Here, we set out to explore the 'wrinkled' microchip cell culture platform, in addition to two additional shrink-based platforms, for the alignment of our embryonic stem cell-derived endothelial cells (ESC-EC).

Results

Characterization of microchip master-mold platforms

A summary of the wrinkle methods and channel dimensions is provided in Figure 1. Nanoscale wrinkles with different dimensions were produced using two different metal deposition thicknesses: 15 nm and 30 nm gold. The 15-nm coated PS sheets generated wrinkles with average wrinkle widths of 320 nm with 280 nm peak-to-valley depths, and the 30-nm coated PS sheets generated wrinkles with average wrinkle widths of 510 nm with 430 nm peak-to-valley depths (see Figure 2). The selectively-crazed acetone-etched samples generated much larger channels of approximately 19 μm ridges and 10.5 μm valley widths (see Figure 2), but only 10.5 μm in depth. The fully-crazed

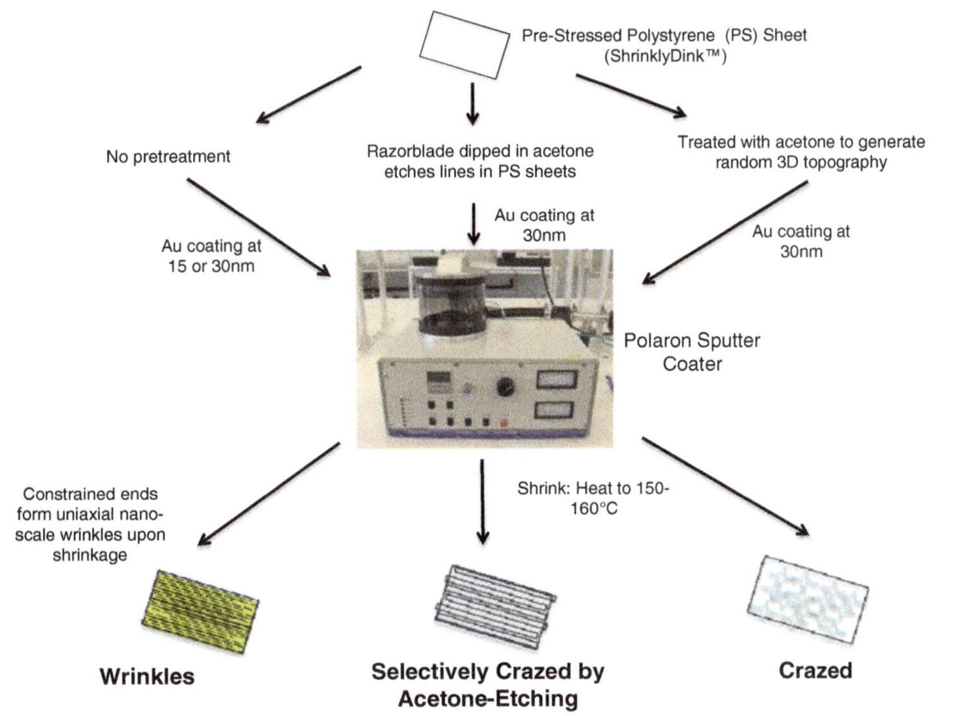

Figure 1 Microchip Fabrication. Flowchart of three fabrication methods for the different topographies. The fabrication methods include: two different sizes of **wrinkles** that generate uniaxial nanoscale channels in the molds, selectively crazed **acetone-etched** method that generates larger micron-sized channels, and a **crazed** 'control' topography that does not generate uniform channels, but does exhibit random three-dimensional hills, valleys and ridges.

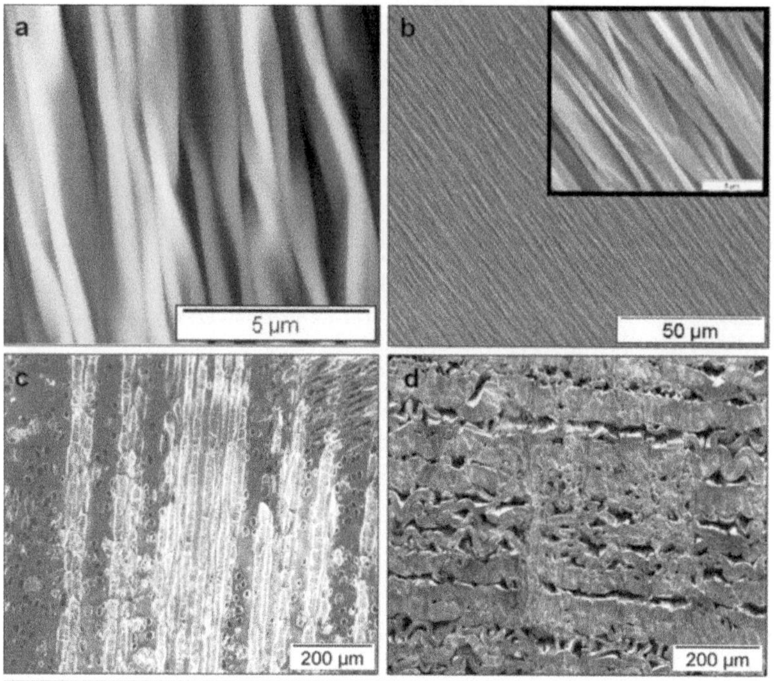

e

Topography of Master-Molds	Pre-Treatment	Sputter-coating Thickness	Shrinking	Average Channel Width	Average Ridge Width	Average Channel Depth
Smaller Nanowrinkles	None	15nm Au	Uniaxial	320nm	570nm	280nm
Larger Nanowrinkles	None	30nm Au	Uniaxial	510nm	890nm	430nm
Selectively Crazed Acetone-Etched	Acetone-etched lines	30nm Au	Biaxial	10.5µm	19µm	10.5µm
Fully Crazed	Acetone-coated	30nm Au	Biaxial	20µm	30µm	65µm

Figure 2 AFM and SEM Images of the surface topography on wrinkled, selectively crazed acetone-etched channels, and crazed microchips. a) AFM image shows the wrinkled topography developed using a 15 nm gold layer having an average wrinkle channel width of about 320 nm. Scale = 5 µm. **b)** SEM image of wrinkled topography produced using a 30 nm gold layer. Scale = 50 µm. The average wrinkle channel width is about 510 nm. Right corner: AFM image of the same, scale = 5 µm. **c)** Acetone-etched topography has an average channel width distance of about 10.5 µm. Etch lines are traveling parallel to each other after biaxial shrinking. Scale = 200 µm. **d)** Crazed topography developed after shrinking with channels about 20 µm in width. Scale = 200 µm. **e)** Table summarizing the dimensions of the four topographical surfaces.

topography contained an average ridge width of about 30 µm and an average valley width of about 20 µm and 65 µm in depth.

Characterization of the ESC-derived EC

The cultured ESC-EC were stained for EC specific markers: Flk-1, Flt-1, Tie-1, and VE-cad, as well as arterial specific: Notch-1, EphrinB2, DLL4, and venous specific: EphB4. The smooth muscle cell (SMC) marker, calponin, was used to verify the absence on that contaminating cell types and all cells were counterstained with DAPI nuclear stain. The expression of EC and SMC markers were then analyzed by laser-scanning confocal microscopy. As seen in Figure 2a-h, the ESC-EC express all endothelial markers (Figure 3a-h), including the arterial-specific [28]: Notch-1, EphrinB2, DLL4 (Figure 3e-g) and the venous-specific [28]: EphB4 (Figure 3h), but do not express the SMC marker (Figure 3i). The percent of cells expressing the arterial-specific: EphrinB2, Notch-1, DLL4, and the venous-specific marker: COUP-TFII [29] were than quantified. The histograms of these markers (Figure 4a-d) depict the unstained cells (red), IgG control (if available;

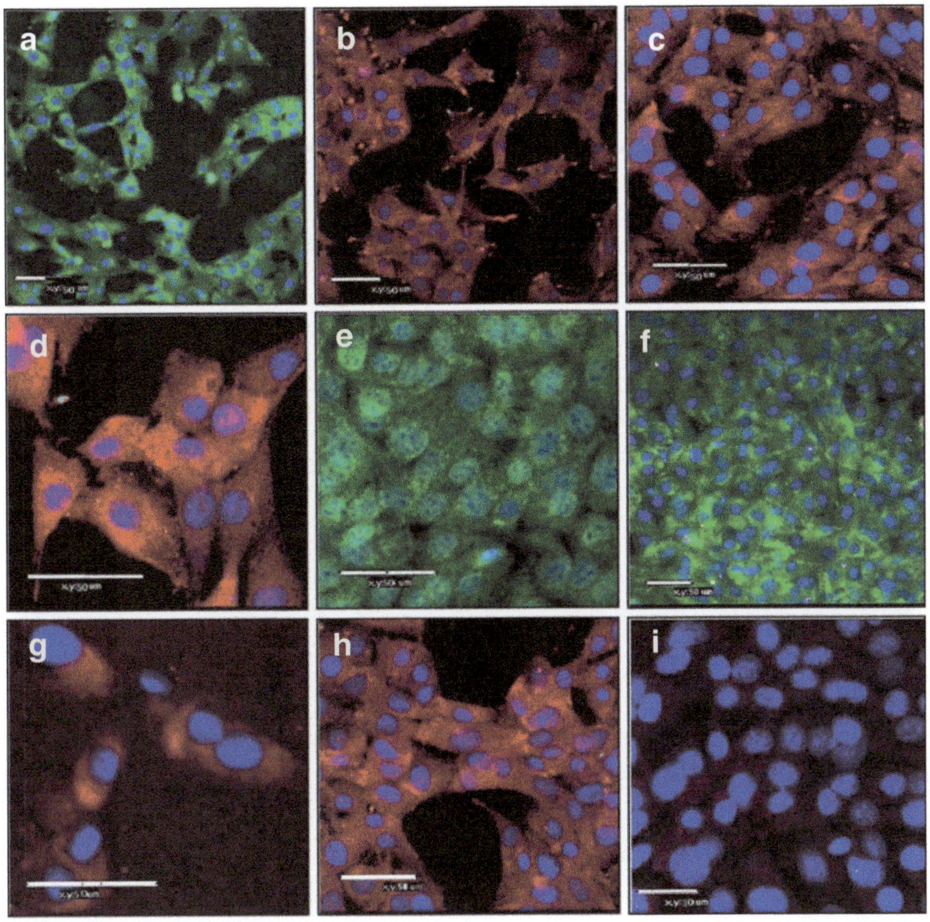

Figure 3 a-i) Immunofluorescent images of ESC-EC. These ESC-EC express endothelial markers: **(a)** Flk-1 (green), **(b)** Flt-1 (red), **(c)** Tie-1 (red), **(d)** VE-cad (red), arterial specific: **(e)** Notch-1 (green), **(f)** EphrinB2 (green), **(g)** DLL4 (red), and venous specific: **(h)** EphB4 (red), but not smooth muscle marker **(i)** calponin. All cells counterstained with DAPI (blue). Scale = 50 μm.

blue) and stained cells (orange). These data were than averaged (N = 3) and presented in a bar graph (Figure 4e). Based on this quantification data, it seems as if over 60% of these ESC-derived EC express arterial-specific markers, but that almost 90% also exhibit venous identity.

Alignment of ESC-EC

One day after plating ESC-EC on the different topographical surfaces (Figure 5) and staining cells for F-actin (green) and DAPI (blue), the alignment of the cells cultured on the nano-scale wrinkles and larger selectively-crazed channels is apparent. However, EC alignment and elongation are not evident on the crazed three-dimensional surface. Note that the cells also preferentially adhere to the ridges, compared with the valleys, on the selectively-crazed acetone-etched and crazed micro-scale topographies. At day 5 (Figure 6), after the ESC-EC have grown to confluence, the ESC-EC remain aligned on the wrinkle and selectively-crazed uniaxial topographies compared with the cobblestone-shaped EC grown on a flat surface (control) and the randomly crazed three-dimensional surface. The initial adhesion of the EC on the flat, nanowrinkled, and acetone-etched surfaces was significantly greater than initial adhesion of the EC on the crazed surface (Figure 7a). However, the proliferation capability of the EC was limited on the acetone-etched surface as well as the crazed surface (Figure 7b). No difference in the proliferation rates was observed (not shown).

The percent of ESC-EC aligning on the different topographies was then quantified. Based on the criterion that the cells were "aligned" if their long-axis was oriented within ±30° with the channel direction, 100% of the cells cultured on the 2 sizes of nano-scale wrinkles were aligned by day 3, and close to 80% of the cells on selectively-crazed channels (Figure 8a). These were all statistically significant compared with the control and crazed topographies. The percent of cells aligned on the nano-scale wrinkles was also statistically greater than the acetone-

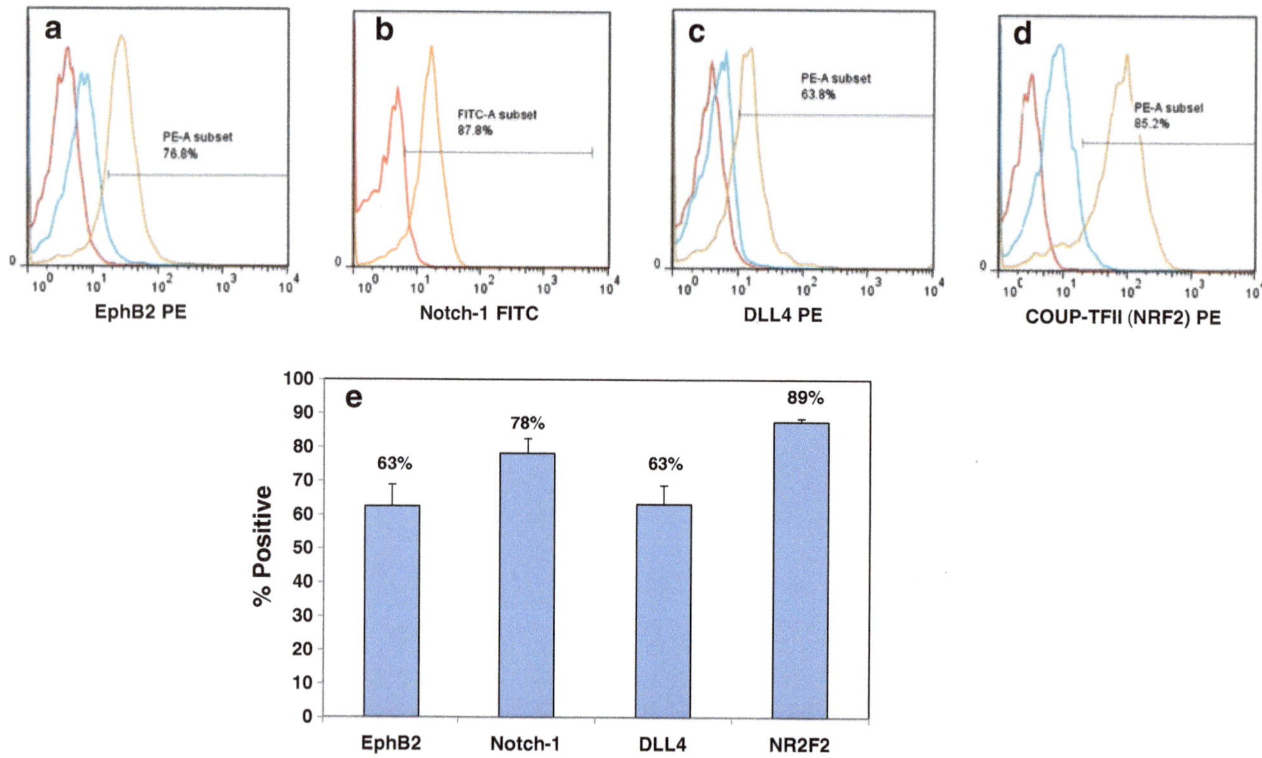

Figure 4 FACS analysis of ESC-EC indicate that these cells express both arterial and venous endothelial markers. The percent of arterial cells indicated by **(a-c)** EprinB2, intracellular Notch-1, and DLL4 staining is slightly lower than the percent of venous-specific EC **(d)** NRF2. **(e)** However, because the cultures contain over 50% of both arterial and venous phenotypes, it seems that these stem cell-derived EC cultures are retaining some plasticity regarding their arterial-venous (a-v) specification.

etched micro-channels, suggesting that the multiple nano-scale cues provide stronger (i.e. more) signals for the cells to align. The alignment on the small nano-scale wrinkles also increases from day 1 to day 3, but then is reduced when the cells become confluent by day 5 (Figure 8b), whereas the percent of aligned cells on the acetone-etched surfaces continue to increase from day 1, to day 3, to day 5.

We also calculated the elongation factors for these cells (N = 50; Figure 8c). On day 1, EC cultured on the smaller, 320 nm, nano-scale wrinkles exhibited the greatest elongation, and this was statistically greater compared with the control EC. Moreover, the EC cultured on the flat control surface and both wrinkled surfaces exhibited the greatest elongation on day 1 with decreased elongation by day 3 and 5. Conversely, the elongation factors for cells on the selectively-crazed channels remained similar over time, and were not significantly elongated compared with control EC. It is important to note that EC cultured on flat surfaces (control) also elongate more at subconfluence, but appear more cobblestone-shaped at confluence (Additional file 1: Figure S1). Based on the comparison with control cells, we know that there is a reduction in elongation from subconfluent to confluent cultures, and that the wrinkles aid the elongation of confluent EC.

Discussion

The stained cells depicted in Figure 2a-h, combined with previously published characterization studies [30-33], illustrate that the ESC-derived EC are composed of largely pure populations of EC-specific cells. However, in our experience, the arterial or venous (a-v) identity of the ESC-EC can vary somewhat from isolation-to-isolation, with most cells predominantly expressing venous EC markers. Therefore, we first set out to characterize the a-v identity of the ESC-EC used for these alignment studies. Ephrin-B2 and EphB4 mark arterial and venous EC, respectively [34] while Notch1 and DLL4 are also considered markers of arterial EC [28] and have been shown to play a role in arterial specification from stem cells [35]. The data indicate that over 50% of the ESC-EC must be expressing both phenotypes simultaneously, as greater than 60% express arterial EC markers, while 90% express the venous EC marker. Because reciprocal signaling between Ephrin-B2 positive arteries and EphB4 positive veins is crucial for morphogenesis of the capillary beds, this data may indicate that our ESC-EC more closely resemble the capillary EC subphenotype compared with an arterial or venous phenotype.

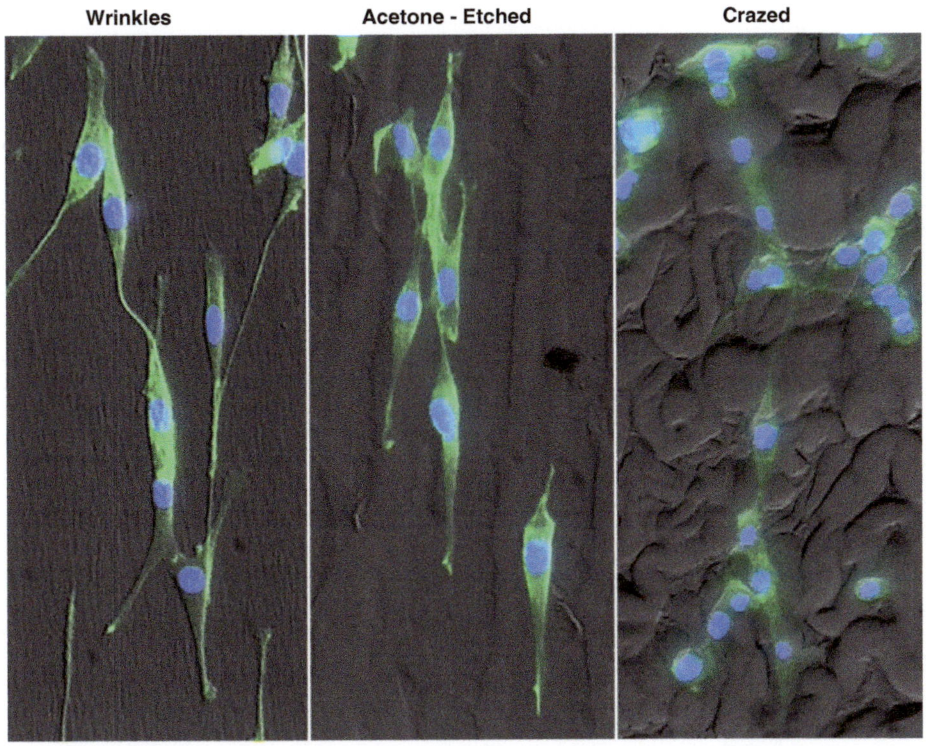

Figure 5 Images of the 3 distinct topographies with cells cultured at low density. The EC do align well with the uniaxial nanoscale wrinkles and larger acetone-etched microchannels, especially at low density. The cells also appear to adhere preferentially to the ridges compared with the valleys in the acetone-etched and crazed topographies. All cells are stained with F-actin (green) and counterstained with DAPI (blue). Magnification = 20×.

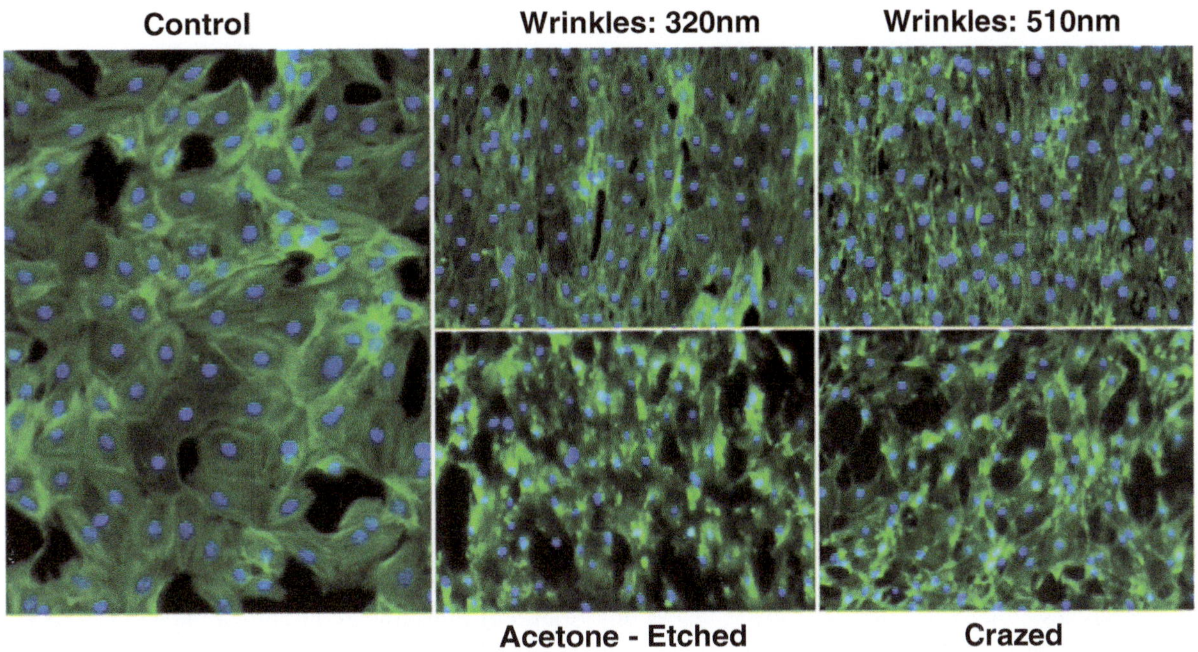

Figure 6 Images of EC cultured on the 5 topographical surfaces at high density. After 5 days of culture, the EC on the different surfaces have become confluent. Compared with the flat and crazed control surfaces, the EC align well on the wrinkles and actetone-etched uniaxial channels. All cells are stained with F-actin (green) and counterstained with DAPI (blue). Magnification = 20× (control) and 10×.

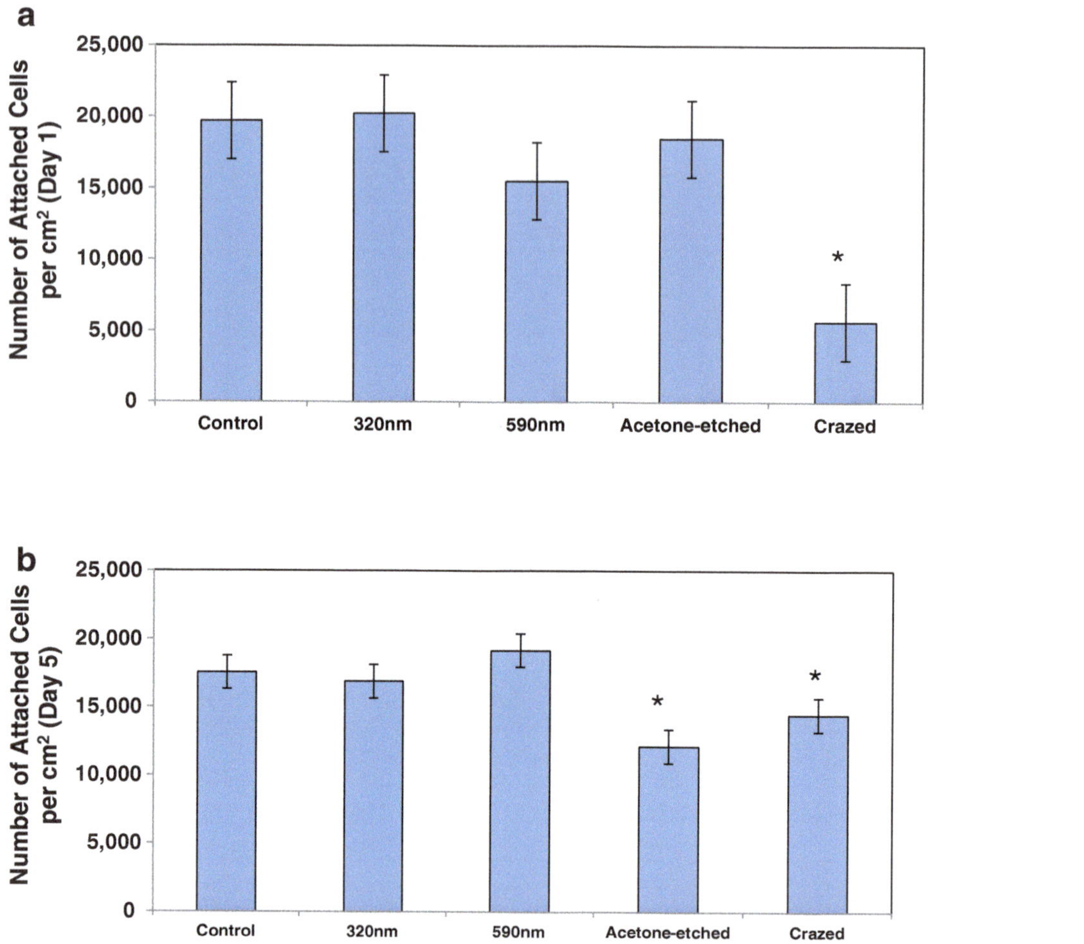

Figure 7 EC adhesion on topographical surfaces. The number of adherent EC per cm^2 was quantified **a)** at day 1, initial adhesion, and **b)** day 5 when the cells approached confluence for that surface. The fully crazed surface significantly interfered with both the initial adhesion and the ability to reach confluence ($P < 0.05$) whereas, the partially crazed etched surface limited only final confluence ($P < 0.05$).

After plating the ESC-EC on gelatin-coated PDMS microchips with the nano- and micro-scale topographies, the ESC-EC align with both 320 nm and 510 nm 'wrinkles' as well as on the ridges (Figures 5 and 6) of the larger 10.5 μm channels formed in the selectively-crazed microchips, with the greatest cell alignment seen on the nano-sized wrinkles compared with the larger channels (Figure 8a). This data agrees closely with one other study that compared four human vascular endothelial cell-types from large and small diameter vessels [20]. This study found that orientation and alignment of human umbilical vein endothelial cells (HUVEC) was the most pronounced on 800 nm anisotropically ordered ridges, but the EC isolated from larger vessels preferred the larger (1200 nm-4000 nm) topographies [20].

Although significant differences in the ESC-EC adhesion and proliferation cultured on the nano-scale wrinkles compared with flat surfaces were not observed, the larger crazed surfaces did limit both adhesion (fully crazed only) and the ability for the cells to proliferate (partially crazed and fully crazed surfaces). This data agrees with a study that found no difference in cell adhesion of adipose-derived stem cells compared with flat substrates [36]. Another study that examined pyramid-shaped features, ranging from 50 nm-2 μm in height, also found that the fewer cells adhered to the rough surfaces and that micro-scaled pyramids also hindered migration [37], like the reduced adhesion we observed on our crazed surface.

It was also seen that as the ESC-EC are cultured on the nano-scale wrinkles, the percent of aligned cells and elongation factor is slightly reduced at day 3 and 5 as the cells also become more confluent (Figure 8b). This is consistent with the elongated (although not directionally oriented) morphology of subconfluent EC and the cobblestone-shaped morophlogy of EC cultured on flat surfaces (Additional file 1: Figure S1). We expect that the cell-to-cell adhesions via cadherins are beginning to compete with cell-substrate integrin attachment sites as

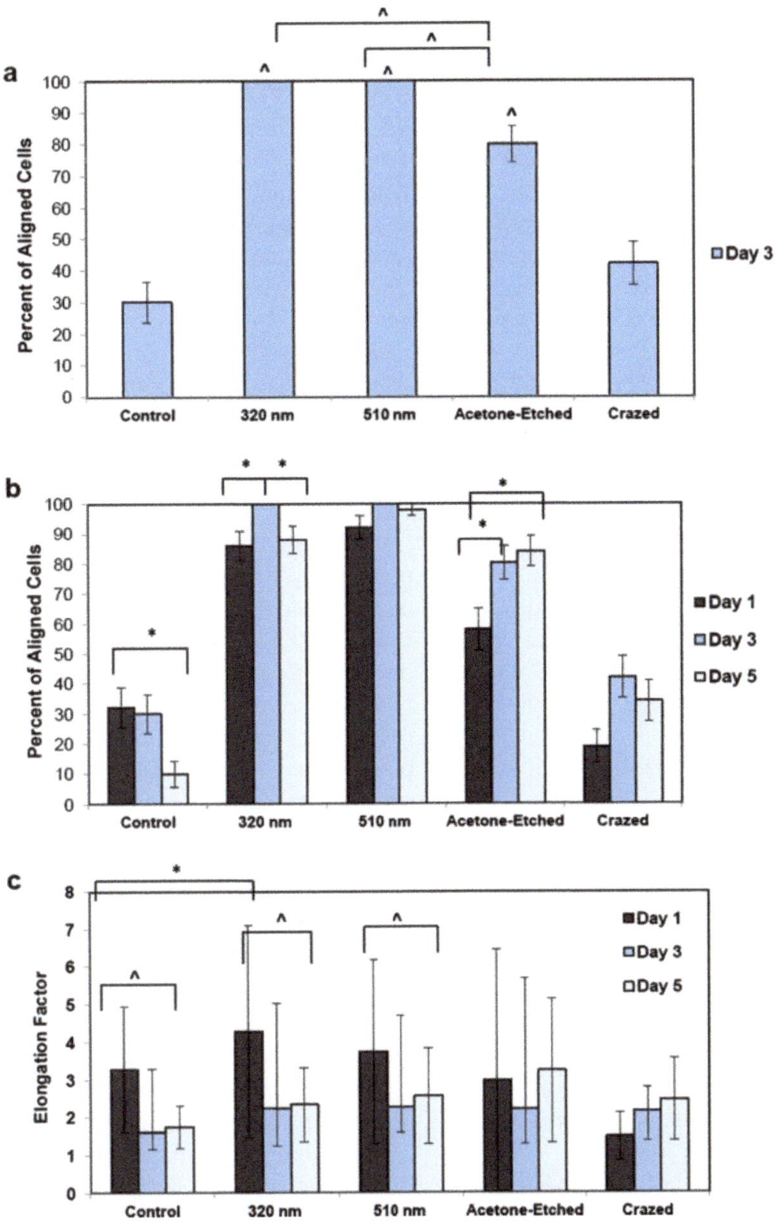

Figure 8 Quantification of the aligned cells. The aligned EC were quantified. **a)** At day 3 of culture, the cells were not yet confluent, but still had ample time to respond to topographical cues. At day 3, 100% of the cells on the two wrinkled surfaces aligned in the wrinkle direction and approximately 80% aligned on the acetone-etched surface, a much higher percentage than the control EC cultured on a flat PDMS microchip, in which only 30% aligned in the wrinkle direction ^ P-value < 0.0001. The alignment on the wrinkles is also statistically greater than the alignment seen on the larger micro-scale etched wrinkles ^, P-value < 0.0001. **b)** The EC alignment was then analyzed over time up to a confluent EC culture on day 5. For the small 390 nm wrinkles, a statistically significant peak in alignment is seen at day 3 compared with day 1 and day 5. For the large micro-channels in the acetone-etched surfaces, the percent of EC aligning on the channels increase over time, whereas the percent of aligned cells decrease for the flat control surface, * P-value < 0.01. **c)** The average elongation factors were also calculated for all topographical surfaces and time points. The greatest EC elongation was seen in the 390 nm wrinkled surface at day 1, and was a statistically significant increase compared with the flat control surface, * P-value < 0.05. Moreover, as the cells became more confluent, they began to lose some of their elongation, ^ P-value < 0.0001.

the cell cultures become confluent. Interestingly, this phenomenon of reduced alignment at confluence is not observed in EC cultured on the larger micro-scale channels. The width of an EC that has elongated along a ridge-edge, D_{min}, is smaller than the 10.5 μm channel width, and therefore, the cell-to-cell interactions across the large channel widths would be more difficult to establish. Moreover, because the ESC-EC, as well as other

EC [38], preferentially adhere and align on the ridges of the topographical features, the percent of aligned ESC-EC on the larger micron-sized channels actually continued to increase (Figure 8b) with time, but probably are less likely to bridge the large channels gaps for generating confluence. Other studies have also confirmed that micron-scale topographies actually interfere with cell migration and have proposed a 2-step model of cell adhesion [37] suggesting that the surface topography guides initial cell contact, but that the engagement of the cell surface receptors is controlled by the surface chemistry (i.e. ligand density).

In addition to using this microchip topographical platform to study the physiologic response to surface-induced cell alignment, the technology also has potential applications in regenerative medicine. Due to the highly thrombotic surface of the disrupted endothelium following coronary angioplasty (stent insertion), approaches for enhancing endothelial cell adhesion via cell elongation are being explored including surface modification for generating micro- and nano- structured surfaces [39]. Moreover, because shear stress signaling on confluent and aligned endothelium is known to impart athero-protective benefits, it is likely that the EC alignment could be advantageous in both cell adhesion and anti-thrombosis in the absence of other mechanical forces.

Conclusions

This work describes the use of distinct 3D topographical cell culture platforms for aligning ESC-derived EC. Consistent with the small anisotropic nano-scale fiber sizes in the BM of vessels, the ESC-EC align most robustly on the anisotropic nano-scale wrinkles. On the larger micron-sized channels, the ESC-EC preferred to align with the ridge edges of the channels and do not seem to form the lateral cell-cell borders that lead to reduced elongation of the EC at confluence. Characterization of the ESC-EC showed that they express a number of EC markers, including both arterial- and venous-specific markers. The dual positive a-v identity may be an indication that these cells are consistent with a microvascular EC subphenotype. The ESC-EC aligning most robustly on the smaller nano-scale topology is also consistent with a micro-vessel subphenotype.

Methods

Generation of microchips with nano- and micro-scale topography

Metal wrinkles were fabricated as previously reported and described in detail [27]. Briefly (Figure 1a), gold was deposited by sputter coating (SEM Sputter Coater; Polaron) at various thicknesses. After deposition, the pre-strained polystyrene (PS) sheets (ShinkyDinks, Inc.) were induced to thermally shrink by heating to a temperature of 150–160°C. When the pre-strained polystyrene was exposed to temperatures above glass transition (T_g for polystyrene ~ 105 Celsius), the polystyrene shrinks and contracts. When constrained at opposite ends, the buckling of the surface layer produces the wrinkled topography. The metal coating treatments allow some control over wrinkle dimensions: 15 nm to 90 nm coating correlates with average wrinkle widths ranging from 800 nm to 1 μm, increasing proportionally with coating thickness [27]. These 'wrinkled' metal surfaces serve as master molds for generating PDMS microchip cell culture platforms. A 10:1 ratio of PDMS and curing agent (Sylgard 184 Silicon Elastomer Kit, Dow Corning) was poured on the metal mold and set to cure at 75°C. The chips were sterilized following standard procedures and then coated with 0.5% gelatin and seeded with cells. For this study, we generated four different surface topographies to attain different wrinkle wavelength dimensions. The first two use the standard procedure described above, depositing 15-nm and 30-nm of gold onto the pre-strained PS sheets. The third topography, called selectively crazed acetone-etched, uses acetone with a razorblade to selectively etch 'lines' onto the pre-strained polystyrene sheets. The width of the etch lines is controlled based on the thickness of the razor blade edge and the depth of the channel is determined by exposure to the solvent prior to thermal shrinking. The final topography, called 'fully crazed' is generated after the entire surface of the PS sheet is treated with acetone, then coated with gold metal and thermally shrunk by heating to 150–160°C without constraining edges of the PS sheet. The crazing of polymer surfaces to produce this type of topography has been well-explained in the literature [40].

Characterization of microchip master-mold platforms

The depth and width features of the nano-scale wrinkles were characterized using the Atomic Force Microscope (AFM) while the dimensions in the acetone-etched and crazed topographies were characterized using Scanning Electron Microscopy (SEM). Cross sections of the platforms were used to verify the height features in the crazed topographies. The AFM was operated in tapping mode and the total scan size was 10 μm × 10 μm. Once images were captured, analysis was conducted using AnalySIS Pro software to determine the range of dimensions.

ESC cell culture

Mouse D3-ESC (American Type Culture Collection, Manassas, VA) were initially maintained on irradiated or Mitomycin C-treated (Sigma) mouse embryonic fibroblast feeder layers in Knockout Dulbecco's Modified Eagle Medium (KO-DMEM; Gibco) containing 15% ES Cell Qualified Fetal Bovine Serum (Gibco), 5% Knockout Serum Replacement (Gibco), 1,000 units per ml of leukemia inhibitory factor (ESGRO; Chemicon International) and 5 ×

10^{-5} M β-mercaptoethanol. Cells were then cultured on 0.1% gelatin (no feeders) for one week before switching to differentiation conditions.

EC derivation from ESC

The EC used in these studies were derived from mouse ESC using previously published protocols [30-33,41]. Briefly, initial induction of EC required 4 days of culture on collagen type IV-coated dishes in media containing FBS and without leukemia inhibitory factor (LIF). Differentiation medium consisted of 93% alpha-Minimal Essential Medium, 5% Fetal Bovine Serum, 1% Penicillin/Streptomycin, 1% L-Glutamine, and 5×10^{-5} M β-mercaptoethanol. The cells expressing Flk-1 were then sorted using magnetic cell sorting (MACS®, Miltenyi Biotech) and allowed to grow for one week on collagen type-IV coated dishes. After one week, the $Flk1^{+}$ positive cells exhibited 2 phenotypes; elongated smooth muscle morphology or cobblestone-like endothelial morphology. The cells exhibiting endothelial morphology were manually selected and fed endothelial cell medium (EGM-2 medium supplemented with EGM-2 Bullet Kit; Clonetics - 10 ml FBS, 0.2 ml hydrocortisone, 2 ml hFGF-β, 0.5 ml VEGF, 0.5 ml R3-IGF-1, 0.5 ml ascorbic acid, 0.5 ml hEGF, 0.5 ml GA-1000, 0.5 ml heparin – plus 5×10^{-5} M β-mercaptoethanol, and an extra 50 ng/ml of recombinant human VEGF; $VEGF_{165}$, R&D Systems). Methods consistently yield 25 population doublings at >95% purity [31].

ESC-EC characterization: confocal imaging

Cells were fixed with 4% paraformaldehyde (PFA) and permeabilized with 0.7% Triton 100× in buffer containing 0.5% bovine serum albumin and 5% donkey serum. Primary rabbit antibodies against EphrinB2 (Santa Cruz Biotech) and calponin (Santa Cruz Biotech), as well as, goat antibodies against Flt-1 (BD Pharmingen), Tie-1 (SC Biotech), EphB4 (Santa Cruz Biotech), and VE-cad (Santa Cruz Biotech), were added at concentrations of 1 μg total per sample and refrigerated overnight. The next day, the cells were rinsed before adding FITC anti-rabbit (Fitzgerald) and PE anti-goat (Santa Cruz Biotech) antibodies which were incubated for an additional 2 hours before imaging. Directly-conjugated antibodies for Flk-1 FITC (BD Pharmingen), Notch-1 FITC (Abcam) as well as PE-conjugated antibodies against Delta-like Ligand 4 (DLL4; Biolegend) were also incubated overnight. Cells were imaged using a laser-scanning confocal microscope (Technical Instruments).

ESC-EC characterization: FACS analysis

Cells were removed with Cell Dissociation Buffer (Invitrogen), fixed with 4% paraformaldehyde (PFA) and permeabilized with 0.7% Triton 100× in buffer with 0.5% bovine serum albumin. Rabbit antibodies against EphrinB2 (Santa Cruz Biotech) and Coup-TFII (NR2F2; Abcam), Notch-1 (Abcam), their IgG controls, and PE conjugated DLL4 (Biolegend) were added at concentrations of 1 μg total per sample and refrigerated overnight. The next day, the cells were rinsed with PBS before adding FITC anti-rabbit (Fitzgerald) or PE anti-goat (Santa Cruz Biotech) antibodies which were incubated for an additional 2 hours before FACS analysis.

ESC-EC on nano- and micro-scale topographies

Each of the PDMS microchips was coated with 0.5% gelatin and then plated with 20,000 ESC-EC per cm^2. At day 1, 3 and 5, the ESC-EC were fixed with 4% paraformaldahyde, stained for F-actin (Atto 488 Phalloidin, Sigma) and counterstained with DAPI. Samples were mounted on a cover glass using mounting medium (Vector Laboratories) and imaged with an inverted fluorescent microscope (Nikon Eclipse TE2000-U) and digital camera (Photometrics Coolsnap). ESC-EC were defined as "aligned" if their major axes are within ± 30° with respect to the wrinkle or channel direction (the "aligned" direction was randomly chosen for control and crazed surfaces with no distinct direction). Based on this criterion, the percentage of cells on the channeled surface was quantified. All comparisons for statistical significance were conducted using a student's *T*-test using the mean percentage and standard error of the mean (SEM) for N = 50 cells. The elongation factor (EF) was also calculated by the ratio of the maximal diameter (D_{max}), length, to the minimal diameter (D_{min}), or length-to-width ratio of the cell ($EF = D_{max}/D_{min}$) [42]. All comparisons for statistical significance of EF were conducted using a Student's *t*-test using the mean EF and standard deviation (SD) for N = 50 cells.

Additional file

Additional file 1: Figure S1. Images of ESC-EC on flat surfaces. The ESC-EC exhibit increased elongation in subconfluent cultures (left) compared with confluent cultures (right).

Competing interests

The authors declare that they have no competing interests.

Author information

Kara E McCloskey: https://eng.ucmerced.edu/people/kmccloskey/KaraMcCloskey

Authors' contributions

The first author, RH, along with DG, characterized the ESC-derived EC used in this the study and initiated all cell culture experiments and data analysis of images. The second author, KM generated and characterized the cell culture surface materials. JL collected the fluorescent images. The principle investigators, VL and KMcC, led the study: reviewed all experimental procedures, data collection methods, and data analyses. All authors read and approved the final manuscript.

Acknowledgements
This work was funded, in part, through an NSF-Science and Technology Center (STC) for the Emergent Behavior of Integrated Biological Systems (EBICS) Award # 0939511, NSF-Integrative Graduate Education and Research Traineeship (IGERT) Award # 0965918, and a New Faculty Award from the California Institute for Regenerative Medicine (CIRM) Award # RN2-00921-1. The authors would like to thank Mike Dunlap and the Imaging and Microscopy Facility at UC Merced for assistance and Yang Liu for help with AFM imaging.

Author details
[1]School of Natural Sciences, University of California, Merced, CA, USA. [2]Graduate Program in Biological Engineering and Small-scale Technologies, University of California, Merced, CA, USA. [3]School of Engineering, University of California, P.O. Box 2039, Merced, CA 95344, USA.

References

1. Dalby MJ, Yarwood SJ, Riehle MO, Johnstone HJ, Affrossman S, Curtis AS: **Increasing fibroblast response to materials using nanotopography: morphological and genetic measurements of cell response to 13-nm-high polymer demixed islands.** *Exp Cell Res* 2002, **276**(1):1–9.
2. Recknor JB, Recknor JC, Sakaguchi DS, Mallapragada SK: **Oriented astroglial cell growth on micropatterned polystyrene substrates.** *Biomaterials* 2004, **25**(14):2753–2767.
3. Sorensen A, Alekseeva T, Katechia K, Robertson M, Riehle MO, Barnett SC: **Long-term neurite orientation on astrocyte monolayers aligned by microtopography.** *Biomaterials* 2007, **28**(36):5498–5508.
4. Webb A, Clark P, Skepper J, Compston A, Wood A: **Guidance of oligodendrocytes and their progenitors by substratum topography.** *J Cell Sci* 1995, **108**(Pt 8):2747–2760.
5. Shimizu K, Fujita H, Nagamori E: **Alignment of skeletal muscle myoblasts and myotubes using linear micropatterned surfaces ground with abrasives.** *Biotechnol Bioeng* 2009, **103**(3):631–638.
6. Au HT, Cheng I, Chowdhury MF, Radisic M: **Interactive effects of surface topography and pulsatile electrical field stimulation on orientation and elongation of fibroblasts and cardiomyocytes.** *Biomaterials* 2007, **28** (29):4277–4293.
7. Badie N, Bursac N: **Novel micropatterned cardiac cell cultures with realistic ventricular microstructure.** *Biophys J* 2009, **96**(9):3873–3885.
8. Biehl JK, Yamanaka S, Desai TA, Boheler KR, Russell B: **Proliferation of mouse embryonic stem cell progeny and the spontaneous contractile activity of cardiomyocytes are affected by microtopography.** *Dev Dyn* 2009, **238**(8):1964–1973.
9. Deutsch J, Motlagh D, Russell B, Desai TA: **Fabrication of microtextured membranes for cardiac myocyte attachment and orientation.** *J Biomed Mater Res* 2000, **53**(3):267–275.
10. Geisse NA, Sheehy SP, Parker KK: **Control of myocyte remodeling *in vitro* with engineered substrates.** *In Vitro Cell Dev Biol Anim* 2009, **45**(7):343–350.
11. Heidi Au HT, Cui B, Chu ZE, Veres T, Radisic M: **Cell culture chips for simultaneous application of topographical and electrical cues enhance phenotype of cardiomyocytes.** *Lab on a chip* 2009, **9**(4):564–575.
12. Luna JI, Ciriza J, Garcia-Ojeda ME, Kong M, Herren A, Lieu DK, Li R, Fowlkes CC, Khine M, McCloskey KE: **Multi-scale biomimetic topography for the alignment of neonatal and embryonic stem cell-derived heart cells.** *Tissue Eng Part C Methods* 2011, **17**(5):579–588.
13. Motlagh D, Hartman TJ, Desai TA, Russell B: **Microfabricated grooves recapitulate neonatal myocyte connexin43 and N-cadherin expression and localization.** *J Biomed Mater Res* 2003, **67**(1):148–157.
14. Thomas SP, Bircher-Lehmann L, Thomas SA, Zhuang J, Saffitz JE, Kleber AG: **Synthetic strands of neonatal mouse cardiac myocytes: structural and electrophysiological properties.** *Circ Res* 2000, **87**(6):467–473.
15. Yeong WY, Yu H, Lim KP, Ng KL, Boey YC, Subbu VS, Tan LP: **Multiscale topological guidance for cell alignment via direct laser writing on biodegradable polymer.** *Tissue Eng Part C Methods* 2010, **16**(5):1011–1021.
16. Zong X, Bien H, Chung CY, Yin L, Fang D, Hsiao BS, Chu B, Entcheva E: **Electrospun fine-textured scaffolds for heart tissue constructs.** *Biomaterials* 2005, **26**(26):5330–5338.
17. Rajnicek AM, Foubister LE, McCaig CD: **Alignment of corneal and lens epithelial cells by co-operative effects of substratum topography and DC electric fields.** *Biomaterials* 2008, **29**(13):2082–2095.
18. Teixeira AI, McKie GA, Foley JD, Bertics PJ, Nealey PF, Murphy CJ: **The effect of environmental factors on the response of human corneal epithelial cells to nanoscale substrate topography.** *Biomaterials* 2006, **27**(21):3945–3954.
19. Stolberg S, McCloskey KE: **Can shear stress direct stem cell fate?** *Biotechnol Prog* 2009, **25**(1):10–19.
20. Liliensiek SJ, Wood JA, Yong J, Auerbach R, Nealey PF, Murphy CJ: **Modulation of human vascular endothelial cell behaviors by nanotopographic cues.** *Biomaterials* 2010, **31**(20):5418–5426.
21. Wood JA, Shah NM, McKee CT, Hughbanks ML, Liliensiek SJ, Russell P, Murphy CJ: **The role of substratum compliance of hydrogels on vascular endothelial cell behavior.** *Biomaterials* 2011, **32**(22):5056–5064.
22. Gasiorowski JZ, Liliensiek SJ, Russell P, Stephan DA, Nealey PF, Murphy CJ: **Alterations in gene expression of human vascular endothelial cells associated with nanotopographic cues.** *Biomaterials* 2010, **31**(34):8882–8888.
23. Candiello J, Balasubramani M, Schreiber EM, Cole GJ, Mayer U, Halfter W, Lin H: **Biomechanical properties of native basement membranes.** *FEBS J* 2007, **274**(11):2897–2908.
24. Liliensiek SJ, Nealey P, Murphy CJ: **Characterization of endothelial basement membrane nanotopography in rhesus macaque as a guide for vessel tissue engineering.** *Tissue eng Part A* 2009, **15**(9):2643–2651.
25. McKee CT, Wood JA, Ly I, Russell P, Murphy CJ: **The influence of a biologically relevant substratum topography on human aortic and umbilical vein endothelial cells.** *Biophys J* 2012, **102**(5):1224–1233.
26. Wood MA: **Colloidal lithography and current fabrication techniques producing in-plane nanotopography for biological applications.** *J R Soc Interface* 2007, **4**(12):1–17.
27. Fu C, Grimes A, Long M, Ferri C, Rich BD, Ghosh S, Ghosh S, Lee LP, Gopinathan A, Khine M: **Tunable nanowrinkles on shape memory polymer sheets.** *Adv Mater* 2009, **21**:4472–4476.
28. Yamashita JK: **Differentiation of arterial, venous, and lymphatic endothelial cells from vascular progenitors.** *Trends Cardiovasc Med* 2007, **17**(2):59–63.
29. You LR, Lin FJ, Lee CT, DeMayo FJ, Tsai MJ, Tsai SY: **Suppression of Notch signalling by the COUP-TFII transcription factor regulates vein identity.** *Nature* 2005, **435**(7038):98–104.
30. Glaser DE, Gower RM, Lauer NE, Tam K, Blancas AA, Shih AJ, Simon SI, McCloskey KE: **Functional characterization of embryonic stem cell-derived endothelial cells.** *J Vasc Res* 2011, **48**(5):415–428.
31. McCloskey KE, Lyons I, Rao RR, Stice SL, Nerem RM: **Purified and proliferating endothelial cells derived and expanded *in vitro* from embryonic stem cells.** *Endothelium* 2003, **10**(6):329–336.
32. McCloskey KE, Smith DA, Jo H, Nerem RM: **Embryonic stem cell-derived endothelial cells may lack complete functional maturation *in vitro*.** *J Vasc Res* 2006, **43**(5):411–421.
33. McCloskey KE, Stice SL, Nerem RM: ***In vitro* derivation and expansion of endothelial cells from embryonic stem cells.** *Methods Mol Biol* 2006, **330**:287–301.
34. Wang HU, Chen ZF, Anderson DJ: **Molecular distinction and angiogenic interaction between embryonic arteries and veins revealed by ephrin-B2 and its receptor Eph-B4.** *Cell* 1998, **93**(5):741–753.
35. Yurugi-Kobayashi T, Itoh H, Schroeder T, Nakano A, Narazaki G, Kita F, Yanagi K, Hiraoka-Kanie M, Inoue E, Ara T, *et al*: **Adrenomedullin/cyclic AMP pathway induces notch activation and differentiation of arterial endothelial cells from vascular progenitors.** *Arterioscler Thromb Vasc Biol* 2006, **26**(9):1977–1984.
36. Shi Z, Neoh KG, Kang ET, Poh CK, Wang W: **Enhanced endothelial differentiation of adipose-derived stem cells by substrate nanotopography.** *J Tissue Eng Regen Med* 2012. PMID: 22628362.
37. Le Saux G, Magenau A, Bocking T, Gaus K, Gooding JJ: **The relative importance of topography and RGD ligand density for endothelial cell adhesion.** *PLoS One* 2011, **6**(7):e21869.

38. Uttayarat P, Toworfe GK, Dietrich F, Lelkes PI, Composto RJ: **Topographic guidance of endothelial cells on silicone surfaces with micro- to nanogrooves: orientation of actin filaments and focal adhesions.** *J Biomed Mater Res* 2005, **75**(3):668–680.
39. Mani G, Feldman MD, Patel D, Agrawal CM: **Coronary stents: a materials perspective.** *Biomaterials* 2007, **28**(9):1689–1710.
40. Kausch HH, Gensler R, Grein C, Plummer CJG, Scaramuzzino P: **Crazing in semicrystalline thermoplastics.** *J Macromol Sci Phys* 1999, **B38**(5–6):803–815.
41. Blancas AA, Lauer NE, McCloskey KE: **Endothelial differentiation of embryonic stem cells.** *Curr Protoc Stem Cell Biol* 2008. Chapter 1: Unit 1F 5.
42. Rao GN, Shaw EL, Arthur EJ, Aquavella JV: **Endothelial cell morphology and corneal deturgescence.** *Ann Ophthalmol* 1979, **11**(6):885–899.

Assembly of eukaryotic algal chromosomes in yeast

Bogumil J Karas[1], Bhuvan Molparia[1], Jelena Jablanovic[1], Wolfgang J Hermann[1], Ying-Chi Lin[1], Christopher L Dupont[2], Christian Tagwerker[1], Isaac T Yonemoto[1], Vladimir N Noskov[3], Ray-Yuan Chuang[3], Andrew E Allen[2], John I Glass[3], Clyde A Hutchison III[1], Hamilton O Smith[1], J Craig Venter[1,2,3] and Philip D Weyman[1*]

Abstract

Background: Synthetic genomic approaches offer unique opportunities to use powerful yeast and *Escherichia coli* genetic systems to assemble and modify chromosome-sized molecules before returning the modified DNA to the target host. For example, the entire 1 Mb *Mycoplasma mycoides* chromosome can be stably maintained and manipulated in yeast before being transplanted back into recipient cells. We have previously demonstrated that cloning in yeast of large (> ~ 150 kb), high G + C (55%) prokaryotic DNA fragments was improved by addition of yeast replication origins every ~100 kb. Conversely, low G + C DNA is stable (up to at least 1.8 Mb) without adding supplemental yeast origins. It has not been previously tested whether addition of yeast replication origins similarly improves the yeast-based cloning of large (>150 kb) eukaryotic DNA with moderate G + C content. The model diatom *Phaeodactylum tricornutum* has an average G + C content of 48% and a 27.4 Mb genome sequence that has been assembled into chromosome-sized scaffolds making it an ideal test case for assembly and maintenance of eukaryotic chromosomes in yeast.

Results: We present a modified chromosome assembly technique in which eukaryotic chromosomes as large as ~500 kb can be assembled from cloned ~100 kb fragments. We used this technique to clone fragments spanning *P. tricornutum* chromosomes 25 and 26 and to assemble these fragments into single, chromosome-sized molecules. We found that addition of yeast replication origins improved the cloning, assembly, and maintenance of the large chromosomes in yeast. Furthermore, purification of the fragments to be assembled by electroelution greatly increased assembly efficiency.

Conclusions: Entire eukaryotic chromosomes can be successfully cloned, maintained, and manipulated in yeast. These results highlight the improvement in assembly and maintenance afforded by including yeast replication origins in eukaryotic DNA with moderate G + C content (48%). They also highlight the increased efficiency of assembly that can be achieved by purifying fragments before assembly.

Keywords: Eukaryote, Chromosome, TAR cloning, *Saccharomyces cerevisiae*, Autonomously replicating sequence (ARS), *Phaeodactylum tricornutum*

Background

Synthetic biology can be described as the iterative process of modeling biological phenomena, building genetic devices to implement these models, and evaluating device performance. This discipline practiced at the scale of entire chromosomes can be defined as synthetic genomics. Synthetic genomic approaches focus on understanding function and regulation of multi-gene pathways, function of chromosome architecture, and strategies for multi-locus engineering. Several recent examples of synthetic genomic approaches include the creation of a bacterial cell from a chemically-synthesized genome [1], the cloning and manipulation of entire bacterial chromosomes in *Bacillus subtilis* [2,3], and efforts to synthesize and shuffle yeast chromosomes [4-6].

We have previously used the synthetic genomic approach with prokaryotic genomes which are typically small, circular, and contained on a single molecule. These qualities allow for efficient maintenance and

* Correspondence: pweyman@jcvi.org
[1]Department of Synthetic Biology and Bioenergy, J. Craig Venter Institute, 10355 Science Center Dr., San Diego, CA 92121, USA
Full list of author information is available at the end of the article

manipulation of the chromosome in yeast [7-10]. Once maintained in yeast, chromosomes can be efficiently manipulated before reintroduction into the host. Expanding synthetic genomic techniques to other bacteria is an active research area, but may also be useful for eukaryotic organisms. Chromosomes from eukaryotic organelles have been synthesized or assembled in yeast including the mitochondrion [11] and the chloroplast [12], but to our knowledge, no entire eukaryotic nuclear chromosome has been cloned in yeast.

Large regions of eukaryotic chromosomes have been cloned in yeast. Human (41% G + C) and mouse (42% G + C) YAC libraries have been made with average insert sizes of 620 kb and 700 kb, respectively [13], and human YAC clones greater than 1 Mb have been reported [14,15]. Small DNA fragments from diverse eukaryotes can function as yeast replication origins including *Neurospora crassa* (49% G + C), *Dictyostelium discoideum* (22% G + C), *Caenorhabditis elegans* (35% G + C), *Drosophila melanogaster* (42% G + C), and *Zea mays* (47% G + C) [16]. Addition of yeast origins of replication was previously shown to improve maintenance of large regions of prokaryotic DNA, especially when the sequence has a higher average G + C content than yeast [17]. DNA with low G + C content will have a greater chance of having A + T-rich sites that function as yeast replication origins while DNA of higher average G + C content will have statistically fewer of these sites (Table 1). Given that some sequences from diverse eukaryotes can function as yeast replication origins, we sought to test whether adding additional yeast replication origins could improve maintenance of entire eukaryotic chromosomes from non-fungal species.

To demonstrate the ability of yeast to maintain entire eukaryotic chromosomes, we investigated the yeast-based assembly of chromosomes from the diatom *Phaeodactylum tricornutum*. This marine diatom has a total genome size of 27.4 Mb and a G + C content of 48%. After the completely closed genomes of *Ostreococcus lucimarinus* [18] and *Cyanidioschyzon merolae* [19], *P. tricornutum* has one of the best-assembled algal genomes with 33 chromosome-sized scaffolds including 12 with telomere-to-telomere coverage [20]. Additionally, some of the telomere-to-telomere scaffolds are relatively small (<600-kb), making them good targets for developing synthetic genomic approaches. Experiments with the *P. tricornutum* system benefit from efficient transformation methods [21] and a long history of physiological research [22]. Because *P. tricornutum* is an emerging model genetic system for marine microalgal biology and biotechnology [23], we chose it to develop the first stage of a synthetic genomic approach: cloning, maintaining, and manipulating chromosomes in yeast. In this paper, we present data on the prospects and limitations associated with assembling, maintaining and manipulating *P. tricornutum* chromosomes in yeast and demonstrate improvements on the yeast assembly process including gel purification by electroelution as a powerful way to concentrate large DNA fragments for assembly.

Results

The 27.4 Mb *P. tricornutum* nuclear genome currently exists as 33 individual scaffolds. These scaffolds are described as chromosomes in reports on the genome sequence [20], although ongoing work to completely close the genome sequence may collapse some of these scaffolds without telomere-to-telomere coverage into single chromosomes. To be consistent with published work describing the genome, we refer to these scaffolds as "chromosomes". We chose *P. tricornutum* chromosomes 25 and 26 (lengths of 497 kb and 441 kb, respectively) as targets to assemble entire eukaryotic chromosomes in yeast because they are the smallest chromosomes for which telomere-to-telomere assembly exists. Chromosomes 25 and 26 each have a single remaining sequencing gap predicted to be 3.7 kb and 17 kb on chromosomes 25 and 26, respectively.

Cloning fragments spanning chromosomes

To clone an entire eukaryotic chromosome in yeast, we used an assembly-based strategy in which the chromosome was first cloned as smaller fragments in yeast and then assembled into the full-length molecule. Previous results cloning large DNA fragments from bacteria with moderately high G + C content (55%) indicated that fragments greater than *ca.* 150 kb were optimally maintained after addition of a yeast replication origin sequence [17]. With the large bacterial fragments, we demonstrated

Table 1 11-base yeast ARS consensus sequence

Organism	Chromosome size (kb)	11 base consensus count (WTTTAYRTTTW)	Average consensus per base pair	Longest region without consensus sequence (kb)
Mycoplasma mycoides	1084	303	0.28	22.5
Acholeplasma laidlawii	1498	179	0.12	65.2
Prochlorococcus MED4	1658	328	0.20	34.5
P. tricornutum chromosome 25	497	12	0.02	235
P. tricornutum chromosome 26	441	8	0.02	188

that adding additional yeast replication origins to the DNA allowed for stable maintenance of a *ca.* 450-kb fragment in yeast. Eukaryotic DNA fragments from sources such as human (41% G + C) and mouse (42% G + C) have been cloned in YACs that exceed 1 Mb, and we sought to test whether *P. tricornutum* DNA fragments (48% G + C) larger than *ca.* 150 kb could be replicated efficiently in yeast because of the higher G + C content.

To test whether additional yeast replication origins would improve the maintenance of larger *P. tricornutum* DNA, we first cloned chromosomes 25 and 26 each as five smaller *ca.* 100-kb fragments. Breaking the chromosome into smaller pieces allowed us to test for toxicity of any genes that may happen to be expressed in yeast and allowed us to further modify the pieces to test whether adding additional yeast origins of replication improved assembly efficiency in yeast.

We cloned the five, large, overlapping fragments that compose a chromosome using a yeast-based method called Transformation-Associated Recombination (TAR) cloning [24-26]. In TAR cloning, a yeast vector containing sequence overlaps with a specific target sequence is co-transformed with genomic DNA containing the target sequence. Yeast performs specific recombination events to recombine the target sequence into the plasmid. The efficiency of this process is greatly improved if the target sequence is linearized near at least one of the sites of recombination. We took advantage of rare-cutting restriction sites present in the *P. tricornutum* chromosomes to design our cloning strategy which would allow us to create double-stranded breaks near at least one of the homologous recombination sites. Digestion sites were selected such that each of the five chromosomal fragments overlapped by 10–40 kb (Table 2, Additional file 1: Figure S1). To clone each fragment, *P. tricornutum* DNA was prepared in agarose plugs to prevent unwanted shearing, and plugs were digested with the appropriate enzyme(s) to yield long DNA fragments (Additional file 1: Figure S1). To clone the *P. tricornutum* chromosome fragments, we designed vectors which had two 200-bp homology regions specific for each fragment. Cloned fragments were selected by the *HIS3* marker on the vector, and, because our vector design required two recombination events resulting in elimination of the *URA3* marker from the vector, empty vector molecules could be eliminated by selecting for the absence of the *URA3* maker using 5-fluoroorotic acid (5FOA, see Methods) [27].

Resulting yeast colonies were screened by PCR using primers designed to amplify specific, small, unique regions (less than 1 kb) located approximately in the middle of each fragment. Clones verified as correct by PCR were purified from yeast and transformed into *E. coli* to obtain higher plasmid yields. The large, cloned, ~100 kb fragments were then separated as supercoiled plasmids on an agarose gel to estimate their size. All fragments from chromosome 25 yielded bands of the expected size (Figure 1 and Table 2). Fragments from chromosome 26 were also of the expected size with the exception of fragments 26–1 and 26–2. For fragment 26–1, we expected a fragment of size 88 kb (including vector sequence) but obtained fragments of around 130 kb (Figure 1). For fragment 26–2, we expected a fragment at around 152 kb (including vector sequence), but we obtained plasmids of two different sizes: one was too small at around 100 kb and one that appeared too large at around 170 kb. Because we worried that the smaller-sized plasmids contained deletions, we chose the larger plasmid (see discussion for further explanation) to proceed with the assembly of chromosome 26.

After all fragments had been cloned in yeast and moved to *E. coli*, a single clone from each of the five fragments was chosen to move forward for assembly of

Table 2 *P. tricornutum* fragments used for assembly of chromosomes 25 and 26

Fragment #	Fragment size (kb) (final sizes with KmArs (frags 2,3,4) and Ura (frag 3))	Restriction sites used to cut genomic DNA for TAR cloning	Release sites (5′ and 3′)	5′ overlap (kb)	3′ overlap (kb)
25-1	105.5	NotI (3′ only)	I-CeuI + I-SceI	0.6	35.4
25-2	126.9	FseI (3′ only)	I-CeuI + I-SceI	35.4	9.6
25-3	122.3	SrfI	I-CeuI + I-SceI	9.6	27.8
25-4	125.2	FseI (5′ only)	I-CeuI + I-SceI	27.8	32.8
25-5	127.5	PmeI (5′ only)	I-CeuI + I-SceI	32.8	0.6
26-1	76.6	AscI (3′ only)	I-CeuI + I-SceI	0.5	26.1
26-2	142.6	SwaI	I-CeuI + I-SceI	26.1	39.5
26-3	129.3	NotI + AscI	I-CeuI + I-SceI	39.5	25.1
26-4	116.6	AsiSI	I-CeuI + I-SceI	25.1	12.1
26-5	83.1	AbsI (5′ only)	I-CeuI + I-SceI	12.1	0.5

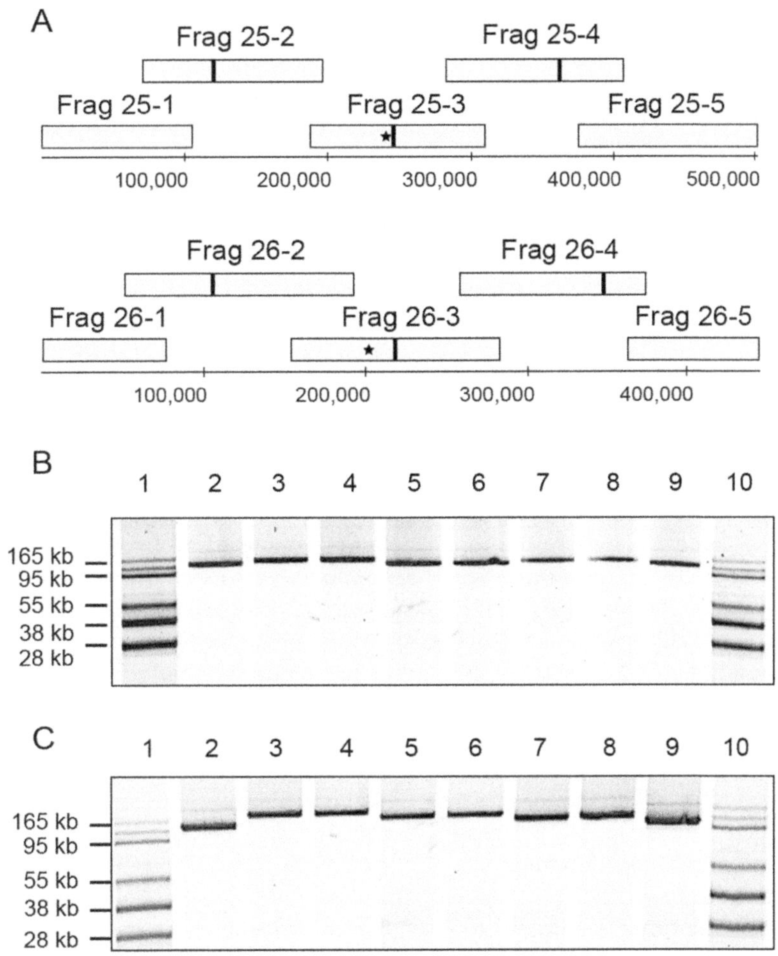

Figure 1 Cloning of *P. tricornutum* chromosomes as five overlapping fragments. A. Chromosomes 25 and 26 were each cloned as five overlapping fragments in yeast using a BAC-YAC vector. Fragments were further modified by addition of the *URA3* selectable marker in fragment 3 (asterisk) and the addition of a yeast replication origin (ARS) in fragments 2, 3, and 4 (vertical bar). **B** and **C.** The cloned fragments were purified from yeast, transformed into *E. coli*, and were purified and separated on agarose gels as supercoiled plasmids. Chromosome 25 **(B)** and 26 **(C)** fragments were each in the following order: BAC-Tracker supercoiled DNA ladder (Epicentre, lane 1), Fragment 1 (lane 2), Fragment 2 (larger version ~ 170 kb, lane 3), Fragment 2-ARS (lane 4), Fragment 3-URA (lane 5), Fragment 3-URA-ARS (lane 6), Fragment 4 (lane 7), Fragment 4-ARS (lane 8), Fragment 5 (lane 9), BAC-Tracker ladder (lane 10).

the entire chromosome (Figure 1). To facilitate chromosome assembly and selection of correctly-assembled molecules in yeast, fragment 3 from each of chromosomes 25 and 26 was modified in yeast with an additional uracil selectable marker (*URA3*, Table 3). Each of fragments 2, 3, and 4 from chromosomes 25 and 26 was modified by addition of a yeast replication origin (also called an autonomously replicating sequence or ARS, Table 3). After modifications, plasmids were extracted and separated by agarose gel electrophoresis to verify that the correct plasmid size was maintained (Figure 1).

To further validate the identity of the cloned fragments, we tested the plasmids containing the large *P. tricornutum* fragments using a higher resolution multiplex PCR. The multiplex primer set was designed to amplify a fragment every ~20 kb. All fragments in chromosome 25 had the expected high resolution multiplex PCR pattern with the exception of 25–3 that lacked one band where the primer binding site was interrupted by the *URA3* insertion (Additional file 1: Figure S2) and a faint band present for 25–2 (MPX2, Additional file 1: Figure S2) that was likely the result of non-specific primer binding (see Additional file 1: Figure S2 for additional explanation). High resolution multiplex PCR was also performed after modification with the ARS regions (Additional file 1: Figure S2). Modification with the ARS regions did not alter the multiplex PCR pattern indicating that all regions of the chromosome were still present after modification.

For chromosome 26, fragments 26–1, 26–3, 26–4, and 26–5 had the expected high resolution multiplex bands PCR (Additional file 1: Figure S1). The larger than

Table 3 Sites of insertion of yeast replication origins (KmR-ARS) and yeast selectable markers (URA) in cloned *P. tricornutum* fragments

Fragment	KmR-ARS insertion position (nt from beginning of chromosome sequence at 5′ site of insertion)	URA insertion position (nt from beginning of chromosome sequence at 5′ site of insertion)
25-2	120000	NA
25-3	241000	239990
25-4	357000	NA
26-2	115973	NA
26-3	230089	213353
26-4	358818	NA

expected size of 26–1 may be explained by the sequencing gap found within this fragment. High resolution multiplex PCR with several larger 26–2 clones indicated that they were each lacking one of the multiplex bands near the 5′ end of the fragment (Additional file 1: Figure S1). We continued with these fragments because the multiplex PCR indicated that there was still sequence overlap between fragments 26–1 and 26–2. Of all the fragments cloned, 26–1 and 26–2 were the only fragments that yielded a discrepancy in either size or DNA content from what was predicted by the published genome sequence, and we discuss possibilities for these observed differences at the end of the paper.

Our cloning strategy involved capturing sequence immediately adjacent to the telomeric repeats found at the ends of fragments 25–1, 25–5, 26–1, and 26–5. To verify that the regions immediately inside of the telomeric regions were cloned, we amplified the vector-insert junctions of the 5′ region of fragments 25–1 and 26–1 and the 3′ region of fragments 25–5 and 26–5. When these amplicons were sequenced, the results confirmed that the junction was seamless between the subtelomeric region and the vector, and all telomeric repeats had been removed in the TAR-cloning process.

Assembly of fragments into chromosomes

To assemble whole chromosomes in yeast, plasmids carrying the ~100 kb *P. tricornutum* fragments (with or without added ARS insertion) were purified from *E. coli* and the cloned fragments were released from the cloning vector by restriction enzyme digestion. After digestion, *P. tricornutum* DNA fragments were separated from the vector sequence by agarose gel electrophoresis, and *P. tricornutum* fragments were electroeluted from the agarose gel. The five *P. tricornutum* fragments spanning an entire chromosome were then assembled into a new vector in yeast using spheroplast transformation in one of two experimental configurations. In the first, the five fragments lacked added yeast origins while in the second, fragments 2, 3, and 4 each contained an additional yeast origin. In both configurations, fragment 3 contained an additional *URA3* gene to assist in selection of correct assemblies. Selectable markers on both the vector and fragment 3 complemented tryptophan and uracil auxotrophies, respectively, and improved selection for complete assemblies.

Resulting yeast colonies were patched several times on selective media. One hundred colonies of each assembly (i.e. with or without additional yeast replication origins) were screened by multiplex PCR in which one amplicon derived from each of the five assembled fragments. For both chromosomes 25 and 26, no complete assemblies were observed without the addition of yeast replication origins (Figure 2). In the chromosome 25 assembly with additional yeast origins of replication, 36% of the yeast colonies contained all five fragments (Figure 2). For the chromosome 26 assembly with added ARS regions, 62% of the clones contained all five fragments (Figure 2). The greatly improved efficiency of assembly using fragments that contain additional yeast origins confirms their utility for improving chromosome-scale assemblies with eukaryotic DNA (Figure 2). We also investigated the use of electroelution as a purification method for large DNA fragments. Results without gel purification of the fragments by electroelution yielded a lower percentage of correct colonies, and only 5 of 100 colonies in the chromosome 25 assembly had all five fragments.

Assembled chromosomes that were positive by PCR for all five fragments were further analyzed by high resolution multiplex PCR. In this analysis, primers were designed to amplify a band every 20 kb. The results from the high resolution multiplex PCR indicated that all regions present in the original five fragments were successfully assembled into a final molecule (Additional file 1: Figure S3). Assembled chromosome plasmids with correct multiplex PCR profiles were transformed to *E. coli* and retested with high resolution multiplex PCR to confirm that the entire plasmid was transferred.

The size of the assembled chromosomes was then analyzed by pulsed field gel electrophoresis. The assembled circular *P. tricornutum* chromosomes were prepared from *E. coli* in agarose plugs to prevent shearing the large circular plasmids and were separated by pulsed field gel electrophoresis (Figure 3). DNA preparations from *E. coli* cells containing chromosome 25 yielded a single band at a size consistent with the expected size of the cloned chromosome 25 (513 kb including vector and other added sequence). Preparations from *E. coli* cells containing chromosome 26 also contained a single band at *ca.* 500 kb. This observed size was slightly larger than expected (457 kb including vector and other added sequence), but the larger observed size of the assembled

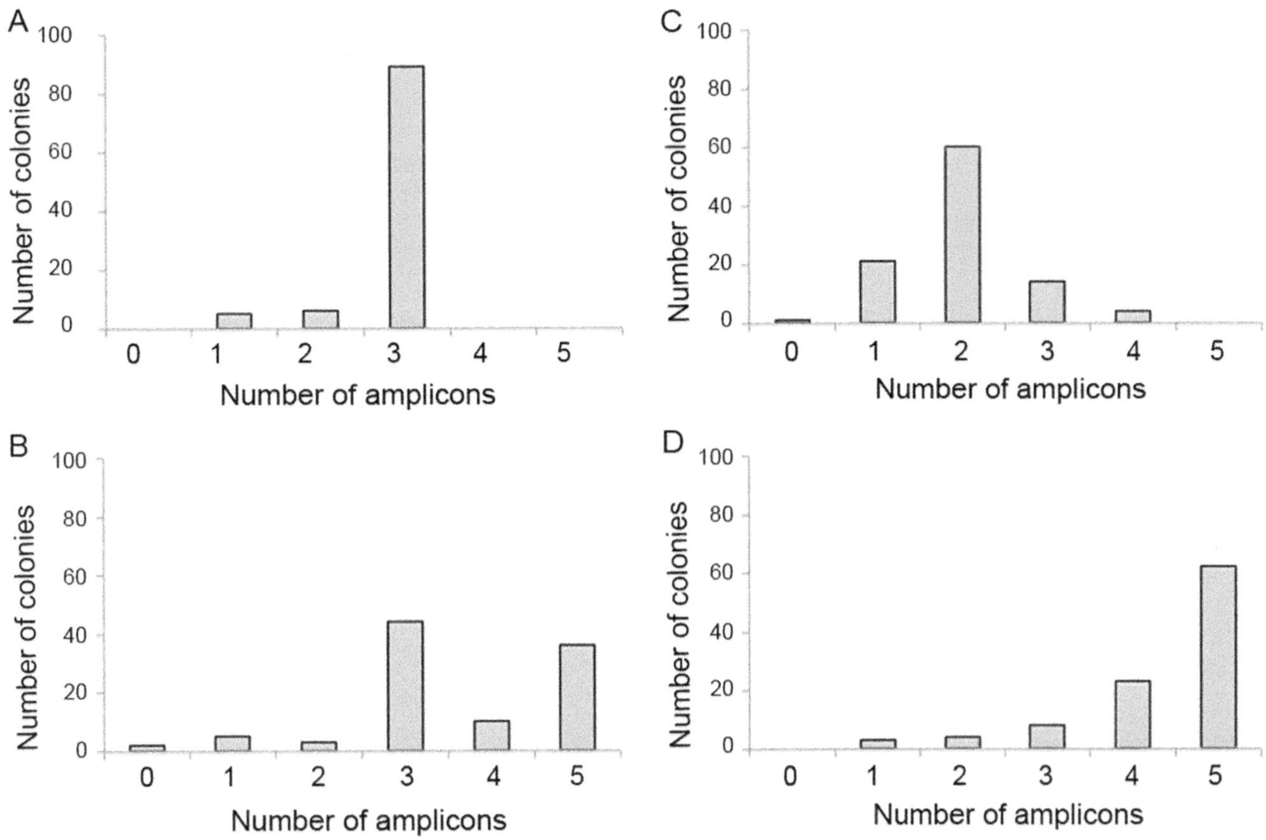

Figure 2 Summary of assembly efficiency for chromosomes 25 and 26. Assembly of each of *P. tricornutum* chromosomes 25 (A and B) and 26 (C and D) from five overlapping fragments was performed using yeast spheroplast transformation. Fragments to assemble were modified with the *URA3* selectable marker in the third fragment only (**A** and **C**) or *URA3* selectable marker in the third fragment and yeast origins of replication in fragments 2, 3, and 4 (**B** and **D**). One hundred resulting colonies were screened by low resolution multiplex PCR (one amplicon per fragment), and the number of fragments cloned for each assembly was plotted.

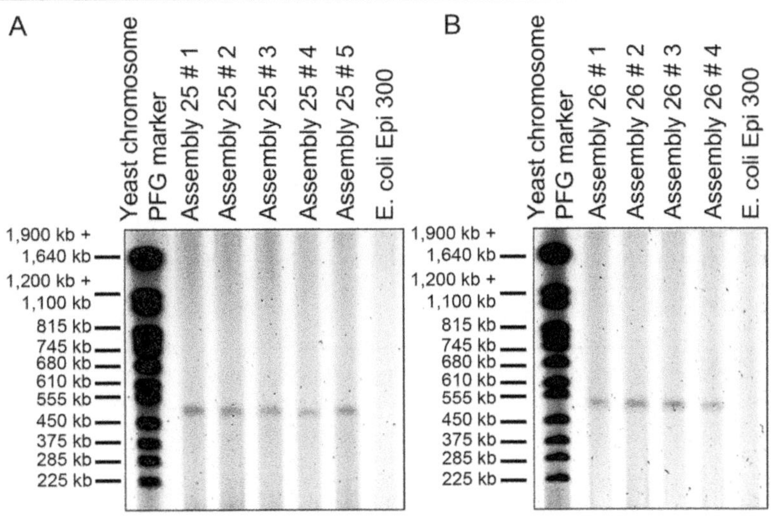

Figure 3 Final assemblies of *P. tricornutum chromosomes.* Assembled chromosomes that tested positive by high resolution multiplex PCR were transferred to *E. coli* and supercoiled plasmids were purified in agarose plugs. DNA purified in plugs from *E. coli* containing assembled chromosome 25 **(A)** or 26 **(B)** was separated by pulsed-field gel electrophoresis.

molecule is consistent with the larger than expected size of fragments 26–1 and 26–2 (Figure 1).

Manipulation of chromosome fragments in yeast

One of the strengths of the synthetic genomic approach outlined above is the ability to manipulate target organism DNA using efficient yeast and *E. coli* genetic techniques. To demonstrate how this might be applied to the cloned *P. tricornutum* DNA, we used a mutational strategy to progressively shorten fragment 26–3. As shown in Figure 4A, a *URA3* cassette was PCR amplified to have 40-bp homology on the 5' end to a position in the cloning vector and at the 3' end of the *URA3* to part of the cloned 26–3 *P. tricornutum* fragment. This fragment was then transformed into yeast carrying the 26–3 plasmid to homologously recombine and replace native sequence with the *URA3* marker. Deletions from 20–100 kb were constructed using this method (Figure 4B). The deletions constructed using this method are not limited to extending from the vector as shown in Figure 4, but can be made anywhere in the fragment. Serial mutations can be easily made using techniques to recycle the *URA3* marker [28]. Once a *URA3* cassette is inserted, replacement of the URA-marked deleted region with a sequence of interest can be easily selected by plating the transformant on media containing 5-FOA [27]. After the fragments are modified as desired, they can be reassembled into entire chromosomes as described above.

Discussion

In this work, we found that the efficiency of yeast-based assembly of entire *P. tricornutum* chromosomes was

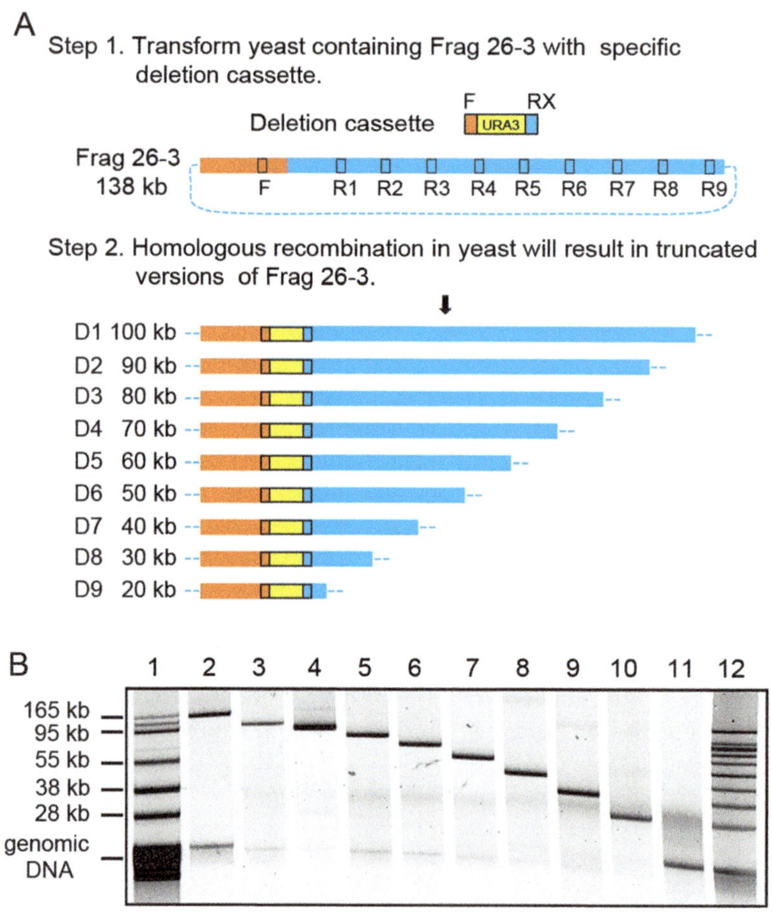

Figure 4 Progressive minimization of *P. tricornutum* fragment 26–3 using yeast homologous recombination. A. Diagram of the deletion strategy. To generate reduced versions of fragment 26–3, we inserted yeast deletion cassettes (which had 40-bp homology at the 5' end (F) common for all deletions, 40-bp homology at the 3' end specific for each fragment (RX where X represents homology sites 1–9), and a URA3 marker). These cassettes were PCR amplified, and yeast transformation was performed using the lithium acetate method (Step 1). Homologous recombination between the deletion cassettes and fragment 26–3 would produce desired fragments (Step 2). URA3 (yellow box), Yeast vector (orange box), *P. tricornutum* DNA representing fragment 26–3 (blue box and blue dotted line). **B.** After creation of the deletion in yeast, the plasmids were transferred to *E. coli* and purified. Supercoiled plasmids were separated by gel electrophoresis in the following order: BAC-Tracker ladder (lane 1), unmodified fragment 26–3 (lane 2), deletion of fragment 26–3 into 100 kb (lane 3), 90 kb (lane 4), 80 kb (lane 5), 70 kb (lane 6), 60 kb (lane 7), 50 kb (lane 8), 40 kb (lane 9), 30 kb (lane 10), 20 kb (lane 11).

improved by adding yeast replication origins to the large fragments before assembly. Although similarly large fragments of eukaryotic DNA have been cloned as YACs in various genome projects, the increased efficiency of assembly and maintenance afforded by adding yeast replication origins may have been related to the 10% higher G + C content of *P. tricornutum* relative to yeast (48% vs. 38% G + C, respectively) (Table 1). Although sequences found in *Zea mays* (47% G + C and similar to *P. tricornutum*) function as yeast origins [16], our finding may highlight the importance of variation from the mean G + C content as well as the mean value itself in predicting whether additional yeast replication origins may improve assembly and maintenance in yeast. For example, while the human genome has an average G + C content of 41%, many small regions were found with G + C contents that vary from 33% to 55% [29], and large YACs spanning hundreds of kilobases of human DNA have been maintained in yeast. Thus, mean G + C content may not be as important as the frequency of deviation from the mean in understanding how well DNA of a particular organism will be maintained in yeast. If frequent regions of low G + C content are found, then large DNA fragments may be more efficiently cloned without addition of yeast origins.

This paper describes an improved chromosome assembly process that uses electroelution to concentrate and purify large DNA fragments to increase the efficiency of assembly. Electroelution has been used extensively to select for large inserts in BAC library construction [30-33]. Initial attempts to assemble the *P. tricornutum* DNA fragments used a "digest and mix" strategy. In this initial method, plasmids were digested to release the *P. tricornutum* fragments from the cloning vector, and the five fragments were subsequently mixed together with a new cloning vector without any purification or removal of old cloning vectors. We found that this method yielded very few colonies of which only a small percentage were correct. In this study, we describe purification of the large, digested DNA fragments by electroelution from agarose gels before assembly in yeast. This modification improved the rate of successful assembly. For electroelution, we used RECO chips (Takara). These small disposable devices consist of a chamber with permeable fabric on one side and impermeable dialysis membrane on the other side. The chip is inserted into an agarose gel after bands of interest have been separated and the DNA of interest is electrophoresed into the chip chamber. The DNA is eluted by a quick low speed centrifugation step. Although the RECO chip product is currently discontinued by the manufacturer, other electroelution techniques should suffice as they have been used successfully in the construction of large BAC libraries [31].

Assembly of chromosome 25 matched the expected size and multiplex PCR profile, and our final assembly of chromosome 26 differed from the expected size and PCR profile based on the published genome sequence. The differences between the expected size and multiplex PCR profiles were likely due to the larger-than-expected sizes of fragments 26–1 and 26–2. In the end, the assembled whole chromosome 26 was consistent with the starting component fragments indicating that the assembly process itself was robust for the *P. tricornutum* DNA. To better understand the differences between chromosomes 25 and 26, we searched for repeated regions in each chromosome using BLAST. The results showed that chromosome 25 had very little repeated DNA while chromosome 26 had a region containing several repeats. These repeats are present in a 30-kb region at the 5' end of fragment 26–2 and contained at least 4 direct repeats each 2–4 kb in length (Additional file 1: Table S1). It is possible that during the process of cloning fragment 26–2 in yeast, different rearrangements occurred resulting in cloned regions of different sizes. However, we also cannot rule out the possibility that the different-sized fragments reflect natural diversity in this region in our *P. tricornutum* cell population. Because the *P. tricornutum* cultures are largely maintained by serial passage for long periods of time, populations of cells within the culture with different genetic characteristics may have resulted. We did not find repeat regions in fragment 26–1; however, the larger than expected size for fragment 26–1 may be explained by a gap in sequence at the 5'end of this fragment. This gap is estimated to be approximately 17 kb; however, it is possible that it is larger than 17 kb which would explain the size of fragment 26–1 which we have cloned.

Despite the differences between our assembled chromosome 26 and the expected version based on the genome sequence, our data support the fact that using the assembly techniques described in this paper we are able to accurately and efficiently assemble the pieces we had cloned. To avoid potential difficulties, future attempts to assemble eukaryotic chromosomes should perform a BLAST search to identify repeated sequences. Such a search can be easily implemented and visualized in the NCBI Genome Workbench software. If repeats are identified, attempts should be made to design a fragment cloning strategy that places the repeated sequences in the middle of the fragment and not at the junctions between two fragments as in fragment 26–2.

Conclusion

Using *P. tricornutum* as our model eukaryote, we have demonstrated a technique for assembly, maintenance, and manipulation of entire eukaryotic chromosomes in yeast. We show that addition of yeast replication origins has improved assembly and maintenance of two eukaryotic chromosomes with moderate G + C content

(47%) and that this is likely to translate to other eukaryotic DNA fragments. Because many other algae have high G + C content including the model green alga *Chlamydomonas reinhardii* (G + C content 64%)[34], application of these techniques may improve genome scale cloning in a variety of organisms. Given the challenges of slow growth and limited genetic tools in a variety of algal species of interest, developing synthetic genomic approaches may improve genetic characterization of these species.

Methods

Strains, media, and growth conditions

For all yeast transformations, we used *Saccharomyces cerevisiae* VL6-48 (ATCC MYA-3666: MATα his3-Δ200 trp1-Δ1 ura3-52 lys2 ade2-1 met14 cir^0). Yeast cells were grown in rich medium (YEPD) or complete minimal (CM) medium lacking histidine and uracil or tryptophan and uracil (Teknova) at 30°C. *Escherichia coli* (Epi300, Epicentre) strains were grown on Luria broth or agar supplemented with chloramphenicol (20 mg L^{-1}) or kanamycin (50 mg L^{-1}) as needed. *Pheodactylum tricornutum* was grown in L1 medium [35] at 18°C under cool white fluorescent lights (50 μE m^{-2} s^{-1}). Sequence information for *P. tricornutum* was taken from the DOE genome sequencing site [36] as described [20].

Preparation of agarose-embedded DNA

DNA from *P. tricornutum* for TAR cloning of the ~100-kb fragments was isolated in agarose plugs using the Bio-Rad CHEF Genomic DNA Plug Kit with a protocol that was adapted for use with *P. tricornutum*. To prepare the plugs, 50 ml of mid-log phase (10^8 cells ml^{-1}) *P. tricornutum* culture was centrifuged at 1500 × g for 5 min at 10°C. Cells were washed once with 50 ml of 1 M sorbitol and were resuspended in 2 mL of SPEM solution (1 M sorbitol, 10 mM EDTA pH 7.5, $Na_2HPO_4 \cdot 7H_2O$ (2.08 g L^{-1}), $NaH_2PO_4 \cdot 1H_2O$ (0.32 g L^{-1})). The cell suspension was incubated for 5 min at 37°C and mixed with an equal volume of 2.0% low-melting-point agarose in 1 × TAE buffer (40 mM Tris, 20 mM acetic acid and 1 mM EDTA) which was equilibrated at 50°C. Aliquots of 100 μl were transferred into plug molds (Bio-Rad, catalog # 170–3713) and allowed to solidify for 10 min at 4°C. Next plugs were removed from the molds into 50 ml conical tube containing 5 ml of protoplasting solution (4.56 mL of SPEM solution, 200 μl Zymolyase-100 T solution (50 mg ml^{-1} dissolved in H_2O), 200 μl lysozyme (25 mg ml^{-1}), 40 μl β-Mercaptoethanol and incubated for 1 h at 37°C. Next, plugs were washed with 25 ml of wash buffer (20 mM Tris, 50 mM EDTA, pH 8.0), and then incubated in 5 ml in Proteinase K buffer (100 mM EDTA (pH 8.0), 0.2% sodium deoxycholate, and 1% sodium lauryl sarcosine, 1 mg ml^{-1} Proteinase K) for 24 hr at 50°C. After proteinase K treatment, plugs were washed according to the Bio-Rad CHEF-Dr III manual.

Construction of cloning vectors used in this study

Vectors used for cloning and assembly of the ~100 kb fragments and entire *P. tricornutum* chromosomes were constructed from the template plasmids pBK-RBYV-HIS3URA3 and pBK-RBYV-TRP1URA3 using yeast assembly methods [17,37]. Maps and sequences for both vectors are available in the Supplementary Information. Both vectors contained the *URA3* gene from pYAC-RC (ATCC 37610). For the vector pBK-RBYV-HIS3URA3, the BACYAC backbone pCC1BACHIS3 was used containing the *HIS3* gene [17]. For pBK-RBYV-TRP1URA3, the HIS3 gene was replaced with the *TRP1* gene from pRS314 (GenBank Accession U03440). Both pBK-RBYV-HIS3URA3 and pBK-RBYV-TRP1URA3 contain the *ShBle* cassette from pAF6 [38] and the conjugative transfer origin (OriT) from pRL2948a (C. P. Wolk, unpublished data). Both vectors were designed to have I-CeuI and XhoI restriction sites at the junction between OriT and *URA3*, and at the junction of the *URA3* and pCC1BAC backbone, XhoI and I-SceI sites were inserted.

Specific vectors for cloning each of the ~100 kb *P. tricornutum* DNA fragments by TAR cloning were assembled in yeast by co-transforming XhoI-digested template plasmids pBK-RBYV-HIS3URA3 and two 200-bp PCR-amplified *P. tricornutum* DNA homology regions. The vector pKB-RBYV-HIS3URA3 was designed so that digestion with XhoI would excise the *URA3* cassette. Our strategy to create 200-bp homology regions on either side of the *URA3* cassette was to reassemble the two pieces of the digested template vector with the 200-bp homology regions inserted on either side of the *URA3* cassette. Each of the 200-bp *P. tricornutum* PCR fragments contained 25–40 bp homologies to the template vector (pBK-RBYV-HIS3URA3) on either side of an XhoI site. The resulting plasmid contained the *URA3* cassette cloned between two 200-bp *P. tricornutum* homology regions that specified the boundaries of the resulting TAR-cloned fragment, and the pBK-RBYV-HIS3 backbone. Primers used to amplify each of the 200-bp *P. tricornutum* homology regions are listed in Additional file 1: Table S4.

The design of the primers used to amplify the 200-bp homology regions was such that after assembly only one of the XhoI sites would be restored and the second XhoI site would be replaced with a PacI site. Digestion with either XhoI or PacI would allow the vector to be linearized while maintaining the *URA3* marker which could be used as negative selection to select for double recombination events in which both homology regions had recombined with their target regions in the *P. tricornutum* chromosome.

Alternatively each vector could be digested with both XhoI and PacI to remove *URA3* marker before TAR-cloning. We used both approaches in this paper and found that the highest cloning efficiency of target genomic fragments that were cut only on one side (e.g. fragment 25–1) was obtained using vector that had been cut at both ends to release *URA3*, and therefore the negative selection could not be applied. For increased yield, vectors verified by PCR as correctly assembled were moved from yeast to *E.coli* and were purified by alkaline lysis before being digested with XhoI and/or PacI [39].

Vectors for assembling entire chromosomes 25 and 26 were constructed in yeast by co-transforming XhoI-digested template pBK-RBYV-TRP1URA3 and two 500-bp PCR amplified homology regions. The first homology region matched the 5' region of the cloned *P. tricornutum* sequence in fragments 25–1 or 26–1. The second homology region matched the 3' region of the cloned *P. tricornutum* sequence in pieces 25–5 or 26–5. The first homology region had 35 bp overlap with the template vector at the XhoI site near the OriT region, and the second 500 bp region had 35 bp overlap at the XhoI site near the pCC1BAC sequence. In addition, the 3' end of homology region one contained 35 bp overlap with 5' end of homology region two so that the two pieces could be assembled together in a vector. The design of the primers used to amplify the homologous regions was such that after assembly an XhoI site would be created at the junction of homology region one and two. This allowed for digestion of the resulting vector with XhoI to linearize and expose the two homology regions before assembly of the entire chromosome. For chromosome 25, the first 500-bp homology region was amplified by primers Pt25-AV15-1 and Pt25-AV15-2new and the second homology region was amplified using primers Pt25-AV15-3new and PT25-AV15-4. For chromosome 26, the first 500-bp homology region was amplified by primers Pt26-av-15-F1 and Pt26-av-15-R1 and the second region was amplified by primers Pt26-av-15-F2 and Pt-av-15-R2 (See supplementary information).

TAR-cloning 100-kb fragments and assembly of full length chromosomes

P. tricornutum DNA for TAR cloning was prepared in agarose plugs as described above and in-plug digestion was performed as previously described [17] using specific enzymes listed in Table 2. After digestion, plugs were washed in TE buffer (pH 8) and were melted in 100 μl TE buffer at 65°C for 10 min. The molten agarose was then equilibrated at 42°C and 2 μl of β-agarase (NEB) was added and incubated for 8 h. 10 μl of DNA from plugs, (typically 100 ng/μl) was mixed with 300–500 ng of vector and the assembly was performed by mixing the DNA with yeast spheroplasts prepared as previously described [17,37]. Resulting colonies were screened by PCR using primers specific to a region in the cloned piece (see multiplex primers below). Fragments 25–1, 25–5, 26–1, and 26–5 were further screened by PCR to verify that the proper sub-telomeric region was cloned using the following primer pairs, respectively: Endchk-F + Pt25-1endchk-R, Endchk-R + Pt25-5endchk-F2, Endchk-F + Pt26-1endchk-R, Endchk-R + Pt26-5endchk-F2 (See Additional file 1: Table S4). Plasmids from positive clones were isolated by alkaline lysis and transformed into *E. coli* for further analysis and confirmation of size by separation with agarose gel electrophoresis.

To assemble entire *P. tricornutum* chromosomes, plasmids containing the cloned fragments were purified from *E. coli* using an alkaline lysis procedure [39]. Each of the five fragments (20–30 μg plasmid) composing either chromosome 25 or 26 was excised from the cloning vector by digest with I-SceI and I-CeuI for 8 h at 37°C in a total volume of 50 μl. These excised fragments were either precipitated and used directly for assembly or subjected to gel purification and electroelution. For gel purification, the entire 50 μl digest was loaded on a 1% agarose gel in 1X Tris-Acetate-EDTA (TAE) buffer and was separated by electrophoresis for 8 h at 80 V at 4°C. The gel was then stained briefly with ethidium bromide, and the excised fragments were each separated from genomic DNA and vector by electroelution using RECO chips (Takara). To capture the large DNA fragments in the chips, the gel was rotated 90° and a RECO chip was inserted into the gel so that the large *P. tricornutum* DNA fragments would electrophorese into the chip for 1 h at 100 V at room temperature. DNA was eluted from the chip by a centrifugation for 10 sec (1,500 × g), and final concentrations of DNA were typically between 20–30 μg μl^{-1}.

Digested fragments were mixed with 500–1000 ng of vector and transformed into yeast spheroplasts as previously described [37]. Typically, 3–5 μg of each fragment was used for assemblies of fragments without electroelution and 100–500 ng of each fragment was used for electroeluted samples. Resulting yeast colonies were screened by colony PCR using a multiplex primer set containing one amplicon on each assembled fragment using the Multiplex PCR Kit (Qiagen). Primers used for multiplex PCR can be found in Additional file 1: Table S4 and are listed as "low resolution multiplex PCR". Plasmids from clones positive for all five amplicons were isolated by alkaline lysis and transformed into *E. coli* strain Epi300. Further analysis of the clones was performed on plasmids isolated from *E. coli* including high resolution multiplex PCR and CHEF gel analysis on the whole assembled chromosomes. High resolution multiplex PCR consisted of amplicons designed every ~20 kb along the chromosome. Primers for this can be found in Additional file 1: Table S4 and were pooled according to Additional file 1: Table S3. The reactions were performed using the Qiagen Multiplex PCR kit according to the manufacturer's instructions. Amplicons were separated by agarose gel electrophoresis using 2% agarose gels made with 1X TAE. CHEF gel analysis of assembled *P. tricornutum* chromosomes was performed as previously described for other cloned large plasmids [17]. Briefly, assembled *P. tricornutum* chromosomes propagated in *E. coli* were isolated in agarose plugs using the CHEF Genomic DNA Plug Kit according to the manufacturer's instructions.

Manipulation of fragments

Yeast replication origins were inserted at defined locations in fragments 2, 3, and 4 of chromosomes 25 and 26 (Additional file 1: Table S2). The *ARSH3* sequence from yeast was fused to a sequence encoding the kanamycin antibiotic resistance marker from pACYC177 and recombined into the cloned *P. tricornutum* fragments using lambda red recombineering as previously described [17]. The KanR-ARS fragment was amplified by PCR using primers described in Additional file 1: Table S2. Positive colonies were screened by PCR using primers flanking the insertion site and the increase in amplicon size based on insertion of the KanR-ARS fragment indicated a correctly-recombined clone (Additional file 1: Table S2). Positive clones were purified, confirmed to have the correct size by agarose gel electrophoresis and used in assembly reactions.

The strategy to modify the fragments in yeast was based on insertion of the *URA3* gene using homologous recombination. This strategy was used to insert the *URA3* gene in the middle of fragments 25–3 and 26–3 to assist in proper assembly, as well as to shorten fragment 26–3 to make a series of deletions. The *URA3* gene with appropriate homology regions was amplified from pYAC-RC vector (ATCC 37610). For the insertions into fragment 25–3 using primers URApt25-3_F + URApt25-3_R, and into fragment 26–3 using primers URApt26-3_F + URApt26-3_R (See Additional file 1: Table S2). For deletions in fragment 26–3, the *URA3* gene was amplified using a common forward primer for each insertion and a unique reverse primer that specified the insertion site and the degree of deletion created (e.g. URA-F + URA-R100, etc., see Additional file 1: Table S4). The amplified *URA3* cassettes were transformed into yeast carrying Fragment 25–3 or 26–3 using lithium acetate transformation [40]. Yeast colonies were selected on media lacking histidine and uracil, and correct clones were identified by PCR at either side of the *URA3* insertion junction. The junction at the 5' end of the insertion (i.e. the "left" side) was common to all plasmids generated in the deletion series and was screened using the primers F-left + R-left). The junction at the 3' end of the URA3 insertion (i.e. the "right" side) was screened using a common forward primer (F-right) and a primer that denoted the final size of the resulting plasmid (100-R-right, etc., See Additional file 1: Table S4). Fragments identified as having the correct insertion by PCR were moved from yeast to *E.coli* and plasmids were purified and further tested by agarose gel electrophoresis to verify correct plasmid size.

Additional file

Additional file 1: Supplemental Figures, Tables, and Plasmid sequences used in this study.

Abbreviations

BAC-YAC: Bacterial artificial chromosome-yeast artificial chromosome; TAR: Transformation-associated recombination.

Competing interests

JCV. is Chief Executive Officer and Co-Chief Scientific Officer of Synthetic Genomics, Inc. HOS. is Co-Chief Scientific Officer and on the Board of Directors of Synthetic Genomics, Inc. CAH. is Chairman of the Synthetic Genomics, Inc. Scientific Advisory Board. All three of these authors and the J. Craig Venter Institute hold Synthetic Genomics, Inc. stock.

Authors' contributions

The following authors designed experiments and analyzed data: BJK, BM, JJ, WJH, YCL, CT, ITY, VNN, RYC, AEA, CLD, JIG, CAH, HOS, JCV, and PDW. Experiments were performed by BJK, BM, JJ, WJH, YCL, CT, ITY, and PDW. The paper was written by BJK and PDW. All authors read and approved the final manuscript.

Acknowledgements

The authors thank Chuck Merryman and Dan Gibson from the J. Craig Venter Institute and Peter DeHoff, Rob Brown, Michele Champagne, Toby Richardson, and Jim Flatt from Synthetic Genomics, Inc. for many helpful suggestions and discussions about this project.

This work was supported by the United States Department of Energy Genomics Science program grant (DE-SC0008593) and by Synthetic Genomics, Inc. (SGI).

Author details

[1]Department of Synthetic Biology and Bioenergy, J. Craig Venter Institute, 10355 Science Center Dr., San Diego, CA 92121, USA. [2]Department of Microbial and Environmental Genomics, J. Craig Venter Institute, 10355 Science Center Dr., San Diego, CA 92121, USA. [3]Department of Synthetic Biology and Bioenergy, J. Craig Venter Institute, 9704 Medical Center Dr., Rockville, MD 20850, USA.

References

1. Gibson DG, Glass JI, Lartigue C, Noskov VN, Chuang R-Y, Algire MA, Benders GA, Montague MG, Ma L, Moodie MM, Merryman C, Vashee S, Krishnakumar R, Assad-Garcia N, Andrews-Pfannkoch C, Denisova EA, Young L, Qi Z-Q, Segall-Shapiro TH, Calvey CH, Parmar PP, Hutchison CA, Smith HO, Venter JC: **Creation of a bacterial cell controlled by a chemically synthesized genome.** *Science* 2010, **329:**52–56.
2. Watanabe S, Shiwa Y, Itaya M, Yoshikawa H: **Complete Sequence of the First Chimera Genome Constructed by Cloning the Whole Genome of Synechocystis Strain PCC6803 into the Bacillus subtilis 168 Genome.** *J Bacteriol* 2012, **194:**7007.

3. Itaya M, Tsuge K, Koizumi M, Fujita K: **Combining two genomes in one cell: Stable cloning of the Synechocystis PCC6803 genome in the Bacillus subtilis 168 genome.** *Proc Natl Acad Sci USA* 2005, **102:**15971–15976.
4. Muller H, Annaluru N, Schwerzman J, Richardson S, Dymond J, Cooper E, Bader J, Boeke J, Chandrasegaran S: **Assembling large DNA segments in yeast.** *Methods Mol Biol* 2012, **852:**133–150.
5. Dymond J, Boeke J: **The Saccharomyces cerevisiae SCRaMbLE system and genome minimization.** *Bioeng Bugs* 2012, **3:**168–171.
6. Dymond JS, Richardson SM, Coombes CE, Babatz T, Muller H, Annaluru N, Blake WJ, Schwerzmann JW, Dai J, Lindstrom DL, Boeke AC, Gottschling DE, Chandrasegaran S, Bader JS, Boeke JD: **Synthetic chromosome arms function in yeast and generate phenotypic diversity by design.** *Nature* 2011, **477:**471–476.
7. Tagwerker C, Dupont CL, Karas BJ, Ma L, Chuang R-Y, Benders G a, Ramon A, Novotny M, Montague MG, Venepally P, Brami D, Schwartz A, Andrews-Pfannkoch C, Gibson DG, Glass JI, Smith HO, Venter JC, Hutchison C a: **Sequence analysis of a complete 1.66 Mb Prochlorococcus marinus MED4 genome cloned in yeast.** *Nucleic Acids Res* 2012, **40:**10375–10383.
8. Karas B, Jablanovic J, Sun L, Stam J, Ma L, Ramon A, Venter J, Weyman P, Gibson D, Glass, JI Hutchison C III, Smith H, Suzuki Y: **Direct transfer of whole genomes from bacteria to yeast.** *Nat Meth* 2013, **10:**410–412.
9. Benders GA, Noskov VN, Denisova EA, Lartigue C, Gibson DG, Assad-Garcia N, Chuang R-Y, Carrera W, Moodie M, Algire MA, Phan Q, Alperovich N, Vashee S, Merryman C, Venter JC, Smith HO, Glass JI, Hutchison CA: **Cloning whole bacterial genomes in yeast.** *Nucleic Acids Res* 2010, **38:**2558–2569.
10. Karas BJ, Tagwerker C, Yonemoto I, Hutchison CA III, Smith HO: **Cloning the Acholeplasma laidlawii PG-8A Genome in Saccharomyces cerevisiae as a Yeast Centromeric Plasmid.** *ACS Synth Biol* 2012, **1:**22–28.
11. Gibson DG, Smith HO, Hutchison CA, Venter JC, Merryman C: **Chemical synthesis of the mouse mitochondrial genome.** *Nat Meth* 2010, **7:**901–903.
12. O'Neill BM, Mikkelson KL, Gutierrez NM, Cunningham JL, Wolff KL, Szyjka SJ, Yohn CB, Redding KE, Mendez MJ: **An exogenous chloroplast genome for complex sequence manipulation in algae.** *Nucleic Acids Res* 2012, **40:**2782–2792.
13. Larin Z, Monaco AP, Lehrach H: **Yeast artificial chromosome libraries containing large inserts from mouse and human DNA.** *Proc Natl Acad Sci USA* 1991, **88:**4123–4127.
14. Albertsen HM, Abderrahim H, Cann HM, Dausset J, Le Paslier D, Cohen D: **Construction and characterization of a yeast artificial chromosome library containing seven haploid human genome equivalents.** *Proc Natl Acad Sci USA* 1990, **87:**4256–4260.
15. Marschall P, Malik N, Larin Z: **Transfer of YACs up to 2.3 Mb intact into human cells with polyethylenimine.** *Gene Ther* 1999, **6:**1634–1637.
16. Stinchcomb DT, Thomas M, Kelly J, Selker E, Davis RW: **Eukaryotic DNA segments capable of autonomous replication in yeast.** *Proc Natl Acad Sci USA* 1980, **77:**4559–4563.
17. Noskov VN, Karas BJ, Young L, Chung R-Y, Gibson DG, Lin Y-C, Stam J, Yonemoto IT, Suzuki Y, Pfannkoch C, Glass JI, Smith HO, CAH III, Venter JC, Weyman PD: **Assembly of large, high G + C bacterial DNA fragments in yeast.** *ACS Synth Biol* 2012, **1:**267–273.
18. Palenik B, Grimwood J, Aerts A, Rouzé P, Salamov A, Putnam N, Dupont C, Jorgensen R, Derelle E, Rombauts S, Zhou K, Otillar R, Merchant SS, Podell S, Gaasterland T, Napoli C, Gendler K, Manuell A, Tai V, Vallon O, Piganeau G, Jancek S, Heijde M, Jabbari K, Bowler C, Lohr M, Robbens S, Werner G, Dubchak I, Pazour GJ, *et al*: **The tiny eukaryote Ostreococcus provides genomic insights into the paradox of plankton speciation.** *Proc Natl Acad Sci USA* 2007, **104:**7705–7710.
19. Maruyama S, Matsuzaki M, Kuroiwa H, Miyagishima S-Y, Tanaka K, Kuroiwa T, Nozaki H: **Centromere structures highlighted by the 100%-complete Cyanidioschyzon merolae genome.** *Plant Signal Behav* 2008, **3:**140–141.
20. Bowler C, Allen AE, Badger JH, Grimwood J, Jabbari K, Kuo A, Maheswari U, Martens C, Maumus F, Otillar RP, Rayko E, Salamov A, Vandepoele K, Beszteri B, Gruber A, Heijde M, Katinka M, Mock T, Valentin K, Verret F, Berges J a, Brownlee C, Cadoret J-P, Chiovitti A, Choi CJ, Coesel S, De Martino A, Detter JC, Durkin C, Falciatore A, *et al*: **The Phaeodactylum genome reveals the evolutionary history of diatom genomes.** *Nature* 2008, **456:**239–244.
21. Apt KE, Kroth-Pancic PG, Grossman AR: **Stable nuclear transformation of the diatom Phaeodactylum tricornutum.** *Mol Gen Genet* 1996, **252:**572–579.
22. Valenzuela J, Mazurie A, Carlson RP, Gerlach R, Cooksey KE, Peyton BM, Fields MW: **Potential role of multiple carbon fixation pathways during lipid accumulation in Phaeodactylum tricornutum.** *Biotechnol Biofuels* 2012, **5:**40.
23. Bozarth A, Maier U-G, Zauner S: **Diatoms in biotechnology: modern tools and applications.** *Appl Microbiol Biotechnol* 2009, **82:**195–201.
24. Kouprina N, Noskov V, Koriabine M, Leem S-H, Larionov V: **Exploring transformation-associated recombination cloning for selective isolation of genomic regions.** *Methods Mol Biol* 2004, **255:**69–90.
25. Larionov V, Kouprina N, Graves J, Chen XN, Korenberg JR, Resnick MA: **Specific cloning of human DNA as yeast artificial chromosomes by transformation-associated recombination.** *Proc Natl Acad Sci USA* 1996, **93:**491–496.
26. Kouprina N, Larionov V: **TAR cloning: insights into gene function, long-range haplotypes and genome structure and evolution.** *Nat Rev Genet* 2006, **7:**805–812.
27. Boeke J, Croute F, Fink G: **A positive selection for mutants lacking orotidine-5′-phosphate decarboxylase activity in yeast: 5-fluoro-orotic acid resistance.** *Mol Gen Genet* 1984, **197:**345–346.
28. Noskov VN, Segall-Shapiro TH, Chuang R-Y: **Tandem repeat coupled with endonuclease cleavage (TREC): a seamless modification tool for genome engineering in yeast.** *Nucl Acids Res* 2010, **38:**2570–2576.
29. Cohen N, Dagan T, Stone L, Graur D: **GC composition of the human genome: in search of isochores.** *Mol Biol Evol* 2005, **22:**1260–1272.
30. Gangalum RK, Jing Z, Nagaoka Y, Jiang M, Bhat SP: **Purification of BAC DNA for high-efficiency transgenesis.** *Biotechniques* 2011, **51:**335–336. 338.
31. Strong SJ, Ohta Y, Litman GW, Amemiya CT: **Marked improvement of PAC and BAC cloning is achieved using electroelution of pulsed-field gel-separated partial digests of genomic DNA.** *Nucleic Acids Res* 1997, **25:**3959–3961.
32. Osoegawa K, Woon PY, Zhao B, Frengen E, Tateno M, Catanese JJ, de Jong PJ: **An improved approach for construction of bacterial artificial chromosome libraries.** *Genomics* 1998, **52:**1–8.
33. Magbanua ZV, Ozkan S, Bartlett BD, Chouvarine P, Saski CA, Liston A, Cronn RC, Nelson CD, Peterson DG: **Adventures in the enormous: a 1.8 million clone BAC library for the 21.7 Gb genome of loblolly pine.** *PLoS One* 2011, **6:**e16214.
34. Montsant A, Jabbari K, Maheswari U, Bowler C: **Comparative genomics of the pennate diatom Phaeodactylum tricornutum.** *Plant Physiol* 2005, **137:**500–513.
35. Hallegraeff GM, Anderson DM, Cembella AD (Eds): *Manual on Harmful Marine Microalgae.* UNESCO Publishing; 2003.
36. *Phaeodactylum Tricornutum v2.0.* http://genome.jgi-psf.org/Phatr2/Phatr2.home.html.
37. Gibson DG: **Gene and genome construction in yeast.** In *Curr Protoc Mol Biol. Volume Chapter 3.* Edited by Ausube FM. ; 2011(April):Unit3.22.
38. Falciatore A, Casotti R, Leblanc C, Abrescia C, Bowler C: **Transformation of nonselectable reporter genes in marine diatoms.** *Mar Biotechnol* 1999, **1:**239–251.
39. Sambrook J, Fritsch EF, Maniatis T: *Molecular Cloning: A Laboratory Manual. Volume 3.* Cold Spring Harbor Laboratory Press; 2001:2344.
40. Gietz RD, Woods RA: **Yeast transformation by the LiAc/SS Carrier DNA/PEG method.** *Methods Mol Biol* 2005, **313:**107.

Pushing the science forward: chitosan nanoparticles and functional repair of CNS tissue after spinal cord injury

Bojun Chen[1†], Debra Bohnert[1†], Richard Ben Borgens[1,2*] and Youngnam Cho[1,3*]

Abstract

Background: We continue our exploration of the large polysaccharide polymer Chitosan as an acute therapy for severe damage to the nervous system. We tested the action of subcutaneously injected nanoparticles (~ 100 – 200 nanometers in diameter; 1 mg per ml) against control injections (silica particle of the same size and concentration) in a standardized *in vivo* spinal cord injury model. These functional tests used standardized physiological measurements of evoked potentials arriving at the sensorimotor cortex subsequent to stimulation of the tibial nerve of the contralateral hindlimb. We further explored the degree of acetylation and molecular weight of chitosan on the success of sealing cell damage using specific probes of membrane integrity.

Results: Not one of the control group showed restored conduction of evoked potentials stimulated from the tibial nerve of the hindleg – through the lesion – and recorded at the sensorimotor cortex of the brain. Investigation if the degree of acetylation and molecular weight impacted "membrane sealing" properties of Chitosan were unsuccessful. Dye - exchange membrane probes failed to show a difference between the comparators in the function of Chitosan in ex vivo injured spinal cord tests.

Conclusions: We found that Chitosan nanoparticles effectively restore nerve impulse transmission through the crushed adult guinea pig spinal cord in vivo after severe crush/compression injury. The tests of the molecular weight (MW) and degree of acetylation did not produce any improvement in Chitosan's membrane sealing properties.

Background

The integrity of the cell membrane is critical for maintaining cell physiological structure and function. For example, immediate function loss, progressive degeneration, and the death of neurons after acute spinal cord injury (SCI) is initiated after cell membrane disruption [1-4]. Spontaneous membrane self-repair is often initiated after the damage, but fails to overcome the overwhelming tissue distortion and physiological derangement, such as unregulated Ca^{2+} influx, reactive oxygen species (ROS) generation, and subsequent lipid peroxidation (LPO) (reviewed in [5] and [6]). Among all therapies, restoring membrane integrity rapidly and effectively after injury would be critical in early stages of Central Nervous System (CNS) damage interfering with progressive secondary injury [5,7-9].

It has been established that water-soluble polymers such as polyethylene glycol (PEG) can fulfill some of the requirements discussed above. PEG and some other synthetic polymers rapid and effectively seal membrane disruption [2,10]. Moreover it can preserve very significant structural, physiological, and behavioral function after SCI, Traumatic Brain Injury (TBI), and even localized peripheral nerve damage [2,11-13]. However, due to the viscosity of high molecular weight (MW) and the toxicity of low MW after the degradation of PEG, its administration must be limited in concentration and in timing after acute clinical neurotrauma [14,15]. Under some circumstances PEG can be quite toxic to CNS tissue [16-19] and see Figure Thirty one; Recovery of Behavioral and Physiological Function In Vivo page 128 in Borgens 2003 [20]. In a parallel line of investigation,

* Correspondence: cpr@purdue.edu; cho22@purdue.edu
†Equal contributors
[1]Center for Paralysis Research, Department of Basic Medical Sciences, College of Veterinary Medicine, Purdue University, West, Lafayette, IN 47907, USA
[2]Weldon School of Biomedical Engineering, Purdue University, West, Lafayette, IN 47907, USA
Full list of author information is available at the end of the article

studies showed that Chitosan also had similar and even more significant sealing actions than PEG [21]. Actually, the effect of chitosan, a non-toxic biodegradable polysaccharide polymer, has already been widely studied and used in biomedical and industrial applications, such as beverage clarification, wound healing, surgical adhesion, and drug delivery [22,23].

The chemical characteristics of Chitosan are mainly determined by two variables: the degree of acetylation (DA) (Figure 1) and MW [24]. The DA determines the number of free amino groups in the chitosan polymer which is inversely proportional to the degree of protonation. On the other hand, the MW determines the length of the main chain of the polymer, which potentially influences the viscosity of the solution and the shape of the polymer when it is presented in solution to targets such as cells and tissues. Previous studies showed that the DA plays a critical role in artificial membrane incorporation, mammalian membrane sealing, and neurophysiological function restoration ex vivo [21,25-28]. Electrostatic interaction between negative charged lipid headgroups on the cell membrane and the primary amines on the cationic deacetylated unit, β-(1–4)-linked D-glucosamine on the chitosan backbone has been proposed as the major physical driving force for its membrane adsorption and incorporation [25-27].

The effect of hydrophobic interaction between the inner non-polar hydrocarbon chain of the membrane and the nonionic acetylation unit, N-acetyl-D-glucosamine on the backbone of chitosan is also claimed to complete membrane insertion after chitosan adsorption onto the membrane surface [25-27]. As mentioned above, MW is closely related to the morphological structure of chitosan presented in the aqueous solution. Previous studies suggested the important role of viscosity and physical shape of chitosan presented in chitosan aqueous solution in biomedical applications [29]. Furthermore we have learned that chitosan in an injectable solution moves throughout the systemic circulation – apparently without issue dependent on viscosity [21]. Thus the route of administration (Intraperitoneal, Intraveneous, or Subcutaneous) does not appear to matter in a manner similar to PEG [10,12]. Nanoparticles and microparticles offer increased targeting to damaged cells coincident with a corresponding decrease in the concentration of the polysaccharide circulating within the body. Whether chitosan nanoparticles are able to reach and repair damaged tissues as has chitosan in aqueous solution in spinal injured animals with vascular trauma has not been tested until this report.

In addition, the effect of MW of chitosan on membrane organization has been discussed, mainly in the field of artificial lipid membranes considered as a simplified cell model. However, the issue was not settled [28,30,31]. Fang reported the enhanced interactions of lipid bilayers with higher MWs of chitosan, indicated by the significantly reduced cooperative units compared with those of lower MWs [30]. In contrast, Quemeneur observed the independent role of chitosan's MW on membrane adsorption in controlled pH environments [28].

Using these observations, we push our laboratory's success using Chitosan as an acutely applied membrane "sealant" forward – and present the first functional test of chitosan nanoparticles (100 – 200 nm diameter) application by subcutaneous injection to adult guinea pigs with crush/compression injury to the mid-thoracic spinal cord.

The use of nanoparticles (> 100 nm) provides a means to deliver Chitosan in large concentrations to only the interface of damaged cells (of course making up damaged tissues) escaping the need to systemically deliver the polymer as we have past reported. By revealing a functional benefit to nanoparticle administration – we further tested the "membrane repair" activity of Chitosan possessing different degrees of acetylation and molecular weight to determine the best composition/fabrication to move forward with further clinical and clinically oriented studies (Table 1). These studies used chemically modified chitosan in solution in tests of membrane integrity. These tests were: A) evaluating the loss of a large enzyme from

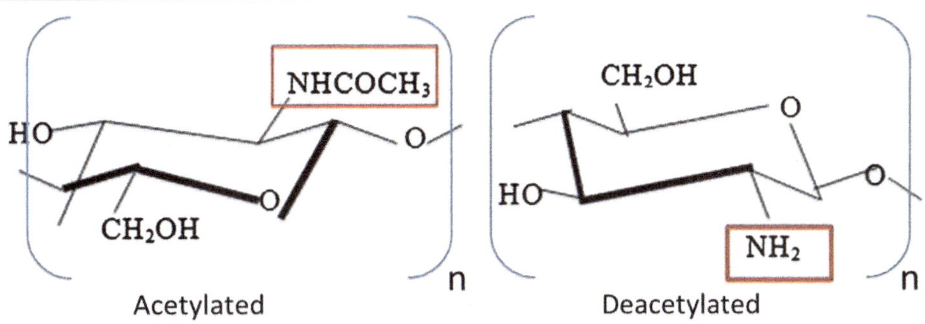

Figure 1 The chemical composition of acetylated and deacetylated chitosan. The amine group of the deacetylated chitosan (on the right), squared in red, is produced by the removal of the acetyl group, -$COCH_3$, from the acetylated chitosan (on the left) during the process of the deacetylation. Therefore, the amine group, squared in red, in the deacetylated chitosan can be protonated in the acidic environment.

Table 1 Different degree of acetylation and molecular weights of chitosan in this study

		Samples	DA (%)	MW (Da)
Molecular Weight (MW)	1 Low	Low MW chitosan	15 ~ 25	100 K
	2 Medium	Medium MW chitosan	15 ~ 25	200 K
	3 Oligo	Oligo MW chitosan	15 ~ 25	5000
Degree of Acetylation (DA)	4 DAO	DAO Chitosan	0	
	5 DA10	DA10 Chitosan	10	
	6 DA20	DA20 Chitosan	20	
	7 DA100	DA100 Chitosan	100	

cytoplasm to the extracellular milieu; and B), the intrusion of a dye molecule into the cytoplasm through compromised membrane. A more extensive discussion and technical detail relative to these "dye exclusion" techniques is detailed in [5,32]. Altogether, we pursue this line of investigation to provide a biodegradable alternative to PEG's use following acute Neurotrauma.

Discussion and conclusion

We provide the first preliminary results in the "whole animal" suggesting subcutaneous injection of Chitosan nanoparticles can dramatically improve nerve impulse conduction through the lesion – and upwards to the cerebral sensory centers after severe crush of mid-thoracic spinal cord of the adult guinea pig. In the absence of this experimental therapy, conduction remained silent. To our knowledge this is the first study of chitosan nanofabrications in any in vivo neurotrauma model. These data fit nicely with a previous in vitro investigation revealing a profound interference with progressive secondary injury mechanisms in acrolein - poisoned cell samples, membrane sealing, and the unique ability of chitosan particles to carry a "drug cargo". This significantly enhances its therapeutic potential [21].

Our results reveal that Chitosan nanoparticles can indeed function as an important intervention in acute spinal cord injury by ameliorating the effects of acute compression/crush to the adult guinea pig spinal cord. The physiological recovery of conduction shown here frames the basis for functional recovery of motor, sensory, and autonomic functioning after neurotrauma (see discussion in [6]). However, being cautious, this promise is not yet fulfilled until clinically meaningful recovery of behavioral outcome measures is shown. Furthermore - by evaluating different compositions of chitosan in controlled solutions ex vivo - we do not believe MW or DA may be a critical factor in developing this use of a safe biodegradable polysaccharide for clinical treatment of SCI or TBI [12]. In other words, the strong membrane sealing effect of chitosan on damaged spinal cord tissues occurred regardless of DA and MW. Specifically, the TMR and LDH study confirmed the significant effect of chitosan on inhibiting i.) the intrusion of an exogenous applied dye into damaged spinal cord cells ($P < 0.05$); ii.) as well as escape of a large endogenous enzyme LDH through membrane defects ($P < 0.05$).

Membrane sealing and repair

While there may be differences in scientific nomenclature between repair and sealing – those differences are moot in the context of trauma. The exchange of ions and macromolecules after membrane compromise is actually the initial event triggering the eventual death and degeneration of cells after injury (reviewed by [5,6]). The only logical means of acutely dampening and even reversing progressive secondary injury, cell degeneration, localized axotomy, and the demise of neurons and their support cells during the steady collapse of the nervous system is *to restore membrane integrity, at a minimum, to block this exchange*. This can be accomplished using various polymers and surface – active agents we have called membrane "sealants". While the mechanisms of action may vary somewhat between inorganic polymers and polysaccharides while accomplishing this feat [32] – the outcome is the same: the membrane defect is acutely "plugged"; the organization of water and ions in and around the defect is normalized; and spontaneous re-assembly of the plasmalemma then proceeds via lipid bilayer components resolving into one another. In the case of Polymers such as PEG – there is an affinity for damaged membranes based on the disruption of the hydrophilic outer leaflet. The membrane fusion is believed mediated by a dehydration effect and volume-exclusion aggregation of membrane lipids bringing adjacent lipids into close physical contact [2,3,10-13,33]. With Polymers such as Chitosan – the "reorganization" of membrane structure is initiated, but not limited to, the edge of the membrane defect. The subsequent continuing aggregation of chitosan on the membrane surface results from the extended conformation of chitosan's main chain [25,26,28,30-32]. We hasten to add that PEG can facilitate a functional reconnection between proximal and distal segments of severed axons [2,20]. This takes many minutes of absolute immobility of the axon segments in the presence of the polymer to accomplish this task, and an unknown, and untested, period of immobilization of the tissue for the repair to become permanent. We do not see why chitosan would not as well induce membrane fusion permitting reconnection of axon segments in an experimental setting. However, in the context of the animal or human suffering from neurotrauma – such immobility would be impossible so this mechanism of action for any polymer-based sealant is a moot point.

The issue of the degree of acetylation

The different degrees of acetylation represent different strengths of electrostatic interaction between the cationic chitosan and anionic lipids. Specifically, the lower the DA, the higher the number of positive ions a chitosan molecule carries. In turn, the higher concentration of surface charges on chitosan enables a more extended conformation due to electrostatic repulsion among charged amines on the chitosan backbone. Therefore, the polarity of the molecules is greatly enhanced. Since the lipid headgroups are negatively charged, molecules with lower DA might have a greater opportunity to diffuse close to the membrane breaches, and thus adsorb and seal the membrane through electrostatic interactions. This mechanism is consistent with previous studies in which chitosan polymers incorporated with artificial membrane films and vesicles where physical attractions and interactions were a result of membrane adsorption [25,26]. Our data however, did not reveal a hint of an improvement in function of chitosan based on the DA.

The issue of molecular weight

Here, equivalent sealing effects using different MW chitosan treatments was suggested in both TMR and LDH tests, and in both transection and compression ex vivo injury models. Based on the fact that different MW chitosan in our study shared a similar DA, one possible reason for the consistency of data found in different MW chitosan treatments might be the predominant role of DA over chitosan MW in initiating membrane repair. A similar result is also observed in the zeta-potential evaluation of the influence of different MWs of chitosan on membrane adsorption [28].

The data shown in Figure 2 reveals that all tests showed a statistically significant reduction in TMR fluorescent intensity. There was little difference between them however - with the exception of DA100 which revealed a possible lower capability for sealing. An alternative thought is that the MW could play a role in membrane sealing [30,31] as it is closely related with the morphological structure of chitosan presented in the aqueous solution. Typically, the two most common structures are 1.) rope-like and 2.) coil-like polymeric shapes. The most likely shape in an oligo-chitosan solution would be rope-like due to the relatively few numbers of monomers existing in the chitosan. The less folded main chains of this polymer require less energy to diffuse close to the cell membrane and fit near or into its breaches. On the other hand, MW is also proportional to the viscosity of the aqueous solution, which reflects the hydrodynamic diameter of chitosan molecules; refer back to Figures 3 and 2 [29]. The degree of the intermingling entanglement of chitosan polymer is proportional to the chain length, in other words - the MW. Considering the higher MW of Medium chitosan which is 40 times the size of the oligomer, this might result in the reduced mobility of the polymer. This

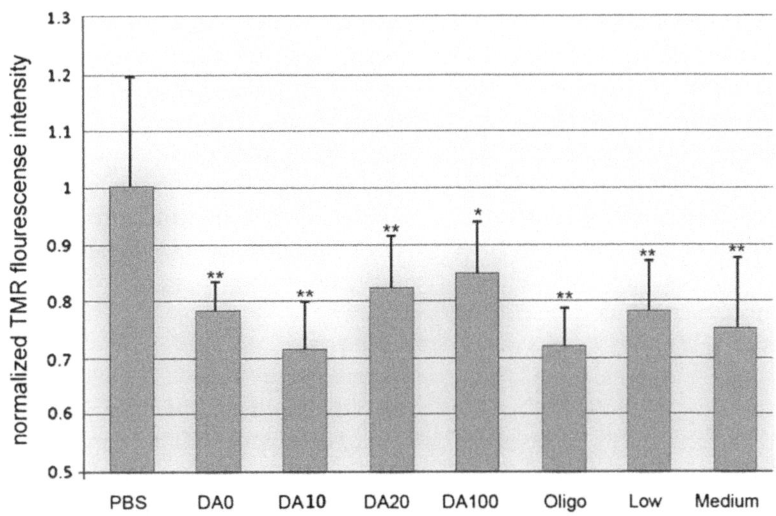

Figure 2 Quantification of normalized fluorescence intensity of TMR uptake as a function of chitosan treatments. The fluorescence intensity in the PBS group was used as the control. Note that all chitosan treatments significantly decreased the fluorescence intensity of TMR uptake following the transection injury, *$P < 0.05$, ** $P < 0.01$ (Dunnett test). A Statistical difference was not observed between chitosan treatments with different DA and MW, $P > 0.05$ (Tukey test). $N = 6$.

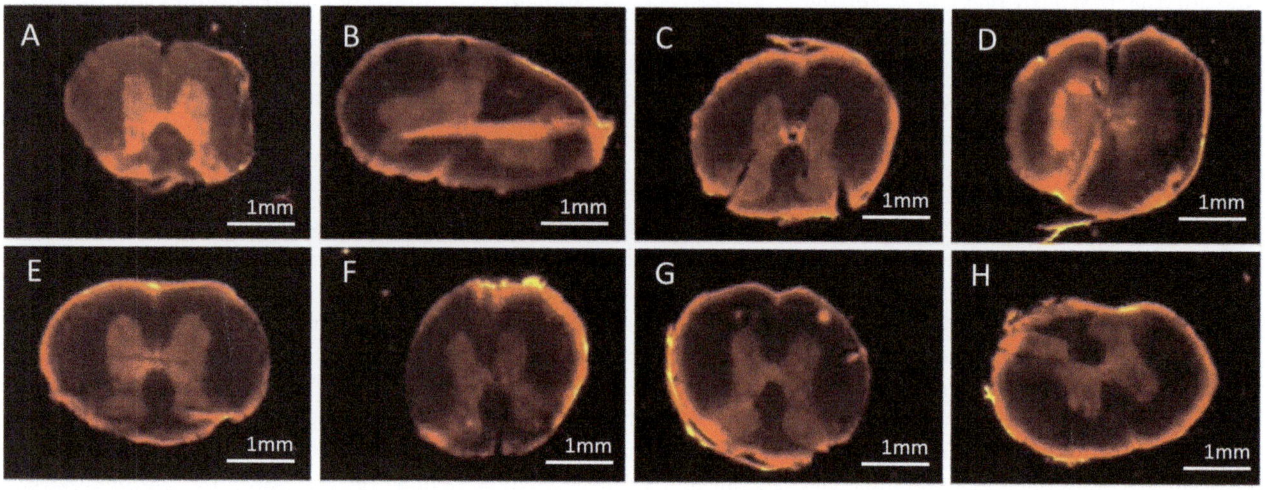

Figure 3 The fluorescence intensity of TMR uptake after transection injury and its inhibition by different chitosan treatments. A) control, **B**) DA100, **C**) DA20, **D**) DA10, **E**) DA0, **F**) Oligo, **G**) Low, **H**) Medium. See test for nomenclature.

is likely to enhance polymer adhesion by increasing the contact area between chitosan molecules and lipids [34].

"Nanomedicine"

The superiority of nano-fabrications in particular over aqueous suspensions of injected polymers, cannot be under appreciated. High molecular weights of PEG are too viscous for facile IV use. Lowering the MW in an effort to produce a clinically easy injection may produce some level of toxicity due to the circulation of polymers which do not degrade and must be removed at the level of the kidney and liver (as discussed and cited above). Chitosan, as a naturally occurring polysaccharide, is not toxic. Detection of it in bodily fluids after systemic administration is unlikely.

Moreover, PEG coated silica nanoparticles required the PEG surface coat fabrication over the silica nanoparticle which is inert. To be further useful as a targeted repair agent and a delivery vehicle requires a third step - the bonding of surface PEG to the drug or cytokine of choice. In chitosan nanofabrications, the chitosan itself is the "sealant" while capturing or surface bonding of a drug/cytokine to be released requires only another single step in the process. Chitosan nanospheres - by themselves - show significant sealing properties producing improved physiological recovery after SCI.

Results

In vivo testing

SSEP responses to injection of chitosan microspheres

As expected, and observed many times before, all 5 silica nanoparticle – injected control animals failed to recover any form of evoked potential (EP) by 3 weeks after compression injury. The lack of conduction was confirmed by clear recordings of EPs elicited from Median nerve "control" stimulation at the same stimulation parameters before and after spinal compression. Of the 11 nanoparticle-treated animal group, 3 of these died prior to the final physiological measurement session. Instead of complicating evaluation by periodic scoring, we statistically compared only those animals that survived until the end of the study. SSEP conduction recovered in every one of these eight healthy animals. This difference between "controls and experimentals" was strikingly statistically significant ($P = 0.007$; Fisher's exact; see Figure 4).

Enzyme leakage from injured and untreated and treated spinal cords

The membrane sealing effect of chitosan was further confirmed by the inhibition of leakage of the intracellular enzyme Lactate Dehydrogenase (LDH) by specific assay (Figure 5). The signal intensity was proportional to the amount of LDH present in the bathing solution and inversely related with the condition of membrane integrity. More than a 60% decrease of the signal intensity after all chitosan treatments strongly indicated the sealing effect of chitosan ($P < 0.05$). Similar to the Tetramethylrhodamine (TMR) test, we did not observe any correlation with the result and the composition (MW and DA) of chitosan.

Methods

In vivo testing

Spinal cord injury in the adult guinea pig

We have used the adult guinea pig model of crush/compression injury to their mid-thoracic spinal cords for over 25 years. Extensive detail of the surgical manipulations and aftercare of spinal–injured guinea pigs, can be found in several previous reports (for examples see; [5,10,34,35]). Briefly: adult female guinea pigs (300 – 400 gms; Hartley strain) were deeply anesthetized, and a midline incision

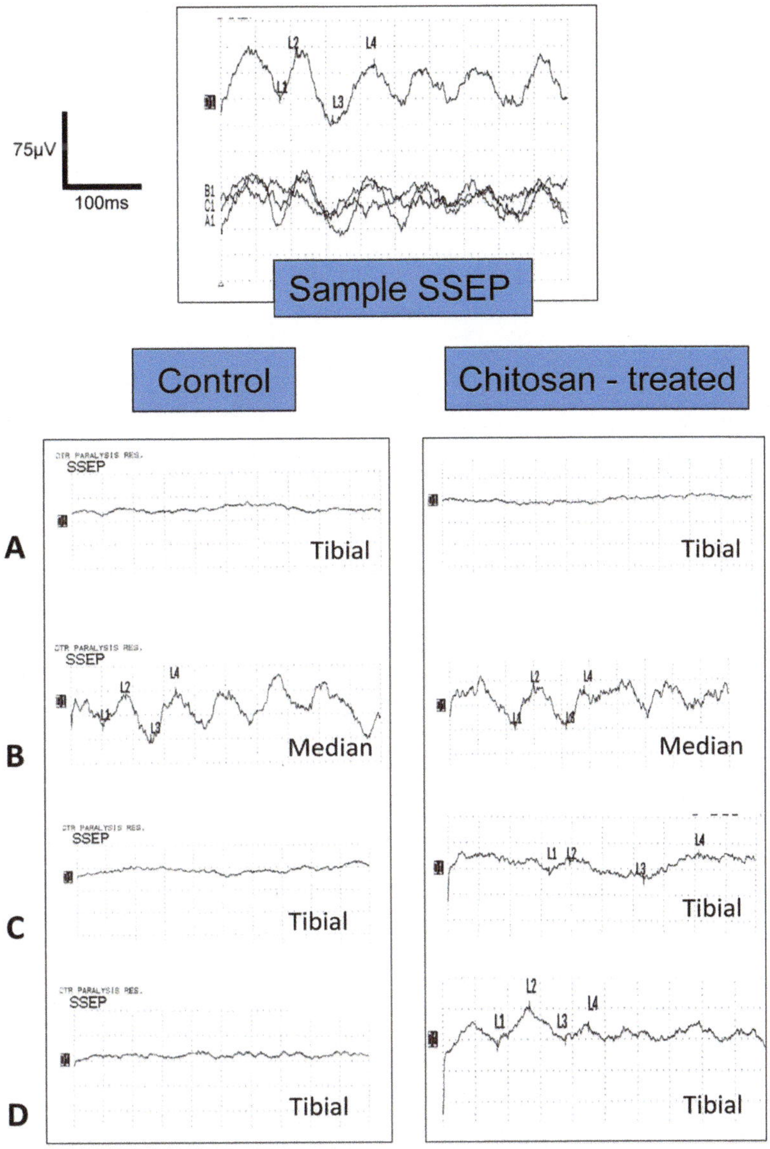

Figure 4 Somatosensory recordings in control and injured guinea pig spinal cords. The uppermost electrical record is a sample to show the single and averaged evoked potential as they are recorded. This record was produced by stimulation of the tibial nerve and recording the arrival of the ascending evoked potentials (EPs) of ~ 40 msec latency at the sensory motor Cortex. The three bottom traces are single records taken consecutively from the same uninjured animal as detailed in the text, while the top record is an average of these three. Any statistical deviation from baseline is automatically annotated. All data presented below are such averaged signals. The left column shows records taken from a Control animal. The right column shows records obtained from a Chitosan Nanoparticle-treated animal. Note that in both samples, compression injury eliminated transmission of EPs through the cord. Immediately below, the clear recordings of SSEP in both animals were obtained from Median Nerve stimulation/recording. This neural circuit from the forelimb to the sensory cortex is unaffected by injury to the midthoracic spinal cord, and is an internal control for the functioning of the system – eliminating the likelihood of "false negatives". Note the appearance of weak EPs in only the Chitosan Nanoparticle – treated animal – while electrophysiological functioning in the Control animal were absent throughout the observation period. Reading down, recordings are; **A**) immediately post injury, **B**) median nerve control, **C**) one week post injury, and **D**) two weeks post injury. The time base is 10 msec per division, with an amplitude of 1.25 microvolts/division.

in the mid-thoracic regions exposed the vertebral column. A hemi-laminectomy procedure was performed exposing the spinal cord, and it was crushed using a laboratory - fabricated instrument which was basically a forceps possessing a détente. In this way the cord compression could be reasonably standardized between animals – and this "consultant displacement" injury is far superior in producing a standardized outcome than any "constant force"/weight drop technique. Animals were kept warm until fully recovered and moved to

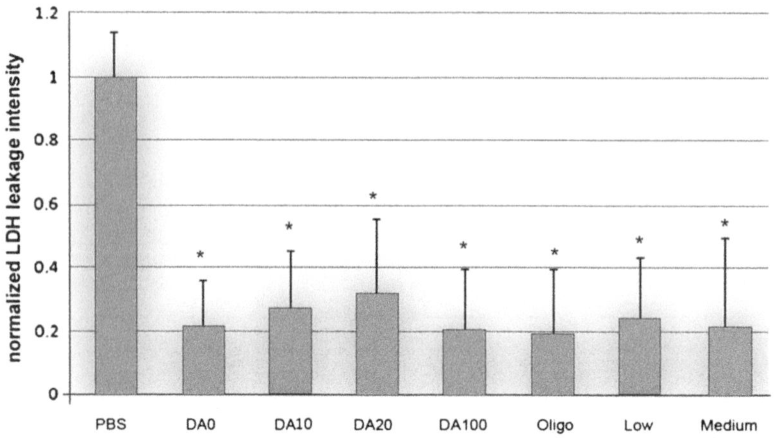

Figure 5 Quantification of the normalized fluorescence intensity of LDH leakage as a function of chitosan treatments. The fluorescence intensity in the PBS group was used as the control. Note that all chitosan treatments significantly decreased the fluorescence intensity of LDH leakage following the compression injury, *$P < 0.05$ (Dunnett test). A statistical difference was not observed between chitosan treatments of different DA and MW, $P > 0.05$ (Tukey test). $N = 6$.

special housing where they received daily care from the technical staff of the Center for Paralysis Research (CPR) and Veterinary Animal Holding Facility.

The measurement of spinal cord evoked potentials in adult guinea pig as a measure of functional recovery

This laboratory pioneered the use of Somatosensory Evoked Potential (SSEP) recording using commercial clinical instruments for the measurement of evoked potentials as a means to determine conduction through guinea pig spinal lesions. We do not evaluate SSEPs to study the nature of subtle neurological conduction in CNS issues – but rather to determine the presence or absence of EP conduction relative to each animal's pre-operative recordings. For over a decade, we have learned this is the *most sensaive* of functional tests performed to judge the animals' recovery from SCI. Sedated guinea pigs were connected to a Nihon Kohden Neuropack 2 Stimulator/Recorder using subcutaneous pin electrodes. Stimulating electrodes were inserted near the tibial nerve of the hindleg and a separate pair near the medial nerve of the ipsilateral foreleg.

The frequency of stimulation of the leg(s) was set at ~ 200 HZ which would produce a clear twitching of the very local region surrounding the stimulating electrodes. The magnitude of the stimulation was then reduced until such movement was nearly undetectable. Recording electrodes were inserted beneath the scalp superior to the region of the sensory cortex on the contralateral side (see the citations above for figures and for more detail). Control stimulation of the foreleg medial nerve was an excellent tool to prove electrode placement (and other variables) did not affect the recordings of ascending Evoked Potentials (EPs). This latter neural circuit was "above" the level of the injury to the cord. On the other hand, SCI completely blocks such conduction to the brain when the tibial nerve is stimulated. A recovery of EPs recorded at the contralateral sensory cortex subsequent to a standardized regimen of electrical stimulation of the hindlegs was taken as a recovery of electrophysiological function in spinal cord injured animals. Recording of normal tibial – mediated SSEPs were always performed on each animal prior to SCI. Median nerve control procedures were a part of every recording period. Replacement or movement of electrodes was unnecessary since the medial nerve circuit set the standard for stimulation/recording, and the use of a laboratory – fabricated "toggle" allowed switching between stimulations of the hind and foreleg without touching the animal or the electrical connections.

Construction, evaluation, and administration of chitosan nanoparticles

We have already provided complete details of fabrication, chemical analysis, imaging, including the loading of drugs into nanoconstructions of Chitosan spheres as used in this study (for examples see [21,36]). We did not perform these analytical procedures again in this study since all features were well standardized, understood, and reported.

The Chitosan nanospheres used here were within 100 – 200 nm in diameter, fabricated using sterile media for use in animals, at a concentration of 1 mg of chitosan particles/ml of injection. 1 ml of this solution was injected subcutaneously in the nape of the neck, within ½ hour after SCI. Control injections of sterile lactated Ringers containing silica particles of the same dimensions and concentration were performed in a separate set of control animals accordingly. Only 5 control animals were studied since their use here was identical to over 50 such saline- injected animals used as controls in 4 individual published prior studies. Animals never recover SSEPs for varying periods of study durations up to – and sometimes exceeding - 1 month post SCI.

All guinea pigs used in this study were handled in accordance with, and prior approval by, the Purdue University Animal Care and Use Committee (PACUC).

Test solutions

0.1% (w/v) chitosan solutions completely dissolved in 1% acetic acid solution by stirring overnight. Figure 1 shows the acetylated and deacetylated forms of chitosan which is that of basic organic chemistry. Detail information on these polymers is provided in Table 1. All samples were bubbled with 95% O_2/5% CO_2 throughout the duration of the experiment.

Considering membrane permeability

The membrane permeability of damaged spinal cord segments with different treatments of different degrees of acetylation and molecular weights of chitosan solutions was evaluated by the interference with Tetramethylrhodamine (TMR; 4000 MW) and the leakage of the large endogenous enzyme Lactate Dehydrogenase (LDH, 140 kDa). The leakage of the relatively large molecular weight enzyme from disrupted membrane not only indicated compromize - but is a gauge of how severe such membrane breaches were [21,36]. The membrane sealing effect of chitosan was measured as a function of the leakage of LDH and the intrusion of TMR.

Briefly, after compression injury was induced, equally weighted segments were incubated immediately into different chitosan and 0.1 $molL^{-1}$ Phosphate buffered saline (PBS) solutions respectively for 15 mins. Then, all segments were quickly rinsed in PBS solution for three times to remove any leaked LDH extracellularly. Segments were incubated in 1 ml fresh PBS solution for additional 45 mins to allow LDH leakage from un-repaired membrane breaches. The level of extracellular LDH in the above 600 μl bathing solution was assayed using the TOX-7 kit (Sigma-Aldrich, Inc, USA). Six replicates were applied in this study. Five measurements were taken for each sample.

Measurement of TMR uptake

The membrane sealing effect of chitosan was measured by the uptake of TMR. Briefly, immediately after transection injury, 1-cm segments of spinal cord were incubated in different chitosan solutions and PBS solution, respectively for 15 mins. Then, all segments were transferred and incubated in 0.1% TMR dextran solution for 15 mins in the dark. Subsequently, segments were quickly rinsed in 0.1 $molL^{-1}$ PBS solution for three times and fixed in 4% paraformaldehyde for 5 h at 4°C in the dark. Segments were then embedded in Tssue –Tek OCT compound, frozen in liquid nitrogen. Sections were cut at 50 μm thickness using a freezing microtome (Thermo Electron, Waltham, MA, USA). Sections were visualized by epifluorescence on an Olympus BX61 microscope with a standard rhodamine cube (545 nm and 590 nm of excitation filter and emission filter, Olympus). The fluorescence of white matter was quantified using Image J software (NIH). Five sections were randomly selected from the injury zone. Six replicates were conducted in this test.

Statistical analysis

The data is expressed as a mean ± SD. One-way ANOVA were used for statistical analyses (SAS 9.2, SAS Institute Inc., NC, USA). Results showing significance between different chitosan treated groups were subjected to the Tukey's test. In vivo evaluations used a test of proportions; Fisher's Exact Test.

Competing interests

The authors declare they have no competing interests. There is no conflict of interest of any sort in the reporting of these data relative to any author.

Authors' contributions

BC drafted the manuscript, YC and RBB designed the experiments. DB and BC performed the experiments. RBB is the Principle Investigator and Director of the CPR and is responsible for all elements of the research. This work is in partial fulfillment of the Doctoral Degree requirements of BC. All authors read and approved final manuscript.

Acknowledgements

We appreciate the professional technical assistance of Michel Schweinsberg with the graphics, and the assistance of Jennifer Danaher for manuscript preparation. We thankfully acknowledge the financial support from the General Funds of the Center for Paralysis Research, The State of Indiana, and a generous endowment from Mrs. Mari Hulman George.

Author details

[1]Center for Paralysis Research, Department of Basic Medical Sciences, College of Veterinary Medicine, Purdue University, West, Lafayette, IN 47907, USA. [2]Weldon School of Biomedical Engineering, Purdue University, West, Lafayette, IN 47907, USA. [3]Present address: New Experimental Therapeutics Branch, National Cancer Center, 809 Madu-1dong, Ilsandong-gu, Goyang-si, Gyeonggi-do 410-769, Korea.

References

1. Luo J, Li N, Robinson JP, Shi R: **The increase of reactive oxygen species and their inhibition in an isolated guinea pig spinal cord compression model.** *Spinal Cord* 2002, **40**:656–665.
2. Borgens RB, Shi R: **Immediate recovery from spinal cord injury through molecular repair of nerve membranes with polyethylene glycol.** *FASEB J* 2000, **14**:27–35.
3. Borgens RB: **Cellular engineering: molecular repair of membranes to rescue cells of the damaged nervous system.** *Neurosurgery* 2001, **49**:370–378. discussion 378–379.
4. Schwab ME: **Repairing the injured spinal cord.** *Science* 2002, **295**:1029–1031.
5. Cho Y, Borgens RB: **Polymer and nano-technology applications for repair and reconstruction of the central nervous system.** *Exp Neurol* 2012, **233**:126–144.
6. Borgens RB, Liu-Snyder P: **Understanding secondary injury.** *Q Rev Biol* 2012, **87**:89–127.
7. Shi Y, Kim S, Huff TB, Borgens RB, Park K, Shi R, Cheng JX: **Effective repair of traumatically injured spinal cord by nanoscale block copolymer micelles.** *Nat Nanotechnol* 2009, **5**:80–87.
8. Liu-Snyder P, Borgens RB, Shi R: **Hydralazine rescues PC12 cells from acrolein-mediated death.** *J Neurosci Res* 2006, **84**:219–227.

9. Liu-Snyder P, Logan MP, Shi R, Smith DT, Borgens RB: **Neuroprotection from secondary injury by polyethylene glycol requires its internalization.** *J Exp Biol* 2007, **210**:1455–1462.
10. Borgens RB, Bohnert D: **Rapid recovery from spinal cord injury following subcutaneously administered polyethylene glycol.** *J Neurosci Res* 2001, **66**:1179–1186.
11. Koob AO, Colby JM, Borgens RB: **Behavioral recovery from traumatic brain injury after membrane reconstruction using polyethylene glycol.** *J Biol Eng* 2008, **2**:9.
12. Koob AO, Duerstock BS, Babbs CF, Sun Y, Borgens RB: **Intravenous polyethylene glycol inhibits the loss of cerebral cells after brain injury.** *J Neurotrauma* 2005, **22**:1092–1111.
13. Donaldson J, Shi R, Borgens R: **Polyethylene glycol rapidly restores physiological functions in damaged sciatic nerves of guinea pigs.** *Neurosurgery* 2002, **50**:147–156.
14. Brent J: **Current management of ethylene glycol poisoning.** *Drugs* 2001, **61**:979–988.
15. Caravati EM, Erdman AR, Christianson G, Manoguerra AS, Booze LL, Woolf AD, Olson KR, Chyka PA, Scharman EJ, Wax PM, *et al*: **Ethylene glycol exposure: an evidence-based consensus guideline for out-of-hospital management.** *Clin Toxicol (Phila)* 2005, **43**:327–345.
16. Schneiderman S, Farber JL, Baserga R: **A simple method for decreasing the toxicity of polyethylene glycol in mammalian cell hybridization.** *Somat Cell Mol Genet* 1979, **5**:263–269.
17. Szabó G, Kiss A, Trón L: **Permeabilization of lymphocytes with polyethylene glycol 1000. Discrimination of permeabilized cells by flow cytometry.** *Cytometry* 1982, **3**:59–63.
18. Webster R, Elliott V, Park BK, Walker D, Hankin M, Taupin P: **PEG and PEG conjugates toxicity: towards an understanding of the toxicity of PEG and its relevance to PEGylated biologicals.** In *PEGylated Protein Drugs: Basic Science and Clinical Application.* Edited by Veronese FM. Switzerland: Birkhäuser Basel; 2009:127–146. Milestones in Drug Therapy.
19. Cole A, Shi R: **Prolonged focal application of polyethylene glycol induces conduction block in guinea pig spinal cord white matter.** *Toxicol Vitr* 2005, **19**:215–220.
20. Borgens RB: *Restoring function to the injured human spinal cord.* Heidelberg, Germany: (Monograph) Springer-Verlag; 2003.
21. Cho Y, Shi R, Borgens RB: **Chitosan produces potent neuroprotection and physiological recovery following traumatic spinal cord injury.** *J Exp Biol* 2010, **213**:1513–1520.
22. Chavasit V, Torres JA: **Chitosan-poly(acrylic acid): mechanism of complex formation and potential industrial applications.** *Biotechnol Prog* 1990, **6**:2–6.
23. Lee DW, Lim H, Chong HN, Shim WS: **Advances in chitosan material and its hybrid derivatives: a review.** *Open Biomater J* 2009, **1**:10–20.
24. Aranaz I, Mengibar M, Harris R, Panos I, Miralles B, Acosta N, Galed G, Heras A: **Functional characterization of chitin and chitosan.** *Curr Chem Biol* 2009, **3**:203–230.
25. Pavinatto FJ, Caseli L, Pavinatto A, Dos Santos DS Jr, Nobre TM, Zaniquelli ME, Silva HS, Miranda PB, De Oliveira ON Jr: **Probing chitosan and phospholipid interactions using Langmuir and Langmuir-Blodgett films as cell membrane models.** *Langmuir* 2007, **23**:7666–7671.
26. Pavinatto FJ, Pavinatto A, Caseli L, Santos DS Jr, Nobre TM, Zaniquelli ME, Oliveira ON Jr: **Interaction of chitosan with cell membrane models at the air-water interface.** *Biomacromolecules* 2007, **8**:1633–1640.
27. Pavinatto FJ, Caseli L, Oliveira ON: **Chitosan in nanostructured thin films.** *Biomacromolecules* 2010, **11**:1897–1908.
28. Quemeneur F, Rinaudo M, Pepen-Donat B: **Influence of molecular weight and pH on adsorption of chitosan at the surface of large and giant vesicles.** *Biomacromolecules* 2008, **9**:396–402.
29. Chattopadhyay DP, Inamdar MS: **Aqueous behaviour of chitosan.** *Int J Polym Sci* 2010.
30. Fang N, Chan V, Mao HQ, Leong KW: **Interactions of phospholipid bilayer with chitosan: effect of molecular weight and pH.** *Biomacromolecules* 2001, **2**:1161–1168.
31. Fang N, Chan V: **Chitosan-induced restructuration of a mica-supported phospholipid bilayer: an atomic force microscopy study.** *Biomacromolecules* 2003, **4**:1596–1604.
32. Chen B, Zuberi M, Borgens RB, Cho Y: **Affinity for, and localization of, PEG-functionalized silica nanoparticles to sites of damage in an ex vivo spinal cord injury model.** *J Biol Eng* 2012, **6**:18.
33. Lentz B: **PEG as a tool to gain insight into membrane fusion.** *Eur Biophys J* 2007, **36**:315–326.
34. Shi R, Borgens RB: **Acute repair of crushed guinea pig spinal cord by polyethylene glycol.** *J Neurophysiol* 1999, **81**:2406–2414.
35. Cho Y, Shi R, Borgens R, Ivanisevic A: **Repairing the damaged spinal cord and brain with nanomedicine.** *Small* 2008, **4**(10):1676–1681.
36. Cho Y, Shi R, Ben Borgens R: **Chitosan nanoparticle-based neuronal membrane sealing and neuroprotection following acrolein-induced cell injury.** *J Biol Eng* 2010, **4**:2.

21

Influence of surface topography on the human epithelial cell response to micropatterned substrates with convex and concave architectures

Mee-Hae Kim[1], Yoshiko Sawada[2], Masahito Taya[2] and Masahiro Kino-oka[1*]

Abstract

Background: Understanding the fundamental mechanisms underlying the cellular response to topographical surface features will extend our knowledge regarding the regulation of cell functions. Analyzing the cellular response to different topographical features, over multiple temporal and spatial scales, is central to understanding and guiding several biological functions. We used micropatterned substrates with convex and concave architectures to evaluate the behaviors of human epithelial cells on these substrates.

Results: Pillar and pit substrates caused heterogeneous spatial growth and distribution, with differences in cell density, over 48 h. Regional densities and distribution were significantly increased at pillar sidewalls, and at pit sidewalls and bottoms compared with those on flat unpatterned areas. Time-lapse observations revealed that different mechanisms of cell migration were dependent upon pillar and pit features. Cells on pillar substrate migrated towards the sidewall, whereas cells on pit substrate tended to move towards the sidewalls and bottom. Cytoskeletal staining of F-actin and vinculin showed that this migration can be attributed to difference in spatial reorganization of actin cytoskeleton, and the formation of focal adhesions at various points on the at the convex and concave corners of pillar and pit substrates. Cells cultured on the pillar substrate had stress fibers with extended filopodia and immature focal contacts at the sidewalls and convex corners, similar to those on the flat unpatterned substrate. Cells at the sidewalls and concave corners of pit substrate had more contractile stress fibers and stable focal contacts compared with cells on the pillar substrate. We also found that the substrate structures affect cell-cell contact formation *via* E-cadherin, and that this was associated with reorganization of the actin cytoskeleton at the sidewall, and at the convex and concave corners of the substrate.

Conclusion: Migration is an important factor affecting spatial growth and distribution. Heterogeneity at various locations was caused by different migratory behaviors at the convex and concave corners of pillar and pit substrates. We propose that this investigation is a valuable method for understanding cell phenotypes and the heterogeneity during spatial growth and distribution of epithelial cells during culture.

Keywords: Human epithelial cells, Convex architecture, Concave architecture, Spatial cell distribution, Stereoscopic cell imaging, Cytoskeletal formation, Cell-cell interaction

* Correspondence: kino-oka@bio.eng.osaka-u.ac.jp
[1]Department of Biotechnology, Graduate School of Engineering, Osaka University, 2-1 Yamadaoka, Suita, Osaka 565-0871, Japan
Full list of author information is available at the end of the article

Introduction

In tissue engineering, various efforts have been made to promote tissue regeneration. Studies have shown that providing an appropriate environment, with chemical or physical support, is important for cellular functions such as cell adhesion, proliferation, migration and differentiation [1-3]. In particular, cell adhesion on scaffold surfaces is a primary step for guiding cellular function and further tissue generation [3]. By understanding the manner in which cells interact with their physical environment, it might be possible to control cellular behavior through the fabrication of substrates with unique physical properties [4]. These approaches could allow researchers to study the dynamic responses of cells to well-defined micropatterned substrates, and the effects of modulating cell behaviors such as cell–cell and cell–substrate interactions with respect to using such constructs for tissue replacement [2,4].

Many approaches to manipulate the cell microenvironment have been conducted on micropatterned surfaces. The behaviors of many cell types have been examined using various micropatterned substrates created by a variety of microlithography and micropatterning techniques [2-16]. Of these modified surfaces, regular micropatterns affected cell responses, controlling cell morphology and function, when compared with cells grown on unpatterned flat substrates [2,3]. Green *et al.* studied the growth rate of human abdominal fibroblasts cultured on substrates patterned with square pillars or pits. They observed that cells were more sensitive to pillars or pits with smaller sizes [14]. The influence of microarchitecture on cell behavior with respect to morphology and functionality is further exemplified by the ability of cells to acutely sense variability in topographic cues. However, there have been few systematic analyses into the impact of micropatterned features on cell adhesion and migration mechanisms. Therefore, analyzing the cell responses to different topographical cues, acting over multiple temporal and spatial scales, is central to understanding and guiding several biological functions.

In this study, micropatterned substrates with convex and concave architectures were established to assess the responses of human epithelial cells to these substrates and to determine their spatial growth and distribution. We investigated the fundamental mechanisms of cell and culture surface interactions with respect to the formation of the actin cytoskeleton, focal adhesion, and cell-cell contacts.

Materials and methods

Fabrication of micropatterned substrates

Micropatterned substrates were provided by Kuraray Co., Ltd. (Kurashiki-shi, Japan). Two different topographic patterns, pillar and pit, were fabricated in polystyrene using a UV-lithographic technique. The fabrication process was completed by growing SiO_2 using a vacuum deposition system on the substrates. A schematic outlining micropatterned surfaces composed of pillar and pit are shown in Additional file 1: Figure S1. The different spatial aspects for the pillar and pit features were: the top surface; the sidewalls; the bottom surface; and gaps. The gaps refer to spaces between adjacent pillar and pit. Fabricated samples were observed with a scanning electron microscope (Additional file 1: Figure S1). The dimensions of the pillar features were 50.8 ± 0.56 μm wide and 25.9 ± 0.18 μm high, with a pitch of 198.0 ± 0.57 μm. The dimensions of pit features were 53.8 ± 0.75 μm wide and 22.6 ± 0.59 μm high, with a pitch of 195.2 ± 3.48 μm. The pitch sizes were set to specific values so as to clarity an individual topography itself effects.

Cells and culture conditions

Infinity telomerase immortalized human epithelial cells (hTERT-HME1; Clontec Laboratories, San Diego, CA, USA) were thawed and incubated in a 25-cm^2 flask (Nunc, Roskilde, Denmark). Unless otherwise stated, the cells were cultivated in serum-free medium containing 10 μg/ml insulin (HuMedia-KG2; Kurabo Industries, Osaka, Japan) at 37°C under a 5% CO_2 atmosphere. For experiments, the seeding density of viable cells (X_0), determined by trypan blue exclusion staining, was 8.0×10^3 cells/cm^2. Culture medium was replaced every 72 h.

Determination of spatial cell distribution

The procedures used for staining of cytoplasm and nuclei were similar to those described previously [15]. Briefly, cells were rinsed twice with phosphate-buffered saline (PBS, Sigma-Aldrich, MO, USA). Samples were incubated at 37°C under an atmosphere with 5% CO_2 in HuMedia-KG2 containing CellTraker Green CM-FDA (Molecular Probes, Eugene, OR, USA) to stain the cytoplasm of living cells. After 45 min, the cells were further incubated for 30 min in HuMedia-KG2 without CellTraker Green CM-FDA and then were rinsed twice with PBS.

Cells were fixed with 3.7% paraformaldehyde in phosphate buffer (Wako Pure Chemical Industries, Osaka, Japan) for 10 min at room temperature and rinsed with PBS. They were then soaked in PBS with 0.25% Triton X-100 for 4 min. Cells were washed three times with PBS and then counterstained with TOPRO-3 (Molecular Probes) for visualization of nuclei. Cells were then observed using a confocal laser-scanning microscope (CLSM, model FV-300; Olympus, Tokyo, Japan) with a 60× objective lens. Two-dimensional (2-D) images were generated by scanning along the longitudinal height of the bottom surface. The resolution of each 2-D image was 256 × 256 pixels, which covered 0.25 mm^2 of the captured area with 256 gray level, ranging from 0 (black) to 255 (white). A 2-D image was captured every 0.6 μm along the *z*-axis. The signal intensities for CellTraker Green CM-FDA and TOPRO-3 were obtained by exciting at the corresponding wavelengths of 488 and 633 nm, respectively.

The experimental procedure to determine the number and spatial distribution of epithelial cells cultured on micropatterned substrates with pillars and pits is outlines in Figure 1. Spatial cell distribution was measured in the vertical and horizontal directions. To measure the cell number and distribution on micropatterned substrates in vertical direction, 2-D images were stacked and stereoscopic images were analyzed to estimate the location of cytoplasms and nuclei using Image-Pro Plus version 6.0 software (Media Cybernetics, Silver Spring, MD, USA). The region of interest (ROI; 235 μm × 235 μm) was defined as the total area (R_{TA}) and was divided into two regions (R_1 and R_2). The densities and distributions of cells in the vertical direction were determined using ROI. We determined the number of adherent cells at 24 and 48 h and calculated the ratio (X_{48}/X_{24}) of adherent cells between these time points.

For reconstruction of a three-dimensional (3-D) image, the 2-D images were stacked, and the stereoscopic image was analyzed to estimate the location of nuclei using Image-Pro Plus version 6.0 software. Stereoscopic images were subjected to an algorithm with serial processes for primary noise removal, particle numbering, secondary noise removal, and 3-D construction.

To define the spatial distribution of epithelial cells on the micropatterned substrates in the horizontal direction, image capturing was conducted using epithelial cells cultured for 48 h; the coordinates for nucleus gravity were estimated. To analyze the local distribution of nuclei, it was assumed that the gravity height of nuclei on the bottom surface (Z) was lower than a given threshold value (Z_t) as follows:

$$Z_t = Z_m + 2SD \quad (1)$$

where the average height of nucleus gravity (Z_m = 4.5 μm) and standard deviation (SD = 1.2 μm) were determined in advance for cells cultured on the micropatterned substrates. The position of the bottom surface ($Z = 0$) was determined by measuring the autofluorescence of the culture vessel. Additionally, the height of the surface (Z_h) was determined by measuring autofluorescence at the top and bottom of the surface (Z_h = 20 μm). The distribution of nuclei on the micropatterned substrate at 24 h can be seen in Figure 1. The distribution of nuclei can be compartmentalized to three topographic regions: the top (T; $Z_h < Z$); the sidewall (S; $Z_t < Z \le Z_t$); and the bottom (B; $Z \le Z_t$). The spatial growth and

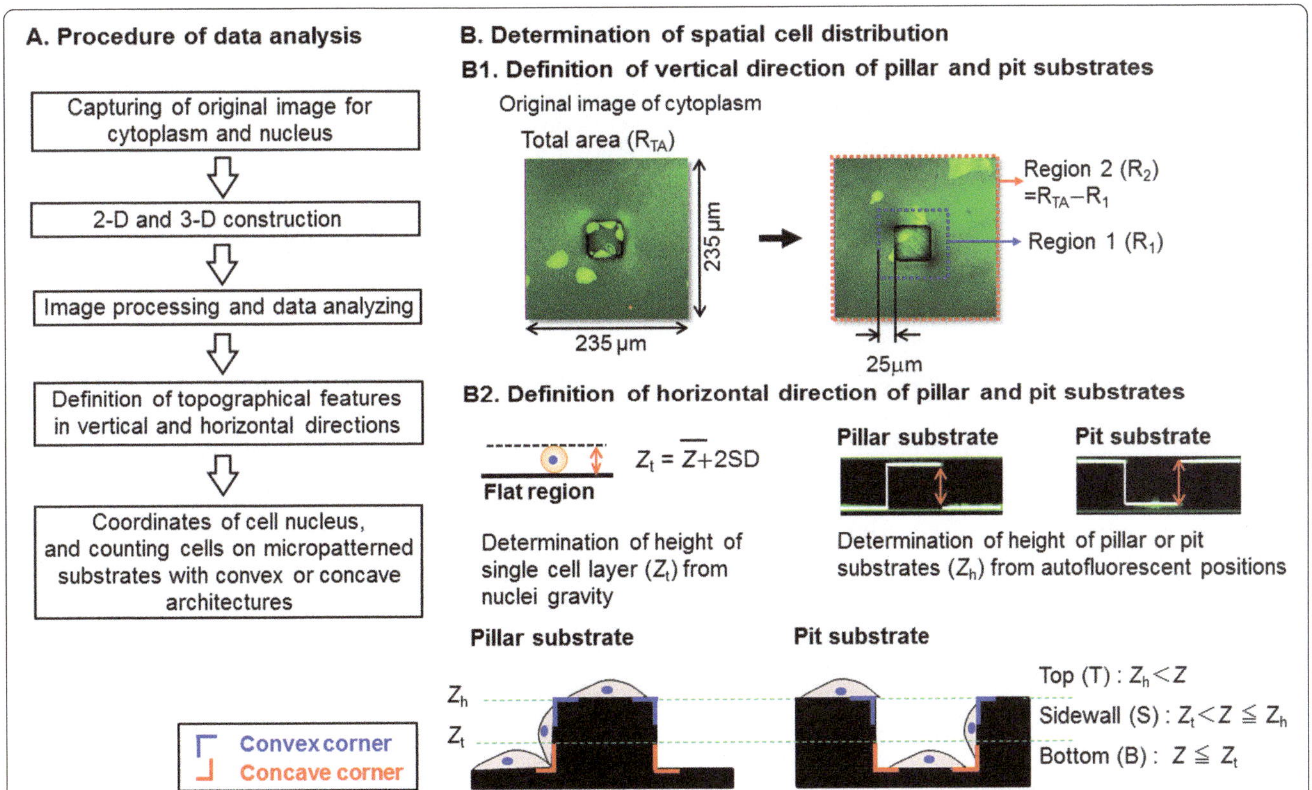

Figure 1 Schematic showing the experimental procedures used for analyzing distribution of spatial cell nuclei. (A) Whole procedure of data analysis and **(B)** determination of spatial cell distribution on micropatterned substrates with convex or concave architectures.

distribution of cell nuclei on micropatterned substrates were determined for the top, sidewall and bottom of pillar and pit substrates.

Observation of cell behaviors

Time-lapse observation for dynamic behavior was conducted by obtaining images every 10 min at several positions using a custom-made observation tool [16]. Time-lapse images of cells at 48 h were traced backward and their dynamic movements analyzed along with division and migratory processes. Growth ability was estimated by comparing the number of dividing cells (R_d) with the total number of cells at 48 h after seeding.

Immunofluorescence staining

The procedure for immunofluorescence staining was similar to that described previously [17]. Cells were fixed with 3.7% paraformaldehyde in phosphate buffer for 10 min at room temperature and rinsed with PBS. They were then soaked in PBS with 0.25% Triton X-100 for 4 min. After masking of non-specific proteins by incubation in Block Ace (Dainippon Sumitomo Pharma Co., Ltd., Osaka, Japan) for 1 h at ambient temperature, cells were treated with a primary antibody at 4°C overnight. The primary antibodies used were against vinculin (Santa Cruz Biotechnology, CA, USA) or anti-E-cadherin (Santa Cruz Biotechnology) and were diluted appropriately in PBS containing 10% Block Ace. Cells were washed with Tris-buffered saline and then incubated with Alexa Fluor 594-conjugated goat anti-mouse IgG (Molecular Probes) for 1 h. F-actin was stained with Alexa Fluor 488 phalloidin (Molecular Probes). Cells were observed using a CLSM with a 60× objective lens.

Statistical analysis

All experiments were conducted at least three times and data were expressed as means with standard deviations. Student's t-test was used to determine the statistical significance among data sets. A p-value less than 0.05 was considered significant.

Results

Spatial growth and distribution of epithelial cells on micropatterned substrates

The cultures of epithelial cells were performed for 48 h on micropatterned substrates with pillar and pit. As shown in Figure 2, the densities of cells in R_1 and R_2 regions on the pillar substrate were respectively 9.6×10^3 and 9.6×10^3 cells/cm^2 at 24 h, similar to the density in R_{TA}. Cell densities were increased for pillar and pit substrates in comparison with those on flat unpatterned substrate (7.1×10^3 cells/cm^2). At 48 h, the cell densities in R_1 and R_2 regions on the pillar substrate were 3.8×10^4 and 1.3×10^4 cells/cm^2, respectively. These were 4- and 1.4-fold enhancements compared with those flat unpatterned substrate.

The growth ability of cells in the R_1 and R_2 regions for the pillar and pit substrates after 48 h is clearly shown in Table 1. Compared with cells on the flat unpatterned substrate, the ratio of adherent cells from 48 h to 24 h (X_{48}/X_{24}) in the R_{TA} region of pillar and pit substrates was lower, but significantly different from that in the R_1 and R_2 regions. The X_{48}/X_{24} corresponding to the R_1 region of pillar and pit substrates was 3.9 ± 1.8 and 2.7 ± 1.4, respectively, which was 2.8 and 1.8-fold enhancements as compared with those in the R_2 region. However, the ratio of dividing cells to the total number of cells (R_d) for 48 h in R_1 region was similar to that for the R_2 region, irrespective of the substrate's topographical feature.

Based on the determined coordinates of nuclei (x, z), the spatial distribution and nucleus density on the micropatterned substrates occurred on the top surface, sidewall, and/or bottom of the substrate. As shown in Figure 3, cells were predominantly restricted to the sidewalls of the pillar substrate, and to the sidewalls and bottoms of pit substrate. For the pillar substrate, nucleus densities were 0.04, 0.37 and 0.56 nuclei/topographic region at the top, sidewall and bottom, respectively after 24 h. After 48 h, the nucleus density at the sidewall was higher than at the top and bottom. For the sidewall,

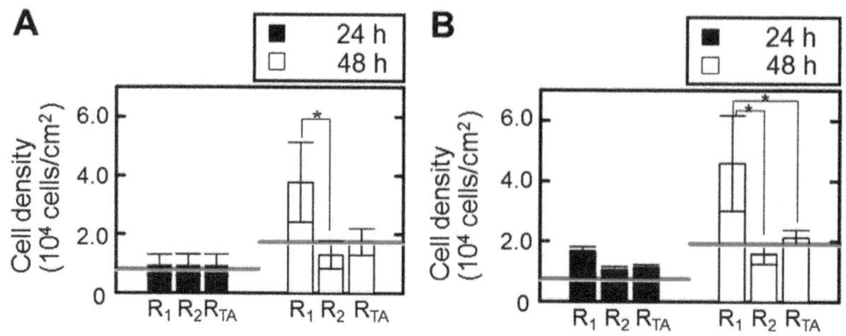

Figure 2 Vertical spatial growth and distribution of epithelial cells cultured for 48 h on pillar (A) and pit (B) substrates. The red lines indicate the data determined on the flat unpatterned substrate. Data are measured from individual pillars and pits of 9. The bar shows the standard deviation obtained from triplicate measurements.

Table 1 Growth ability of epithelial cells cultured for 48 h on micropatterned substrates with pillar and pit

Locational classification of convex or concave in vertical direction	Pillar substrate		Pit substrate	
	Ratio of increased cell number, X_{48}/X_{24} (−)	Ratio of dividing cells, R_d (10^{-2})	Ratio of increased cell number, X_{48}/X_{24} (−)	Ratio of dividing cells, R_d (10^{-2})
R_{TA} region	1.8 ± 0.8	91.3 ± 7.3	1.8 ± 0.5	94.0 ± 3.9
R_1 region	3.9 ± 1.8	90.9 ± 8.4	2.7 ± 1.4	89.2 ± 8.0
R_2 region	1.4 ± 1.0	91.4 ± 7.1	1.5 ± 0.4	94.3 ± 5.6
Flat Region	2.1 ± 0.7	84.0 ± 8.4	2.6 ± 0.6	96.1 ± 2.1

Measurements were made in an individual 9 pillars and pits. Data are expressed as means ± standard deviation from triplicate measurements and analyzed using Student's *t*-test. There were no statistically significant differences among the data sets.

nucleus density was 2.6 nuclei/topographic region, which was 13.5- and 5.5-fold higher than those at the top and bottom. For the pit substrate, similar trends regarding nucleus density were observed, with 0.61, 0.85 and 0.24 nuclei/topography for the top, sidewall and bottom at 24 h. The densities of nuclei at the sidewall after 48 h were 1.8-fold higher than those at those at the top and the bottom.

Behavior of epithelial cells on micropatterned substrates

To understand the dynamic behavior of epithelial cells on the micropatterned substrates, we conducted time-lapse observations of representative cells on the pillar and pit substrates as well as the flat unpatterned substrate. As time elapsed, cells exhibited active division and migration, irrespective of topographical feature (Additional file 2: Movie S1, Additional file 3: Movie S2, and Additional file 4: Movie S3). Cells on the pillar substrate migrated towards the sidewall, or stayed at the top of the surface. However, cells on pit substrate tended to move towards the sidewalls and bottom, or confined at the bottoms of the surface. Additionally, cells in the gaps tended to cling to attaching cells at the sidewalls, and to have extended protrusions to the sidewalls. In contrast, cells on the flat unpatterned substrate formed a cluster of cells through division and migration, and rotated persistently in constrained spaces.

Cytoskeletal formation of epithelial cells on micropatterned substrates

To confirm the cytoskeletal organization and focal contact of epithelial cells cultured on the micropatterned substrates for 48 h, staining of F-actin and vinculin was conducted. For cells on the pillar substrate, stress fibers assembled along straight edges, with extended lamellipodia and filopodia at the sidewalls and convex corners of the structure; this was also seen for cells on flat unpatterned substrate (Figure 4A1 and Figure 4C1). Immature vinculin spots were detected at the cytoplasm and periphery of cells at the sidewalls and convex corners of pillar substrate (Figure 4A3(a) and Figure 4A3(b)). In contrast, the formation of the contractile stress fibers with lamellipodia and filopodia were seldom seen at the concave corners of the pit substrate (Figure 4B1). The formation of the distinct spots of vinculin became pronounced appeared (Figure 4B3(c)).

Cell-cell contacts of epithelial cells on micropatterned substrates

To investigate contribution of E-cadherin to cell adhesion, the expression of E-cadherin in the cells cultured on the micropatterned substrates with pillar and pit for 144 h was observed. As shown in Figure 5, expression of E-cadherin was observed in almost all cells on the pillar substrate, similarly to those on the flat unpatterned

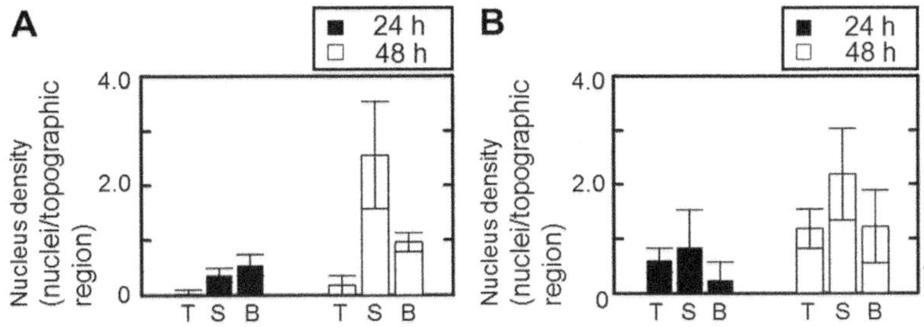

Figure 3 Horizontal spatial growth and distribution of epithelial cells cultured for 48 h on pillar (A) and pit (B) substrates. Data are measured from individual pillars and pits of 9. The bar shows the standard deviation obtained from triplicate measurements.

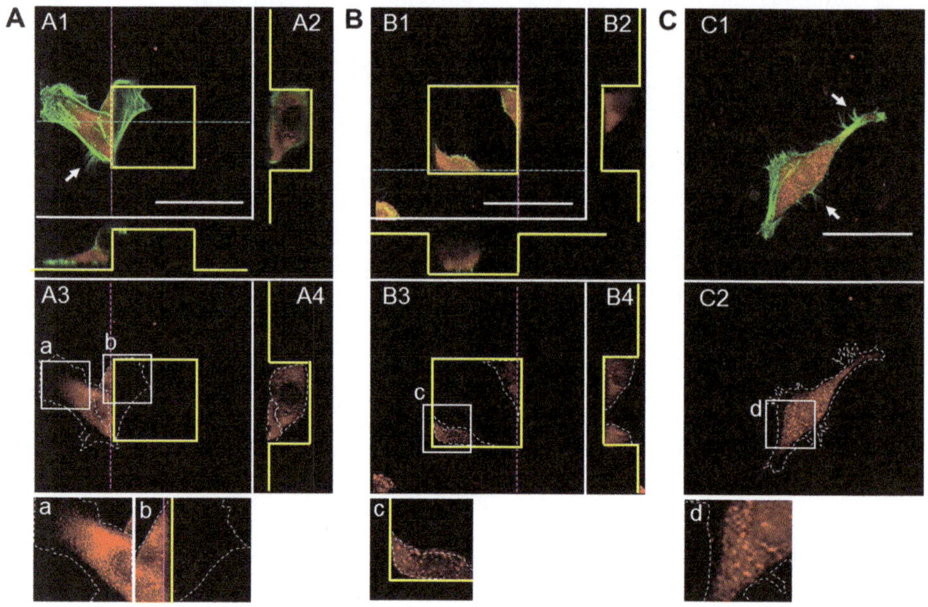

Figure 4 Immunostaining of vinculin (red) and F-actin (green) for epithelial cells cultured on pillar (A), pit (B) and flat (C) substrates. The scale bars show 50 μm. The yellow lines outline the pillar or pit features. The white dotted lines indicate the formation of actin stress fibers. Panels A2, A4, B2, and B4 show the cross-sectional views (dotted pink and blue lines) of the images in panels A1, A3, B1, and B3, respectively. Panels a–d show merged enlargements of the corresponding boxed areas in panels A3, B3, and C2. The asterisks indicate the formation of filopodia.

substrate. However, cells on the pit substrate dispersed as confluent monolayers with low levels of E-cadherin expression at the bottom (Figure 5B2(b)).

Discussion

In this study, we have described cellular responses to 3-D topographical surface features. We compared spatial growth and distribution on two micropatterned substrates with well-defined anisotropic topographical cues. The topographical features of pillar and pit substrates with convex and concave architectures caused heterogeneous spatial growth and distribution in vertical and horizontal directions (Figures 2 and 3). At 24 h, the regional densities and distribution on the pillar substrate was similar to those on pit substrate, regardless of their topographical region. At 48 h, the highest cell densities were observed at the sidewalls of pillar substrate, and at the bottom and sidewalls of pit substrate, respectively. However, there were no significant differences between the topographical regions in the growth ability with number of dividing cells from 24 h to 48 h (Table 1). These results suggest that the topographical features of pillar and pit possibly limit the ability of cells to migrate vertically and horizontally.

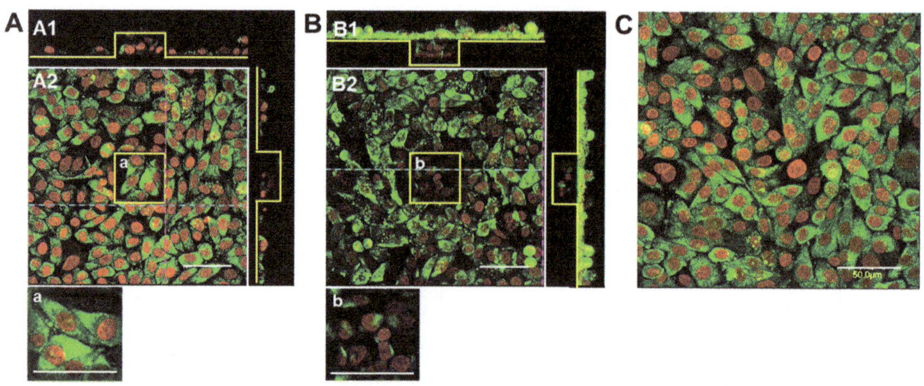

Figure 5 Immunostaining of E-cadherin (green) for epithelial cells cultured on pillar (A), pit (B) and flat (C) substrates. Nuclei are stained in red. The scale bars show 50 μm. The yellow lines outline the pillar or pit features. Panels A1 and B1 show the cross-sectional views (dotted pink and blue lines) of the images in panels A2 and B2. Panels a and b show merged enlargements of the corresponding boxed areas in panels A2 and B2.

Surface topography is critical to guide cellular behaviors such as adhesion, spreading and migration [2,8]. Responding to the topographical features, cells protrude their leading edge. The extension of membranes toward the direction of motility, including both lamellipodia and filopodia, brings attachment and thus the traction force to the substrate, resulting in a counter-force on the cell to promote cell migration [1,7]. With these steps of cell migration in mind it follows that surface topography may modulate direction of motility through contact guidance, as the reaction of cells to topographical features leads to cell polarization, lamellipodial and filopodial protrusion, actin bundle alignment, and focal adhesion formation preferentially along these surface features. In the present study, we address the spatial responses of epithelial cells to the micropatterned substrates with pillar and pit. We propose that the heterogeneity of spatial cell distribution, induced by directional migration, is caused by the convex and concave corners of pillar and pit substrates in Figure 6. Cells on pillar substrate responded to convex corners by altering morphology and migrating from the top to the bottom of the surface (Additional file 2: Movie S1), suggesting that they might be predominantly located in sidewalls connecting the top and bottom surfaces. For cells grown on the pit substrate, cells migrated to the sidewalls and bottom of the pit substrate, often bridging corners and attaching to both the bottom and sidewalls. Cells at bottom had an apparent preference for settling at the bottom of the concave cavity, suggesting that they might be sensitive to changes at concave corners (Additional file 3: Movie S2). It indicates that the convex and concave corners possibly limit the ability of cells to migrate vertically and horizontally. It is therefore possible to selectively influence either cell adhesion or morphology via surface topography. These observations indicate that orientation, migration, and morphology of the cells appeared to be governed by topographical features of the convex and concave corners of the pillar and pit substrates. It is not clear, however, whether morphological polarity of the cell itself at convex and concave corners of pillar and pit substrates can determine the direction of movement. The different patterns of movement during migration take place at the convex and concave corners of pillar and pit substrates, resulting in heterogeneity of spatial cell distribution.

Although a detailed mechanism for the transduction of such topographic signals is unclear, one possibility is that mechanical forces are transmitted through integrins. This might cause an associated sensing protein on the cytoplasmic side to alter its conformation and enzyme-substrate activities [3,4]. Consequently, cytoskeletal organization and adhesion to a substrate alters the way in which cells sense and respond to their microenvironment, thereby affecting cell-cell contact [18]. The cytoskeleton of a moving cell's protrusion is the property that determines such reactions to topography. We found

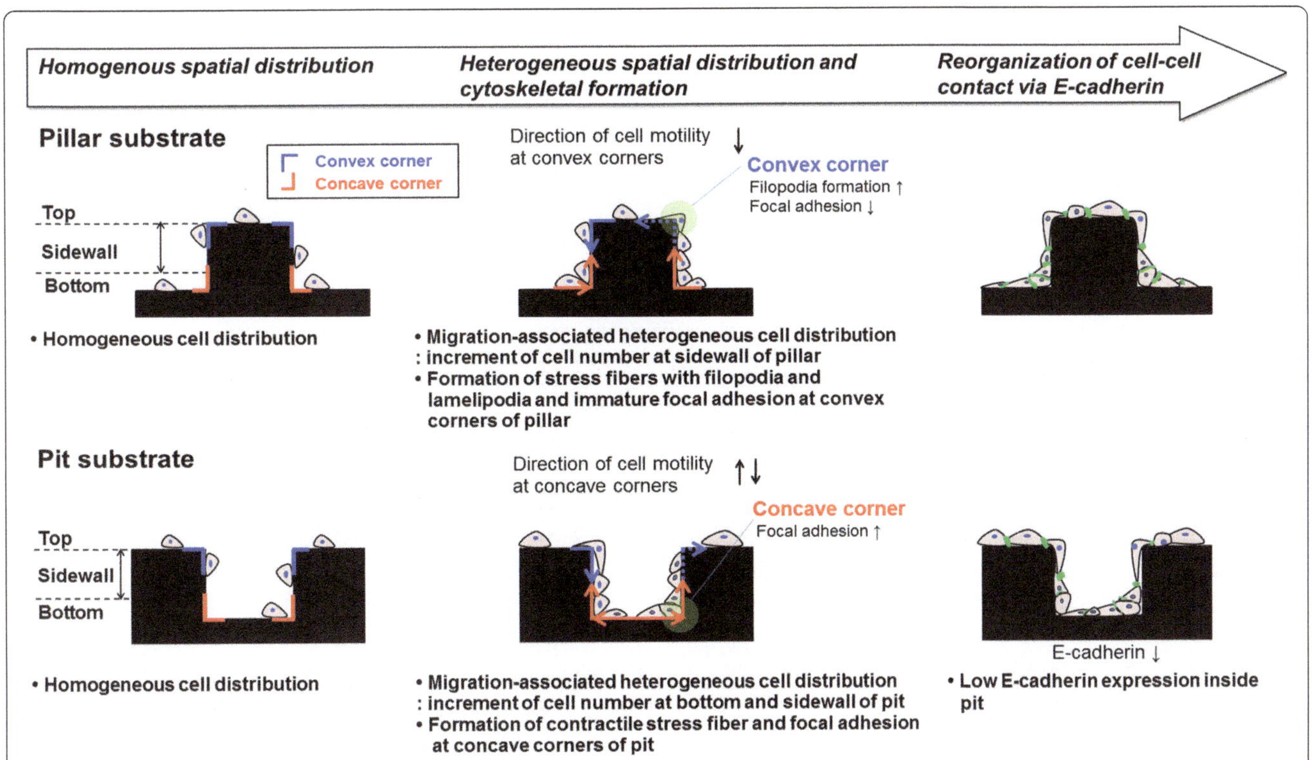

Figure 6 Schematic summarizing effect of micropatterned substrates with pillar and pit on cell behaviors observed in this study.

that the response of cells upon encountering a topographical feature could be predicted. Convex and concave topographies have attracted significant attention with regards to providing insights into the mechanisms of cell migration [2]. Spatial organization and dimensions of these micropatterns have been shown to affect the migration of cells [2,7]. Ghibaudo *et al.* investigated the role of substrate topography in cell adhesion and migration [18]. The topographical features of convex substrate induced changes in cellular morphology that were caused by alterations in the cytoskeleton in response to focal adhesion formation. Focal adhesions are known to serve as membrane sensing entities that control local and global adhesion mediated by Rho family GTPases signaling [19-23]. Rac1 and cadherin appear to be the major players in the maintenance of epithelial cell morphology [18,19]. Previous studies have demonstrated that this switching mechanism occurs in cells moving from 2-D to 3-D environments because of lower Rac1 activity in 3-D cell cultures [18]. Thus, the location and patterns of adhesive sites imposed by substrate topography most probably drive the direction of cell migration by modulating RhoA and Rac1 signaling pathways, and thus, cell polarity, adhesion and traction forces [18]. Furthermore, Rac1 signaling has emerged as a key regulator of E-cadherin-mediated cell–cell adhesion [23]. In this study, we found that variations in the spatial distribution and cytoskeletal structure of cells on the micropatterned substrates were dependent upon altered migration. Cytoskeletal staining for F-actin and vinculin showed that cells cultured on pillar substrate had stress fibers with extended filopodia and immature focal contacts at the sidewalls and bottom corners, similar to those seen on the flat unpatterned substrate (Figure 4A and Figure 4C). However, cells at the sidewalls and concave corners of the pit substrate had contractile stress fibers and more stable focal contacts than convex corners of the pillar substrate (Figure 4A and Figure 4B). Overall there were more developed filopodia and immature adhesions at the convex corners of pillar substrate compared with concave corners of pit substrate, possibly accounting for increased migration. This suggests that cells followed the convex corners and migrated toward the sidewalls of pillar substrate. Based on our observations it is plausible that the location for focal contacts on the micropatterned substrate is specified by spatial restrictions where cells can favorably form contact points with the surface. In addition to cytoskeletal formation, E-cadherin expression was seen for cells attached to the top of the pillar substrate. This expression was reduced in cells toward the bottom of the pit substrate (Figure 5b). When stable cell adhesion occurs inside a concave structure, stress in the actin cytoskeleton is induced. This disrupts the cytoskeleton during persistent migration, limiting the intracellular pathways responsible for cell-cell contact formation. We propose that epithelial phenotypes at the convex and concave corners of micropatterned substrates are regulated by the actin cytoskeletal architecture through cell–cell and cell–substrate interactions. These, in turn, elicit different contractile forces and levels of adhesion to stimulate cell migration. Taken together, our results demonstrate that anisotropic topographical features are important factors that affect spatial growth and distribution. Heterogeneity was caused by differences in migratory behavior at the convex and concave corners of pillar and pit substrates.

Conclusion

Our study supports the notion that the evaluation of spatial distribution in culture of epithelial cells can help in understanding the proliferation and migration potentials of cells grown on the pillar and pit substrates. Variations in orientation, migration, and morphology of cells were dependent upon substrate topography. We found that this altered migration could be attributed to spatial reorganization of the actin cytoskeleton, and formation of focal adhesions at various points along the convex and concave corners of pillar and pit substrates. Cells cultured on the pillar substrate had stress fibers with extended filopodia and immature focal contacts at the sidewalls and convex corners of substrate. Cells at the sidewalls and concave corners of pit substrate had contractile stress fibers, and formed more distinct focal contacts than cells on the pillar substrate. The convex and concave corners of pillar and pit substrates possibly created patterns of mechanical forces that modulate the level and direction of intracellular forces. The anisotropic topographical features of micropatterned substrates affected the spatial growth and distribution, thereby changing cell-cell and cell-surface interactions. Our investigations could lead to an effective approach for the design of the materials to control cell migration, and in the design of implants for tissue engineering applications.

Additional files

Additional file 1: Figure S1. Substrate topography. (A) Schematic showing two micropatterned substrates with pillar and pit. (B) Scanning electron microscopy images of micropatterned substrates with pillar (B1) and pit (B2).

Additional file 2: Movie S1. Dynamic behavior of epithelial cells cultured on pillar substrate.

Additional file 3: Movie S2. Dynamic behavior of epithelial cells cultured on pit substrate.

Additional file 4: Movie S3. Dynamic behavior of epithelial cells cultured on flat substrate.

Competing interests
The authors declare that they have no competing interests.

Authors' contributions
Conceived and designed the experiments: MHK, YS, MT, MK. Performed the experiments: MHK, YS. Analyzed and interpreted the data: MHK, YS, MT, MK. Wrote the paper: MHK, MT, MK. All authors read and approved the final manuscript.

Acknowledgment
We are grateful to Mr. T. Tazaki (Kuraray Co., Ltd., Kurashiki-shi, Japan) for preparation and characterization of micropatterned substrates. This work was supported in part by the MEXT project, "Creating Hybrid Organs of the future" at Osaka University and the project of "Development of cell manufacturing and processing system for industrialization of regenerative medicine" (No. P14006) commissioned by the New Energy and Industrial Technology Development Organization (NEDO).

Author details
[1]Department of Biotechnology, Graduate School of Engineering, Osaka University, 2-1 Yamadaoka, Suita, Osaka 565-0871, Japan. [2]Division of Chemical Engineering, Graduate School of Engineering Science, Osaka University, 1-3 Machikaneyama-cho, Toyonaka, Osaka 560-8531, Japan.

References
1. Discher DE, Janmey P, Wang Y-L: **Tissue cells feel and respond to the stiffness of their substrate.** *Science* 2005, **310:**1139–1143.
2. Wilkinson CDW, Riehle M, Wood MA, Gallagher J, Curtis ASG: **The use of materials patterned on a nano- and micrometric scale in cellular engineering.** *Mater Sci Eng* 2002, **19:**263–269.
3. Nikkhah M, Edalat F, Manoucheri S, Khademhosseini A: **Engineering microscale topographies to control the cell-substrate interface.** *Biomaterials* 2012, **33:**5230–5246.
4. le Digabel J, Ghibaudo M, Trichet L, Richert A, Ladoux B: **Microfabricated substrates as a tool to study cell mechanotransduction.** *Med Biol Eng Comput* 2010, **48:**965–976.
5. Flemming RG, Murphy CJ, Abrams GA, Goodman SL, Nealey PF: **Effects of synthetic micro- and nano-structured surfaces on cell behavior.** *Biomaterials* 1999, **20:**573–588.
6. Brunette DM, Chehroudi B: **The effects of the surface topography of micromachined titanium substrata on cell behavior in vitro and in vivo.** *J Biomech Eng* 1999, **121:**49–57.
7. Berry CC, Campbell G, Spadiccino A, Robertson M, Curtis ASG: **The influence of microscale topography on fibroblast attachment and motility.** *Biomaterials* 2004, **25:**5781–5788.
8. Martínez E, Engel E, Planell JA, Samitier J: **Effects of artificial micro- and nano-structured surfaces on cell behaviour.** *Ann Anat* 2009, **191:**126–135.
9. Karuri NW, Liliensiek S, Teixeira AI, Abrams G, Campbell S, Nealey PF, Murphy CJ: **Biological length scale topography enhances cell substratum adhesion of human corneal epithelial cells.** *J Cell Sci* 2004, **117:**3153–3164.
10. Liliensiek SJ, Campbell S, Nealey PF, Murphy CJ: **The scale of substratum topographic features modulates proliferation of corneal epithelial cells and corneal fibroblasts.** *J Biomed Mater Res A* 2006, **79:**185–192.
11. den Braber ET, de Ruijter JE, Smits HTJ, Ginsel LA, von Recum AF, Jansen JA: **Quantitative analysis of fibroblast morphology on microgrooved surfaces with various groove and ridge dimensions.** *Biomaterials* 1996, **17:**2037–2044.
12. Mirzadeh H, Shokrolahi F, Daliri M: **Effect of silicon rubber crosslink density on fibroblast cell behavior in vitro.** *J Biomed Mater Res A* 2003, **67:**727–732.
13. Ng CK, Yu KN: **Proliferation of epithelial cells on PDMS substrates with micropillars fabricated with different curvature characteristics.** *Biointerphases* 2012, **7:**21.
14. Green AM, Jansen JA, van der Waerden JP, von Recum AF: **Fibroblast response to microtextured silicone surfaces: texture orientation into or out of the surface.** *J Biomed Mater Res* 1994, **28:**647–653.
15. Kim MH, Tsubakino N, Kagita S, Taya M, Kino-oka M: **Characterization of spatial cell distribution in multilayer sheet of human keratinocytes through a stereoscopic cell imaging system.** *J Biosci Bioeng* 2011, **112:**289–291.
16. Kino-oka M, Agatahama Y, Hata N, Taya M: **Evaluation of growth potential of human epithelial cells by motion analysis of pairwise rotation under glucose-limited condition.** *Biochem Eng J* 2004, **19:**109–117.
17. Kim MH, Kino-oka M, Morinaga Y, Sawada Y, Kawase M, Yagi K, Taya M: **Morphological regulation and aggregate formation of rabbit chondrocytes on dendrimer-immobilized surfaces with D-glucose display.** *J Biosci Bioeng* 2009, **107:**196–205.
18. Ghibaudo M, Trichet L, Le Digabel J, Richert A, Hersen P, Ladoux B: **Substrate topography induces a crossover from 2D to 3D behavior in fibroblast migration.** *Biophys J* 2009, **97:**357–368.
19. Delon I, Brown NH: **Integrins and the actin cytoskeleton.** *Curr Opin Cell Biol* 2007, **19:**43–50.
20. Frey MT, Tsai IY, Russell TP, Hanks SK, Wang YL: **Cellular responses to substrate topography: role of myosin II and focal adhesion kinase.** *Biophys J* 2006, **90:**3774–3782.
21. Saez A, Ghibaudo M, Buguin A, Silberzan P, Ladoux B: **Rigidity-driven growth and migration of epithelial cells on microstructured anisotropic substrates.** *Proc Natl Acad Sci U S A* 2007, **104:**8281–8286.
22. Galbraith CG, Yamada KM, Sheetz MP: **The relationship between force and focal complex development.** *J Cell Biol* 2002, **159:**695–705.
23. Kaibuchi K, Kuroda S, Fukata M, Nakagawa M: **Regulation of cadherin-mediated cell–cell adhesion by the Rho family GTPases.** *Curr Opin Cell Biol* 1999, **11:**591–596.

Effect of tumor shape, size, and tissue transport properties on drug delivery to solid tumors

Mostafa Sefidgar[1], M Soltani[1,2*], Kaamran Raahemifar[3], Hossein Bazmara[1], Seyed Mojtaba Mousavi Nayinian[1] and Majid Bazargan[1]

Abstract

Background: The computational methods provide condition for investigation related to the process of drug delivery, such as convection and diffusion of drug in extracellular matrices, drug extravasation from microvessels or to lymphatic vessels. The information of this process clarifies the mechanisms of drug delivery from the injection site to absorption by a solid tumor. In this study, an advanced numerical method is used to solve fluid flow and solute transport equations simultaneously to investigate the effect of tumor shape and size on drug delivery to solid tumor.

Methods: The advanced mathematical model used in our previous work is further developed by adding solute transport equation to the governing equations. After applying appropriate boundary and initial conditions on tumor and surrounding tissue geometry, the element-based finite volume method is used for solving governing equations of drug delivery in solid tumor. Also, the effects of size and shape of tumor and some of tissue transport parameters such as effective pressure and hydraulic conductivity on interstitial fluid flow and drug delivery are investigated.

Results: Sensitivity analysis shows that drug delivery in prolate shape is significantly better than other tumor shapes. Considering size effect, increasing tumor size decreases drug concentration in interstitial fluid. This study shows that dependency of drug concentration in interstitial fluid to osmotic and intravascular pressure is negligible.

Conclusions: This study shows that among diffusion and convection mechanisms of drug transport, diffusion is dominant in most different tumor shapes and sizes. In tumors in which the convection has considerable effect, the drug concentration is larger than that of other tumors at the same time post injection.

Keywords: Drug concentration, Solid tumor, Tumor shape and size, Tissue transport properties

Background

Cancer, the main cause of death in developed countries, is the second leading cause of death in developing countries [1]. Solid tumors account for 85% of human cancers [2]. Chemotherapy is one of the ways widely used for cancer treatment. Based on the findings from clinical applications, most cancer treatments with drugs fail to eliminate solid tumors completely [3]. The computational method can investigate why systemic administration cannot distribute drug uniformly in tumors. The drug exchange between microvessels and extracellular matrices, drug removal by lymphatic system, drug diffusion and convective transport in extracellular matrices should be included by mathematical simulation. Computational fluid dynamics (CFD) can model the whole drug delivery process and clarify the mechanisms of drug delivery from the injection site to absorption by a solid tumor.

Baxter and Jain, based on the theoretical framework in their 1D mathematical method, found the effective factors on drug delivery such as microvessel permeability, interstitial fluid pressure (IFP), and interstitial fluid velocity (IFV) [4-7].

Improving the 1D model of Baxter et al. [5,6,8] and Saltzman et al. [9], Wang et al. [10-12] developed a simulation framework of drug delivery to tumors by

* Correspondence: msoltani@jhu.edu
[1]Department of Mechanical Engineering, K. N. T. University of Technology, Tehran, Iran
[2]Division of Nuclear Medicine, Department of Radiology and Radiological Science, School of Medicine, Johns Hopkins University, Baltimore, MD, USA
Full list of author information is available at the end of the article

considering the complex 3D geometry. Wang and Li [10] used modified MRI images for tumor geometry. They considered interstitial fluid flow with blood and lymphatic drainage in their model. Wang et al. [11] studied the effect of elevated interstitial pressure, convective flux, and blood drainage on the delivery of specified solute to brain tumors.

The study of tissue transport property effect on drug delivery is considered in recent studies. Zhao et al. [13] used a 3D computational model to predict the distribution of IFV, IFP, and solute transport through a tumor. Arifin et al. [14] studied the sensitivity of drug distribution to physiochemical properties in realistic models of brain tumors. A specific tumor captured by MRI is used by Pishko et al. [15] for modeling drug distribution in tissue with spatially-varying porosity and vascular permeability. The sensitivity of solute distribution to tumor shape and size is not considered in above mentioned works.

In our previous work [2], tumor shape and size effect on drug delivery is investigated by modeling interstitial fluid flow and assuming that drug particles flow with the interstitial fluid. In the present work, by adding the solute transport equation to the previous developed model in our group [16-20], new governing equations are investigated to find drug concentration in interstitial fluid (DCIF). Solving fluid flow and solute transport equations simultaneously, the effects of tumor shape, size, and tissue transport properties on drug delivery to solid tumor are also investigated.

Spherical and non-spherical tumors and their surrounding normal tissue are modeled with assumption of rigid porous media. The vasculature as a source term and lymphatic vessel as a sink term vary spatially. In the following parts of this paper, the sensitivity analysis provides a better understanding of the effects of tissue transport parameters on drug delivery.

Results

Simulation of interstitial fluid flow for baseline values (Table 1) predicts that IFP has its greatest value in the tumor center. IFP is non-dimensionalized by effective pressure. The effective pressure, P_{eff}, is a parameter defined by intravascular pressure, plasma osmotic pressure, and interstitial osmotic pressure by Equation (1). The non-dimensionalized pressure is defined by Equation (2)

$$P_{eff} = P_B^t - \sigma_s^t\left(\pi_B^t - \pi_i^t\right) \qquad (1)$$

$$P_n = \frac{P_i}{P_{eff}} = \frac{P_i}{P_b^t - \sigma_s^t\left(\pi_b^t - \pi_i^t\right)} \qquad (2)$$

where t stands for tumor tissue. Parameters used in Equation (1) and (2) are introduced in the method section.

Table 1 Interstitial transport properties used in numerical simulations

Parameter	Baseline value	Reference
L_P[cm/mmHg s]		
Normal	0.36×10^{-7}	[16]
Tumor	2.80×10^{-7}	
K[cm^2/mmHg s]		
Normal	8.53×10^{-9}	[16]
Tumor	4.13×10^{-8}	
S/V[cm^{-1}]		
Normal	70	[16]
Tumor	200	
P_B[mmHg]		
Normal	15.6	[16]
Tumor	15.6	
π_B[mmHg]		
Normal	20	[16]
Tumor	20	
π_i[mmHg]		
Normal	10	[16]
Tumor	15	
σ		
Normal	0.91	[16]
Tumor	0.82	
P_L		
Normal	0	[15]
L_{PL} S_L/V [1/mmHg s]		
Normal	1.33×10^{-5}	[15]

Non-dimensionalized IFP along transverse and vertical lines (shown in Figure 1), are illustrated in Figures 2 and 3. The maximum value of IFP occurs in the tumor center (Figure 2). This value is equal to P_{eff}, 1.53 kPa, except for the case of $\lambda = 0.1$ (Prolate shape). IFP has uniform distribution at tumor region and in the inner boundary (for more detail see Figure 1 and boundary condition section) IFP falls down sharply to around 150 Pa as shown in Figure 2 for all shapes except for prolate one ($\lambda = 0.1$). In normal tissue, pressure has uniform distribution close to outer boundary (for more detail, see Figure 1 and boundary condition section) then it decreases smoothly to peripheral pressure. For prolate shape as shown in Figure 2, IFP reduces smoothly from tumor center to the outer boundary, and IFP has a non-zero gradient in whole domain. Figure 2 shows the IFP along vertical line. Results are very similar to IFP results in Figure 2. Only for a prolate shape IFP has different pattern from what is observed in Figure 2. Maximum value of IFP in the prolate shape does not occur in the tumor

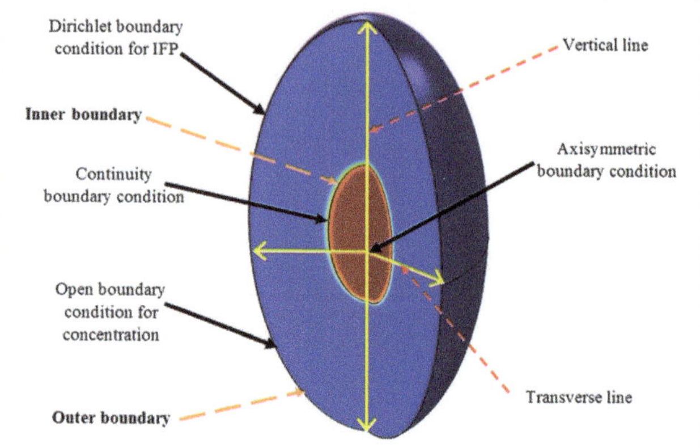

Figure 1 Schematic of considered geometry and boundary conditions. The transverse and vertical lines are used to show results along them.

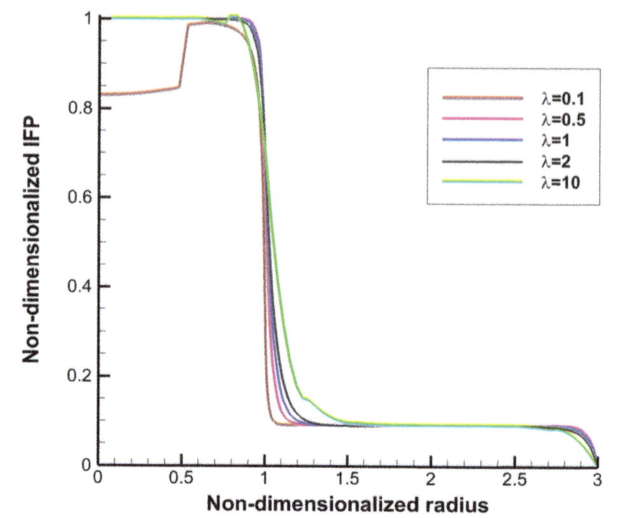

Figure 3 Interstitial fluid pressure profile along vertical line. The 1 in "x" axis shows the boundary of tumor (inner boundary).

center and takes place somewhere between the tumor center and tumor periphery.

IFV distribution along the vertical and transverse lines is presented in Figures 4 and 5. Maximum value of IFV occurs close to the inner boundary. Also, in normal tissue, IFV reaches zero far from the inner boundary. However, for the prolate shape, velocity is not zero, especially along transvers line (Figure 4).

DCIF is simulated in two cases of injection. In the first case, the continuous injection which leads to constant plasma concentration (C_p = constant) is considered and in the second case, the bolus injection in which the plasma concentration decreases with time exponentially ($C_p = C_p^0 e^{-\ln(0.5)t/\tau}$) is studied, in which τ is the drug half-life in plasma (Table 2). DCIF are non-dimensionalized by C_P for continuous injection and C_P^0 for bolus injection, respectively. Average of non-dimensionalized DCIF for two injection cases in different times is shown in Figures 6 and 7. DCIF of prolate shape ($\lambda = 0.1$) has the maximum value and DCIF of oblate shape ($\lambda = 10$) has the minimum value. Other considered shapes show the similar transient behavior.

Non-dimensionalized DCIF along two lines (transverse and vertical) is illustrated in Figures 8 and 9. The bolus injection results are presented in 8 hr post injection in which the concentration is maximum based on Figure 7 and for continuous injection is presented at final time of simulation, in 72 hr post injection. The profiles of two types of injections are similar in spite of different values of DCIF. As observed in IFV and IFP profiles, the prolate

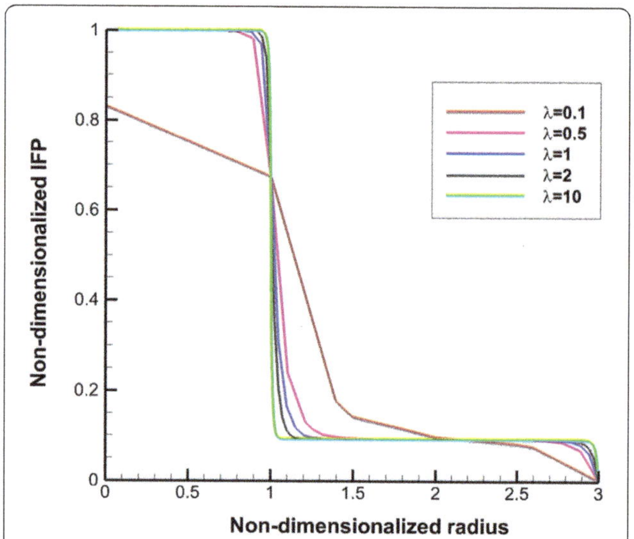

Figure 2 Interstitial fluid pressure profile along transverse line. The 1 in "x" axis shows the boundary of tumor (inner boundary).

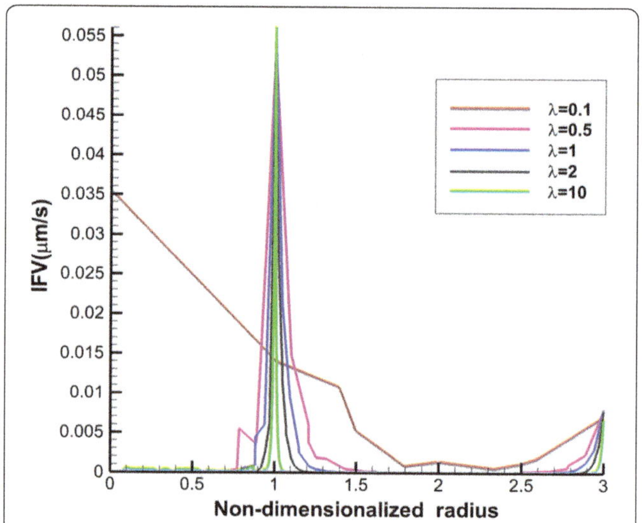

Figure 4 Interstitial fluid velocity profile along transverse line. The 1 in "x" axis shows the boundary of tumor (inner boundary).

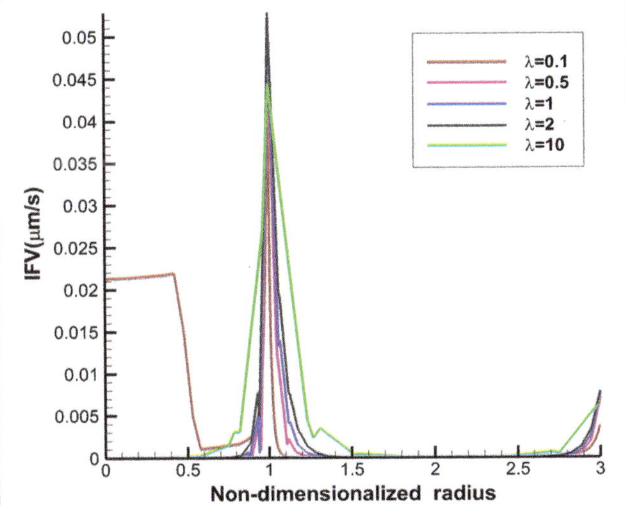

Figure 5 Interstitial fluid velocity profile along vertical line. The 1 in "x" axis shows the boundary of tumor (inner boundary).

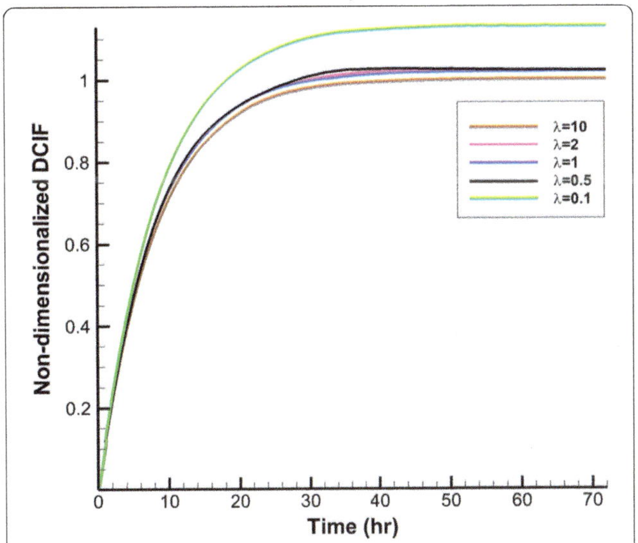

Figure 6 Average non-dimensionalized DCIF in tumor region in during time for continuous injection.

shape is different from the other shapes in DCIF distribution. A bump is observed in DCIF curves at the inner boundary of all tumor shapes. The normal tissue has uniform distribution of DCIF except near the boundaries (inner and outer). DCIF distribution in normal tissue is the same for all tumor shapes.

2D contours of DCIF in tumor region for bolus injection in 8 hr post injection are shown in Figure 10 for all tumor shapes. Results show that DCIF for tumors with $\lambda = 0.1$ and $\lambda = 10$ has different distributions. Generally, in the inner boundary, DCIF has its maximum value. In Figures 11 and 12, Peclet number distribution along two lines for continuous injection is shown. Peclet number demonstrates the ratio of convection to diffusion across the microvessel wall. Results show that in the tumor region Peclet number is zero except for prolate shape. Peclet number for the prolate shape especially along short radius (transverse line) is greater than zero.

Some of tissue transport parameters mentioned (effective pressure, hydraulic conductivity, and tumor size) in Table 1 are investigated for sensitivity analysis. The values of these parameters are selected near the ranges reported in the literature [2,15,21].

Figure 13 shows the influence of changing effective pressure, P_{eff}, on numerical results of tumor shapes with the same volume. Maximum value of IFP in tumor region (Figure 13a) linearly increases when P_{eff} increases. Only in

Table 2 Parameters of solute transport model used in numerical simulations

Parameter	Baseline value	Reference
σ_f		
Normal	0.9	[8]
Tumor	0.9	
D_{eff} [m^2/s]		
Normal	0.16×10^{-12}	[8]
Tumor	2.0×10^{-12}	
P[m/s]		
Normal	$2.2\times 10^{-10*}$	[8]
Tumor	$17.3\times 10^{-10*}$	
τ [hr]**		
Plasma	6.1	[8]

*- the microvessel permeability coefficient is 10% of effective permeability introduced in [8].
**- the drug half-life in plasma introduced in results section.

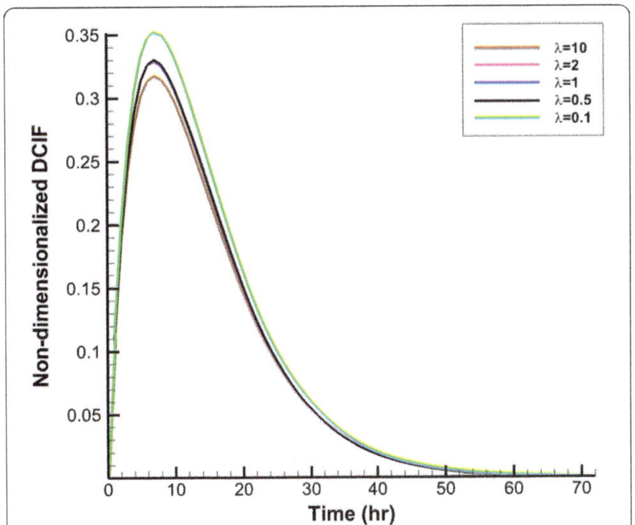

Figure 7 Average non-dimensionalized DCIF in tumor region in during time for bolus injection.

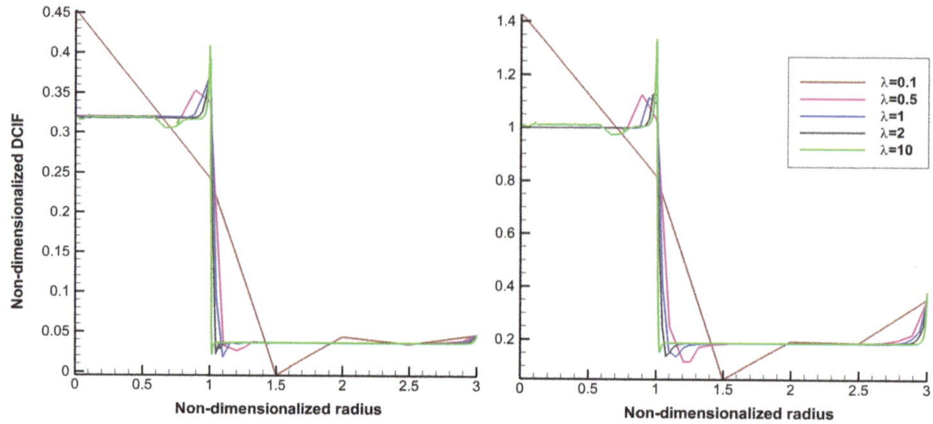

Figure 8 Non-dimensionalized DCIF profile along transverse line for normal. Left for bolus injection and right for continuous injection. The 1 in "x" axis shows the boundary of tumor (inner boundary).

the prolate shape the pressure increases more than other tumor shapes. Average of IFV on inner boundary also has the same pattern and linearly changes with P_{eff} (Figure 13b). The changes in average of DCIF in tumor region for two cases of injection are shown in Figure 13c and d. For these two cases of injection, the prolate shape is more sensitive to P_{eff} changes than other tumor shapes and DCIF changes around 20% when the P_{eff} changes from 500 Pa to 2500 Pa. The other shapes do not show significant variations in DCIF for these ranges of P_{eff}.

Different tumor sizes are studied with changes in their volume. One of important metric of disease development and response to tumor therapy with drug is volume of tumors [22-25]. To quantify response to therapeutic regimens and also assess disease progression, tumor volume is used as a metric in many studies, such as Char et al. [26], Jensen et al. [27] and Gass et al. [28].

As shown in Figure 14a, IFP has less value than P_{eff} when the tumor volume is smaller than 1 cm^3. When the tumor volume is in the order of 1 cm^3, IFP reaches P_{eff} and by increasing the tumor radius, IFP remains constant. Average of IFV on the inner boundary generally decreases with increasing tumor size, Figure 14b. Only in prolate shape in small radii, IFV increases with the tumor size. As shown in Figure 14c and d, the mean values of DCIF have the greatest value in the smallest tumor. Also, the prolate shape has the highest value of DCIF in all studied tumors.

Figure 15 shows the sensitivity of IFF parameters and DCIF to hydraulic conductivity changes. Results show that in all tumor shapes, if hydraulic conductivity increases, the maximum value of IFP remains constant and then decreases sharply. Average of IFV increases by increasing hydraulic conductivity. DCIF increases by increasing hydraulic conductivity and then reaches a constant value in spite of increasing hydraulic conductivity.

Discussion

This study presents DCIF, IFP, and IFV in solid tumors surrounded by normal tissue in two types of injection; bolus and continuous one. The model used in this study

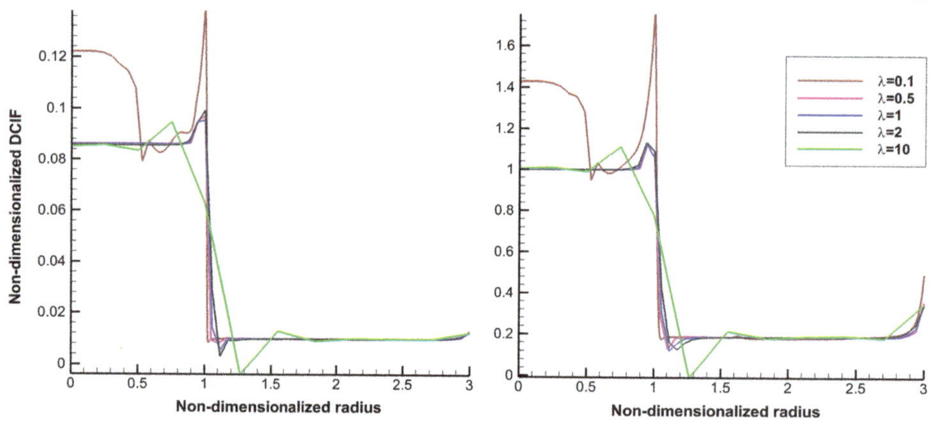

Figure 9 Non-dimensionalized DCIF profile along vertical line. Left for bolus injection and right for continuous injection. The 1 in "x" axis shows the boundary of tumor (inner boundary).

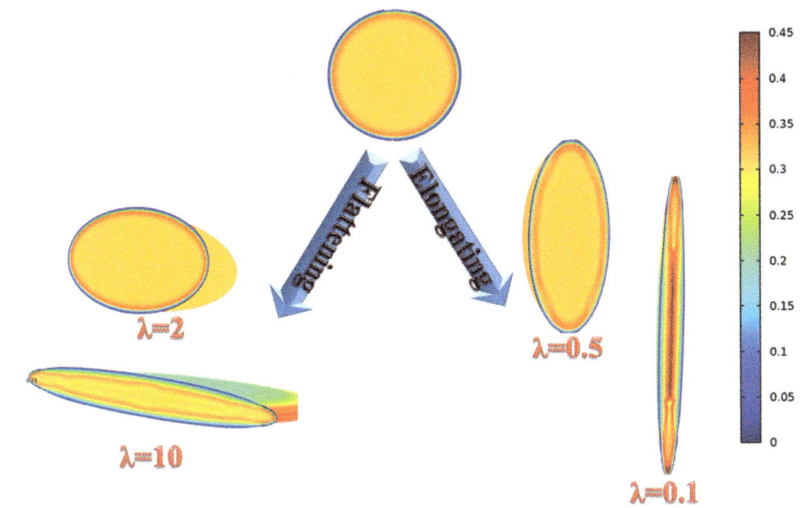

Figure 10 2D contour of DCIF in tumor region for bolus injection in 8 hr post injection for all tumor shapes.

investigates the effect of two characteristics of tumors on concentration distribution; tumor shape and size.

Results of high IFP in tumors are discussed in our previous studies [2,16,17] and in the experimental results of Arifin et al. [29] and Huber et al. [30]. Maximum value of IFP in spherical tumors (1 cm radius) for baseline values in Table 1 is 1529.5 Pa which is close to the studies of Jain et al. [31], Chauhan et al. [32], and Arfin et al. [33]. The current results are verified by experimental data of IFP measured by Nielsen et al. [34]. In their work, the wick-in-needle technique is used to measure IFP in two types of tumors with the same volume as tumors in the current study. They found IFP in the range of 1400 Pa to 1600 Pa.

IFV on tumor boundary in spherical tumors with baseline values is around 0.05 $\mu m/s$ which is at the same order of the prediction of Jain et al. [31] and experimental observation of Hompland et al. [35]. Also, the profile of drug concentration for simulation with baseline values for spherical tumor in different times (Figures 6 and 7) agrees well with Baxter and Jain's predictions [8]. Results show that DCIF for prolate shape ($\lambda = 0.1$) always has the greatest value. Results of transient DCIF (Figures 6 and 7) show that drug delivery is much easier in the prolate tumors. Also, the oblate shape ($\lambda = 10$) has the most resistance to drug delivery. Following of this section, the reason of these phenomena is investigated.

The uniform values of IFP for all tumor shapes except prolate one (Figures 2 and 3) is equal to P_{eff}. Non-uniform IFP in prolate shape results in the maximum value of

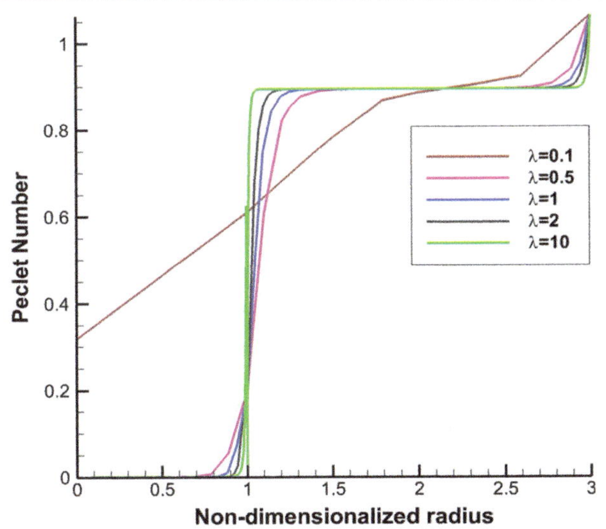

Figure 11 Peclet number (ratio of convection to diffusion across the microvessel wall) along transverse line.

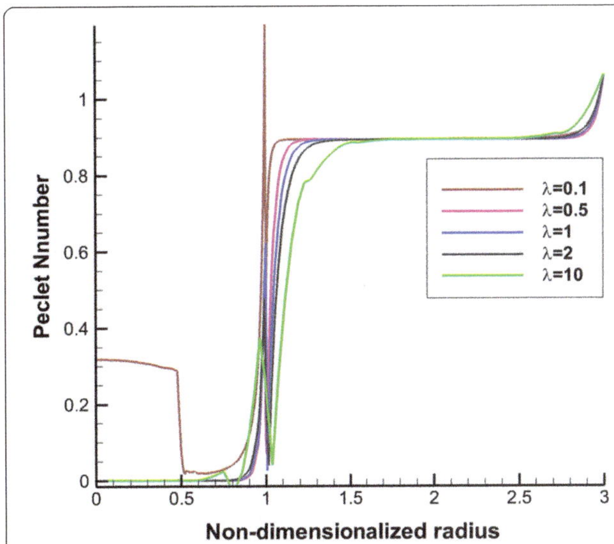

Figure 12 Peclet number (ratio of convection to diffusion across the microvessel wall) along vertical line.

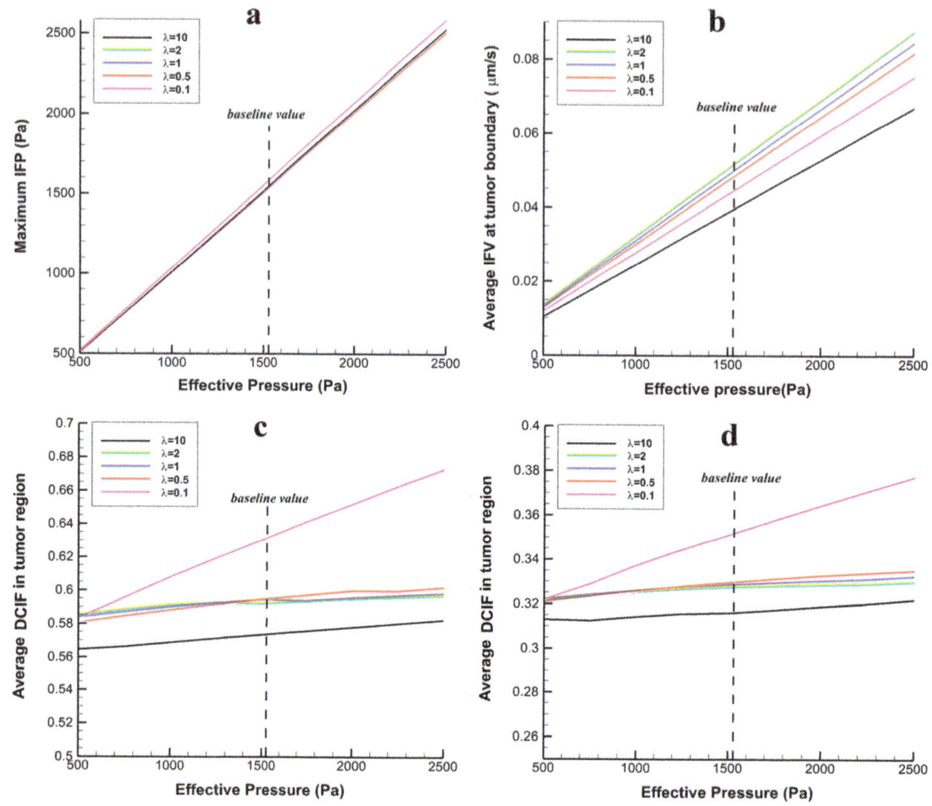

Figure 13 The effect of different values of effective pressure on IFP, IFV and DCIF. a) The variation of maximum IFP in tumor region. **b)** The variation of averaged IFV at tumor boundary. **c)** Average of DCIF in tumor region for continuous injection. **d)** Average of DCIF in tumor region for bolus injection.

DCIF among other tumor shapes with the same volume. Equation (17) is able to legitimize this behavior of tumor shapes. The source term (the last term in the right hand side of Equation (17)) includes diffusion and convection terms. The convection term depends on the differences between IFP and P_{eff} based on starling's law. This pressure difference is close to zero (Figures 2 and 3) for all tumor shapes except for the prolate one and therefore the convection term only for prolate shape has non-zero value. Because of the non-zero values of Peclet number for prolate shape in tumor region (Figures 11 and 12), the concentration for this shape is affected not only by diffusion rate from vessels but also by convection rate from vessels. The convection rate leads to higher level of DCIF in prolate shape than that of other tumor shapes (Figures 6, 7, 8, 9 and 10). The non-zero values of Peclet number for prolate shape are seen in an image based work of Zhao et al. as well [13]. In normal tissue, Peclet number shows that drug delivery from microvessel to tissue is done by both mechanisms of transfer, convection and diffusion.

The other effect of uniform pressure is zero pressure gradients in all tumor shapes except for prolate one. Because of zero IFP gradient and based on Darcy's equation, IFV is close to zero in tumor tissue except for prolate shape. The close to zero value of IFV is predicted in a few numerical studies such as Welter and Rieger [36]; Roy and Riahi [37]; and experimental results of Hompland et al. [35]. Zero IFV results in a negligible convection transport (the second term of the right hand side of Equation (17)) and consequently the convection transport does not affect drug distribution; and the diffusion transport (the first term of the right hand side of Equation (17)) is the only reason of drug transport in all tumor shapes except for prolate one. Therefore, non-zero IFV in prolate shape, also seen in Zhao et al. [13], is another reason of higher DCIF values in this tumor shape with respect to other shapes.

The sharp pressure gradient (Figures 2 and 3) and highest value of IFV (Figures 4 and 5) in the inner boundary for all tumor shapes increases drug transport and make a bump at this boundary for DCIF.

The sensitivity analysis of effective pressure shows that the effective pressure does not have too much effect on DCIF. In all tumor shapes, DCIF for a wide range of effective pressure changes smoothly; however, in prolate shape, this change is sharper and increases by effective pressure (Figure 13c and d). As mentioned, in the tumor region for all tumor shapes except for prolate one the convection rate, which depends on P_{eff}, has a negligible role

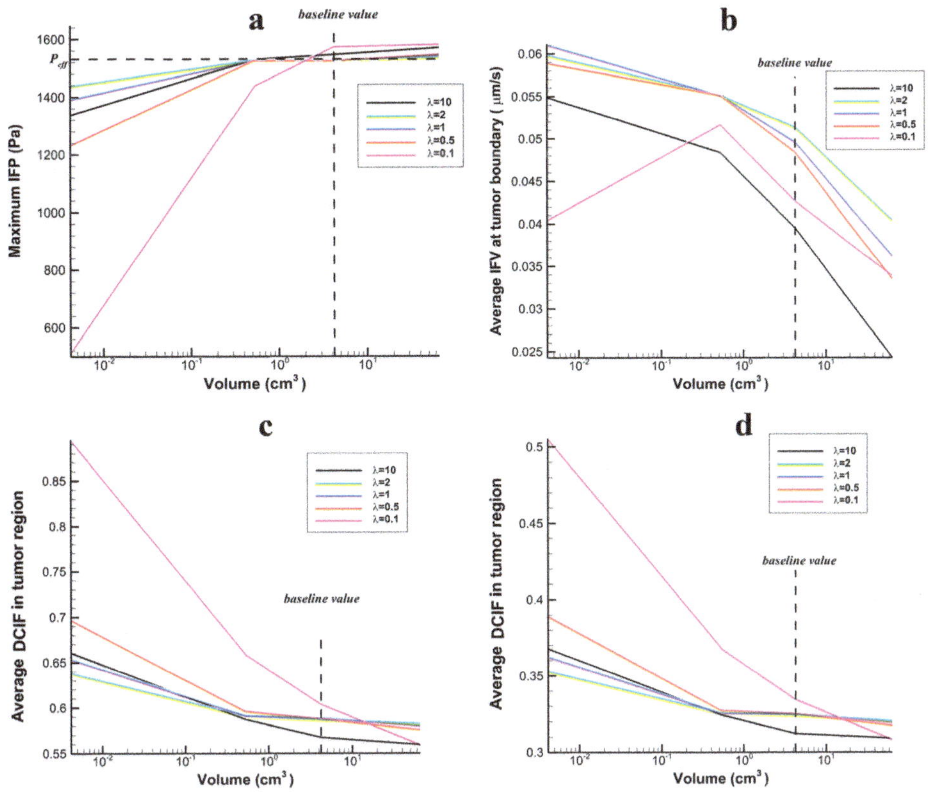

Figure 14 The effect of different values of tumor size on IFP, IFV and DCIF. a) The variation of maximum IFP in tumor region. **b)** The variation of averaged IFV at tumor boundary. **c)** Average of DCIF in tumor region for continuous injection. **d)** Average of DCIF in tumor region for bolus injection.

on drug distribution; therefore, P_{eff} cannot have significant effect on drug concentration. However, in prolate shape, increasing P_{eff} increases pressure difference between IFP and P_{eff} and consequently the convection rate from vessels in the tumor region; therefore, DCIF in prolate shape is sensitive to P_{eff} changes.

The tumor volume shows more effects on IFP, IFV, and DCIF than other investigated parameters such as effective pressure and hydraulic conductivity. The increasing tumor volume increases significantly IFP (Figure 14a). The dependency of IFP to tumor volume is observed in experimental study of Gutmann et al. [38], Hompland et al. [39], and Leguerney et al. [40], as well. When the tumor volume is in the order of 1 cm^3, the sensitivity of IFP to tumor size decreases. The independency of IFP to tumor volume in in large tumors is observed in the study of Leguerney et al. [40], as well. In their work, IFP changes very slowly with tumor volume. Since the high IFP is introduced as the main barrier of drug delivery to tumors, IFP increasing with tumor volume leads to DCIF decrease in these tumors. This reduction of DCIF with tumor size is observed in Au et al. [41], as well. Since IFP reaches around the effective pressure with increasing tumor volume, the convection rate is vanished and the diffusion rate reaches a constant value, and consequently the sensitivity of DCIF to tumor size reduces. Lower IFP in the small tumor sizes leads to increase the convection rate of source term in Equation (17). Therefore, it is expected to have a better drug distribution in small tumors.

Results show that IFV, IFP, and DCIF are sensitive to tissue hydraulic conductivity changes. The hydraulic conductivity is appeared only in Darcy's law. This parameter has a direct effect on IFF and indirect effect on DCIF. Theoretical analysis shows that the $\frac{P_i}{P_{eff}}$ in tumor region is proportional to $1-\frac{1}{\sqrt{\kappa}}/\sinh\left(\frac{1}{\sqrt{\kappa}}\right)$ [8] (κ is hydraulic conductivity, see material section). In low values of κ, the $\frac{1}{\sqrt{\kappa}}/\sinh\left(\frac{1}{\sqrt{\kappa}}\right)$ is negligible and P_i is close to P_{eff} (Figure 15a). Increasing κ, increases the $\frac{1}{\sqrt{\kappa}}/\sinh\left(\frac{1}{\sqrt{\kappa}}\right)$ and leads to sharp decrease in IFP. This dependency is also observed by McCarty and Johnson [42]. For high values of κ, IFP is very low and negligible in comparison to P_{eff}. IFP reduction from effective pressure increases IFV around 5 times (Figure 15b). The hydraulic conductivity affects DCIF by convection rate from vessels (as mentioned this value depends on difference between IFP and P_{eff}). In low values of κ, since IFP is equal to P_{eff}, the

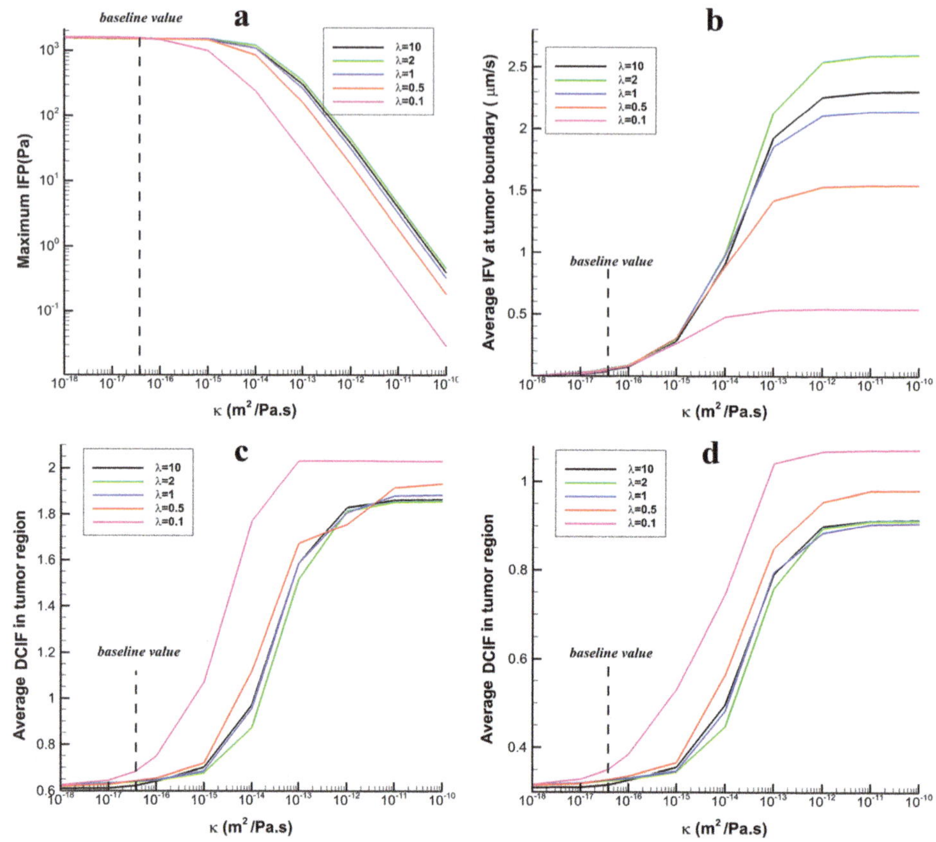

Figure 15 The effect of different values of hydraulic conductivity on IFP, IFV and DCIF. a) The variation of maximum IFP in tumor region. **b)** The variation of averaged IFV at tumor boundary. **c)** Average of DCIF in tumor region for continuous injection. **d)** Average of DCIF in tumor region for bolus injection.

effect of convection rate is not significant and DCIF remains constant. Increasing κ increases the convention rate and consequently DCIF.

When hydraulic conductivity increases two to three orders of magnitude, the mean values of DCIF are two times greater than the average of DCIF for baseline values in Table 1. However, after a specific value of hydraulic conductivity, DCIF changes smoothly and reaches a constant value because IFP is very low and convection rate only depends on P_{eff}.

Conclusions

A numerical approach which couples the mathematical model of the lymphatic system and the interstitial flow with the mathematical model of solute transport demonstrates that DCIF is affected by two transport mechanisms, convection and diffusion.

Drug convection and drug transport from microvessel to tumor are blocked by high interstitial pressure (IFP) which is uniformly distributed in most part of the tumor. The large pressure gradient results in an outward convective flow that washes out the drug extravasated from microvessels at the tumor periphery. This study shows that when there is IFP gradient in the tumor region instead of uniform IFP distribution which occurs in some tumor shapes, DCIF is greater than that of the uniform one.

As the effects of osmotic and intravascular pressure on convection rate are negligible in most of tumor shapes, the dependency of DCIF to these parameters is very low.

The hydraulic conductivity which is another considered parameter in sensitivity analysis has significant effect on drug distribution since it increases the convection rate from vessels.

Method

Mathematical model of interstitial fluid transport

This section introduces the mathematical model of interstitial fluid transport in macroscopic scale [2,16]. Since normal and tumor tissue have characteristics the same as porous media, fluid flow behavior is defined by coupling the fluid flow governing equations. The mass balance or continuity equation for steady state incompressible fluid in the porous media with source and sink of mass is [16]:

$$\nabla . v_i = \phi_B - \phi_L \tag{3}$$

where

v_i: The interstitial fluid velocity,

ϕ_B: The source term, extravasation from microvessels, and

ϕ_L: The drainage term, elimination by lymphatic system.

In biological tissues, the fluid source is evaluated through Starling's law as follows [16]:

$$\phi_B = \frac{L_P S}{V}(P_B - P_i - \sigma_s(\pi_b - \pi_i)) \tag{4}$$

where

P_i : Interstitial fluid pressure,

P_B : Blood pressure in microvessel,

$\frac{S}{V}$: The surface area per unit volume of tissue for transport in the interstitium,

π_B: Microvessel oncotic pressure,

π_i : Interstitial oncotic pressure,

L_p : The hydraulic conductivity of the microvessel wall, and

σ_s: Osmotic reflection coefficient.

and the lymphatic system is related to the pressure difference between the interstitium and the lymphatic vessels and is considered only for normal tissues [43]:

$$\phi_L(r) = \begin{cases} \frac{L_{PL} S_L}{V}(P_i - P_L) & \textit{Normal tissue} \\ 0 & \textit{Tumor tissue} \end{cases} \tag{5}$$

where

ϕ_L: The volumetric flow rate into the lymphatic,

$\frac{L_{PL} S_L}{V}$: The lymphatic filtration coefficient, and

P_L: The hydrostatic pressure of the lymphatic.

The momentum balance equation in its general form can be written as Equation (6) [44]:

$$\rho\left(\frac{\partial v_i}{\partial t} + (v_i.\nabla)v_i\right) = \nabla\cdot\left[-P_i + \mu\left(\nabla v_i + (\nabla v_i)^T\right) - \frac{2\mu}{3}(\nabla\cdot v_i)\right] - \left(\frac{\mu}{K}\right)v_i + F \tag{6}$$

where

K: The permeability of the porous medium,

ρ : The density,

μ: The viscosity, and

F: The volume forces.

Since interstitial fluid is a Newtonian fluid and has low velocity through the tissues, Equation (6) in the steady state condition is simplified to Darcy's law [16]:

$$\nabla P_i = -\left(\frac{\mu}{K}\right)v_i \tag{7}$$

The K/μ is defined as hydraulic conductivity of the interstitium, κ:

$$v_i = -\kappa \nabla P_i \tag{8}$$

Combination of momentum equation (Equation (8)) and the continuity equation (Equation (3)) results in

$$-\nabla\cdot(\kappa \nabla P_i) = \phi_B - \phi_L \tag{9}$$

When κ is constant, the interstitial pressure can be expressed by

$$-\kappa \nabla^2 P_i = \phi_B - \phi_L \tag{10}$$

By substituting Equations (4) and (5) into Equation (10):

$$-\kappa\nabla^2 p = \begin{cases} \frac{L_P S}{V}(P_B - P_i - \sigma_s(\pi_b - \pi_i)) - \frac{L_{PL} S_L}{V}(P_i - P_L) & \textit{Normal tissue} \\ \frac{L_P S}{V}(P_B - P_i - \sigma_s(\pi_b - \pi_i)) & \textit{Tumor tissue} \end{cases} \tag{11}$$

Macroscopic solute transport

Molecular transport in tumors is based on the conservation laws for mass and momentum. The interstitial transport of drug is governed by the convection diffusion equation; therefore, the general equation for the mass balance of solutes can be written as:

$$\frac{\partial C}{\partial t} = \nabla\cdot(D_{eff}\cdot\nabla C) - \nabla\cdot(v_i C) + (\Phi_B - \Phi_L) \tag{12}$$

where

C : The solute concentration based on tissue volume,

Φ_B : The rate of solute transport per unit volume from microvessel into the interstitial space,

Φ_L : The rate of solute transport per unit volume from the interstitial space into lymphatic vessels, and

D_{eff} : The effective diffusion tensor.

For an isotropic and uniform diffusion in tissues, equation (12) can be written as:

$$\frac{\partial C}{\partial t} = D_{eff}\nabla^2 C - \nabla\cdot(v_f C) + (\Phi_B - \Phi_L) \tag{13}$$

The solute transport rate across the lymphatic vessels can be considered as [15]

$$\Phi_L = \begin{cases} \phi_L C & \textit{Normal Tissue} \\ 0 & \textit{Tumor Tissue} \end{cases} \tag{14}$$

The solute transport rate across the microvessel can be represented by Patlak equation [45]:

$$\Phi_B = \phi_B(1 - \sigma_f)C_P + \frac{PS}{V}(C_P - C)\frac{Pe}{e^{Pe} - 1} \tag{15}$$

where

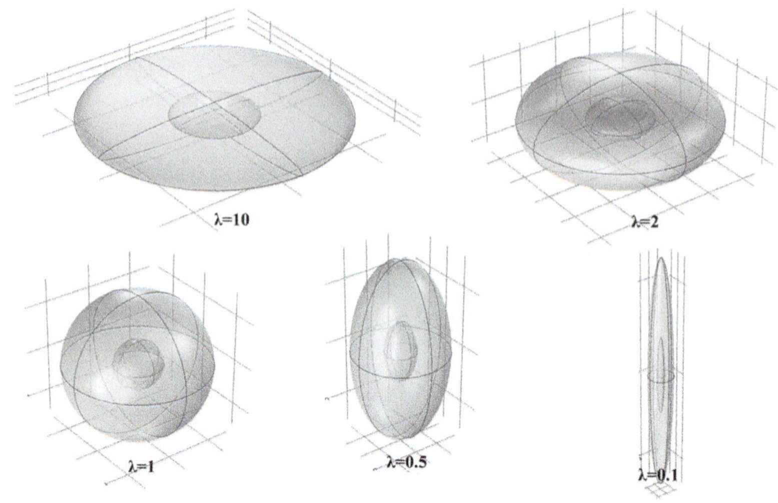

Figure 16 Different shapes of solid tumors considered in this study. The five ellipsoid shapes changed from oblate to prolate are studied.

$$Pe = \frac{\phi_B(1-\sigma_f)V}{PS} \tag{16}$$

ϕ_B: The fluid flow rate per unit volume of tissue across the microvessel wall,

σ_f : The filtration reflection coefficient,

P : The microvessel permeability coefficient,

S/V : The microvessel surface area per unit volume of tissue,

C_p : Solute concentration in the plasma.

By substituting Equations (15) and (14) in to Equation (13):

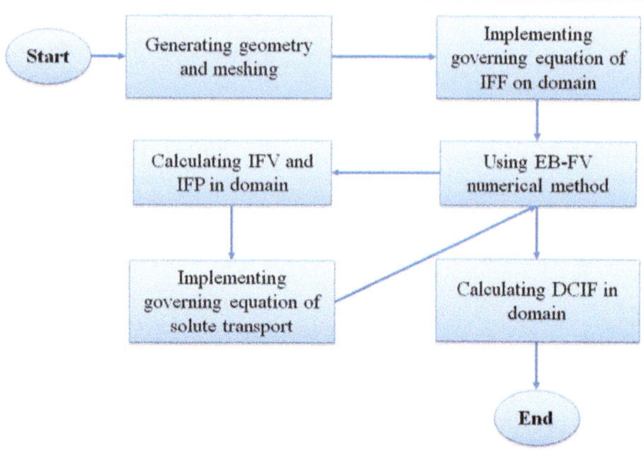

Figure 17 Algorithm of numerical simulation used for calculating interstitial fluid flow parameter (IFV, IFP) and solute transport parameter (DCIF).

$$\begin{cases} \dfrac{\partial C}{\partial t} = D_{eff}\nabla^2 C - \nabla\cdot(v_i f C) \\ \quad + \left(\phi_B(1-\sigma_f)C_P + \dfrac{PS}{V}\left(C_P - \dfrac{C}{K_{AV}}\right)\dfrac{Pe}{e^{Pe}-1} - \phi_L C\right) \quad \textit{Normal region} \\ \dfrac{\partial C}{\partial t} = D_{eff}\nabla^2 C - \nabla\cdot(v_i f C) \\ \quad + \left(\phi_B(1-\sigma_f)C_P + \dfrac{PS}{V}\left(C_P - \dfrac{C}{K_{AV}}\right)\dfrac{Pe}{e^{Pe}-1}\right) \quad \textit{Tumor region} \end{cases} \tag{17}$$

Boundary conditions

A tumor surrounded by normal tissue is considered in this study. The tumor shape is considered ellipsoid. A 2D section of the geometry and boundaries are shown in Figure 1. The three boundaries are indicated for geometry: a) the center of tumor, b) the boundary between tumor and normal tissue, named inner boundary, c) the boundary at the outer edge of geometry, named outer boundary. The appropriate boundary conditions are implemented for Equations (11) and (17). The no flux boundary condition is applied at the center of the tumor; i.e. [16],

$$\nabla P_i = 0 \; for \; r = 0 \tag{18}$$

$$D_{eff}\nabla C + v_i C = 0 \; for \; r = 0 \tag{19}$$

The continuity of pressure and velocity for Equation (11) and concentration and its flux for Equation (17) are considered as appropriate boundary conditions for inner boundary:

$$\begin{aligned} -\kappa_t \nabla P_i|_{\Omega^-} &= -\kappa_n \nabla P_i|_{\Omega^+} \\ P_i|_{\Omega^-} &= P_i|_{\Omega^+} \end{aligned} \tag{20}$$

$$\left(D^t_{eff}\nabla C + v_i C\right)|_{\Omega^-} = \left(D^n_{eff}\nabla C + v_i C\right)|_{\Omega^-} \qquad (21)$$
$$C|_{\Omega^-} = C|_{\Omega^+}$$

where Ω^- and Ω^+ indicate the tumor and normal tissue at the inner boundary.

For outer boundary, that the interstitial pressure is constant; the Dirichlet type of boundary condition must be applied [16]:

$$P_i = \textit{for outer region} \qquad (22)$$

And for concentration, the open boundary condition is used in the outer region [46]. The Open Boundary is used to set up mass transport across boundaries where both convective inflow and outflow can occur and defined by Equation (23):

$$-\mathbf{n}\cdot\nabla C = 0 \qquad (23)$$

where **n** is the normal vector.

Geometry

To investigate the effect of tumor shape on drug delivery, 5 different shapes are considered. The 3D fundamental shape of tumors is assumed to be an ellipsoid in different studies. The assumption of considering ellipsoid tumor shape is investigated in many research such as breast cancer [47], prostate cancer [48,49] cervical cancer [50], glioma cancer [51], and others [52]. Based on this mentioned reason, ellipsoidal shapes of tumors are considered in this study. Different shapes are produced by changing ratio of two radii of ellipsoid shown in Figure 1. This ratio ($\lambda = b/a$) is changed from 0.1 (prolate) to 10 (oblate). In all shapes shown in Figure 16 the volume of the tumor is remained constant and equal to the volume of spherical tumor with radius R. The baseline value of R is 1 cm. The values of R used in sensitivity analysis are changed from 0.1 cm to 2.5 cm. This range is obtained from the literature and is close to the experimental observations [40].

Model parameterization

The interstitial transport properties for normal and tumor tissue are listed in Table 1. These values are used as baseline and some of them are investigated and changed in specified ranges for sensitivity analysis.

The parameters of solute transport model taken from Baxter and Jain [8] are listed in Table 2. Although, the numerical model is applicable for any type of solute, in present study the properties of Fragment antigen-binding ($F(ab')_2$) as a sample is used.

Numerical solution

A systematic flow chart is illustrated in Figure 17 to clarify the computational techniques involved in this paper. The criterion for the convergence of iterative solution based on element–based finite volume (EB-FV) method is to reduce the residual by 6 orders of magnitudes. The details of numerical method is mentioned in our previous works [2,16,53].

Abbreviations

CFD: Computational fluid dynamics; IFP: Interstitial fluid pressure; IFV: Interstitial fluid velocity; DCIF: Drug concentration in interstitial fluid; EB-FV: Element based- finite volume.

Competing interests

The authors declare that they have no competing interests.

Authors' contributions

MS and MS carried out the numerical simulations and drafted the manuscript. HB participated in the numerical simulations. MM and MB and KR helped to draft the manuscript and discussed the results. MS and MS conducted the numerical simulation, discussed the results and reviewed the manuscript. All authors read and approved the final manuscript.

Author details

[1]Department of Mechanical Engineering, K. N. T. University of Technology, Tehran, Iran. [2]Division of Nuclear Medicine, Department of Radiology and Radiological Science, School of Medicine, Johns Hopkins University, Baltimore, MD, USA. [3]Electrical & Computer Department of Ryerson University, Toronto, Ontario.

References

1. Siegel R, Naishadham D, Jemal A: **Cancer statistics.** *A Cancer J Clin* 2013, **63**:11–30.
2. Soltani M, Chen P: **Effect of tumor shape and size on drug delivery to solid tumors.** *J Biol Eng* 2012, **6**:4.
3. Blakeslee S: **Impenetrable tumors found to block even the newest cancer agents.** In [http://www.nytimes.com/1989/06/27/science/impenetrable-tumors-found-to-block-even-the-newest-cancer-agents.html]
4. Baxter LT, Jain RK: **Transport of fluid and macromolecules in tumors IV: A microscopic model of the perivascular distribution.** *Microvasc Res* 1991, **41**:252–272.
5. Baxter LT, Jain RK: **Transport of fluid and macromolecules in tumors (II): role of heterogeneous perfusion and lymphatics.** *Microvasc Res* 1990, **40**:246–263.
6. Baxter LT, Jain RK: **Transport of fluid and macromolecules in tumors (iii): role of binding and metabolism.** *Microvasc Res* 1991, **41**:5–23.
7. Jain RK, Baxter LT: **Mechanisms of heterogeneous distribution of monoclonal antibodies and other macromolecules in tumors: significance of elevated interstitial pressure.** *Cancer Res* 1988, **48**:7022–7032.
8. Baxter LT, Jain RK: **Transport of fluid and macromolecules in tumors (I): role of interstitial pressure and convection.** *Microvasc Res* 1989, **37**:77–104.
9. Saltzman WM, Radomsky ML: **Drugs released from polymers: diffusion and elimination in brain tissue.** *Chem Engng Sci* 1991, **46**:2429–2444.
10. Wang CH, Li J: **Three-dimensional simulation of IgG delivery to tumors.** *Chem Eng Sci* 1998, **53**:3579–3600.
11. Wang CC, Li J, Teo CS, Lee T: **The delivery of BCNU to brain tumors.** *J Control Release* 1999, **61**:21–41.
12. Tan WHK, Wang F, Lee T, Wang C-H: **Computer simulation of the delivery of etanidazole to brain tumor from PLGA wafers: comparison between linear and double burst release systems.** *Biotechnol Bioeng* 2003, **82**:278–288.
13. Zhao J, Salmon H, Sarntinoranont M: **Effect of heterogeneous vasculature on interstitial transport within a solid tumor.** *Microvasc Res* 2007, **73**:224–236.
14. Arifin DY, Lee KYT, Wang C-H: **Chemotherapeutic drug transport to brain tumor.** *J Control Release* 2009, **137**:203–210.
15. Pishko GL, Astary GW, Mareci TH, Sarntinoranont M: **Sensitivity analysis of an image-based solid tumor computational model with heterogeneous vasculature and porosity.** *Ann Biomed Eng* 2011, **39**:2360–2373.
16. Soltani M, Chen P: **Numerical modeling of fluid flow in solid tumors.** *PLoS One* 2011, **6**:1–15.

17. Soltani M, Chen P: **Numerical modeling of interstitial fluid flow coupled with blood flow through a remodeled solid tumor microvascular network.** *PLoS One* 2013, **8**:e67025.
18. Soltani M: *Numerical modeling of drug delivery to solid tumor microvasculature.* Waterloo, Ontario, Canada: PhD thesis, Chem. Eng. (Nanotechnology); 2012.
19. Sefidgar M, Soltani M, Bazmara H, Mousavi M, Bazargan M, Elkamel A: **Interstitial flow in cancerous tissue: effect of considering remodeled capillary network.** *J Tissue Sci Eng* 2014, **4**:1–8.
20. Sefidgar M, Raahemifar K, Bazmara H, Bazargan M, Mousavi SM, Soltani M: **Effect of remodeled tumor-induced capillary network on interstitial flow in cancerous tissue.** In *2nd Middle East Conf:17-20 February 201;Doha, Qatar.* Biomed. Eng. IEEE; 2014:212–215.
21. Netti PA, Berk DA, Swartz MA, Grodzinsky AJ, Jain RK: **Role of extracellular matrix assembly in interstitial transport in solid tumors.** *Cancer Res* 2000, **60**:2497–2503.
22. Mayr NA, Yuh WTC, Zheng J, Ehrhardt JC, Sorosky JI, Magnotta VA, Pelsang RE, Hussey DH: **Tumor size evaluated by pelvic examination compared with 3-D MR quantitative analysis in the prediction of outcome for cervical cancer.** *Int J Radiat Oncol Biol Phys* 1997, **39**:395–404.
23. Kim HJ, Kim W: **Method of tumor volume evaluation using magnetic resonance imaging for outcome prediction in cervical cancer treated with concurrent chemotherapy and radiotherapy.** *Radiat Oncol J* 2012, **30**:70–77.
24. Ayers GD, Mckinley ET, Zhao P, Fritz JM, Metry RE, Deal BC, Adlerz KM, Coffey RJ, Manning HC: **Volume of preclinical xenograft tumors is more accurately assessed by ultrasound imaging than manual caliper measurements.** *J Ultrasound Med* 2010, **29**:891–901.
25. Narayan K, Fisher R, Bernshaw D: **Significance of tumor volume and corpus uteri invasion in cervical cancer patients treated by radiotherapy.** *Int J Gynecol Cancer* 2006, **16**:623–630.
26. Char D, Kroll S, Phillips T: **Uveal melanoma: growth rate and prognosis.** *Arch Ophthalmol* 1997, **115**:1014–1018.
27. Jensen PK, Hansen MK: **Ultrasonographic, three-dimensional scanning for determination of intraocular tumor volume.** *Acta Ophthalmol* 1991, **69**:178–186.
28. Gass J, Donald M: **Comparison of uveal melanoma growth rates with mitotic index and mortality.** *Arch Ophthalmol* 1985, **103**:924–931.
29. Arifin DY, Lee LY, Wang C-H: **Mathematical modeling and simulation of drug release from microspheres: implications to drug delivery systems.** *Adv Drug Deliv Rev* 2006, **58**:1274–1325.
30. Huber PE, Bischof M, Heiland S, Peschke P, Saffrich R, Gro H, Lipson KE, Abdollahi A: **Trimodal cancer treatment: beneficial effects of combined antiangiogenesis, radiation, and chemotherapy.** *Cancer Res* 2005, **65**:3643–3655.
31. Jain RK, Tong RT, Munn LL: **Effect of vascular normalization by antiangiogenic therapy on interstitial hypertension, peritumor edema, and lymphatic metastasis: insights from a mathematical model.** *Cancer Res* 2007, **67**:2729–2735.
32. Chauhan VP, Stylianopoulos T, Martin JD, Chen O, Kamoun WS, Bawendi MG, Fukumura D, Jain RK: **Normalization of tumour blood vessels improves the delivery of nanomedicines in a size-dependent manner.** *Nat Nanotechnol* 2012, **7**:383–388.
33. Arifin DY, Wang C, Smith KA: **Patient-specific chemotherapeutic drug delivery to brain tumors.** *Mimics Innov Award* 2007, **1**:1–9.
34. Nielsen T, Murata R, Maxwell RJ, Stødkilde-Jørgensen H, Ostergaard L, Ley CD, Kristjansen PEG, Horsman MR: **Non-invasive imaging of combretastatin activity in two tumor models: association with invasive estimates.** *Acta Oncol* 2010, **49**:906–913.
35. Hompland T, Ellingsen C, Øvrebø KM, Rofstad EK: **Interstitial fluid pressure and associated lymph node metastasis revealed in tumors by dynamic contrast-enhanced MRI interstitial fluid pressure and associated lymph node.** *Cancer Res* 2012, **72**:4899–4908.
36. Welter M, Rieger H: **Interstitial fluid flow and drug delivery in vascularized tumors: a computational model.** *PLoS One* 2013, **8**:e70395.
37. Roy R, Riahi DN: **Modeling blood flow in a brain tumor treated concurrently with radiotherapy and chemotherapy.** *Appl Math Phys* 2013, **1**:67–77.
38. Gutmann R, Leunig M, Feyh J, Goetz AE, Messmer K, Kastenbauer E, Jain RK: **Advances in brief interstitial hypertension in head and neck tumors in patients: correlation with tumor size.** *Cancer Res* 1992, **52**:1993–1995.
39. Hompland T, Ellingsen C, Rofstad EK: **Preclinical evaluation of Gd-DTPA and gadomelitol as contrast agents in DCE-MRI of cervical carcinoma interstitial fluid pressure.** *BMC Cancer* 2012, **12**:2–10.
40. Leguerney I, Lassau N, Koscielny S, Rodrigues M, Massard C, Rouffiac V, Benatsou B, Thalmensi J, Bawa O, Opolon P, Peronneau P, Roche A: **Combining functional imaging and interstitial pressure measurements to evaluate two anti-angiogenic treatments.** *Invest New Drugs* 2012, **30**:144–156.
41. Au JL, Jang SH, Zheng J, Chen CT, Song S, Hu L, Wientjes MG: **Determinants of drug delivery and transport to solid tumors.** *J Control Release* 2001, **74**:31–46.
42. Mccarty WJ, Johnson M: **The hydraulic conductivity of Matrigel.** *Biorheology* 2007, **44**:303–317.
43. Wu M, Frieboes HB, McDougall SR, Chaplain MAJ, Cristini V, Lowengrub J: **The effect of interstitial pressure on tumor growth: coupling with the blood and lymphatic vascular systems.** *J Theor Biol* 2013, **320**:131–151.
44. Vafai K: *Porous Media, Application in Biological systems and biotechnology.* Boca Raton: CRC Press; 2011.
45. Curry F: *Mechanics and thermodynamics of transcapillary exchange, Handbook of Physiology, Section 2: The Cardiovascular System.* Bethesda: Amer Physiol Soc; 1984.
46. Orlanski I: **A simple boundary condition for unbounded hyperbolic flows.** *J Comput Phys* 1976, **21**:251–269.
47. Wapnir IL, Barnard N, Wartenberg D, Greco RS: **The inverse relationship between microvessel counts and tumor volume in breast cancer.** *Breast J* 2001, **7**:184–188.
48. Sosna J, Rofsky NM, Gaston SM, DeWolf WC, Lenkinski RE: **Determinations of prostate volume at 3-Tesla using an external phased array coil: comparison to pathologic specimens.** *Acad Radio* 2003, **10**:846–853.
49. Egorov V, Ayrapetyan S, Sarvazyan AP: **Prostate mechanical imaging: 3-D image composition and feature calculations.** *IEEE Trans Med Imaging* 2008, **25**:1329–1340.
50. Mayr NA, Taoka T, Yuh WTC, Denning LM, Zhen WK, Paulino AC, Gaston RC, Sorosky JI, Meeks SL, Walker JL, Mannel RS, Buatti JM: **Method and timing of tumor volume measurement for outcome prediction in cervical cancer using magnetic resonance imaging.** *Int J Radiat Oncol Biol Phys* 2002, **52**:14–22.
51. Schmidt K, Ziu M, Ole Schmidt N, Vaghasia P, Cargioli T, Doshi S, Albert M, Black P, Carroll R, Sun Y: **Volume reconstruction techniques improve the correlation between histological and in vivo tumor volume measurements in mouse models of human gliomas.** *J Neurooncol* 2004, **68**:207–215.
52. James K, Eisenhauer E, Christian M, Terenziani M, Vena D, Muldal A, Therasse P: **Measuring response in solid tumors: unidimensional versus bidimensional measurement.** *J Natl Cancer Inst* 1999, **91**:523–528.
53. Soltani M, Chen P: **Shape design of internal flow with minimum pressure loss.** *Adv Sci Lett* 2009, **2**:347–355.

Comparison of lignin derivatives as substrates for laccase-catalyzed scavenging of oxygen in coatings and films

Kristin Johansson[1], Thomas Gillgren[2], Sandra Winestrand[2], Lars Järnström[1] and Leif J Jönsson[2*]

Abstract

Background: Lignin derivatives are phenylpropanoid biopolymers derived from pulping and biorefinery processes. The possibility to utilize lignin derivatives from different types of processes in advanced enzyme-catalyzed oxygen-scavenging systems intended for active packaging was explored. Laccase-catalyzed oxidation of alkali lignin (LA), hydrolytic lignin (LH), organosolv lignin (LO), and lignosulfonates (LS) was compared using oxygen-scavenging coatings and films in liquid and gas phase systems.

Results: When coatings containing lignin derivatives and laccase were immersed in a buffered aqueous solution, the oxygen-scavenging capability increased in the order LO < LH < LA < LS. Experiments with coatings containing laccase and LO, LH or LA incubated in oxygen-containing gas in air-tight chambers and at a relative humidity (RH) of 100% showed that paperboard coated with LO and laccase reduced the oxygen content from 1.0% to 0.4% during a four-day period, which was far better than the results obtained with LA or LH. LO-containing coatings incubated at 92% RH also displayed activity, with a decrease in oxygen from 1.0% to 0.7% during a four-day period. The oxygen scavenging was not related to the content of free phenolic hydroxyl groups, which increased in the order LO < LS < LH < LA. LO and LS were selected for further studies and films containing starch, clay, glycerol, laccase and LO or LS were characterized using gel permeation chromatograpy, dynamic mechanical analysis, and wet stability.

Conclusions: The investigation shows that different lignin derivatives exhibit widely different properties as a part of active coatings and films. Results indicate that LS and LO were most suitable for the application studied and differences between them were attributed to a higher degree of laccase-catalyzed cross-linking of LS than of LO. Inclusion in active-packaging systems offers a new way to utilize some types of lignin derivatives from biorefining processes.

Keywords: Lignin derivatives, Laccase, Coating, Film, Oxygen scavenger

Background

Endeavors to decrease human impact on the climate and the environment include using resources more efficiently, changing from fossil to renewable resources, and processing resources in more efficient ways. Food production and consumption has a vast impact on climate and environment. In 2009, 230 megatons of plastics were produced worldwide, and 40% of the plastics went to packaging [1]. Since most plastics are made of petroleum, which is not a renewable resource, our massive use of plastic packages runs contrary to a sustainable development. Another environmental issue is wastage in the food-production chain. It is estimated that one third of the food produced is lost or uneaten [2]. For no good reasons, this wastage demands large amounts of electricity, fresh water, fossil fuels, and fertilizers. One possible contribution to both the package waste problem and the food wastage problem would be to use renewable and biodegradable biopolymers to produce active packages. Active packages contain active components that maintain the quality of the food that is packaged.

Lignin is a renewable biopolymer and one of the main components of wood. Although lignin is the second most abundant terrestrial biopolymer [3], it is poorly exploited and not much used in advanced material applications. Whereas the main component of processed wood, cellulose, is used for paper, paperboard, viscose and many other

* Correspondence: leif.jonsson@chem.umu.se
[2]Department of Chemistry, Umeå University, SE-901 87 Umeå, Sweden
Full list of author information is available at the end of the article

products, lignin often goes to energy recovery [4]. Since native lignin is recalcitrant and hard to process, it needs to be partially degraded and extracted in order to be useful. Hence, the lignin available from e.g. the pulp and paper industry is in the form of lignin derivatives. Enzymatic catalysis is an environmentally friendly way to modify the properties of chemicals and materials. Enzymes are used under mild reaction conditions and are generally highly selective catalysts. Enzymes can be utilized as catalysts of oxygen scavenging. Oxygen scavengers consume freely available oxygen in order to decrease the oxygen availability in a specified volume. In particular, oxygen scavengers can be used as an active ingredient in packaging. Food degradation processes are most often oxygen dependent, and keeping oxygen from the food is therefore a way of maintaining the quality and increasing the shelf-life of the food.

In a recent study, the enzyme laccase from the white-rot fungus *Trametes versicolor* was used to catalyze oxygen scavenging in prototype packages: coatings on cardboard and foils, and free-standing films [5]. Laccase carries out a one-electron oxidation of free phenolic hydroxyl groups present in lignin derivatives. Cross-linked products or quinones are then formed from the phenoxy radicals. In the previous investigation, lignosulfonates (LS) were used as substrate for the oxidation [5]. LS are by-products from pulp mills and wood biorefineries using sulfite-based processes. Using paperboard coatings with LS and laccase, the oxygen content in the headspace of a trial packaging was successfully reduced by up to almost 80% [5]. Furthermore, it was found that the oxidation of LS increased the storage modulus (E') of the films, and increased their water resistance. The latter is a weak spot of biopolymer materials. The altered material properties were tentatively attributed to cross-linking, but this was not further analyzed [5].

Since interesting results were obtained with LS and laccase in coatings and films, it is of interest to compare LS with other lignin derivatives and to analyze the effects of the enzyme-catalyzed reaction on the material properties of the films. Hence, in this investigation, we also studied the oxygen-scavenging potential of alkali lignin (LA), hydrolytic lignin (LH) and organosolv lignin (LO). Alkali lignin is separated from black liquor, which is a by-product from the kraft process, by acid treatment. The kraft process is a more common industrial process than sulfite pulping, which generates LS, and the abundance of alkali lignin motivates its inclusion in the study. Hydrolytic lignin is formed by degradation of wood polysaccharides through hydrolysis, and this type of lignin may become a large co-product of lignocellulosic biorefineries [6,7]. Organosolv lignin is recovered from organosolv pulping, which is a pulping technique where organic solvents are used to extract lignin and hemicellulose. The organosolv process is studied as a pretreatment method for bioconversion of lignocellulose [8-10]. Thus, hydrolytic lignin and organosolv lignin are potential co-products of lignocellulosic biorefineries, which motivates their inclusion in the present study. Lignin co-products can be combusted for generation of energy and heat, but utilization of some part of the lignin for making high value-added products, such as active packaging, would be desirable.

The present study is focused on the properties of coated paper board, coated foil, and free-standing films containing starch and different lignin derivatives. Furthermore, by using analytical techniques such as GPC (Gel Permeation Chromatography) we investigated molecular transformations effected by laccase-catalyzed oxidation of lignin derivatives in films. Research in this area may lead to new applications for biomacromolecules such as lignin derivatives and starch, and improves the understanding of the potential of different types of lignin derivatives as well as of laccase-catalyzed reactions in solid media.

Results and discussion

Oxygen scavenging

The lignin derivatives LO, LA, LH and LS were evaluated as potential components of oxygen-scavenging coatings containing laccase from the white-rot fungus *Trametes versicolor* as catalyst. Initial experiments were performed to compare coatings with LO, LA and LH and to select the most promising candidate for further investigations, which also included coatings with LS.

The oxygen-consumption rate of pieces of aluminum foil coated with laccase and different lignin derivatives was determined in buffered aqueous solutions by using an oxygen electrode. Coatings with LS gave the highest value (Figure 1). Coatings with LA and LH performed similarly and reached about 50% of the oxygen-consumption rate determined for LS-containing coatings. Coatings with LO displayed slightly lower oxygen-consumption rate than the ones with LA and LH, and reached approx. 40% of that of coatings with LS. Thus, all four lignin preparations worked relatively well as substrates for laccase in coatings immersed in buffer.

The oxygen-scavenging ability of coated boards with lignin derivatives and laccase was studied in air-tight chambers with an RH of 100% (Figure 2). As expected, coatings without enzyme did not scavenge oxygen (Figure 2B). With respect to enzyme-containing coatings, oxygen was consumed much faster by the coating containing LO than by coatings containing LA or LH. After only three days (before the re-flushing) the oxygen concentration decreased from 1% to 0.7%. This can be compared to results obtained with LS, where the oxygen concentration after three days was 0.9% [5].

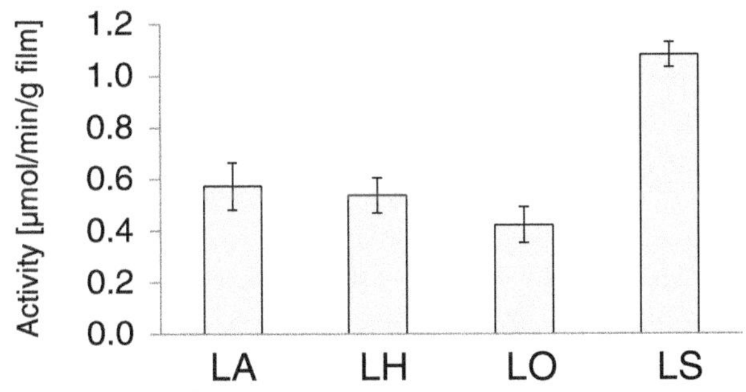

Figure 1 Oxygen-consumption rate of lignin derivatives and laccase coated on aluminum foil. Coated pieces of aluminum foil were immersed in a buffered aqueous solution (23°C and pH 6.5). The coatings contained LA, LH, LO, or LS. Error bars indicate standard deviations of three replicates.

After re-flushing with gas containing 1.0% oxygen and three additional days of incubation, the coating containing LS had decreased the oxygen concentration to 0.3% [5]. At day six in the experiment of the present study, three days after re-flushing with gas with 1.0% oxygen, the coating containing LO had decreased the oxygen concentration to 0.6% (Figure 2). On day seven, four days after re-flushing, the oxygen concentration was down to 0.4% (Figure 2). The coat weight of the coating containing LO was about three times as high as that of the coating containing LS, which means that for LS less substrate and enzyme was available for the reaction with oxygen. To enable a direct comparison between the LO- and LS-containing coatings, the oxygen-scavenging capacity was calculated as cm^3 scavenged oxygen per g coating (Figure 3). The comparison indicates that after three days there was no significant difference between the LO and the LS coatings. However, after six days the coating containing LS had consumed three times more oxygen than the coating containing LO. It is likely that the plasticizing effect of water was highest for the hydrophilic LS coating, which could possibly result in a higher molecular mobility [11]. Water uptake may also explain the difference in oxygen consumption after six days between coatings with LO (Figure 2) and with LS [5]. The presence of water has been found to be of great importance for the oxygen-scavenging ability of glucose oxidase incorporated into paper coatings [12].

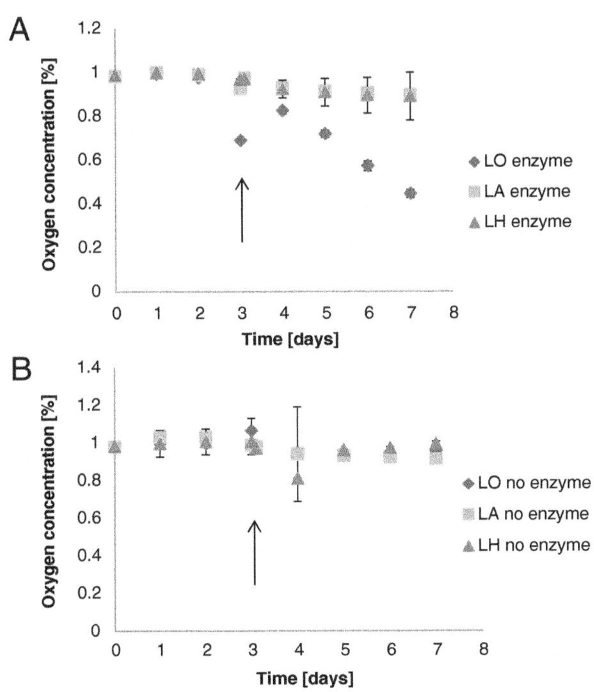

Figure 2 Oxygen content in chambers with coated boards with lignin derivatives and with (A) or without (B) laccase. The chambers were flushed with fresh gas mixture (1% oxygen and 99% nitrogen) after three days (indicated by arrows). Error bars indicate standard deviations of three replicates.

Since laccase oxidizes phenolic hydroxyl groups, it was of interest to see if there was a correlation between oxygen scavenging and the content of phenolic hydroxyl groups. The content of phenolic groups (given as mmol per kg of dry substance) was: LA 1390; LH 890; LS 620; LO 300. LS, which had the second lowest content of phenolic groups, showed the highest oxygen-scavenging capacity. LO, which contained the lowest content of phenolic groups, had higher oxygen-scavenging ability than both LA and LH (Figure 2). Thus, the oxygen-scavenging ability was not reflected by the content of phenolic groups in the lignin derivatives. As the coatings with LA and LH did not consume much oxygen, they were not studied further in subsequent experiments.

When the RH was decreased from 100% to 85%, no oxygen scavenging was detected for coatings with LS and laccase [5]. The oxygen-scavenging ability of

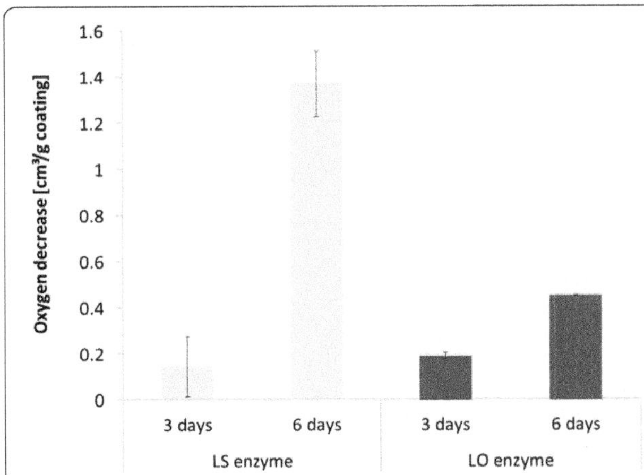

Figure 3 Comparison of the amount of consumed oxygen per g of coating for LS [5] and LO (this study). Error bars indicate standard deviations of three replicates.

LO-containing coatings was therefore tested using an RH in between 85% and 100%. The results are shown in Figure 4. It is clear that 92% RH was sufficient to sustain the reaction, although the rate was lower than at 100% RH. After three days at 92% RH, the oxygen concentration was still about 1.0%, and during the following four-day period it decreased to 0.7%. This result opens up for the possibility to use lignin derivatives and laccase as an oxygen-scavenging system for foodstuffs with a water activity that is higher than 0.9, which would include bread, cheese, meat, and various fruits [13].

Laccase-catalyzed oxidation of lignin derivatives in cast starch-based films

The effects of a possible formation of a three-dimensional network upon oxidation of LO by laccase were evaluated by examining the properties of free films with respect to wet stability and mechanical properties. Films containing denatured enzyme were used as controls.

Figure 5 shows the effect of active enzyme on the wet stability of free films. The absorbance at 380 nm gives a measure of the amount of LO migrating out of the film. The difference between films with active and denatured enzyme was most obvious for samples taken at an early stage (5 and 15 min) (Figure 5A). After a longer incubation, such as 24 h, there was no longer any effect of the enzymatic reaction (Figure 5B). The results indicate that there is a positive correlation between the enzymatic reaction and the wet stability of the films. The effect can be attributed to cross-linking of lignin fragments, which would lead to increased degree of polymerization and stabilization of the films. Films containing LO showed less difference between active and denatured enzyme than reported for the corresponding LS-containing films. After 24 h, the absorption was reduced by 88% in LS films containing active enzyme [5]. This observation suggests more extensive formation of three-dimensional network in the LS films than in the LO films.

Mechanical properties

The mechanical properties of the cast starch-based films with laccase and LO were investigated by determination of the E' modulus at low frequencies. A higher E' modulus indicates that the material is stiffer with a higher cross-linking density [14]. Although the average E' modulus of the films with active laccase was higher than that of the films with inactivated enzyme, the difference was not significant (Figure 6). Similar films as these with an average thickness of 0.13 mm, with inactive enzyme but containing LS instead of LO, displayed even lower E' modulus [5]. A plausible explanation is that LS acted as a plasticizer for the starch in the films. This explanation is supported by several investigations regarding the interaction between starch and LS [15-18]. The increase in modulus due to the active enzyme in the LS films was about 300% [5]. The results from both the mechanical testing and the wet stability testing of the starch-based films suggest that the oxidation of LO did not give rise to as extensive cross-linking as was the case for the LS-containing films. There may also be an incompatibility between the hydrophobic lignin and the hydrophilic starch. It is likely that the hydrophobic LO lignin exists as dispersed particles within the hydrophilic starch matrix [19].

Hydrophilic LS are known to be able to form complexes with starch polymers through hydrogen bonding without any oxidation reaction [20]. LS may hence facilitate the continuity of the starch network during stress, as opposed to LO which rather disrupts the starch network structure.

Py-GC-MS analysis

The S/G (syringyl/guaiacyl) ratios of LO and LS were 2.5 and 0.012, respectively. This indicates that hardwood was used as raw material for LO, and that softwood was used for production of LS.

GPC analysis

Figure 7A shows the chromatograms for LO and LS freeze-dried powder samples (from reactions in solution).

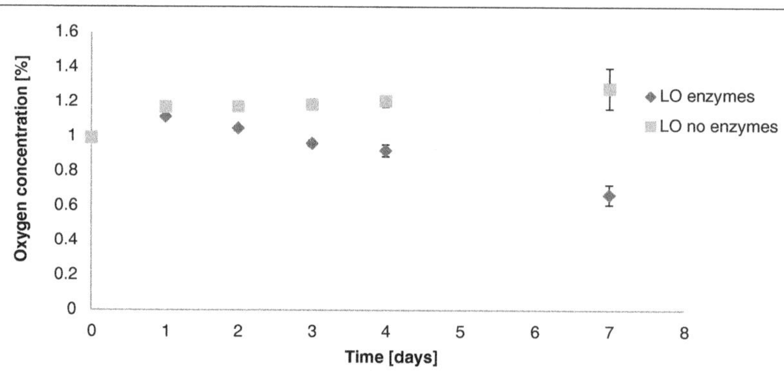

Figure 4 Oxygen concentrations in air-tight chambers measured after addition of board coated with LO. The RH was 92%. In these experiments, there was no flushing with fresh gas mixture after three days. Error bars indicate standard deviations of three replicates.

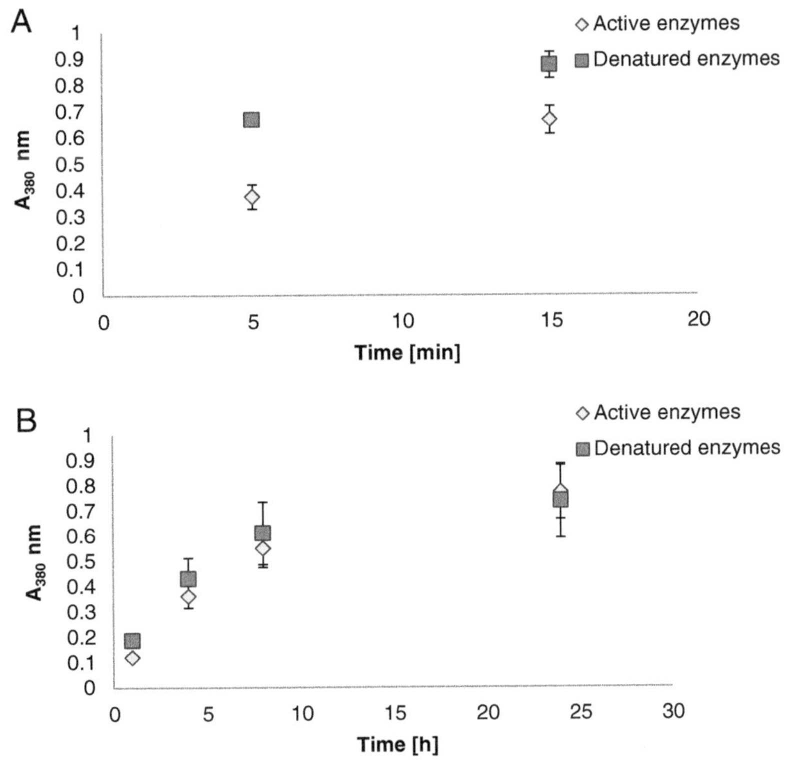

Figure 5 Wet stability of free starch-based films containing LO with and without active laccase. Samples taken early **(A)** were measured without dilution, while samples taken after a longer incubation time **(B)** were diluted 1:9. Error bars indicate standard deviations of three replicates.

The main fraction of LO was divided into two peaks that were not well separated and which had very low peak molecular weights (M_p, the most common molecular weight among the polymers in the sample) (Figure 7A). Enzymatic treatment during 96 h shifted the ratio between these peaks, as the peak with shorter elution time grew relatively the one with longer elution time. In terms of M_w (the weight average molecular weight) for the whole main fraction, this meant an increase from 1,500 to 2,500 as a consequence of the enzymatic activity (Figure 8). Both the

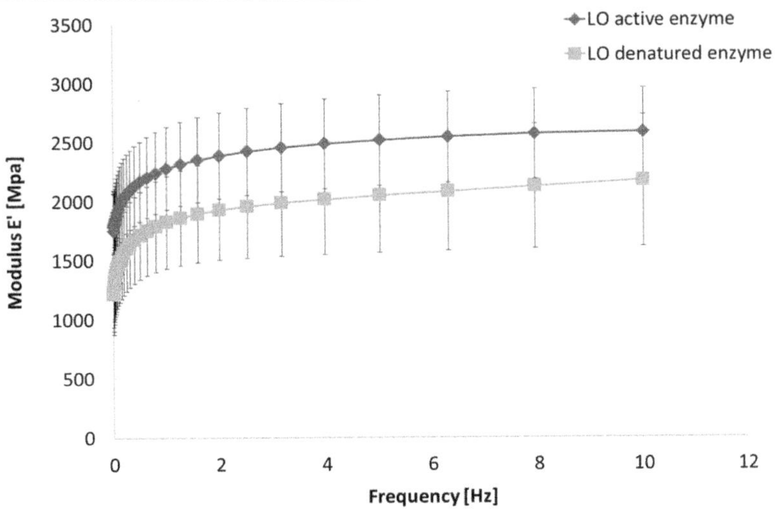

Figure 6 Mechanical properties of starch-based films with LO and active or inactivated laccase. Error bars indicate standard deviations of six replicates.

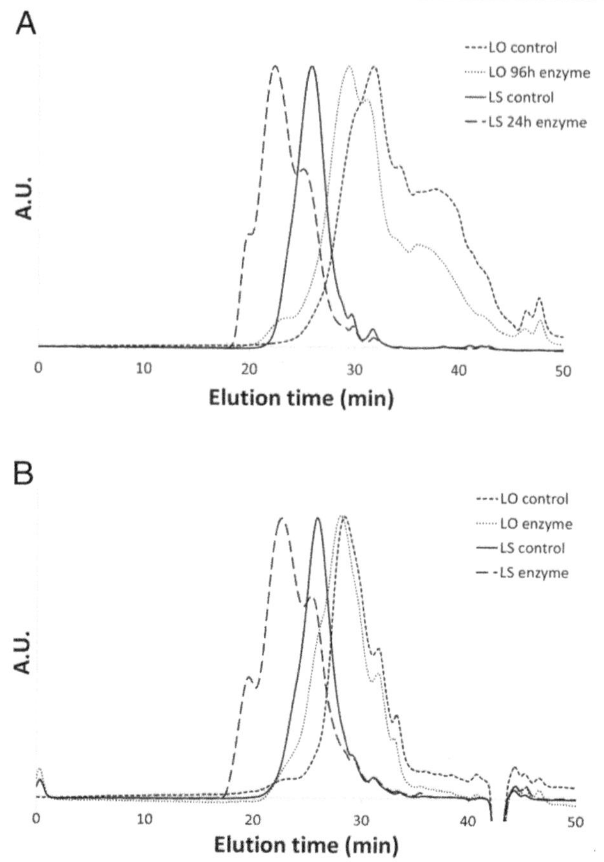

Figure 7 GPC chromatogram of (A) LS and LO powder samples, and (B) LS and LO film samples. Curves are normalized to match approximately the same maximum height.

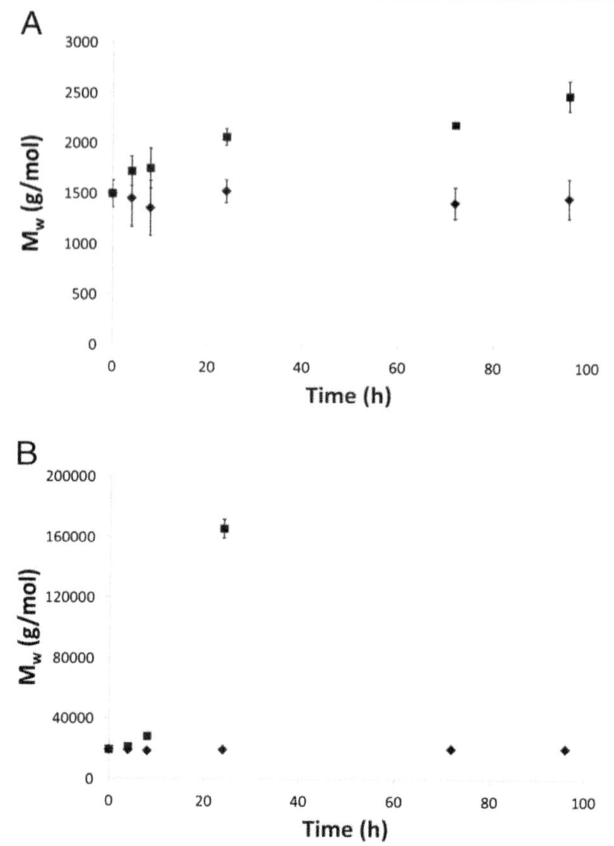

Figure 8 Weight average molecular weights (M_w) of (A) LO and (B) LS powders after incubation with active laccase (squares), and denatured laccase as controls (diamonds), at different times. Error bars indicate standard deviations of four replicates.

enzymatically treated and the control LO powder samples showed a low-molecular-weight fraction (Figure 7A) that was absent in the films (Figure 7B). This effect was attributed to the different preparation methods used for the powder and the film samples. The initial M_p of the LS powder sample was 12,000, but after enzymatic treatment during 24 h it had split into three different sub-peaks with M_p values of 21,000, 120,000 and 620,000 (Figure 7A). The initial M_w was 20,000. The three sub-peaks in the analysis of the 24-h sample had M_w values of 13,000, 160,000, and 880,000, respectively. The reaction resulted in an eight-fold overall increase in M_w (Figure 8) after only 24 h. Samples with longer reaction times were not soluble and could hence not be analyzed. The lack of solubility was believed to be due to extensive cross-linking. Furthermore, according to the time courses shown in Figure 8, the molecular weight of LS increased in an exponential-like mode, which indicates that the cross-linking progressed very rapidly. In contrast, the M_w increase for LO was rather constant throughout the time period studied. It is worth noticing that the LS preparation possessed an initial M_w that was about ten-fold higher than that of LO. The lack of molecular weight increase among the control samples confirmed that it was in fact the enzyme activity that generated the molecular-weight increase of the samples incubated with active enzyme (Figure 8).

Just as for the freeze-dried LS powder samples from the oxidations reactions in water, the analysis of the LS-containing films displayed three sub-peaks with different M_p values (Figure 7B). The M_w of the sub-peaks were 7,500, 130,000, and 1,600,000, values that correspond fairly well with the M_w values of the LS powder samples. The M_w of the two sub-peaks with highest molecular weight increased roughly 9 and 110 times compared to the M_w of the control sample. It was most likely these high M_w fractions that formed the three-dimensional network responsible for the extraordinary improvements in water resistance and storage modulus of the LS-containing films.

The M_p of the LO film sample also increased after the enzymatic reaction (Figure 7B). The laccase-catalyzed oxidation of LO in films resulted in a five-fold increase in M_w. Apparently, there were no high molecular weight

LO fractions in the LO films. This is in agreement with the observations from the water-resistance assessment and the DMA. The more extensive cross-linking of LS (as compared to that of LO) can probably be attributed to the sulphonate groups. Since sulphonate groups make lignin more polar, they will make LS more compatible with the starch matrix. LO, on the other hand, may exist in the form of aggregates. For example, the conformation of lignin from black liquor has been found to be highly dependent on the pH in the solution, and the lignin tends to aggregate in water [21]. Compatibility with starch as a result of higher polarity of LS makes it likely that compared to LO more phenolic groups were exposed to the surroundings, and thereby they would be possible for laccase to oxidize. Furthermore, since LS is known to form complexes with starch [17], it was probably more evenly spread in the starch matrix. It has been reported that co-polymers of LS and starch can be produced through laccase-induced grafting of LS polymers onto the starch [22]. Thereby, cross-linking of LS is more likely to lead to a continuous LS network throughout the films. In contrast, LO molecules, supposedly being accumulated in aggregates, probably cross-link with other LO molecules in their vicinity. Whereas this may lead to strongly bound aggregates, it will not benefit the formation of a continuous network.

An additional contributing factor is the importance of the formation of productive couplings. Areskogh et al. [23] suggested that the presence of a sulphonate group on a lignin model compound directs the laccase-catalyzed oxidation reaction towards productive couplings, enabling further participation in cross-linking of the involved molecules. This may contribute to a more extensive cross-linking of the LS. On the other hand, oxidation of LO (which lacks sulphonate groups) may result in unproductive coupling, making further reactions more difficult. The S/G ratio may also affect the ability to form productive cross-coupling of a lignin derivative, as guaiacyl, compared to syringyl, has one additional site on the benzene moiety available for cross-linking. Further investigations are needed to unravel the roles of polarity, sulphonation and S/G content in cross-linking of lignin derivatives.

Conclusions

Although this was not evident from experiments in buffered aqueous media, LS and LO proved to be superior to LA and LH in experiments with laccase-containing coatings incubated in oxygen-containing gas. This indicates the importance of performing this type of investigations using air-tight reaction chambers and under conditions where the humidity is well controlled. In further experiments with LS and LO, LS-containing coatings performed better with regard to water resistance and mechanical properties. GPC analysis revealed that the laccase-catalyzed reaction resulted in far more extensive cross-linking of LS as indicated by an exceptional increase in molecular weight. It seems plausible that this polymerization of LS resulted in a three-dimensional network, which was responsible for increased storage modulus and improved water resistance. The comparison suggests that sulphonation is a more important feature of lignin derivatives for making them suitable for inclusion in oxygen-scavenging coatings than high content of phenolic hydroxyl groups or low molecular weight. Another important finding is that the relative humidity does not necessarily have to be as high as 100% to obtain laccase-catalyzed oxygen scavenging with lignin derivatives, as activity was detected also at 92% RH using LO-containing coatings. This enables for use of laccase and lignin derivatives as oxygen scavengers in packages for foods such as cheese, bread, fish and fruits.

Methods

Materials

Laccase from the white-rot fungus *Trametes* (syn. *Coriolus*, *Polyporus*) *versicolor* was purchased from Jülich Fine Chemicals GmbH (Jülich, Germany). Technical lignin preparations [alkali lignin (LA), hydrolytic lignin (LH) and organosolv lignin (LO)] were obtained from Sigma-Aldrich (St Louis, MA, USA). The lignosulfonates (LS) were kindly provided by Domsjö Fabriker (Örnsköldsvik, Sweden). Styrene-butadiene latex (SB-latex) was supplied by Styron Europe GmbH (Horgen, Switzerland). The SB-latex had a glass transition temperature (Tg) of 6°C. The dry-solids content (SCAN-P 39:80) was 50% and the pH was 5.5. Kaolin clay (Barrisurf LX) was provided by Imerys Minerals Ltd. (Cornwall, UK). Starch (Perlcoat55, a hydroxypropylated and oxidized starch derived from potatoes) was supplied by Lyckeby Industrial AB (Kristianstad, Sweden). Glycerol (99.5%) was supplied by Karlshamns Tefac AB (Karlshamn, Sweden). Potassium nitrate and 3-(*N*-morpholino)propanesulfonic acid (MOPS) were purchased from Merck (Darmstadt, Germany) and Sigma-Aldrich, respectively. Coated boards were prepared using a three-ply packaging barrier board supplied by Stora Enso (Imatra, Finland). The board had a layer of polyethylene (PE) on the top side, and a layer of PE/ethylene vinyl alcohol (EVOH) on the reverse side. Aluminum foil (Skultuna Aluminium Folie FRYS) was obtained from Skultuna Folie AB (Skultuna, Sweden). One side of the foil is coated with polypropylene (PP). The foil was used for preparation of latex-based coatings that were used to determine the activity of immobilized enzyme. Dimethylacetamide (DMAc, HPLC grade) and lithium bromide (LiBr, ReagentPlus grade) were purchased from Sigma-Aldrich. Sodium hydroxide (NaOH) was purchased from Eka Chemicals (Bohus, Sweden). Sodium polystyrene sulfonate (PSS) standards were purchased from American Polymer Standards Corporation (Mentor, OH, USA) and Sigma-Aldrich.

Phenolic content

The content of phenolic groups was determined by MoRe Research AB (Örnsköldsvik, Sweden) using a method based on the study by Lai et al. [24]. The relative standard error was estimated to 10%.

Reactions with lignin derivatives in buffered aqueous solution

Reaction mixtures contained 100 mg/ml lignin preparation, 2.5 U/ml laccase, and 100 mM MOPS buffer (pH 6.5). LO was first dispersed in 100 mM alkaline MOPS buffer and subsequently adjusted to pH 6.5 using 4 M HCl. LS was added directly to the MOPS buffer (pH 6.5). One unit (U) of laccase equals the amount of enzyme required to catalyze the oxidation of 1 µmol of pyrogallol per min in a reaction performed at pH 6.5 and at a temperature of 25°C. Seven ml of the reaction mixture were transferred to five 50 ml Falcon tubes with pierced caps, and the tubes were then incubated with stirring at 23°C and 50% RH for 4, 8, 24, 72 and 96 h. The reaction mixtures were then freeze-dried (Heto Drywinner, Heto-Holten S/S, Allerød, Denmark). The 0 h sample was prepared by freeze drying seven ml of the reaction mixture immediately after enzyme addition. As controls, identical reaction mixtures as described above were prepared with the exception that the enzyme had been denatured by boiling for 12 min prior to addition. The freeze-dried samples were used for subsequent analyses using GPC (Gel Permeation Chromatography).

Preparation of coated board and aluminum foil

A mixture of starch (20% dry-solids content) and water was prepared by vigorous stirring in a boiling water bath for 45 min. The mixture was then rapidly cooled to room temperature. The clay was dispersed in deionized water (conductivity 1 µS/cm) to a final solids content of 63%. Coating color was prepared by mixing latex, clay, freshly cooked starch, and lignin derivative in the proportions described in Table 1 (latex-based coating). The pH was adjusted to 6.5 with a 1 M solution of NaOH. The coating color was subsequently left to stir for 15 min before laccase was added (0.91 U laccase per g wet coating color). Coating color without laccase was prepared as a control. Before coating, the board was corona-treated (60 Wmin/m^2) by using a laboratory-scale corona equipment (Corona-Plus, Vetaphone, Kolding, Denmark) to obtain sufficient adhesion between the PE-laminated board and the coating color. The treatment was performed with ceramic electrodes and an aluminum roll with a perimeter speed of 50 m/min.

Table 1 Coating-color recipes

Component	Latex-based coating*	Starch-based coating**
Latex	100	-
Starch	10	100
Clay	55	55
Lignin derivative	30	20
Glycerol	-	30
Enzyme preparation	6	6

*Parts of the component per hundred parts (by weight) of dry latex.
**Parts of the component per hundred parts (by weight) of dry starch.

The corona-treated PE/EVOH-side of the board was double-coated with a bench coater (K202 Control Coater, RK Print-Coat Instruments Ltd., Royston, UK) equipped with a wire-wound bar giving a nominal wet deposit of 60 µm. The PP-side of the aluminum foil was double coated with a bar giving a nominal wet deposit of 24 µm. To ensure complete film formation, the coated board was dried at 105°C for 30 s and then at 30°C for 24 h after the first and second application of coating color.

Casting of films

The effects of the laccase-catalyzed reaction on the water stability of the films, the mechanical properties of the films, and the size distribution of the lignin derivatives were evaluated using free films casted in Petri dishes with an inner diameter of 8.7 cm. The coating color was prepared by dispersing lignin derivatives in alkalized deionized water followed by mixing with the starch and glycerol in the proportions indicated in Table 1 (starch-based coating). Glycerol was used as a plasticizer for the starch-based films. The pH was subsequently adjusted to 6.5 using 0.5 M HCl prior to addition of the clay as it precipitates under alkaline conditions. After that, the coating color was stirred for 15 min prior to addition of laccase. A similar mixture, but containing enzyme that had been denatured through boiling for 12 min, was prepared as a control. An amount of dispersion corresponding to a total of about 1 g of dry matter was added to each Petri dish, and the films were then dried in dishes without lids for seven days at 23°C and 50% RH (relative humidity). The thickness of the films was measured prior to analysis using an STFI Thickness Tester M201 (TJT Teknik AB, Järfälla, Sweden), on average the thickness was found to be 0.12 mm.

Oxygen scavenging in buffered aqueous solutions

The enzyme activity of coatings on aluminum foil was evaluated in a buffered solution at 25°C (20 mM MOPS, pH 6.5) using an oxygen electrode (Hansatech Ltd., King's Lynn, UK). Two cm^2 of the coated aluminum film were weighed and added to the reaction chamber containing the pre-heated buffered solution. The decrease in oxygen concentration was monitored during 10 min and the activity was calculated as µmol oxygen/min/g film. All measurements were performed in triplicates.

Oxygen scavenging in air-tight chambers

Two dm^2 of board coated with latex-based coating colors with lignin derivatives were cut into strips and placed inside an air-tight chamber with a volume of 128 ml. Coated boards were analyzed in triplicates. Board coated with a coating color without enzyme was used as a control. The RH inside the chambers was varied and controlled by using a saturated salt solution of potassium nitrate giving an RH of 92% and pure water giving an RH of 100%. The RH was determined experimentally with a relative humidity tester (Testo 625, Testo GmbH & Co., Lenzkirch, Germany). The atmosphere inside the chambers was modified by flushing the chambers for 1.5 min with a gas mixture consisting of 1% oxygen and 99% nitrogen (AGA Gas AB, Enköping, Sweden). The decrease in oxygen concentration was monitored during three days before the chambers were re-flushed with 1% oxygen and the decrease was monitored for an additional period of four days. The concentration of oxygen was measured with a Checkmate II (PBI Dansensor A/S, Ringsted, Denmark) by removing an aliquot of the head-space gas for analysis using a zirconium-based sensor. The oxygen concentration was analyzed during seven days and the total decrease in oxygen concentration was calculated as cm^3/g coating.

Wet stability of cast starch-based films immersed in water

Cut pieces (0.6 × 1.0 cm) of films were placed in 15 ml plastic tubes containing 5 ml of the MOPS buffer. The tubes were incubated with rotation (22 rpm), and samples were taken after 5 min, 15 min, 1 h, 4 h, 8 h, and 24 h. The samples taken after 15 min were diluted 1:9 with the MOPS buffer. Following the method described by Johansson et al. [5], the migration of the lignin derivative from the film to the water phase was evaluated by measuring the increase in absorbance at 380 nm (Abs_{380}) using a spectrophotometer (UV-2101PC, Shimadzu, Kyoto, Japan). Measurements were performed in triplicates.

Mechanical properties of cast starch-based films

The films were cut into rectangular pieces (0.5 × 3.0 cm) and the thickness of each individual piece was measured using the STFI Thickness Tester. The storage modulus (E') of the films was evaluated by performing an oscillation sweep from 0.1 to 10 Hz in tension mode (DMA/SDTA861, Mettler Toledo, GmbH, Schwerzenbach, Switzerland) at constant amplitude 0.029%, 23°C, and 50% RH. An amplitude sweep was performed in order to ensure that the amplitude was within the linear region. The measurement was repeated six times for both types of samples studied (with denatured enzyme and with active enzyme).

Gel Permeation Chromatograhy (GPC)

A 1:2 (v/v) H_2O:DMAc mixture containing 0.1% (w/w) lithium bromide (LiBr) was used as eluent. Freeze-dried LO and LS powders prepared from samples taken after 0, 4, 8, 24, 72 and 96 h reaction were dissolved in the eluent mixture prior to analysis (1 mg sample/ml eluent). LS samples taken after 72 and 96 h were not soluble and could therefore not be analyzed. Starch-based LO and LS films were dissolved in a 10 mM solution of sodium hydroxide (20 mg film sample/ml solution) at 50°C during 7 h by shaking the samples every hour. The film samples were subsequently centrifuged during 10 min at 14,100 *g* using a MiniSpin Plus centrifuge (Eppendorf, Hamburg, Germany). Then, 75% of the liquid phase of the film samples were evacuated into a pipette and mixed with DMAc in a ratio of 3:20 (v/v). Thermogravimetric analysis (results not showed) confirmed that the solid phase of the dissolved film samples consisted of the clay incorporated in the films. The GPC system used comprised two 300 × 7.5 mm PolarGel M columns with a 50 × 7.5 mm PolarGel guard, and an ultraviolet (UV) detector (wavelength 254 nm) in a Polymer Laboratories PL-GPC 50 Plus instrument (Agilent, Santa Clara, CA, USA). The eluent flow-rate was set to 0.5 ml/min. The system was calibrated using 12 PSS standards with molecular masses ranging from 208 Da to 2,600 kDa. Two subsamples of each specimen were taken and analyzed in duplicates.

Pyrolysis – gas chromatography – mass spectroscopy (Py-GC-MS)

The syringyl to guaiacyl (S/G) ratios of LO and LS were determined using Py-GC-MS. The analysis was performed according to a previously described method [25].

Competing interests

The author(s) declare that they have no competing interests.

Authors' contribution

KJ and SW prepared coatings, films and lyophilized reaction mixtures, and carried out experiments on oxygen scavenging. KJ performed analysis of wet stability and mechanical properties. TG performed GPC analyses. LJ and LJJ supervised the work and took part in its design and coordination. LJJ conceived the study. KJ, TG and LJJ drafted the manuscript, which was read, revised and approved by all authors.

Authors' information

KJ is a doctoral student with focus on enzymes for applications in active packaging. TG is a postdoctoral researcher working on biopolymers from renewable resources. SW is a postdoctoral researcher working on enzyme technology in the biorefinery area. LJ is a professor with focus on paper surface treatment. LJJ is a professor with focus on biotechnology for biorefining of lignocellulosic biomass.

Acknowledgments
This work was supported through the Bio4Energy program (<www.bio4energy.se>), the Swedish Energy Agency (P35367-1), and by the Kempe Foundations. Py-GC-MS analysis was performed at the Plant Cell Wall and Carbohydrate Analytical Facility of the Umeå Plant Science Centre.

Author details
[1]Department of Chemical Engineering, Karlstad University, SE-651 88 Karlstad, Sweden. [2]Department of Chemistry, Umeå University, SE-901 87 Umeå, Sweden.

References

1. **Plastics Europe.** http://www.plasticseurope.org.
2. Gustavsson J, Stenberg C, Sonesson U, van Otterdijk R, Meybeck A: *Global Food Losses and Food Waste.* Food and Agriculture Organization of the United Nations: Rome; 2011.
3. Boerjan W, Ralph J, Baucher M: **Lignin biosynthesis.** *Annu Rev Plant Biol* 2003, **54:**519–46.
4. Dahlquist E: *Biomass as Energy Source: Resources, Systems and Applications.* Leiden: CRC Press; 2013.
5. Johansson K, Winestrand S, Johansson C, Järnström L, Jönsson LJ: **Oxygen-scavenging coatings and films based on lignosulfonates and laccase.** *J Biotechnol* 2012, **161:**14–18.
6. Lynd LR, Laser MS, Bransby D, Dale BE, Davison B, Hamilton R, Himmel M, Keller M, McMillan JD, Sheehan J, Wyman CE: **How biotech can transform biofuels.** *Nat Biotechnol* 2008, **26:**169–172.
7. Pu Y, Kosa M, Kalluri UC, Tuskan GA, Ragauskas AJ: **Challenges of the utilization of wood polymers: how can they be overcome?** *Appl Microbiol Biotechnol* 2011, **91:**1525–1536.
8. Pan X, Arato C, Gilkes N, Gregg D, Mabee W, Pye K, Xiao Z, Zhang X, Saddler J: **Biorefining of softwoods using ethanol organosolv pulping: preliminary evaluation of process streams for manufacture of fuel-grade ethanol and co-products.** *Biotechnol Bioeng* 2005, **90:**473–481.
9. Wörmeyer K, Ingram T, Saake B, Brunner G, Smirnova I: **Comparison of different pretreatment methods for lignocellulosic materials. Part II: Influence of pretreatment on the properties of rye straw lignin.** *Bioresour Technol* 2011, **102:**4157–4164.
10. Panagiotopoulos IA, Chandra RP, Saddler JN: **A two-stage pretreatment approach to maximise sugar yield and enhance reactive lignin recovery from poplar wood chips.** *Bioresour Technol* 2013, **130:**570–577.
11. Vieth WR: *Diffusion in and Through Polymers: Principles and Applications.* Münich: Hanser; 1999:73–106.
12. Johansson K, Christophliemk H, Johansson C, Jönsson L, Järnström L: **The effects of coating structure and water-holding capacity on the oxygen-scavenging ability of enzymes embedded in the coating layer.** *Tappi J* 2013, **12:**43–52.
13. Taoukis PS, Richardson M: In *Water Activity in Foods: Fundamentals and Applications.* Edited by Barbosa-Cánovas GV, Fontana AJ, Schmidt SJ, Labuza TP. New Jersey: Wiley-Blackwell; 2008:273–312.
14. Schlesing W, Buhk M, Osterhold M: **Dynamic mechanical analysis in coatings industry.** *Prog Org Coat* 2004, **49:**197–208.
15. Baumberger S, Lapierre C, Monties B, Lourdin D, Colonna P: **Preparation and properties of thermally moulded and cast lignosulfonate-starch blends.** *Ind Crop Prod* 1997, **6:**253–258.
16. Lepifre S, Froment M, Cazaux F, Houot S, Lourdin D, Coqueret X, Lapierre C, Baumberger S: **Lignin incorporation combined with electron beam irradiation improves the surface water resistance of starch films.** *Biomacromolecules* 2004, **5:**1678–1686.
17. Richardsson G, Sun Y, Langton M, Hermansson A-M: **Effects of Ca- and Na-lignosulfonate on starch gelatinization and network formation.** *Carbohyd Polym* 2004, **57:**369–377.
18. Morais LC, Curvelo AAS, Zambon MD: **Thermoplastic starch–lignosulfonate blends. 1. Factorial planning as a tool for elucidating new data from high performance size-exclusion chromatography and mechanical tests.** *Carbohyd Polym* 2005, **62:**104–112.
19. Baumberger S, Lapierre C, Monties B, Della Valle G: **Use of kraft lignin as filler for starch films.** *Polym Degrad Stabil* 1997, **59:**273–277.
20. Baumberger S: **Starch-Lignin Films.** In *Chemical Modification Properties and Usage of Lignin.* Edited by Hu TQ. New York: Kluwer Academic/Plenum Publishers; 2002:1–20.
21. Gidh AV, Decker SR, See CH, Himmel ME, Williford CW: **Characterization of lignin using multi-angle laser light scattering and atomic force microscopy.** *Anal Chim Acta* 2006, **555:**250–258.
22. Shogren RL, Biswas A: **Preparation of starch-sodium lignosulfonate graft copolymers via laccase catalysis and characterization of antioxidant activity.** *Carbohyd Polym* 2013, **91:**581–585.
23. Areskogh D, Nousiainen P, Li J, Gellerstedt G, Sipilä J, Henriksson G: **Sulfonation of phenolic end groups in lignin directs laccase-initiated reactions towards cross-linking.** *Ind Biotechnol* 2010, **6:**50–59.
24. Lai Y-Z, Guo X-P, Situ W: **Estimation of phenolic hydroxyl groups in wood by a periodate oxidation method.** *J Wood Chem Technol* 1990, **10:**365–377.
25. Gerber L, Eliasson M, Trygg J, Moritz T, Sundberg B: **Multivariate curve resolution provides a high-throughput data processing pipeline for pyrolysis-gas chromatography/mass spectrometry.** *J Anal Appl Pyrol* 2012, **95:**95–100.

Permissions

All chapters in this book were first published in JBE, by BioMed Central; hereby published with permission under the Creative Commons Attribution License or equivalent. Every chapter published in this book has been scrutinized by our experts. Their significance has been extensively debated. The topics covered herein carry significant findings which will fuel the growth of the discipline. They may even be implemented as practical applications or may be referred to as a beginning point for another development.

The contributors of this book come from diverse backgrounds, making this book a truly international effort. This book will bring forth new frontiers with its revolutionizing research information and detailed analysis of the nascent developments around the world.

We would like to thank all the contributing authors for lending their expertise to make the book truly unique. They have played a crucial role in the development of this book. Without their invaluable contributions this book wouldn't have been possible. They have made vital efforts to compile up to date information on the varied aspects of this subject to make this book a valuable addition to the collection of many professionals and students.

This book was conceptualized with the vision of imparting up-to-date information and advanced data in this field. To ensure the same, a matchless editorial board was set up. Every individual on the board went through rigorous rounds of assessment to prove their worth. After which they invested a large part of their time researching and compiling the most relevant data for our readers.

The editorial board has been involved in producing this book since its inception. They have spent rigorous hours researching and exploring the diverse topics which have resulted in the successful publishing of this book. They have passed on their knowledge of decades through this book. To expedite this challenging task, the publisher supported the team at every step. A small team of assistant editors was also appointed to further simplify the editing procedure and attain best results for the readers.

Apart from the editorial board, the designing team has also invested a significant amount of their time in understanding the subject and creating the most relevant covers. They scrutinized every image to scout for the most suitable representation of the subject and create an appropriate cover for the book.

The publishing team has been an ardent support to the editorial, designing and production team. Their endless efforts to recruit the best for this project, has resulted in the accomplishment of this book. They are a veteran in the field of academics and their pool of knowledge is as vast as their experience in printing. Their expertise and guidance has proved useful at every step. Their uncompromising quality standards have made this book an exceptional effort. Their encouragement from time to time has been an inspiration for everyone.

The publisher and the editorial board hope that this book will prove to be a valuable piece of knowledge for researchers, students, practitioners and scholars across the globe.

List of Contributors

Arthur T Johnson
Fischell Department of Bioengineering, University of Maryland, College Park, MD 20742, USA

Susanna Zucca
Dipartimento di Ingegneria Industriale e dell'Informazione, Università degli Studi di Pavia, via Ferrata 3, Pavia, Italy
Centro di Ingegneria Tissutale, Università degli Studi di Pavia, via Ferrata 3, Pavia, Italy

Lorenzo Pasotti
Dipartimento di Ingegneria Industriale e dell'Informazione, Università degli Studi di Pavia, via Ferrata 3, Pavia, Italy
Centro di Ingegneria Tissutale, Università degli Studi di Pavia, via Ferrata 3, Pavia, Italy

Nicolò Politi
Dipartimento di Ingegneria Industriale e dell'Informazione, Università degli Studi di Pavia, via Ferrata 3, Pavia, Italy
Centro di Ingegneria Tissutale, Università degli Studi di Pavia, via Ferrata 3, Pavia, Italy

Maria Gabriella Cusella De Angelis
Centro di Ingegneria Tissutale, Università degli Studi di Pavia, via Ferrata 3, Pavia, Italy

Paolo Magni
Dipartimento di Ingegneria Industriale e dell'Informazione, Università degli Studi di Pavia, via Ferrata 3, Pavia, Italy
Centro di Ingegneria Tissutale, Università degli Studi di Pavia, via Ferrata 3, Pavia, Italy

Asif Rahman
Department of Biological Engineering, Utah State University, 4105 Old Main Hill, Logan 84322-4105, UT, USA

Elisabeth Linton
Department of Biological Engineering, Utah State University, 4105 Old Main Hill, Logan 84322-4105, UT, USA

Alex D Hatch
Department of Biological Engineering, Utah State University, 4105 Old Main Hill, Logan 84322-4105, UT, USA

Ronald C Sims
Department of Biological Engineering, Utah State University, 4105 Old Main Hill, Logan 84322-4105, UT, USA

Charles D Miller
Department of Biological Engineering, Utah State University, 4105 Old Main Hill, Logan 84322-4105, UT, USA

Victor K Champagne
U.S. Army Research Laboratory, Aberdeen Proving Ground, MD, USA

Dennis J Helfritch
Dynamic Science, Aberdeen, MD, USA

Meltem Avci-Adali
Department of Thoracic, Cardiac, and Vascular Surgery, University Hospital Tuebingen, Calwerstr. 7/1, 72076 Tuebingen, Germany

Andreas Behring
Department of Thoracic, Cardiac, and Vascular Surgery, University Hospital Tuebingen, Calwerstr. 7/1, 72076 Tuebingen, Germany

Timea Keller
Department of Thoracic, Cardiac, and Vascular Surgery, University Hospital Tuebingen, Calwerstr. 7/1, 72076 Tuebingen, Germany

Stefanie Krajewski
Department of Thoracic, Cardiac, and Vascular Surgery, University Hospital Tuebingen, Calwerstr. 7/1, 72076 Tuebingen, Germany

Christian Schlensak
Department of Thoracic, Cardiac, and Vascular Surgery, University Hospital Tuebingen, Calwerstr. 7/1, 72076 Tuebingen, Germany

Hans Peter Wendel
Department of Thoracic, Cardiac, and Vascular Surgery, University Hospital Tuebingen, Calwerstr. 7/1, 72076 Tuebingen, Germany

Cong Liu
Department of Chemical Engineering, Imperial College London, South Kensington Campus, London SW7 2AZ, UK

Krishnan
Department of Chemical Engineering, Imperial College London, South Kensington Campus, London SW7 2AZ, UK
Centre for Process Systems Engineering and Institute for Systems and Synthetic Biology, Imperial College London, South Kensington, London SW7 2AZ, UK

Xiao Yun Xu
Department of Chemical Engineering, Imperial College London, South Kensington Campus, London SW7 2AZ, UK

Nobuo Fukuda
Biomedical Research Institute, National Institute of Advanced Industrial Science and Technology (AIST), Higashi, Tsukuba, Ibaraki 305-8566, Japan

Shinya Honda
Biomedical Research Institute, National Institute of Advanced Industrial Science and Technology (AIST), Higashi, Tsukuba, Ibaraki 305-8566, Japan

Brendan M Ryback
Systems and Synthetic Biology, Wageningen University, Wageningen, Netherlands

Dorett I Odoni
Systems and Synthetic Biology, Wageningen University, Wageningen, Netherlands

Ruben GA van Heck
Systems and Synthetic Biology, Wageningen University, Wageningen, Netherlands

Youri van Nuland
Systems and Synthetic Biology, Wageningen University, Wageningen, Netherlands

Matthijn C Hesselman
Systems and Synthetic Biology, Wageningen University, Wageningen, Netherlands

Vítor AP Martins dos Santos
Systems and Synthetic Biology, Wageningen University, Wageningen, Netherlands

Mark WJ van Passel
Systems and Synthetic Biology, Wageningen University, Wageningen, Netherlands
Centre for Zoonoses and Environmental Microbiology, Centre for Infectious Disease Control, National Institute for Public Health and the Environment (RIVM), Bilthoven, Netherlands

Floor Hugenholtz
Laboratory of Microbiology, Wageningen University, Wageningen, Netherlands
Netherlands Consortium for Systems Biology, University of Amsterdam, Amsterdam, Netherlands

Timothy N Snyder
Bioengineering Department, University of Colorado, Anschutz Medical Campus, Mail Stop 8607, 12700 East 19th Avenue, Aurora, CO 80045, USA
Regenerative Sciences, 403 Summit Blvd, Suite 201, Broomfield, CO 80021, USA

Krishna Madhavan
Bioengineering Department, University of Colorado, Anschutz Medical Campus, Mail Stop 8607, 12700 East 19th Avenue, Aurora, CO 80045, USA

Miranda Intrator
Bioengineering Department, University of Colorado, Anschutz Medical Campus, Mail Stop 8607, 12700 East 19th Avenue, Aurora, CO 80045, USA

Ryan C Dregalla
Regenerative Sciences, 403 Summit Blvd, Suite 201, Broomfield, CO 80021, USA

Daewon Park
Bioengineering Department, University of Colorado, Anschutz Medical Campus, Mail Stop 8607, 12700 East 19th Avenue, Aurora, CO 80045, USA

Zhanar Abil
Department of Biochemistry, University of Illinois at Urbana-Champaign, Urbana, IL 61801, USA

Carl A Denard
Department of Chemical and Biomolecular Engineering, University of Illinois at Urbana-Champaign, Urbana, IL 61801, USA

Huimin Zhao
Department of Biochemistry, University of Illinois at Urbana-Champaign, Urbana, IL 61801, USA
Department of Chemical and Biomolecular Engineering, University of Illinois at Urbana-Champaign, Urbana, IL 61801, USA
Department of Bioengineering, Department of Chemistry, Center for Biophysics and Computational Biology, and Institute for Genomic Biology, University of Illinois at Urbana-Champaign, Urbana, IL 61801, USA

Marcello Pilia
Department of Biomedical Engineering, University of Texas, San Antonio, TX, USA
Joint Program in Biomedical Engineering, University of Texas Health Science Center at San Antonio/University of Texas, San Antonio, TX, USA

Teja Guda
Department of Biomedical Engineering, University of Texas, San Antonio, TX, USA
Joint Program in Biomedical Engineering, University of Texas Health Science Center at San Antonio/University of Texas, San Antonio, TX, USA

Stefanie M Shiels
Department of Biomedical Engineering, University of Texas, San Antonio, TX, USA

Mark R Appleford
Department of Biomedical Engineering, University of Texas, San Antonio, TX, USA
Joint Program in Biomedical Engineering, University of Texas Health Science Center at San Antonio/University of Texas, San Antonio, TX, USA

Kumar Selvarajoo
Institute for Advanced Biosciences, Keio University, Tsuruoka, Yamagata, Japan
Systems Biology Program, Graduate School of Media and Governance, Keio University, Fujisawa, Japan

Matthew Schmerer
Center for Computational Biology and Bioinformatics, The University of Texas at Austin, Austin, TX, USA

Ian J Molineux
Department of Molecular Biosciences, The University of Texas at Austin, Austin, TX, USA
Institute for Cellular andMolecular Biology, The University of Texas at Austin, Austin, TX, USA

Dilara Ally
Bayer Crop Science - Biologics, 1540 Drew Ave, Unit 170, Davis, CA, USA

Jabus Tyerman
Total New Energies USA Inc., 5858 Horton Street, Suite 253, Emeryville, CA, USA

Nicole Cecchini
Department of Integrative Biology, The University of Texas at Austin, Austin, TX, USA

James J Bull
Center for Computational Biology and Bioinformatics, The University of Texas at Austin, Austin, TX, USA
Institute for Cellular andMolecular Biology, The University of Texas at Austin, Austin, TX, USA
Department of Integrative Biology, The University of Texas at Austin, Austin, TX, USA

Clemens Lattermann
AVT.Biochemical Engineering, RWTH Aachen University, Worringer Weg 1, 52074 Aachen, Germany

Matthias Funke
Lonza Group Ltd, Münchensteinerstraße 38, 4002 Basel, Switzerland

Sven Hansen
Evonik Industries AG, Rodenbacher Chaussee 4, 63457 Hanau-Wolfgang, Germany

Sylvia Diederichs
AVT.Biochemical Engineering, RWTH Aachen University, Worringer Weg 1, 52074 Aachen, Germany

Jochen Büchs
AVT.Biochemical Engineering, RWTH Aachen University, Worringer Weg 1, 52074 Aachen, Germany

Martin Kunze
AVT-Chair for Biochemical Engineering, RWTH Aachen University, Worringerweg 1, 52074 Aachen, Germany

Clemens Lattermann
AVT-Chair for Biochemical Engineering, RWTH Aachen University, Worringerweg 1, 52074 Aachen, Germany

Sylvia Diederichs
AVT-Chair for Biochemical Engineering, RWTH Aachen University, Worringerweg 1, 52074 Aachen, Germany

Wolfgang Kroutil
Department of Chemistry (Organic and Bioorganic Chemistry), University of Graz, Heinrichstraße 28/II, 8010 Graz, Austria

Jochen Büchs
AVT-Chair for Biochemical Engineering, RWTH Aachen University, Worringerweg 1, 52074 Aachen, Germany

Nikolai Eroshenko
Department of Chemical and Life Science Engineering, Virginia Commonwealth University, Richmond, VA, USA
Current Address: School of Engineering and Applied Sciences, Harvard University, Boston, MA, USA

Rukmani Ramachandran
Department of Chemical and Life Science Engineering, Virginia Commonwealth University, Richmond, VA, USA

Vamsi K Yadavalli
Department of Chemical and Life Science Engineering, Virginia Commonwealth University, Richmond, VA, USA

Raj R Rao
Department of Chemical and Life Science Engineering, Virginia Commonwealth University, Richmond, VA, USA

Linda J Kahl
Bioengineering Department, Stanford University, Y2E2 Room 269C, 473 Via Ortega, Stanford, CA 94305-4201, USA

Drew Endy
Bioengineering Department, Stanford University, Y2E2 Room 269C, 473 Via Ortega, Stanford, CA 94305-4201, USA

Rachel Hatano
School of Natural Sciences, University of California, Merced, CA, USA

Kevin Mercurio
Graduate Program in Biological Engineering and Small-scale Technologies, University of California, Merced, CA, USA

Jesus Isaac Luna
Graduate Program in Biological Engineering and Small-scale Technologies, University of California, Merced, CA, USA

Drew E Glaser
Graduate Program in Biological Engineering and Small-scale Technologies, University of California, Merced, CA, USA

Valerie J Leppert
Graduate Program in Biological Engineering and Small-scale Technologies, University of California, Merced, CA, USA
School of Engineering, University of California, P.O. Box 2039, Merced, CA 95344, USA

Kara E McCloskey
Graduate Program in Biological Engineering and Small-scale Technologies, University of California, Merced, CA, USA
School of Engineering, University of California, P.O. Box 2039, Merced, CA 95344, USA

Bogumil J Karas
Department of Synthetic Biology and Bioenergy, J. Craig Venter Institute, 10355 Science Center Dr., San Diego, CA 92121, USA

Bhuvan Molparia
Department of Synthetic Biology and Bioenergy, J. Craig Venter Institute, 10355 Science Center Dr., San Diego, CA 92121, USA

Jelena Jablanovic
Department of Synthetic Biology and Bioenergy, J. Craig Venter Institute, 10355 Science Center Dr., San Diego, CA 92121, USA

Wolfgang J Hermann
Department of Synthetic Biology and Bioenergy, J. Craig Venter Institute, 10355 Science Center Dr., San Diego, CA 92121, USA

Ying-Chi Lin
Department of Synthetic Biology and Bioenergy, J. Craig Venter Institute, 10355 Science Center Dr., San Diego, CA 92121, USA

Christopher L Dupont
Department of Microbial and Environmental Genomics, J. Craig Venter Institute, 10355
Science Center Dr., San Diego, CA 92121, USA

Christian Tagwerker
Department of Synthetic Biology and Bioenergy, J. Craig Venter Institute, 10355 Science Center Dr., San Diego, CA 92121, USA

Isaac T Yonemoto
Department of Synthetic Biology and Bioenergy, J. Craig Venter Institute, 10355 Science Center Dr., San Diego, CA 92121, USA

Vladimir N Noskov
Department of Synthetic Biology and Bioenergy, J. Craig Venter Institute, 9704 Medical Center Dr., Rockville, MD 20850, USA.

Ray-Yuan Chuang
Department of Synthetic Biology and Bioenergy, J. Craig Venter Institute, 9704 Medical Center Dr., Rockville, MD 20850, USA

Andrew E Allen
Department of Microbial and Environmental Genomics, J. Craig Venter Institute, 10355
Science Center Dr., San Diego, CA 92121, USA

John I Glass
Department of Synthetic Biology and Bioenergy, J. Craig Venter Institute, 9704 Medical Center Dr., Rockville, MD 20850, USA

Clyde A Hutchison III
Department of Synthetic Biology and Bioenergy, J. Craig Venter Institute, 10355 Science Center Dr., San Diego, CA 92121, USA

Hamilton O Smith
Department of Synthetic Biology and Bioenergy, J. Craig Venter Institute, 10355 Science Center Dr., San Diego, CA 92121, USA

J Craig Venter
Department of Synthetic Biology and Bioenergy, J. Craig Venter Institute, 10355 Science Center Dr., San Diego, CA 92121, USA
Department of Microbial and Environmental Genomics, J. Craig Venter Institute, 10355
Science Center Dr., San Diego, CA 92121, USA
Department of Synthetic Biology and Bioenergy, J. Craig Venter Institute, 9704 Medical Center Dr., Rockville, MD 20850, USA

Philip D Weyman
Department of Synthetic Biology and Bioenergy, J. Craig Venter Institute, 10355 Science Center Dr., San Diego, CA 92121, USA

Bojun Chen
Center for Paralysis Research, Department of Basic Medical Sciences, College of Veterinary Medicine, Purdue University, West, Lafayette, IN 47907, USA

Debra Bohnert
Center for Paralysis Research, Department of Basic Medical Sciences, College of Veterinary Medicine, Purdue University, West, Lafayette, IN 47907, USA

Richard Ben Borgens
Center for Paralysis Research, Department of Basic Medical Sciences, College of Veterinary Medicine, Purdue University, West, Lafayette, IN 47907, USA
Weldon School of Biomedical Engineering, Purdue University, West,Lafayette, IN 47907, USA

Youngnam Cho
Center for Paralysis Research, Department of Basic Medical Sciences, College of Veterinary Medicine, Purdue University, West, Lafayette, IN 47907, USA
Present address: New Experimental Therapeutics Branch, National Cancer Center, 809 Madu-1dong, Ilsandong-gu, Goyang-si, Gyeonggi-do 410-769, Korea

Mee-Hae Kim
Department of Biotechnology, Graduate School of Engineering, Osaka University, 2-1 Yamadaoka, Suita, Osaka 565-0871, Japan

Yoshiko Sawada
Division of Chemical Engineering, Graduate School of Engineering Science, Osaka University, 1-3 Machikaneyama-cho, Toyonaka, Osaka 560-8531, Japan

Masahito Taya
Division of Chemical Engineering, Graduate School of Engineering Science, Osaka University, 1-3 Machikaneyama-cho, Toyonaka, Osaka 560-8531, Japan

Masahiro Kino-oka
Department of Biotechnology, Graduate School of Engineering, Osaka University, 2-1 Yamadaoka, Suita, Osaka 565-0871, Japan

Mostafa Sefidgar
Department of Mechanical Engineering, K. N. T. University of Technology, Tehran, Iran

M Soltani
Department of Mechanical Engineering, K. N. T. University of Technology, Tehran, Iran
Division of Nuclear Medicine, Department of Radiology and Radiological Science, School of Medicine, Johns Hopkins University, Baltimore, MD, USA

Kaamran Raahemifar
Electrical & Computer Department of Ryerson University, Toronto, Ontario

Hossein Bazmara
Department of Mechanical Engineering, K. N. T. University of Technology, Tehran, Iran

Seyed Mojtaba Mousavi Nayinian
Department of Mechanical Engineering, K. N. T. University of Technology, Tehran, Iran

Majid Bazargan
Department of Mechanical Engineering, K. N. T. University of Technology, Tehran, Iran

Kristin Johansson
Department of Chemical Engineering, Karlstad University, SE-651 88 Karlstad, Sweden

Thomas Gillgren
Department of Chemistry, Umeå University, SE-901 87 Umeå, Sweden

Sandra Winestrand
Department of Chemistry, Umeå University, SE-901 87 Umeå, Sweden

Lars Järnström
Department of Chemical Engineering, Karlstad University, SE-651 88 Karlstad, Sweden

Leif J Jönsson
Department of Chemistry, Umeå University, SE-901 87 Umeå, Sweden